Communications
in Computer and Information Science   43

Dominic Palmer-Brown   Chrisina Draganova
Elias Pimenidis   Haris Mouratidis (Eds.)

Engineering Applications of Neural Networks

11th International Conference, EANN 2009
London, UK, August 27-29, 2009
Proceedings

 Springer

Volume Editors

Dominic Palmer-Brown
London Metropolitan University
Faculty of Computing
London, UK
E-mail: d.palmer-brown@londonmet.ac.uk

Chrisina Draganova
University of East London
School of Computing, IT and Engineering
London, UK
E-mail: c.draganova@uel.ac.uk

Elias Pimenidis
University of East London
School of Computing, IT and Engineering
London, UK
E-mail: e.pimenidis@uel.ac.uk

Haris Mouratidis
University of East London
School of Computing, IT and Engineering
London, UK
E-mail: h.mouratidis@uel.ac.uk

Library of Congress Control Number: 2009933042

CR Subject Classification (1998): F.1, I.2, H.5, I.7, I.5, I.2.7, I.4.8

ISSN	1865-0929
ISBN-10	3-642-03968-5 Springer Berlin Heidelberg New York
ISBN-13	978-3-642-03968-3 Springer Berlin Heidelberg New York

Typesetting: Camera-ready by author, data conversion by Scientific Publishing Services, Chennai, India
Printed on acid-free paper   SPIN: 12744978   06/3180   5 4 3 2 1 0

Preface

A cursory glance at the table of contents of EANN 2009 reveals the amazing range of neural network and related applications. A random but revealing sample includes: reducing urban concentration, entropy topography in epileptic electroencephalography, phytoplanktonic species recognition, revealing the structure of childhood abdominal pain data, robot control, discriminating angry and happy facial expressions, flood forecasting, and assessing credit worthiness. The diverse nature of applications demonstrates the vitality of neural computing and related soft computing approaches, and their relevance to many key contemporary technological challenges. It also illustrates the value of EANN in bringing together a broad spectrum of delegates from across the world to learn from each other's related methods. Variations and extensions of many methods are well represented in the proceedings, ranging from support vector machines, fuzzy reasoning, and Bayesian methods to snap-drift and spiking neurons.

This year EANN accepted approximately 40% of submitted papers for full-length presentation at the conference. All members of the Program Committee were asked to participate in the reviewing process. The standard of submissions was high, according to the reviewers, who did an excellent job. The Program and Organizing Committees thank them. Approximately 20% of submitted papers will be chosen, the best according to the reviews, to be extended and reviewed again for inclusion in a special issue of the journal *Neural Computing and Applications*.

We hope that these proceedings will help to stimulate further research and development of new applications and modes of neural computing.

August 2009   Dominic Palmer-Brown

Organization

EANN 2009 was organized by the School of Computing, IT and Engineering, University of East London in cooperation with the Faculty of Computing, London Metropolitan University.

Organizing Committee

Chair: Dominic Palmer-Brownn	London Metropolitan University, UK
Chrisina Draganova	University of East London, UK
Elias Pimenidis	University of East London, UK
Haris Mouratidis	University of East London, UK
Sin Wee Lee	University of East London, UK
Manolis Christodoulakis	University of East London, UK
Miao Kang	University of East London, UK
Frank Ekpenyong	University of East London, UK
Terry Walcott	University of East London, UK
Administrative Support: Linda Day	University of East London, UK
Administrative Support: Farhanaz Begum	University of East London, UK
Administrative Support: Carole Cole	University of East London, UK

Program Committee

Chair: Dominic Palmer-Brownn	London Metropolitan University, UK
Graham Ball	Nottingham Trent University, UK
Chris Christodoulou	University of Cyprus, Cyprus
Manolis Christodoulakis	University of East London, UK
Chrisina Draganova	University of East London, UK
Frank Ekpenyong	University of East London, UK
Alexander Gegov	University of Portsmouth, UK
Lazaros Iliadis	Democritus University of Thrace, Greece
Miao Kang	University of East London, UK
Sin Wee Lee	University of East London, UK
James Allen Long	London South Bank University, UK
Liam McDaid	University of Ulster, UK
Haralambos Mouratidis	University of East London, UK
Elias Pimenidis	University of East London, UK

Kathleen Steinhofel	King's College, UK
Ioannis Vlahavas	Aristotle University of Thessaloniki, Greece
Carlo Francesco Morabito	University Mediterranea of Reggio C. - Italy, and SIREN (Italian Society of Neural Networks)
Christos N. Schizas	University of Cyprus, Cyprus
Terry Walcott	University of East London, UK
Abdullah Al-Zakwani	Muscat, Sultanate of Oman
Diego G. Loyola R.	Remote Sensing Technology Institute, Germany
Antony Browne	University of Surrey, UK
Kaizhu Huang	University of Bristol, UK
Richard Dybowski	University of East London, UK
Aloysius Edoh	University of East London, UK
Marco Ramoni	Harvard University, USA
Stephen Rees	Aalborg University, Denmark
Bernadette Sharp	Staffordshire University, UK
Peter Weller	City University, UK
Stefan Wermter	Sunderland University, UK
Hassan Kazemian	London Metropolitan University, UK
J. Fernandez de Canete	University of Malaga, Spain
Karim Ouazzane	London Metropolitan University, UK
Kenneth Revett	University of Westminster, UK

Sponsoring Institutions

University of East London, UK
London Metropolitan University, UK
International Neural Network Society INNS, USA

Table of Contents

Intelligent Agents Networks Employing Hybrid Reasoning: Application in Air Quality Monitoring and Improvement

Lazaros S. Iliadis and A. Papaleonidas

Democritus University of Thrace, 193 Padazidou st., 68200, Nea Orestiada, Greece
liliadis@fmenr.duth.gr

Abstract. This paper presents the design and the development of an agent-based intelligent hybrid system. The system consists of a network of interacting intelligent agents aiming not only towards real-time air pollution monitoring but towards proposing proper corrective actions as well. In this manner, the concentration of air pollutants is managed in a real-time scale and as the system is informed continuously on the situation an iterative process is initiated. Four distinct types of intelligent agents are utilized: Sensor, Evaluation, Decision and Actuator. There are also several types of Decision agents depending on the air pollution factor examined. The whole project has a Hybrid nature, since it utilizes fuzzy logic – fuzzy algebra concepts and also crisp values and a rule based inference mechanism. The system has been tested by the application of actual air pollution data related to four years of measurements in the area of Athens.

Keywords: Fuzzy Logic, Intelligent Agents Networks, Hybrid Intelligent Agents, Pollution of the Atmosphere.

1 Introduction

1.1 Aim of This Project

The present work aims firstly to design and develop a network of intelligent distributed agents that employ hybrid inference techniques in order to offer real-time monitoring and to address the air pollution problem by suggesting corrective actions mainly related to human activities. Intelligent agents are software programs with an attitude. An agent is autonomous because it operates without the direct intervention of humans or others and has control over its actions and internal state. An agent is social because it cooperates with humans and with other agents in order to achieve its tasks. An agent is reactive since it perceives the environment and it responds to changes that occur in it [3]. A multiagent system can model complex cases and introduce the possibility of agents having common or conflict goals [3]. Hybrid intelligent systems are becoming popular due to their capabilities in handling many real world complex problems. They provide us with the opportunity to use both, our knowledge and row data to solve problems in a more interesting and promising way. This multidisciplinary research field is in continuous expansion in the artificial intelligence research community [11].

D. Palmer-Brown et al. (Eds.): EANN 2009, CCIS 43, pp. 1–16, 2009.
© Springer-Verlag Berlin Heidelberg 2009

The research effort described here is characterized as a Hybrid one. This is due to the fact that it utilizes fuzzy logic and fuzzy sets in one hand, but also crisp data in the other, fuzzy "IF-THEN" rules where both the antecedents and the consequents take linguistic values but also rule sets with crisp numerical values and at the same time it utilizes the basic properties of a distributed network of interacting intelligent agents. Though they are state of the art technology, several intelligent agents systems have been developed with very useful application domains [14],[10].

The multiagent network is informed on the air pollution situation in a scale of seconds and it is trapped in an iterative process. The whole model is based in a simple and very crucial principle. The received data and the inference engine determine the corrective actions and vise versa. The proposed and implemented system is realized by employing a distributed architecture, since the network comprises of several interacting independent agents of four distinct types. This offers the advantage of parallel execution of several instances of the same agent type. The system has been tested with actual air pollution data gathered from the site of Pathsion street which is located right in the center of Athens in an altitude of 105 meters.

1.2 Air Pollution

There are several air pollutants that have harmful effect in the health of the citizens of major cities. Some of them do not have direct effect but they initiate secondary harmful processes. This research considers the carbon oxide CO, carbon dioxide CO_2, nitric oxide NO, nitrogen dioxide NO_2, the particular matter PM and ozone O_3 which is one of the most pervasive and potentially harmful air pollutants especially in major metropolitan centers like Athens Greece. It is a critical atmospheric species, which drives much of the tropospheric photochemistry. It is also considered responsible for regulating the tropospheric oxidation capacity and it is the main ingredient of the photochemical smog. It is formed when volatile organic compounds (VOCs), nitric oxide and nitrogen dioxide react chemically under the influence of heat and sunlight [17]. Various medical studies have revealed that ozone can be blamed for inflammation and irritation of the respiratory tract, particularly during heavy physical activity, as well as ocular diseases [6], [7], [19].Various artificial neural networks have been developed in order to model the ground-level O_3 concentrations [2], [6], [16], [20].

Also meteorological data are considered, namely: Temperature, Relative Humidity, and the NW-SE direction wind component (u') and the SW-NE direction wind component (v') are used as input to the system. The selection of the u' and v' components instead of the conventional ones, u (W-E) and v (S-N), was considered necessary as u' is almost parallel to the Saronic Gulf coast and v' to the direction of the sea breeze circulation and the main axis of the Athens Basin.

2 Theoretical Background

Fuzzy logic is a "real world" approximator and it has been used in a wide scale towards risk modeling and especially in the case of environmental or natural hazards [13].

2.1 Risk Evaluation Using Fuzzy Logic

Human inference is very approximate. Our statements depend on the contents and we describe our physical world in rather vague terms. Imprecisely defined "classes" are an important part of human thinking [13]. The term *"risky"* due to an involved parameter is both imprecise and subjective and it is determined by a membership function that might have dissimilar shape. The choice of a shape for each particular linguistic variable is both subjective and problem-dependent [13]. Any function $\mu_s(X) \to [0,1]$ describes a membership function associated with some fuzzy set. A trapezoidal and a Triangular membership function is a special case of the following functions 2 and 3 respectively [13].

$$\mu_s(X) = \begin{cases} 0 \text{ if } X < a \\ (X-a)/(m-a) \text{ if } X \in [a,m] \\ 1 \text{ if } X \in [m,n] \\ (b-X)/(b-n) \text{ if } X \in (n,b] \\ 0 \text{ if } X > b \end{cases} \quad (2) \qquad \mu_s(X) = \begin{cases} 0 \text{ if } X < a \\ (X-a)/(c-a) \text{ if } X \in [a,c) \\ (b-X)/(b-c) \text{ if } X \in [c,b) \\ 0 \text{ if } X > b \end{cases} \quad (3)$$

For example in the case of the air pollution due to CO at least three fuzzy sets can be formed and they could have either of the potential following types:

$$\tilde{FS}_1 = \{Risky\_momentum\_due\_to\_CO\}$$

$$\tilde{FS}_2 = \{Very\_Risky\_momentum\_due\_to\_CO\}$$

$$\tilde{FS}_3 = \{Extremely\_Risky\_momentum\_due\_to\_CO\}.$$ Every temporal instance under examination, belongs to all above three fuzzy sets with a different degree of membership (in the closed interval [0,1]) which is determined by the triangular membership function 2. So actually each fuzzy set is a vector of ordered pairs of the form: (C_i, μ_i) where C_i is the crisp value and μ_i is the degree of membership of the value C_i to the corresponding fuzzy set. The fuzzy set with the highest degree of membership indicates the actual *"Linguistic"* value that can be assigned to a temporal instance. For example if μ_i ($\tilde{FS}_3$)=0.9 then the situation is rather critical, since the temporal instance is characterized as extremely risky due to CO. In this case, special actions have to be taken. Another advantage of this approach is the fact that an *"extremely risky"* Linguistic might have a different batch of actions compared to another if the degrees of membership have a significant differentiation.

A major key issue is the determination of the overall degree of risk when two or more parameters are involved. It can be faced by the use of the fuzzy conjunction operators which are called T-Norms. So let's suppose that we need to estimate the unified degree of membership to a fuzzy set $\tilde{FS}_4$ that is the fuzzy conjunction between two other fuzzy sets. $\tilde{FS}_4 = (\{Extremely\_Risky\_momentum\_due\_to\_CO\}$ AND ($\{Risky\_momentum\_due\_to\_NO\}$)

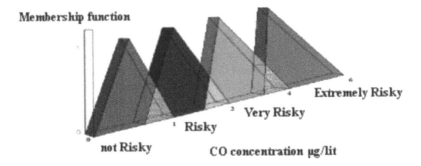

Fig. 1. Fuzzy determination of the triangular CO pollution degree of risk

In this case, the $\mu_i(\tilde{FS}_4)$ will be determined by the application of one of the following T-Norms functions 3,4,5,6,7 [13].

$$Minimum\ Approach\ = MIN(\mu_A(X), \mu_B(X)) \tag{3}$$

$$Algebraic\ Product\ = \mu_A(X) * \mu_B(X) \tag{4}$$

$$Drastic\ Product = MIN(\mu_A(X), \mu_B(X)).if..MAX(\mu_A(X), \mu_B(X)) = 1$$
$$otherwise\ URI = 0 \tag{5}$$

$$Einstein\ Product = \mu_A(X) * \mu_B(X) / (2-(\mu_A(X) + \mu_B(X) - \mu_A(X) * \mu_B(X))) \tag{6}$$

$$Hamacher\ Product = \mu_A(X) * \mu_B(X) / (\mu_A(X) + \mu_B(X) - \mu_A(X) * \mu_B(X)) \tag{7}$$

Such fuzzy intelligent systems have been developed by Iliadis et al. [12]. When weighted conjunction is required according to the importance of each parameter (which is estimated empirically) the following function 8 can be employed. The Aggregation function can be any of the above mentioned T-Norms [12].

$$\mu_{\underset{S}{\sim}}(x_i) = Agg\left(f\left(\mu_{\underset{A}{\sim}}(x_i), w_1 \right), f\left(\mu_{\underset{A}{\sim}}(x_i), w_2 \right), ..., f\left(\mu_{\underset{A}{\sim}}(x_i), w_n \right) \right) \tag{8}$$

3   The Agent-Based System

3.1   System's Architecture

The system is *Hardware independent* since both the multiagent development platform used (Jade) and the implemented agents have been programmed in *Java*. Jade is a software platform providing middleware-layer functionalities independent of the specific application. It was released as open source in 2000 by Telecom Italia under the LGPL (Library Ginu Public Licence) [3]. Though there are also other agent communication languages like ARPA KSI or Knowledge Query and Manipulation Language (KQML) [1] Jade was chosen due to the fact that it is open source and Java based.

The developed system consists of four types of intelligent agents, namely the **Sensor** agents, the **Evaluation** ones, the **Decision making** agents and finally the **Actuators**. Before describing the system's architecture it should be specified that the message exchanged when two intelligent agents interact are specified and encoded according to the FIPA (*Foundation for Intelligent Physical Agents*) standardization. FIPA is dominated by computer and telecommunications companies and it is focused on agent-level issues [4]. Thus *INFORM, CONFIRM, AGREE, ACCEPT_PROPOSAL, REFUSE, REQUEST, SUBSCRIBE* are some characteristic types of exchanged messages. The following figure 2 depicts the general architecture of the system that has been developed in this study.

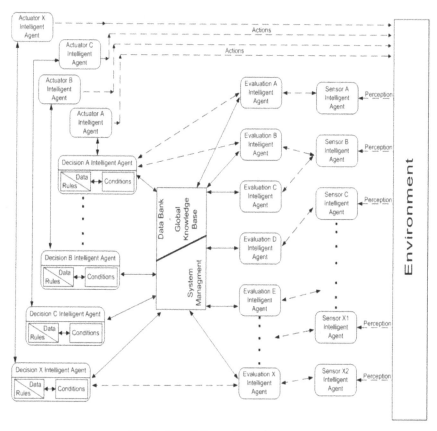

Fig. 2. Overall System's Architecture

3.2 Sensor Agents Architecture

The above figure 3 makes a detailed description of the operational and structural aspects of the sensor intelligent agents,

A Sensor Type intelligent agent, employees three types of behaviors. The "*setup behavior*" is performed once during the initial creation of the agent in order to specify the startup arguments namely (Data Input file, input time interval, Graphical user

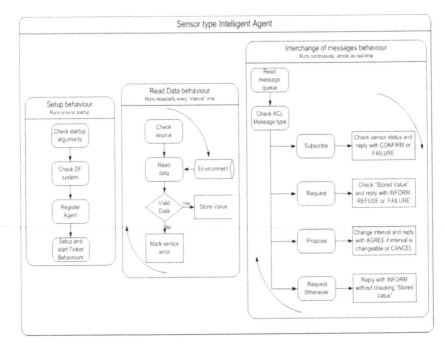

Fig. 3. Architecture of a sensor intelligent agent

interface and changeability of the temporal interval). It also checks the existence of the Directory Facilator (DF) which (under the JADE environment) stores several details regarding each intelligent agent (e.g. name, owner, PC where the agent is connected, ontologies, the type of the agent). The setup ensures that there is no identical agent in the system and it registers the agent in the DF. After the registration, the other two behaviors are activated.

The "*Read Data behavior*" is executed when a specific amount of time passes (the interval time) and it is responsible for the input of the parameters' values. During the execution of each iteration, several checks are made for potential input errors in the input file. If errors are encountered the agent is marked as "*non operational*".

The third behavior is the "*Interchange of messages*" which undertakes the communication of the agent with the rest of the system (other agents). It checks in the message queue of the agent for incoming messages. The following message types are executed: *Subscribe messages* that aim in checking and updating the system's status. In each subscribe request the agent responds with a *confirm* or *failure* subscribe message depending on the functionality of the agent as it is defined by "Read Data behavior". "*Request*" *messages* that aim in updating the system with the last crisp measurement that the agent holds. The sensor agent sends a reply of the type *inform, refuse* or *failure* depending on the time elapsed from the previous request and of course from its status. Messages of "*Propose*" type where the agent is asked to change its time interval. Depending on it setup arguments it answers the proposal with agree or with cancel. The last type of message it accepts is the "*Request whenever*" where other agents require the last data input to a sensor agent regardless its operational status. The answer is of the type *inform*.

3.3 Evaluation Agents

The risk estimation model described above has been used in the development of the intelligent agent network in order to determine the seriousness of the situation. More specifically the following two semi-trapezoidal functions 9 and 10 and the triangular membership function 11 are applied. The boundaries of the following membership functions for the air pollutants were estimated by the European Union instructions 199/30/EC, [18] 200/69/EC [5] and COM(2000)613 [9] and for meteorological factors they were specified according to the Heat index and thermal comfort models SO Standard 7730/1984 [15]. They are all presented in the following table 1.

$$\mu_s(x;a,b) = \begin{cases} 1, & x \le a \\ \dfrac{b-x}{b-a}, & a < x < b \\ 0, & b \le x \end{cases} \tag{9}$$

$$\mu_s(x;f,g) = \begin{cases} 0, & x \le f \\ \dfrac{x-f}{g-f}, & f < x < g \\ 1, & g \le x \end{cases} \tag{10}$$

$$\mu_s(x;c,d,e) = \begin{cases} 0, & x \le c \\ \dfrac{x-c}{d-c}, & c < x \le d \\ \dfrac{e-x}{e-d}, & d < x < e \\ 0, & e \le x \end{cases} \tag{11}$$

Table 1. Boundaries of the fuzzy membership functions

Parameter	a	b	c	d	e	f	g	unit
Temperature	20	30	20	30	38	30	38	°C
CO	1	2.5	1	2.5	4	2.5	4	mgr^{-3}
NO	18	83	18	83	148	83	148	μgr^{-3}
NO_2	50	130	50	130	200	130	250	μgr^{-3}
O_3	80	120	80	120	210	120	280	μgr^{-3}
pm10	30	67	30	67	79	67	90	μgr^{-3}
Air Pressure	950	1025	950	1025	1200	1026	1200	hPa
R. Humidity	30	50	30	50	70	50	70	%
Solar radiation	250	530	250	530	810	530	810	Wm^{-2}
Wind Speed	1.8	3.6	1.8	3.6	5.4	3.6	5.4	ms^{-1}
Wind u'	20	28	20	28	36	28	36	ms^{-1}
Wind v'	20	28	20	28	36	28	36	ms^{-1}

The following figure 4 presents the structure of the employed three functions.

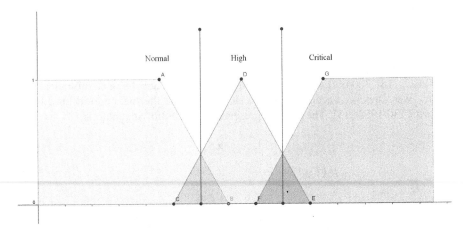

Fig. 4. The three functions applied for the determination of the three potential situations

The *Evaluation* agents are responsible for the assessment of the input values that they accept from the sensor agents by applying their internal logic. They feed the system with fuzzy risk values according to their *fuzzy perspective*. From the architectural point of view, each evaluation intelligent agent is logically connected to a sensor intelligent agent whereas a sensor agent can feed several evaluation agents. There are three behaviors related to an evaluation agent: The "setup behavior" which acts in a similar manner with the corresponding behavior of the sensor agent. It is executed once while the agent is created and it is responsible for the registration of the agent on the DF, for the synchronization of the time interval of an evaluation Intelligent Agent with the time interval of its connected sensor agent. The *"Read Data behavior"* is executed periodically after an interval of time passes and it is responsible for the input of the value of the connected sensor agent, its evaluation and for providing the system with the produced output. In every iteration, it sends a *request type* message to the connected sensor agent. Depending on the answer it receives from the request it updates the system. If it receives an *"inform"* type of answer it produces the fuzzification processes and it updates the system for the degree of membership of the input value to each predefined fuzzy set. If a critical situation is met, or if the situation is getting worse from one measurement to another the evaluation agent might send a *Proposal message* to the sensor agent, proposing the intensification of the time interval which will respond based on its own rules. If a *"refuse"* type answer is received this means that the value of the sensor agent has not changed due to the fact that the interval has not passed. This will make the evaluation agent in a *"hold on"* status. A *"failure"* answer will mean that the sensor agent is not working properly. In this case a time interval *reset* command will be sent and executed to the sensor agent.

The last behavior is the *"Reply to messages"* which handles the messages that the agent receives from its environment. It is executed *periodically*. For every message in the agent's queue its type is checked (based on the FIPA ACL Message type). The first message type that can be accepted is the subscribe message. In each subscribe request accepted by the evaluation agent, it creates and sends a subscribe message to the connected sensor agent. Depending on the operational status of the sensor agent as it is defined by the "Read Data Behavior" it will answer the system with "*confirm*" or with "*failure subscribe*" message. In this way the system is notified on the statuses of both the evaluation and the sensor agents. Messages of type "*inform*" which aim in the *time synchronization* between the agents. In every "*inform*" message the system's timestamp is compared to the time interval of the agent and if it is higher then it informs for a potential malfunction of the system. Finally messages of the type "*Requet whenever*" are employed that are used to ask for the update of the Decision Making agents with the actual crisp values that are also used together with the fuzzy values by the decision making process of the system. This is the Hybrid part of the system's Inference engine. The answer in this type of message is of the type "*Inform*". The following figure 5 is a detailed description of an evaluation agent and its basic behaviors.

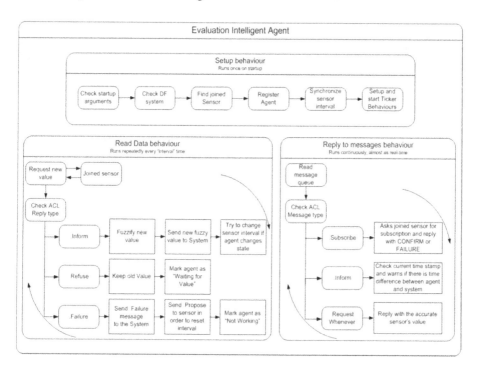

Fig. 5. Architecture of an Evaluation Intelligent Agent

3.4 System's Agents

The System's Intelligent Agent is responsible for the proper synchronization of the other agents and for the storage of the fuzzy output of the evaluation agents. The system's agent is the most complex of all and it comprises of five behaviors. The first is the "*setup behavior*" which is similar with the corresponding behavior of the other agent types. The "*Time synchronize behavior*" is executed periodically for random and short time intervals. In every execution it checks the DF for connected agents and it sends to all of the connected agents messages of the *"inform"* type with the current timestamp of the system in order to enable its comparison to the timestamp of each agent. Another periodically executed behavior is the *"Discover Agents"* which is operating in a *"ping-pong"* fashion. It aims the connection status of the agents to the system and it removes from the DF all agents that are disconnected. It sends *"Subscribe"* messages to all agents that were expected to be connected but they are not. If it receives a "confirm" message it keeps the agents in the DF and in the list of active agents. If it does not receive an answer for more than a certain amount of time (e.g. 2 sec) it considers that the answer is *"failure"*. The *"Reply to requests"* behavior answers to registered and authorized agents in the DF and informs them on the situation of the environment sending actual fuzzy data from the Databank with a message of the type *"Inform"* or *"Failure"* otherwise. The *"Accept Value Inform"* (periodically executed) behavior reads messages sent by the Evaluation Agents regarding the situation of the environment and it informs the DataBank of the system. In each of its iterations it uses messages of the *"Inform"* type from the messages' queue ad it controls if the message sender (agent) is registered in the DF and authorized. In this case it informs the Databank with the new values.

The following figure 6 presents the actual structure and the basic operations performed in a System's Intelligent Agent.

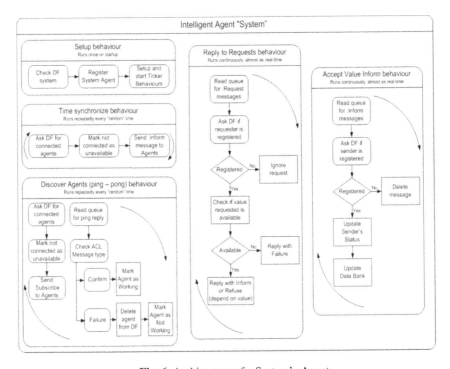

Fig. 6. Architecture of a System's Agent

3.5   Decision Agents

The "*Decision*" type Intelligent Agents perform (each one based on its reasoning and on its knowledge base) assessment of the environmental condition, regarding the parameters examined by each one and they decide on the direction of changing the environmental conditions by proposing interventions and actions towards the improvement of the situation. Each one of the *Decision* type Intelligent Agents are characterized by their particular distinct properties which support the decision making process, they vary in the degree of significance of each one and on the number of checks required. The above particularities combined with the need for the construction of an open system in which the users groups will be able to add custom *Decision* type agents led to various forms of Decision agents that use the platform to obtain data vectors and then they act independently to output proper actions and results. All of the *Decision* agents are compatible to the FIPA protocol of communication and they contact systems agent or the evaluation agents using the messages described above. They have 3 behaviors. The following description is related to the Temperature and Relative Humidity *Decision* agents. The first is the "*setup behavior*". The "*keep connected to Evaluation Agents*" behavior allows the decision agents to control (in random time intervals) the evaluation agents if they are connected to the system and if they operate properly. In each iteration the operational statuses of the evaluation agents related to Temperature and to Relative Humidity are checked. A *main behavior* is the (periodic) "*Decide and Evaluate*" that performs the assessment of the data and the assessment of the examined case and leads to the determination of the proper actions. It reads from the system agent the fuzzy degrees of membership for the Temperature and the Relative Humidity of the Environment.

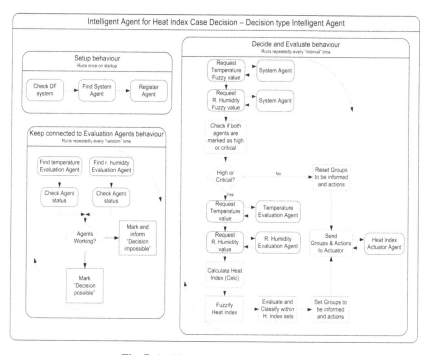

Fig. 7. Architecture of a Heat Index Agent

When both parameters belong to the fuzzy set ***normal*** then the Heat Index is in the scale "***Comfortable***" where the agent sets the *inform groups* and the *proposed actions* to zero informing the actuator agent.

In the system's agent sends values higher than Normal (high ή critical) then the agent sends a message of type "***request whenever***" to the corresponding evaluation agents in order to obtain the actual crisp Temperature and Relative Humidity values. Based on these values it estimates the Heat Index which is then translated to proper fuzzy Linguistic values. The Linguistics used are namely: *slightly uncomfortable, uncomfortable, very uncomfortable, intolerable and Death danger* [21].

Depending on the characterization of the case the vulnerable groups of the population are picked and all the necessary measures are proposed and sent to the connected actuator agent.

3.6 Actuators

The ***Actuators*** are receiving messages from the connected ***Decision Agents*** and if there is no reason for blocking their execution (any kind of contradiction to other actions proposed by other agents) it sends the decisions to policy makers in order to be executed. For example if none of the evaluation agents are in a critical situation in any of the air pollution factors, then the actuators propose to the decision makers the opening of the city centre ring to the cars. The opposite action is taken if the situation is characterized as critical at least for one air pollution parameters.

4 Graphical User Interface

The user interface has been designed to operate in a flexible and friendly manner. Below we present a small sample of the graphical user interface for some specific agents.

Screen 1. Graphical User Interface of an evaluation agent

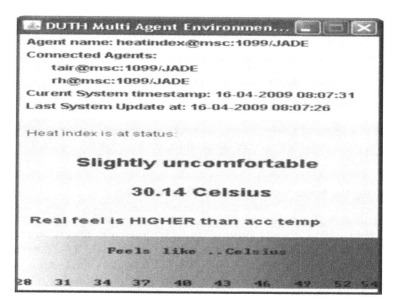

Screen 2. Graphical User Interface of a Decision *"Heat Index"* agent

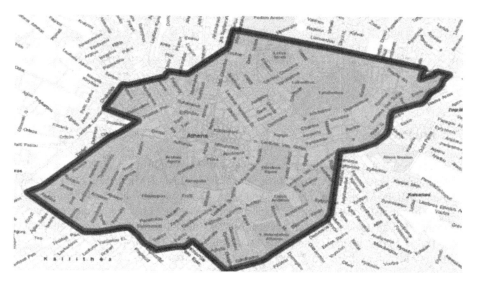

Screen 3. Graphical User Interface of an Actuator *"Ring closing"* Agent

5   Pilot Application of the System

5.1   Running the Agent Network for Heat Index

The agent for the estimation and evaluation of the heat index has checked the situation of the environment 4233 times running in an iterative mode. From the 4233 iterations the agent has calculated the heat index in 4123 cases since in the rest 109 the two determining factors (temperature and relative humidity) were characterized as been in a *normal state*.

The heat index in 1958 cases has proven to be higher than the actual temperature whereas in 2615 instances it was lower. This can be easily explained due to the dry climate of the Attica area. Totally in 2748 cases the heat index was below the temperature of 30 degrees C and thus no action was required. In 1256 instances the heat index was characterized as slightly uncomfortable, whereas 119 times it was characterized as uncomfortable with immediate need for actions, directed especially towards the protection of the most vulnerable population groups.

5.2 Running the Agent Network for Air Pollutants

The agent network (based on the existing data) has checked the situation of the air quality in terms of air pollutants' concentration. Five times the situation became "very risky" for the NO_2 concentration with a duration extended to 2,1,2,4,3 hours respectively (the official limit is 250mg for 3 hours). In four cases there was a *"critical situation"* regarding the concentration of the NO_2 (the official limit is 360mg for 3 hours). The temporal duration of the critical condition was 1,4,2,4 hours respectively.

Of course the critical incidents would be much more if we also had night and winder data in our disposal. Unfortunately the data that we have gathered are daily and they are related only to summer months. For the particular matter pm_{10} the situation became the situation became "very risky" (the official limit is 90mg for 5 continuous days) whereas in 3 cases the situation became "critical" for 7 continuous days (the official limit is 110mg for 7 continuous days). Once we had a red alert for the PM_{10} and for the NO_2 simultaneously. In every risky situation the map of the city center ring turns to orange color, whereas in the case of the critical conditions it turns into red.

6 Conclusions

As it has been already mentioned the system is Hardware independent. Due to its source-internal support of various Java versions like Java 2 Micro Edition, Personal Java and MIDP -Mobile Information Device Profile- Java, (beyond the J2SE version) and also due to their distribution and their division in smaller ones they can be executed in smaller mobile devices like cell phones, personal digital assistants (PDA) and palmtops. This is a great advantage and it shows the potential portability of the system.

Its hybrid nature employing fuzzy algebra, towards the flexible evaluation (with proper linguistics) of the air quality and of the heat index situation, enhances the intelligent and rapid inference capabilities of the agent network.

So it can be characterized as a real-time innovation that can not only monitor the air quality and the heat index for a major urban centre like Athens but moreover it can interact with environmental policy makers towards the deployment of corrective actions. Its portability to any urban center all over the globe without any adjustments is a very serious advantage. The only thing remaining after the first pilot application in Athens city center is its experimental adaptation by the civil protection services towards a better quality of life for all citizens.

References

1. Alonso, F., Fernández, R., Frutos, S., Soriano, F.J.: Engineering agent conversations with the DIALOG framework. In: Fischer, K., Timm, I.J., André, E., Zhong, N. (eds.) MATES 2006. LNCS, vol. 4196, pp. 24–36. Springer, Heidelberg (2006)
2. Balaguer Ballester, E., Camps Valls, G., Carrasco Rodriguez, J.L., Soria Olivas, E., Del Valle, T.S.: Effective 1-day ahead prediction of hourly surface ozone concentrations in eastern Spain using linear models and neural networks. Ecological Modelling 156, 27–41 (2002)
3. Bellifemine, F., Caire, G., Greenwood, D.: Developing mutli-agent systems with Jade. John Wiley and Sons, USA (2007)
4. Bigus, J., Bigus, J.: Constructing Intelligent Agents using Java, 2nd edn. J. wiley and sons, Chichester (2001)
5. Borrego, C., Tchepel, O., Costa, A.M., Amorim, J.H., Miranda, A.I.: Emission and dispersion modelling of Lisbon air quality at local scale Atmospheric Environment. 37(3) (2003)
6. Brauer, M., Brook, J.R.: Ozone personal exposures and health effects for selected groups residing in the Fraser Valley. Atmospheric Environment 31, 2113–2121 (1997)
7. Burnett, R.T., Brook, J.R., Yung, W.T., Dales, R.E., Krewski, D.: Association between Ozone and Hospitalization for Respiratory Diseases in 16 Canadian Cities. Environmental Research 72, 24–31 (1997)
8. Chaloulakou, A., Saisana, M., Spyrellis, N.: Comparative assessment of neural networks and regression models for forecasting summertime ozone in Athens. The Science of the Total Environment 313, 1–13 (2003)
9. Commission of the European Communities, Directive of the european parliament and of the council relating to ozone in ambient air (2000),
   http://aix.meng.auth.gr/AIR-EIA/EU/en_500PC0613.pdf
10. Content, J.M., Gechter, F., Gruer, P., Koukam, A.: Application of reactive multiagent system tolinear vehicle platoon. In: Proceedings of the IEEE ICTAI 2007 International Conference, Patras, vol. 2, pp. 67–71 (2007)
11. Corchado, E., Corchado, J.M., Abraham, A.: Hybrid Intelligence for bio Medical Informatics S.I. In: 2nd International workshop on Hybrid Artificial Intelligence Systems, Spain (2007)
12. Iliadis, L.: Intelligent Information systems and Applications in risk estimation. Stamoulis A., publishing, Thessaloniki, Greece (2007)
13. Kecman: Learning and Soft Computing. MIT Press, Cambridge (2001)
14. Kokawa, T., Takeuchi, Y., Sakamoto, R., Ogawa, H., Kryssanov, V.: An Agent Based System for the Prevention of Earthquake-Induced Disasters. In: Proceedings of the IEEE ICTAI 2007 International Conference, Patras, vol. 2, pp. 55–62 (2007)
15. Labaki, L., Barbosa, M.P.: Thermal comfort evaluation in workplaces in Brazil: the case of furniture industry. In: Proceedings of Clima 2007 WellBeing Indoors (2007)
16. Melas, D., Kioutsioukis, I., Ziomas, I.: Neural Network Model for predicting peak photochemical pollutant levels. Journal of the Air and Waste Management Association 50, 495–501 (2000)
17. Paschalidou, A.K., Kassomenos, P.A.: Comparison of air pollutant concentrations between weekdays and weekends in Athens, Greece for various meteorological conditions. Environmental Technology 25, 1241–1255 (2004)

18. Sixth Environment Action Programme (6th EAP) scoreboard (Date of validation: 26/10/2005),
    `http://ec.europa.eu/environment/newprg/pdf/`
    `6eap_scoreboard_oct2005.pdf` (data until 30/09/2005)
19. Smith-Doron, M., Stieb, D., Raizenne, M., Brook, J., Dales, R., Leech, J., Cakmak, S., Krewski, D.: Association between ozone and hospitalisation for acute respiratory diseases in children less than 2 years of age. American Journal of Epidemiology 153, 444–452 (2000)
20. Soja, G., Soja, A.M.: Ozone indices based on simple meteorological parameters: potential and limitations of regression and neural network models. Atmospheric Environment 33, 4229–4307 (1999)
21. Steadman, R.G.: The Assessment of Sultriness. Part I: A Temperature-Humidity Index Based on Human Physiology and Clothing Science. Journal of Applied Meteorology 18(7), 861–873 (1979)

Neural Network Based Damage Detection of Dynamically Loaded Structures

David Lehký and Drahomír Novák

Institute of Structural Mechanics, Faculty of Civil Engineering, Brno University
of Technology, Veveří 95, 602 00 Brno, Czech Republic
{lehky.d,novak.d}@fce.vutbr.cz

Abstract. The aim of the paper is to describe a methodology of damage detection which is based on artificial neural networks in combination with stochastic analysis. The damage is defined as a stiffness reduction (bending or torsion) in certain part of a structure. The key stone of the method is feed-forward multilayer network. It is impossible to obtain appropriate training set for real structure in usage, therefore stochastic analysis using numerical model is carried out to get training set virtually. Due to possible time demanding nonlinear calculations the effective simulation Latin Hypercube Sampling is used here. The important part of identification process is proper selection of input information. In case of dynamically loaded structures their modal properties seem to be proper input information as those are not dependent on actual loading (traffic, wind, temperature). The methodology verification was carried out using laboratory beam.

Keywords: Artificial neural network, damage, identification, eigenfrequencies, mode shapes, statistical simulation.

1 Introduction

Civil engineering structures such as bridges must be periodically inspected to ensure structural integrity since many of these structures have achieved their service life and some damage may occur. Visual inspection and local non-destructive evaluation as the conventional approaches are expensive, subjective, inconsistent, labor and time intensive and need easy access to the damage zone. That is the reason for recently extended research and development of structural health monitoring (SHM) techniques [1]. Among these, modal based techniques have been extensively investigated due to their global nature and simplicity. They can be used for automated damage localization and result in consistent damage assessment. From the practical point of view, damage assessment can be categorized into four levels: 1) detecting if the structure is damaged; 2) finding the location of damage; 3) estimating the magnitude of damage and 4) evaluating the remaining service life of the structure.

Modal based techniques use ambient vibration measurements and can be used for structures in usage. A typical result of the experimental measurements is the dynamic response of the structure in form of time series (accelerations, velocities). Consequently, modal properties (mode shapes and corresponding eigenfrequencies – "characteristic" frequencies at which a system vibrates), damping characteristics and assurance criteria

D. Palmer-Brown et al. (Eds.): EANN 2009, CCIS 43, pp. 17–27, 2009.

MAC, COMAC, DLAC and so on, are evaluated (e.g. [2], [3]). The subject of research of both academic and industrial research groups during last decade is utilization of this kind of structural response information for damage localization and structural health assessment. The task is based on the fact that damaged structure has smaller stiffness in some parts – and this difference will affect vibration and modal properties. The comparison of vibration of virgin (undamaged) structure and damaged structure can be used for the detection of damaged parts (localization of damage).

Efficiency of identification procedure increases with proper sensors placing on the structure. Sensors placed closer to damaged part of the structure shows higher sensitivity [4]. Besides influence of structural damage also so called operative conditions (e.g. temperature change) should be taken into account during inverse analysis [5].

"Model updating method" is the term frequently used in connection with SHM and damage detection [6], [7], [8], [9], [10], [1]. "Updating" means that individual parameters of FEM model are iteratively changed in order to minimize the difference between experimentally measured and calculated response. A sensitivity of the response on model parameters is frequently used and can be directly utilized for efficient identification [11], [12].

The aim of the research is to work out a methodology of dynamic damage detection based on the coupling of Monte Carlo type simulation and artificial neural networks (ANN). It extends a methodology of inverse analysis developed and applied for fracture/mechanical parameters identification [13], [14], [15], [16].

Important part of damage detection procedure is proper selection of input information. For that purpose authors carried out several modal properties studies (numerical and experimental) using simple laboratory beams as well as real bridge structures [17], [18]. The aim of those studies was to find out which eigenfrequencies, mode shapes or assurance criterions are affected by change of stiffness in certain position on the structure. Results show that if the damage is reasonably large, eigenfrequencies are sufficient for damage detection as it is seen later in this paper. Lower frequencies are effected more then higher ones but not constantly along the structure. Their shift corresponds to mode shapes. That is the reason why for detection of damage in some positions on the structure higher eigenfrequencies must be used.

If mode shapes are available their utilization can be helpful for identification – mode shapes itself or modal assurance criterion (MAC) [2]. It is essential that in comparing with frequencies higher mode shapes are more affected by stiffness change than lower ones. Unfortunately it is not easy or even impossible to obtain higher mode shapes from ambient vibration measurement.

Extensive research has been done by other authors to suggest other quantities which can be used for damage detection using various methods. Among them, other assurance criterions (COMAC [2], DLAC, MDLAC [3]), rank ordering of eigenfrequency shifts [19], damage index (DI) [2] and so on, can be mentioned.

2   Methodology of Damage Detection

Proposed method for damage detection is based on artificial neural networks (ANN) in combination with stochastic analysis. The damage is defined as a stiffness reduction (bending or torsion) in certain part of a structure. ANN then serves as an approximation for following inverse task: what damage has caused the given change of structural response? The whole procedure of inverse analysis can be itemized as follows:

1. The computational model of a particular problem has to be first developed using the dynamic finite element method (FEM) software. In case of dynamic damage identification identified parameters (IP) are usually values of stiffness varied along the structure, often Young modulus of elasticity. Measured data (MD) are modal parameters (eigenfrequencies, mode shapes).

2. IP of the computational model are considered as random variables described by a probability distribution; the rectangular distribution is a "natural choice" as the lower and upper limits represent the bounded range of the physical existence of IP. However, also other distributions can be used, e.g. the Gaussian one. IP are simulated randomly based on the Monte Carlo type simulation, the small-sample simulation LHS is recommended. The results are random realizations of IP (vector **y**, see Fig. 1). A statistical correlation between some parameters may be taken into account too.

3. A multiple calculation (simulation) of FEM model using random realizations **y** of IP is performed, a statistical set of the virtual response **p** (see Fig. 1) is obtained. The selection of number of simulations is driven by many factors, mainly by complexity of the problem (computational demand), structure of ANN and variability of IP.

4. Random realizations **y** (outputs of ANN) and the random responses from the computational model **p** (inputs of ANN) serve as the basis for the training of an appropriate ANN. This key point of the whole procedure is sketched in Fig. 1 (here for the FEM model response in the form of eigenfrequencies).

5. The trained neural network is ready to give an answer to the key task: To select the best parameters IP so that the calculation may result in the best agreement with MD, which is performed by means of the network simulation using MD as an input. This results in an optimal set of parameters $\mathbf{y}_{opt}$.

6. The last step is the results verification – the calculation of the computational model using optimal parameters $\mathbf{y}_{opt}$.

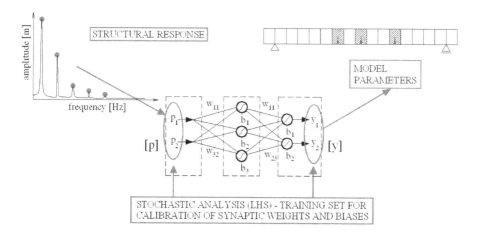

Fig. 1. A scheme of ANN based dynamic damage detection

2.1 Artificial Neural Network

In structural mechanics the classical problem is defined in the way that for given structural, material, loading, environmental ... data (vector **y**) the corresponding structural response is obtained – experimentally or numerically (vector **p**):

$$\mathbf{p} = f(\mathbf{y}). \tag{1}$$

In case of damage detection of dynamically loaded structures we can simplify it that for stiffness distribution along the structure the structural response in form of modal properties is obtained. But the task which is being solved here is opposite to the classical one: what damage has caused the given change of structural response? This inverse task is defined as (see Fig. 1):

$$\mathbf{y} = f^{-1}(\mathbf{p}) = T(\mathbf{p}). \tag{2}$$

Instead of finding inversion to function f in analytical form the transformation T using artificial neural network is used.

Final step is to calculate structural response using active phase of artificial neural network using identified parameters $\mathbf{y}_{opt}$.

$$\mathbf{p} = f(\mathbf{y}_{opt}). \tag{3}$$

Suitable type of neural network for this task is feed-forward multi-layer perceptron. Such a network consists set of neurons arranged in several layers – input layer (sometimes called zero layer), a number of hidden layers and output layer. Number of hidden layers and neurons in it is driven by complexity of problem which is being solved. Considering Kolmogorov's theorem about the third Hilbert's problem (e.g. [20]) the maximum needful number of hidden layers is two with necessary amount of neurons in each of them. For damage detection the one hidden layer proves to be sufficient quantity (see application further).

To determine proper amount of neurons in hidden layers is not a simple task. As a recommendation for a first rough estimation the following formula for network with one hidden layer can be used [21]:

$$N_{hidden} = \sqrt{N_{inp} \cdot N_{out}}, \tag{4}$$

where N_{inp} is the number of inputs and N_{out} is the number of neurons in output layer.

As a transfer (activation) function of neurons in hidden layer hyperbolic tangent function is used; output neurons have linear transfer function.

2.2 Stochastic Analysis

ANN must be trained first using appropriate training set for correct approximation of inverse task (2). Such a training set consists of ordered pairs input–output [$\mathbf{p}_i,\mathbf{y}_i$] (see Fig. 1) where inputs are modal data (eigenfrequencies, mode shapes) and outputs are stiffness distributions along the structure for various damage situations. It is impossible to obtain appropriate training set for the real structure in usage (we are not allowed to damage structure), therefore stochastic analysis using numerical model is carried out to get training set virtually. For that purpose an appropriate numerical

FEM model has to be first developed and checked to be in good agreement with real structural behavior. Parameters of the model are than randomized (stiffness in certain parts is reduced) and stochastic analysis is performed.

Numerical FEM analyses can be very time consuming, especially in non-linear cases, therefore efficient simulation technique should be used. In proposed methodology of damage detection a simulation method Latin Hypercube Sampling (LHS) is used. It is Monte Carlo type simulation method with stratified sampling scheme. The whole multi-dimensional space of IP is covered perfectly by relatively small number of simulations [22], [23]. Efficiency of this method for training of the ANN was proved by authors in [13].

An important task in the inverse analysis is to determine the significance of parameters which are subject to identification. With respect to the small-sample simulation techniques described above, the straightforward and simple approach can be used based on the non-parametric rank-order statistical correlation between the basic random variables and the structural response variables by means of the Spearman correlation coefficient or Kendall tau [23]. The sensitivity analysis is obtained as an additional result of LHS, and no additional computational effort is necessary.

Training of ANN means setting up synaptic weights and biases of neurons. That is optimization problem where the following error of the network is minimized:

$$E = \frac{1}{2}\sum_{i=1}^{N}\sum_{k=1}^{K}\left(y_{ik}^{v} - y_{ik}^{*}\right),$$ (5)

where N is a number of ordered pairs input–output in training set, y_{ik}^{*} is required output value of k-th output neuron for i-th input and y_{ik}^{v} is the real output value (for the same input). In below described application Levenberg–Marquardt optimization method [24] and evolution strategy method [25] were used.

3 Software Tools

The authors combined efficient techniques of statistical simulation of Monte Carlo type with artificial neural network and FEM structural dynamics analysis to offer an advanced tool for the damage identification of concrete structures. The combination of all parts (structural analysis, statistical simulation, inverse analysis and reliability assessment) is presented together in a package as the Relid software system (see Fig. 2). It includes:

- FReET [26] – the probabilistic engine based on LHS simulation; First, it is used for preparation of training set for artificial neural network. Second, when damage is identified, it can be consequently used for reliability assessment of given problem using the results from inverse analysis.
- DLNNET [27] – artificial neural network software; the key stone of inverse analysis which has to communicate with FReET and SOFiSTik finite element software.
- SOFiSTiK [28] – commercial FEM software which performs structural dynamics analysis of a particular problem.
- Relid [29] – software shell which manages a flexible communication among programs mentioned above in a user-friendly environment.

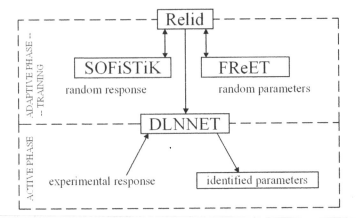

Fig. 2. Software communication scheme within Relid software shell

4 Application – Cantilever Beam

One part of the development of methodology for dynamic damage detection is dynamic laboratory experiments using simple structures. As a first testing configuration a simple cantilever beam made of timber was chosen. Selection was driven by the premise of further experimental testing, existence of analytical solution and non-symmetry of such structure (in the contrary to e.g. simply supported beam).

Dynamic experiment using five specimens ($c11,\dots, c15$) made of ash wood was carried out. Proportions of specimen were as follows: length = 1 m, width = 30 mm and height = 7 mm. Fixed support was realized by satisfactory rigid gripping in length 100 mm; then resulting length of cantilever beam was 0.9 m (Figs. 3 and 4).

Measurement was recorded using five accelerometers Brüel & Kjær 4508-B-001 which were connected to Brüel & Kjær 3560-B-140 analyzer. Fig. 3 shows testing configuration and how fixed support was realized. Loading was implemented with the view of repeatability by fall of plastic bullet with 6 mm diameter and 0.25 g weight from the height of about 32 cm on free end of cantilever beam. Only one touch between bullet and beam was accepted.

Fig. 3. Cantilever beam: a) testing configuration, b) realization of fixed support

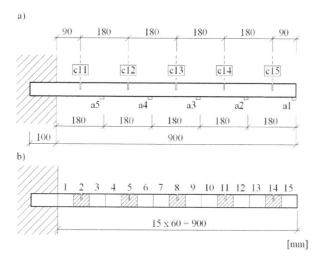

Fig. 4. a) Scheme of cantilever beam – positions of accelerometers and cuts for all five specimens; b) numerical model divided into 15 parts

Table 1. Comparison of results of dynamic measurements for undamaged and damaged beams

Specimen	Frequency [Hz]	Undamaged	Damaged	Relative change [%]
c11	f_1	5.191	4.923	5.15
	f_2	33.712	33.090	1.84
	f_3	96.663	94.188	2.56
	f_4	185.738	184.363	0.74
	f_5	303.588	299.263	1.42
c12	f_1	6.319	6.047	4.32
	f_2	38.062	37.387	1.77
	f_3	107.212	101.938	4.92
	f_4	208.112	205.312	1.35
	f_5	331.512	329.412	0.63
c13	f_1	6.060	5.955	1.72
	f_2	35.040	33.638	4.00
	f_3	96.562	96.837	-0.28
	f_4	186.337	180.112	3.34
	f_5	302.037	303.137	-0.36
c14	f_1	6.060	6.057	0.04
	f_2	36.014	34.662	3.75
	f_3	100.262	94.563	5.68
	f_4	193.662	191.012	1.37
	f_5	312.712	309.062	1.17
c15	f_1	5.944	5.932	0.20
	f_2	36.136	36.111	0.07
	f_3	100.837	99.313	1.51
	f_4	193.662	187.112	3.38
	f_5	313.862	302.812	3.52

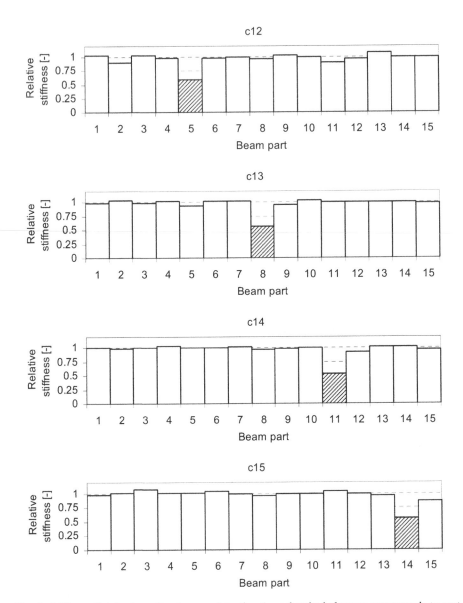

Fig. 5. Stiffness distribution after damage detection (crosshatched element corresponds to part where beam was cut)

When measurements using undamaged beams were carried out all five specimens were cut in certain part (see scheme in Fig. 4). Those damaged beams were then tested again and modal properties were evaluated. Comparison of results of dynamic measurements for undamaged and damaged beams is introduced in Table 1.

Relative changes of all five eigenfrequencies were then used for damage detection using proposed ANN based method. For that purpose a numerical dynamic model in

SOFiSTiK software [28] was created. Beam was divided into 15 parts with independent bending stiffness $EI_1,..., EI_{15}$ (see Fig. 4).

ANN consists of 1 hidden layer with 10 nonlinear neurons and an output layer with 15 linear neurons (15 stiffness values). There are also 5 inputs of the network (5 eigenfrequencies). Training set contains 120 samples and was divided into two parts, 115 samples were used directly for training the network, while 5 samples served for testing of network overfitting. After the network was trained, the eigenfrequencies from experimental testing for damage state were used for the simulation of ANN. The output of ANN is a spatial distribution of stiffness along the girder (15 values, see Fig. 5). Because of inconsistency between experimental and numerical results of specimen c11, the beam was omitted from damage detection. It was probably caused by some testing errors along with high vibration sensitivity to damage located close to fixed end of cantilever beam.

5 Conclusions

Efficient techniques of stochastic simulation methods were combined with artificial neural networks in order to offer an advanced tool for damage identification of dynamically loaded structures. The methodology was tested using wood beams virtually damaged. The efficiency was verified using this artificial laboratory experiment. A multipurpose software tool for damage identification was developed. The methodology was already applied for real structures (bridges) [18], [30] with satisfactory results. These applications are not included here, as the main target of this paper was to present the theory of proposed damage identification and verification using "clear experiment' without any significant "contamination", which is usual in case of real structures.

Acknowledgments. Financial support within the grant of the Grant Academy of Czech Republic No.103/07/P380 and the Research Centre project No. 1M0579 (CIDEAS) are greatly appreciated.

References

1. Wenzel, H., Pichler, D.: Ambient vibration monitoring. John Wiley & Sons Ltd, West Sussex (2005)
2. Salgado, R., Cruz, P.J.S., Ramos, L.F., Lourenço, P.B.: Comparison between damage detection methods applied to beam structures. In: Third International Conference on Bridge Maintenance Safety and Management, Porto, Portugal, CD-ROM (2006)
3. Koh, B.H., Dyke, S.J.: Structural health monitoring for flexible bridge structures using correlation and sensitivity of modal data. Computers and Structures 85, 117–130 (2007)
4. Spencer Jr., B.F., Gao, Y., Yang, G.: Distributed computing strategy for damage monitoring employing smart sensors. In: Shenzhen, C., Ou, L., Duan (eds.) Structural Health Monitoring and Intelligent Infrastructure, pp. 35–47. Taylor & Francis Group, London (2005)

5. Feltrin, G.: Temperature and damage effects on modal parameters of a reinforced concrete bridge. In: 5th European Conference on Structural Dynamics (EURODYN), Munich, Germany (2002)
6. Link, M.: Updating of analytical models – basic procedures and extensions. In: Silva, J.M.M., Maia, N.M.M. (eds.) Modal Analysis and Testing. NATO Science Series. Kluwer Academic Publ, Dordrecht (1999)
7. Teughels, A., Maeck, J., De Roeck, G.: Damage assessment by FE model updating using damage functions. Computers and Structures 80(25), 1869–1879 (2002)
8. Deix, S., Geier, R.: Updating FE-models using experimental modal analysis for damage detection and system identification in civil structures. In: Third European Conference on Structural Control (3ECSC). Vienna University of Technology, Vienna (2004)
9. Fang, X., Luo, H., Tang, J.: Structural damage detection using neural network with learning rate improvement. Computers and Structures 83, 2150–2161 (2005)
10. Huth, O., Feltrin, G., Maeck, J., Kilic, N., Motavalli, M.: Damage identification using modal data: Experiences on a prestressed concrete bridge. Journal of Structural Engineering, ASCE 131(12), 1898–1910 (2005)
11. Strauss, A., Lehký, D., Novák, D., Bergmeister, K., Santa, U.: Probabilistic response identification and monitoring of concrete structures. In: Third European Conference on Structural Control (3ECSC). Vienna University of Technology, Vienna (2004)
12. Strauss, A., Bergmeister, K., Lehký, D., Novák, D.: Inverse statistical FEM analysis – vibration based damage identification of concrete structures. In: International Conference on Bridges, Dubrovnik, Croatia, pp. 461–470 (2006)
13. Novák, D., Lehký, D.: ANN Inverse Analysis Based on Stochastic Small-Sample Training Set Simulation. Engineering Application of Artificial Intelligence 19, 731–740 (2006)
14. Novák, D., Lehký, D.: Inverse analysis based on small-sample stochastic training of neural network. In: 9th International Conference on Engineering Applications of Neural Networks (EAAN2005), Lille, France, pp. 155–162 (2005)
15. Lehký, D., Novák, D.: Probabilistic inverse analysis: Random material parameters of reinforced concrete frame. In: 9th International Conference on Engineering Applications of Neural Networks (EAAN2005), Lille, France, pp. 147–154 (2005)
16. Strauss, A., Bergmeister, K., Novák, D., Lehký, D.: Stochastische Parameteridentifikation bei Konstruktionsbeton für die Betonerhaltung. Beton und Stahlbetonbau 99(12), 967–974 (2004)
17. Frantík, P., Lehký, D., Novák, D.: Modal properties study for damage identification of dynamically loaded structures. In: The Third International Conference on Structural Engineering, Mechanics and Computation, Cape Town, South Africa, pp. 703–704 (2007)
18. Lehký, D., Novák, D., Frantík, P., Strauss, A., Bergmeister, K.: Dynamic damage identification of Colle Isarco viaduct. In: 4th International Conference on Bridge Maintenance, Safety and Management (IABMAS 2008), Seoul, Korea, pp. 2549–2556 (2008)
19. Armon, D., Ben-Haim, Y., Braun, S.: Crack detection in beams by rank-ordering of eigenfrequencies shifts. Mechanical Systems and Signal Processing 8(1), 81–91 (1994)
20. Kůrková, V.: Kolmogorov's theorem and multilayer neural networks. Neural Networks 5(3), 501–506 (1992)
21. Šnorek, M.: Neuronové sítě a neuropočítače (Neural networks and neurocomputers). Vydavatelství ČVUT, Prague, Czech Republic (2002) (in Czech)
22. McKay, M.D., Conover, W.J., Beckman, R.J.: A Comparison of Three Methods for Selecting Values of Input Variables in the Analysis of Output from a Computer Code. Technometrics 21, 239–245 (1979)

23. Novák, D., Teplý, B., Keršner, Z.: The role of Latin Hypercube Sampling method in reliability engineering. In: Proc. of ICOSSAR 1997, Kyoto, Japan, pp. 403–409 (1998)
24. Singh, V., Gupta, I., Gupta, H.O.: ANN-based estimator for distillation using Levenberg-Marquardt approach. Engineering Applications of Artificial Intelligence 20, 249–259 (2007)
25. Schwefel, H.P.: Numerical optimization for computer models. Wiley, Chichester (1991)
26. Novák, D., Vořechovský, M., Rusina, R.: FReET v.1.5 – program documentation. Userś and Theory Guides. Brno/Červenka Consulting, Czech Republic (2008),
    `http://www.freet.cz`
27. Lehký, D.: DLNNET – program documentation. Theory and User's Guides, Brno, Czech Republic (in preparation, 2009)
28. Sofistik, A.G.: SOFiSTiK Analysis Programs, version 21.0, Oberschleissheim, Germany (2004), `http://www.sofistik.com`
29. Lehký, D.: Relid – program documentation. User's Guide, Brno, Czech Republic (in preparation, 2009)
30. Lehký, D., Novák, D., Frantík, P., Strauss, A., Bergmeister, K.: Dynamic damage identification based on artificial neural networks, SARA – part IV. In: The 3rd International Conference on Structural Health Monitoring of Intelligent Infrastructure, Vancouver, British Columbia, Canada, vol. 183 (2007)

Reconstruction of Cross-Sectional Missing Data Using Neural Networks

Iffat A. Gheyas and Leslie S. Smith

Department of Computing Science and Mathematics
University of Stirling Stirling FK9 4LA, Scotland, UK
{iag,lss}@cs.stir.ac.uk

Abstract. The treatment of incomplete data is an important step in
the pre-processing of data. We propose a non-parametric multiple imputation algorithm (GMI) for the reconstruction of missing data, based on
Generalized Regression Neural Networks (GRNN). We compare GMI
with popular missing data imputation algorithms: EM (Expectation
Maximization) MI (Multiple Imputation), MCMC (Markov Chain Monte
Carlo) MI, and hot deck MI. A separate GRNN classifier is trained and
tested on the dataset imputed with each imputation algorithm. The imputation algorithms are evaluated based on the accuracy of the GRNN
classifier after the imputation process. We show the effectiveness of our
proposed algorithm on twenty-six real datasets.

Keywords: Missing values, imputation, multiple imputation, Generalized Regression Neural Networks.

1 Introduction

Real-world datasets are characterized by the very large dimensionality of the
input (explanatory) space often comprising thousands of attributes: in addition,
it is very often the case that, for some cases, the values of one or more explanatory variables are missing. These are incomplete datasets: datasets with
missing values. Most data mining algorithms cannot work directly with incomplete datasets. If missing data are randomly distributed across cases, we could
even end up with no valid cases in the dataset, because each of them will have
at least one missing data element. Missing value imputation is widely used, by
necessity, for the treatment of missing values. A major focus of research is to develop an imputation algorithm that preserves the multivariate joint distribution
of input and output variables. Much of the information in these joint distributions can be described in terms of means, variances and covariances. If the joint
distributions of the variables are multivariate normal, then the first and second
moments completely determine the distributions.

The practice of filling in a missing value with a single replacement is called
single imputation (SI). A major problem with SI is that this approach cannot reflect sampling and imputation uncertainty about the actual value. Rubin
[1] proposed multiple-imputation (MI) to solve this problem. MI replaces each

D. Palmer-Brown et al. (Eds.): EANN 2009, CCIS 43, pp. 28–34, 2009.
© Springer-Verlag Berlin Heidelberg 2009

missing value in a dataset with $m > 1$ (where m is typically small, e.g. 3-10) statistically plausible values. A detailed summary of MI is given in [1,3,4].

Little and Rubin [2] and Schafer [4] classify missing data into three categories: Missing Completely at Random (MCAR), Missing at Random (MAR), and Missing Not At Random (MNAR). MCAR and MAR data are recoverable, where MNAR is not. Various methods are available for handling MCAR and MAR missing data. The most common imputation procedure is mean substitution (MS), replacing missing values with the mean of the variable. The major advantage of the method is its simplicity. However, this method yields biased estimates of variances and covariances.

The most sophisticated techniques for the treatment of missing values are model based. A key advantage of these methods is that they consider interrelations among variables. Model-based methods can be classified into two categories: explicit model based algorithms and implicit model based algorithms. Explicit model based algorithms (such as least squares imputation, expectation maximization and Markov Chain Monte Carlo) are based on a number of assumptions [7,8,9]: however, if the assumptions are violated, the validity of the imputed values derived from applying these techniques may be in question.

Implicit model based algorithms are usually semi-parametric or non-parametric. These methods make few or no distributional assumptions about the underlying phenomenon that produced the data. The most popular implicit model based algorithm is hot deck imputation. This procedure replaces missing values in incomplete records using values from similar, but complete records of the dataset. Past studies suggest that this is promising [7]. A limitation of this method is the difficulty in defining similarity [8]. In addition, it treats all K neighbours in a similar way without consideration of the different distances between the query instance and its neighbours [9]. Recently, a number of studies applied multilayer perceptron (MLP) and radial basis function (RBF) neural networks to impute missing values [10]. However, creation of MLP or RBF networks is complex and requires many parameters. In this paper, we present a novel algorithm for the imputation of missing values. The remainder of this paper is organized as follows: the new algorithm in section 2 (with an overview of GRNN in section 2.1 and details of the proposed algorithm in section 2.2), details of how we assess the new technique in section 3, results and discussions in section 4, followed by summary and conclusions in section 5.

2   Developing a New Algorithm

We propose a simple imputation algorithm (GMI), based on Generalized Regression Neural Networks (GRNN) (described below), to reconstruct probabilistic distributions of multivariate random functions from the incomplete dataset. GMI is a multiple imputation algorithm. Like other multiple imputation algorithms, it has the advantage of taking into account the variability due to sampling and due to non-response and imputation. Three aspects of our approach are novel. Firstly GMI is based on a clustering algorithm and thus avoids distributional assumptions. This is important if the distribution of the data is skewed.

It can handle data from different distributions appropriately. Secondly only one parameter (the smoothing factor) needs to be adjusted for the proposed algorithm. However our empirical observations indicate that the performance of the algorithm is not very sensitive to the exact setting of this parameter value and that the default value of the parameter is almost always a good choice. The inherent model-free characteristics avoid the problem of model mis-specification and parameter estimation errors. Thirdly, although the proposed imputation algorithm closely resembles the hot deck imputation scheme wherein the donor is selected from a neighbourhood comprised of similar records, it does not suffer from the limitations associated with hot-deck imputation because GMI assigns different weights to the k nearest example neighbours according to Gaussian kernel functions and, in this algorithm, all observations participate (according to their Euclidean weights) in the estimation of missing value.

2.1   The Modified GRNN (Generalized Regression Neural Networks) Algorithm

In GRNN [11] each observation in the training set forms its own cluster. When a new input pattern x is presented to the GRNN for the prediction of the output value, each training pattern y_i assigns a membership value h_i to x based on the Euclidean distance $d = d(x, y_i)$ as in equation 1. The Euclidean distance function is used in the standard GRNN,

$$h_i = \frac{1}{\sqrt{2\pi\sigma^2}} \exp(-\frac{d^2}{2\sigma^2}) \tag{1}$$

where σ is a smoothing function parameter (we specify a default value, $\sigma = 0.5$). Finally, GRNN calculates the output value z for the pattern x as in equation 2.

$$z = \frac{\sum_i (h_i \times \text{ output of } y_i)}{\sum_i h_i} \tag{2}$$

If the output variable is binary, the GRNN calculates the probability of event of interest. If the output variable is continuous, then it estimates the value of the variable.

2.2   The GMI Algorithm

Our proposed algorithm (GMI) estimates the conditional mean and conditional variance of each missing value. Each case is replicated a number of times (here, 100). The estimates of missing values are generated based on the estimates of conditional means and total conditional variances (including both sampling and imputation variance) of the missing items. The variance will result in slightly different imputed values for each replica of a record. Hence, the replicas of a record will differ in imputed values but not in observed values.

The pseudo code of our proposed algorithm is as follows: we write D_0 for the original dataset. It has N_r rows and N_c columns.

- Create D_N from D_0 by normalization (column by column) so that all values are between 0 and 1. Write d_{ij} for the element in the i'th row and the j'th column.
- Suppose d_{ij} is missing. We then consider that the i'th pattern of D_N is the test pattern, and the j'th variable is the output variable. Write $C = D - d_{i*}$: that is, D, less the i'th pattern.
- Create $I = [I_{ij}]$ with $I_{ij} = 1$ if d_{ij} is missing and 0 otherwise.
- Write $C1 = C - C_{*j}$: that is C but with the j'th column omitted.
- Use forward stepwise selection [12] to identify independent predictive factors from the pool of candidate input variables. Extract a new dataset C_{new} from $C1$ with the selected input and output data, and delete cases with missing values.
- Set M, the number of imputations to some value, here 100.
- Construct M GRNN networks $G_k^{(\text{mean})}$ and M GRNN networks $G_k^{(\text{var})}$ for estimating the conditional mean and the conditional variance of d_{ij} respectively (where $k = 1 \ldots M$).
- For $k = 1 \ldots M$
  - Create a new training set C_k^{mean} by randomly drawing out 70% of the data from C_{new}
  - Train the k'th GRNN net $G_k^{(\text{mean})}$ on the training set C_k^{mean}
  - Create a training set C_k^{var} for the network $G_k^{(\text{var})}$ by using input patterns of C_k^{mean} as inputs and r_k (squared residuals) as outputs.
  - Train the k'th GRNN net $G_k^{(\text{var})}$ on C_k^{var} in order to teach the net to correctly approximate the conditional variance of the missing value. The neural networks $G_k^{(\text{mean})}$ and $G_k^{(\text{var})}$ are now ready for use
  - Present the test pattern to the trained net $G_k^{(\text{mean})}$ for predicting the conditional mean $\hat{Q}_k$
  - Present the test pattern to the trained network $G_k^{(\text{var})}$ for predicting conditional variance $\hat{U}_k$:
- end For
- The conditional mean of the missing value d_{ij} is the average of the single estimates $\bar{Q} = \frac{1}{M} \sum_{i=1}^{M} \hat{Q}_i$:
- Estimate the within-imputation variance $\bar{U} = \frac{1}{M} \sum_{i=1}^{M} \hat{U}_i$
- Estimate the between-imputation variance $B = \frac{1}{M-1} \sum_{i=1}^{M} (\hat{Q}_i - \bar{Q})^2$
- The total variance T of the missing value d_{ij} is $T = \bar{U} + (1 + \frac{1}{M})B$

We now know the mean and variance of the missing value d_{ij}. Perform exactly the same procedure for estimation of conditional means and variance of other missing values. We then replicate each record of the dataset D_N a number (here 5) times, and impute the missing values setting them to $\bar{Q} + \sqrt{T}R$ where R is a random number between -1 and +1.

3   Assessing the New Technique

We compare our algorithms with the popular imputation algorithms: MCMC MI (Markov Chain Monte Carlo Multiple Imputation), EM MI (Expectation Maximization Multiple Imputation), and Hot Deck MI on 26 real datasets, using 100-fold cross-validation. The public real-world datasets on which we tested the algorithms are (1) Abalone, (2) Adult, (3) Annealing, (4) Arcene, (5) Arrhythmia, (6) Balance Scale, (7) Breast Cancer Wisconsin, (8) Congressional voting records, (9) Dermatology, (10) Diabetes, (11) Heart disease, (12) Hepatitis, (13) Hill-Valley dataset, (14) Iris, (15) Internet Advertisements, (16) Medelon data set, (17) Mushroom, (18) Parkinson, (19) Pima Indians, (20) Poker hand dataset, (21)Post operative patient, (22) Soybean (large), (23) SPECHT Heart, (24) Thyroid disease, (25) Wine, and (26) Yeast. The datasets are obtained from UCI Machine Learning Repository [13]. We artificially remove data using MCAR and MAR mechanisms at different rates of missing values into the training set. Then the imputation algorithms were applied separately to each incomplete dataset for imputation of missing values. Missing data inevitably affect classifier performance. Hence, a classifier (we used a GRNN classifier as classification paradigm) was trained with the imputed training dataset and tested with the testing set. We assess the relative merits of imputation algorithms by evaluating the accuracy of the classifier. The Friedman test is used to test the null hypothesis that the performance is the same for all algorithms. After applying the Friedman test and noting it is significant multiple comparison tests (details are available in [14], p.180) were performed in order to test the (null) hypothesis that there is no significant difference between any pair of the four algorithms.

We simulate missing data by deleting values from the complete training data to simulate MCAR and MAR missing mechanisms. For MCAR missing data pattern, we generate uniformly distributed random number in the interval $(0, 1)$ for each observation and specify a range of values within the interval $(0, 1)$ depending on the percentage of data to be removed. We then remove the observation if the corresponding random number lies within the range. For MAR missing data pattern we have to remove data in such a way that removed values of variable x_k depends on x_m and x_n. To simulate MAR data, we define a model for the non-responsiveness:

$$ p(x_{ki}) = \frac{1}{1 + \exp -(\beta_0 + \beta_m x_{mi} - \beta_n x_{ni})} \tag{3} $$

where $p(x_{ki})$ is the probability of the removal of x_k in the i'th observation.

We arrange the instances in the order of the probability that this element of data should be missing variable x_k. The ordered dataset is then divided into equal sized parts. If the total percentage of missing values is p, remove different percentages of data from different subsets. For example, the percentages of missing values in each one of the two subsets will be $0.2p$ and $0.8p$. We generate uniformly distributed random numbers for each observation of the variable x_k in the interval $(0, 1)$. In subset 1, we remove the value of if the corresponding random number is less than or equal to 0.2. In subset 2, we remove the value of x_k if the corresponding random number is less than or equal to 0.8.

4 Results and Discussion

We compare our proposed imputation algorithm (GMI) against the conventional imputation algorithms (MCMC MI, EM MI, and Hot Deck MI) based on the accuracy of a classifier on the resulting imputed dataset. Table 4 summarizes the results.

Table 1. Summary of results for different algorithms

Missing	GMI				Hot Deck MI				MCMC MI				EM MI			
(%)	Mean	±	Max	Min	Mean	±	Max	Min	Mean	±	Max	Min	Mean	±	Max	Min
5	97	2	100	94	93	3	98	85	89	3	98	76	89	7	100	79
10	95	4	99	86	88	6	100	5	85	7	100	69	83	8	97	67
20	92	5	100	84	84	6	97	72	86	6	96	73	85	7	97	70
30	91	5	99	78	72	8	90	53	80	7	93	70	78	8	94	62
40	82	10	98	61	63	11	84	55	66	7	84	55	66	8	84	50
50	70	7	83	59	53	2	58	50	59	4	66	49	57	6	69	45
60	56	3	63	51	52	2	56	48	53	2	58	50	53	2	58	50
70	54	2	57	50	53	1	57	50	53	2	58	49	53	2	58	49

The key findings are as follows:

- For studies with roughly 5-60% missing values, the performance of GMI is significantly better than the other algorithms at the level of 5% significance.
- The performance of imputation algorithms deteriorates as the rate of missing values increases. The performance of GMI deteriorates relatively slowly. The performance of hot deck MI deteriorated most rapidly.
- The hot deck MI does best when the percentage of missing values is low. The performance of hot deck MI was the second best when the percentage of missing data was up to 10%.
- The performance of MCMC MI and EM MI are very similar.

5 Summary and Conclusion

We presented a non-parametric multiple imputation algorithm (GMI) for imputing missing data. The algorithm is based on the concept of generalized regression neural network. We tested our algorithm on twenty-six real datasets. We compare GMI with three multiple imputation procedures: Markov Chain Monte Carlo, Expected Maximization and Hot Deck impuation. The performance of the algorithms was assessed in terms of classification performance at different percentage of missing values. GMI algorithm appears to be superior to other imputation algorithms.

References

1. Rubin, D.B.: Multiple imputation for non response in surveys. Wiley, New York (1987)
2. Little, R.J.A., Rubin, D.B.: Statistical Analysis with missing data. Wiley, New York (1987)
3. Rubin, D.B., Schenker, N.: Multiple imputation for interval estimation from simple random values with ignorable nonresponse. Journal of the American Statistical Association 81(394), 366–374 (1986)
4. Schafer, J.: Analysis of incomplete multivariate data. Chapman and Hall, London (1997)
5. Bo, T.H., Dysvik, B., Jonassen, I.: LSimpute: accurate estimation of missing values in microarray data with least squares method. Nucleic Acids Research 32(3) (2004)
6. Carlo, G., Yao, J.: A multiple-imputation metropolis version of the EM algorithm. Biometrika 90(3), 643–654 (2003)
7. Lokupitiya, R.S., Lokupitiya, E., Paustian, K.: Comparison of missing value imputation methods for crop yield data. Environmetrics 17(4), 339–349 (2006)
8. Iannacchione, V.: Weighted sequential hot deck imputation macros. In: Proceedings of the Seventh Annual SAS Users Group International Conference, San Francisco, pp. 759–763 (1982)
9. Dan, L., Deogun, J.S., Wang, K.: Gene function classification using fuzzy k-nearest neighbour approach. In: IEEE International Conference on Granular Computing, November 2-4, pp. 644–644 (2007)
10. Schioler, H., Hartmann, U.: Mapping neural network derived from the Parzen window estimator. Neural Networks 2(6), 903–909 (1992)
11. Specht, D.F: A General Regression Neural Network. IEEE Transactions on Neural Networks. 2(6), pp. 568-576 (1991).
12. Kuncheva, L.I.: A stability index for feature selection. In: Proceedings of the 25th IASTED International Multi-Conference: artificial intelligence and applications, pp. 390–395. ACTA Press, Anaheim (2007)
13. UCI Machine Learning Repository: Centre for Machine Learning and Intelligent Systems, http://archive.ics.uci.edu/MI/
14. Siegel, S., Castellan Jr, N.J.: Nonparametric statistics: for the behavioural sciences, 2nd edn. McGraw-Hill, New York (1988)

Municipal Creditworthiness Modelling by Kernel-Based Approaches with Supervised and Semi-supervised Learning

Petr Hajek and Vladimir Olej

Institute of System Engineering and Informatics
Faculty of Economics and Administration
University of Pardubice
Studentska 84, 53210 Pardubice
Czech Republic
{petr.hajek,vladimir.olej}@upce.cz

Abstract. The paper presents the modelling possibilities of kernel-based approaches on a complex real-world problem, i.e. municipal creditworthiness classification. A model design includes data pre-processing, labelling of individual parameters' vectors using expert knowledge, and the design of various support vector machines with supervised learning and kernel-based approaches with semi-supervised learning.

Keywords: Municipal creditworthiness, support vector machines, kernel, supervised learning, semi-supervised learning.

1 Introduction

Municipal creditworthiness is an independent expert evaluation based on a complex analysis of all known municipal creditworthiness parameters. Municipal creditworthiness evaluation is currently being realized by methods combining mathematical-statistical methods and expert opinion. Recently, the models based on hierarchical structures of fuzzy inference systems [1], unsupervised (supervised) methods [2], [3], and neuro-fuzzy systems [4] have been designed for municipal creditworthiness evaluation. The output of the methods is represented by an assignment of the i-th object $o_i \in O$, $O=\{o_1,o_2, \ldots,o_i, \ldots,o_n\}$ to the j-th class $\omega_{i,j} \in \Omega$, $\Omega=\{\omega_{1,j},\omega_{2,j}, \ldots,\omega_{i,j}, \ldots,\omega_{n,j}\}$ [1], [2], [3], [4]. Based on the mentioned facts, it is possible to state that the methods capable of processing and learning expert knowledge, enabling their user to generalize and properly interpret have proved to be most suitable for municipal creditworthiness modelling.

In general, any kernel-based algorithm is composed of two modules, a general purpose learning machine and a problem specific kernel function [5]. Support Vector Machines (SVMs) represent an essential kernel-based method. Many variants of SVMs have been developed since SVMs were proposed [6], e.g. least squares SVMs, robust SVMs, etc. Moreover, conventional methods have been extended to be used in high-dimensional feature space, e.g. kernel least squares,

D. Palmer-Brown et al. (Eds.): EANN 2009, CCIS 43, pp. 35–44, 2009.

kernel Principal Component Analysis, kernel Self-organizing Feature Maps, etc. Concerning classification problem, other classification methods (neural networks, fuzzy systems, etc.) have been also extended to incorporate maximizing margins and mapping to a feature space [6]. Support Vector Machines [7], [8] are affiliated to neural networks and create the category of so-called kernel machines. Support Vector Machines represent methods for supervised, unsupervised, as well as semi-supervised learning [9]. The main principle of SVMs consists in the creation of the decision hyperplane between classes $\omega_{i,j} \in \Omega$ so that the margin between positive and negative patterns is maximized. This feature of SVMs comes from statistical learning theory [9]. More precisely, SVMs represent approximate implementation of structural risk minimization method [5], [7]. This principle is based on the fact that the testing error is limited with the sum of training error and the expression depending on Vapnik-Chervonenkis dimension [7]. The central point in the construction of SVMs algorithm is the inner product kernel between support vector $\mathbf{v}_i$ and vector $\mathbf{x}$ from input space. Support vectors $\mathbf{v}_i$ consist of a small subset of training data O_{train} extracted by the algorithm. Depending on how the inner product kernel is generated, different learning machines can be constructed which are characterized with their own non-linear decision hyperplane [7].

The paper presents the design of parameters vector $\mathbf{x}=(x_1,x_2, \ldots,x_k, \ldots,x_m)$ for municipal creditworthiness evaluation [1], [2], [3], [4]. Further, based on the data analysis, the data representation by means of data matrix $\mathbf{X}$ is proposed and formalized. Then it is possible to design the model for municipal creditworthiness evaluation by SVMs with supervised learning and kernel-based approaches with semi-supervised learning. It consists of data pre-processing, labelling of objects $o_i \in O$ with classes $\omega_{i,j} \in \Omega$ based on expert opinion, and the classifiers of objects $o_i \in O$ into classes $\omega_{i,j} \in \Omega$. The final part of the paper includes an analysis of the results with the aim of comparison of the classifiers based on kernel-based approaches.

2   Municipal Creditworthiness Problem Description

In [1], [2], [3] and [4] there are mentioned common categories of parameters, namely economic, debt, financial and administrative categories. The economic, debt and financial parameters are pivotal. Economic parameters (x_1,x_2,x_3,x_4) affect long-term credit risk. The municipalities with more diversified economy and more favourable social and economic conditions are better prepared for the economic recession. Debt parameters (x_5,x_6,x_7) include the size and structure of the debt. Financial parameters $(x_8,x_9,x_{10},x_{11},x_{12})$ inform about the budget implementation. Their values are extracted from the municipal budget. The design of parameters vector $\mathbf{x}=(x_1,x_2, \ldots,x_k, \ldots,x_m)$, m=12, based on previous correlation analysis and recommendations of notable experts, can be realized as presented in Table 1. The parameters x_3 and x_4 are defined in the r-th year and parameters x_5 to x_{12} as the average value of the r-th and (r-1)th years.

As the descriptions of classes $\omega_{i,j} \in \Omega$ are known (Table 2), the classes $\omega_{i,j} \in \Omega$ have been assigned to Czech municipalities $o_i \in O$ (micro-region Pardubice, the

Table 1. Municipal creditworthiness parameters design

Parameters

Economic $x_1 = PO_r$, PO_r is population in the r-th year.

$x_2 = PO_r/PO_{r-s}$, PO_{r-s} is population in the year r-s, and s is the selected time period.

$x_3 = U$, U is the unemployment rate in a municipality.

$x_4 = \sum_{i=1}^{k}(EPO_i/EIN)^2$, EPO_i is the employed population of the municipality in the i-th economic sector, i=1,2, ...,k, EIN is the total number of employed inhabitants, k is the number of the economic sector.

Debt $x_5 = DS/PR$, $x_5 \in <0,1>$, DS is debt service, PR are periodical revenues.

$x_6 = TD/PO$, TD is total debt.

$x_7 = SD/TD$, $x_7 \in <0,1>$, SD is short-term debt.

Financial $x_8 = PR/CE$, $x_8 \in R^+$, CE are current expenditures.

$x_9 = OR/TR$, $x_9 \in <0,1>$, OR are own revenues, TR are total revenues.

$x_{10} = CAE/TE$, $x_{10} \in <0,1>$, CAE are capital expenditures, TE are total expenditures.

$x_{11} = CAR/TR$, $x_{11} \in <0,1>$, CAR are capital revenues.

$x_{12} = LA/PO$, [Czech Crowns], LA is the size of the municipal liquid assets.

Czech Republic, n=452) by local experts. Based on the presented facts, the following data matrix **X** can be designed

$$\mathbf{X} = \begin{array}{c} \\ o_1 \\ \cdots \\ o_i \\ \cdots \\ o_n \end{array} \begin{vmatrix} x_1 & \cdots & x_k & \cdots & x_m \\ x_{1,1} & \cdots & x_{1,k} & \cdots & x_{1,m} \\ \cdots & \cdots & \cdots & \cdots & \cdots \\ x_{i,1} & \cdots & x_{i,k} & \cdots & x_{i,m} \\ \cdots & \cdots & \cdots & \cdots & \cdots \\ x_{n,1} & \cdots & x_{n,k} & \cdots & x_{n,m} \end{vmatrix} \begin{vmatrix} \omega_{1,j} \\ \cdots \\ \omega_{i,j} \\ \cdots \\ \omega_{n,j} \end{vmatrix}, \quad (1)$$

where $o_i \in O$, $O=\{o_1,o_2, \ldots ,o_i, \ldots ,o_n\}$ are objects (municipalities), x_k is the k-th parameter, $x_{i,k}$ is the value of the parameter x_k for the i-th object $o_i \in O$, $\omega_{i,j} \in \Omega$, $\Omega=\{\omega_{1,j},\omega_{2,j}, \ldots ,\omega_{i,j}, \ldots ,\omega_{n,j}\}$ is the j-th class assigned to the i-th object $o_i \in O$, $\mathbf{x_i}=(x_{i,1},x_{i,2}, \ldots ,x_{i,k}, \ldots ,x_{i,m})$ is the i-th pattern, and $\mathbf{x}=(x_1,x_2, \ldots ,x_k, \ldots ,x_m)$ is the parameters vector.

3 Basic Notions of Support Vector Machines and Learning

The design of SVMs [5], [6], [7], [8] depends on the non-linear projection of the input space $\varXi$ into multidimensional space Λ, and on the construction of an optimal hyperplane. This operation is dependent on the estimation of inner product kernel referred to as kernel function.

Let **x** be a vector from input space $\varXi$ of dimension m. Next, let $g_j(\mathbf{x})$, j=1,2, ...,m be a set of non-linear transformations from input space $\varXi$ into

Table 2. Descriptions of classes $\omega_{i,j} \in \Omega$

Description

$\omega_{i,1}$ High ability of a municipality to meet its financial obligation. Very favourable economic conditions, low debt and excellent budget implementation.

$\omega_{i,2}$ Very good ability of a municipality to meet its financial obligation.

$\omega_{i,3}$ Good ability of a municipality to meet its financial obligation.

$\omega_{i,4}$ A municipality with stable economy, medium debt and good budget implementation.

$\omega_{i,5}$ Municipality meets its financial obligation only under favourable economic conditions.

$\omega_{i,6}$ A municipality meets its financial obligations with difficulty, the municipality is highly indebted.

$\omega_{i,7}$ Inability of a municipality to meet its financial obligation.

multidimensional space Λ of dimension q. Then the hyperplane can be defined as a decision surface as follows [5], [6], [7]

$$\sum_{j=1}^{m} \mathbf{w}_j g_j(\mathbf{x}) + b = 0, \tag{2}$$

where $\mathbf{w}_j$, j=1,2, ...,m is the vector of weights connecting the q-dimensional space Λ with the n-dimensional output space Π, b is bias.

In linearly separable case, the algorithm of SVMs tries to find the separating hyperplane with the widest margin [5], [6], [7], [8]. If patterns are linearly separable, it is always possible to find weights $\mathbf{w}$ and bias b so that the inequalities bounding this optimization problem are satisfied. If the algorithm for separable data is used for inseparable data, it does not find an acceptable solution. Then this fact is realized by means of bounding this optimization problem in the following way: $\mathbf{x}.\mathbf{w}+b \geq +1-\xi_i$ for $y_i=+1$, $\mathbf{x}.\mathbf{w}+b \geq -1-\xi_i$ for $y_i=-1$, where ξ_i, i=1,2, ...,u, and (.) represents the dot product of two vectors $\mathbf{x}$ and $\mathbf{w}$. Lots of $g_j(\mathbf{x})$ represent input supported by weight $\mathbf{w}_j$ through the q-dimensional space Λ. Further, let vector $\mathbf{g}(\mathbf{x})=[g_0(\mathbf{x}),g_1(\mathbf{x}), \dots,g_m(\mathbf{x})]^T$ be defined. Then the vector $\mathbf{g}(\mathbf{x})$ represents the image derived in the q-dimensional space Λ related to the input vector $\mathbf{x}$. Hence, decision hyperplane $\mathbf{w}^T\mathbf{g}(\mathbf{x})=0$ can be defined based on this image. Then based on [5], [6], [7], after the application of optimization condition into the Lagrange equation, the following form of weights is obtained

$$\mathbf{w} = \sum_{i=1}^{N} \alpha_i d_i \mathbf{g}(\mathbf{v}_i), \tag{3}$$

where the vector $\mathbf{g}(\mathbf{v}_i)$ from the q-dimensional space Λ corresponds to the i-th support vector $\mathbf{v}_i$, α_i are Lagrange multipliers determined in the optimization process, and d_i represents the shortest distance separating hyperplane from the nearest positive or negative patterns. Support vectors $\mathbf{v}_i$ consist of small subset of training data extracted by the algorithm. Then as equation (3) is substituted into $\mathbf{w}^T\mathbf{g}(\mathbf{x})=0$, the decision hyperplane is computed in the q-dimensional space $\varLambda$. Inner product of two vectors $\mathbf{g}^T(\mathbf{v}_i)\mathbf{g}(\mathbf{x})$ is derived in the q-dimensional space $\varLambda$ in relation to input vector $\mathbf{x}$ and support vector $\mathbf{v}_i$. Now, the inner product kernel $k(\mathbf{x},\mathbf{v}_i)$ can be defined this way

$$k(\mathbf{x}, \mathbf{v}_i) = \mathbf{g}^T(\mathbf{v}_i)\mathbf{g}(\mathbf{x}) = \sum_{j=0}^{m} g_j(\mathbf{v}_i)g_j(\mathbf{x}), i = 1, 2, \ldots, N. \qquad (4)$$

As obvious from this equation, the kernel function $k(\mathbf{x},\mathbf{v}_i)$ is a symmetrical function with respect to its arguments. The most important fact is that the kernel function $k(\mathbf{x},\mathbf{v}_i)$ can be used in order to construct an optimal hyperplane in the q-dimensional space $\varLambda$ without considering separate q-dimensional space $\varLambda$ in explicit form. Based on given facts, it is possible to find linear separators in the q-dimensional space $\varLambda$ so that $(\mathbf{x},\mathbf{v}_i)$ is replaced by kernel function $k(\mathbf{x},\mathbf{v}_i)$. Accordingly, the process of learning can be realized so that only kernel functions $k(\mathbf{x},\mathbf{v}_i)$ can be computed instead of full list of attributes for each data point. Evidently, the found linear separators can be transformed back into the original space $\varXi$. This way any non-linear boundaries between positive and negative patterns can be obtained. Various kernel functions $k(\mathbf{x},\mathbf{v}_i)$ representing different spaces can be used for modelling, e.g. linear, polynomial, radial basis function (RBF), and sigmoid kernel function [5], [6], [7]. Then the output $f(\mathbf{x}_t)$ of SVMs is defined this way

$$f(\mathbf{x}_t) = \sum_{i=1}^{N} \alpha_i y_i k(\mathbf{v}_i, \mathbf{x}_t) + b, \qquad (5)$$

where $\mathbf{x}_t$ is the evaluated pattern, N is the number of support vectors, $\mathbf{v}_i$ are support vectors, α_i are Lagrange multipliers determined in the optimization process, $k(\mathbf{v}_i,\mathbf{x}_t)$ is actual kernel function $k(\mathbf{x},\mathbf{v}_i)$. Support vectors $\mathbf{v}_i$ represent the component of classifiers' structure, and their number N is cut during the optimization process [5], [6], [7].

Further, let $O_{train}=\{(\mathbf{x}_1,\omega_{1,j}),(\mathbf{x}_2,\omega_{2,j}),\ldots,(\mathbf{x}_r,\omega_{r,j})\}$ be a training set. Then, the goal of the supervised learning lies in the assignment of the j-th class $\omega_{i,j}\in\varOmega$, $\varOmega=\{\omega_{1,j},\omega_{2,j},\ldots,\omega_{i,j},\ldots,\omega_{n,j}\}$ to the i-th pattern $\mathbf{x}_i=(x_{i,1},x_{i,2},\ldots,x_{i,k},\ldots,x_{i,m})$. In this case, $\omega_{i,j}$ are called the labels or targets of the patterns $\mathbf{x}_i$. Support vector machines are based on a direct decision function, i.e. the classification of patterns $\mathbf{x}_i$ into multiclass is not realized directly but there are several formulations of this problem.

Semi-supervised learning is defined e.g. in [9], [10], [11], [12]. This learning represents halfway between supervised and unsupervised learning. In addition to unlabelled data, the algorithm is provided with some supervision information, but not necessarily for all patterns $\mathbf{x}_i$. Consider a classification problem with n classes $\omega_{n,j}$. We will assume that the training set O_{train} consists of two subsets $O_{train}=\{O_{trainS},O_{trainT}\}$, where $O_{trainS}=\{(\mathbf{x}_1,\omega_{1,j}),(\mathbf{x}_2,\omega_{2,j}),\ldots,(\mathbf{x}_s,\omega_{s,j})\}$ is the labelled subset, $O_{trainT}=\{(\mathbf{x}_{s+1}),(\mathbf{x}_{s+2}),\ldots,(\mathbf{x}_{s+t})\}$ is the unlabelled subset, s and t are the numbers of examples in the labelled and unlabelled sets, respectively, and $\mathbf{x}$ is the m-dimensional feature vector.

A comprehensive review of current semi-supervised methods is presented by [12]. Key categories of these methods are represented, for example, expected maximization with generative mixture models, self-training, co-training, transductive SVMs, and graph-based methods represent . For kernel-based approaches with semi-supervised learning, the difference between supervised and semi- supervised learning amounts to the difference between Structural Risk Minimization and Overall Risk Minimization. First access attempts to minimize the sum of the hypothesis space complexity and the empirical error [10]. Also, it claims that a larger separation margin will correspond to a better classifier. Second access attempts to assign class labels to the unlabelled data that would result in the largest separation margin.

4   Modelling and Analysis of the Results

Local experts assign classes $\omega_{i,j} \in \Omega$ to municipalities in the credit analysis process. The data matrix $\mathbf{X}$ is used as the input of the SVMs with supervised learning and kernel-based approaches with semi-supervised learning which realize the classification process. Based on presented facts, the model is designed for the classification of municipalities $o_i \in O$ into classes $\omega_{i,j} \in \Omega$.

The frequencies f of municipalities $o_i \in O$ in classes $\omega_{i,j} \in \Omega$ are presented in Fig. 1. Classes $\omega_{i,2}, \omega_{i,3}$, and $\omega_{i,4}$ have the highest frequencies, i.e. the municipalities prevail which are moderate regarding economic, debt and financial parameters.

4.1   Modelling by SVMs with Supervised Learning

In the process of modelling, number of supervised SVMs with linear, polynomial, sigmoid, and RBF kernel functions has been designed along with various values of input parameters. Two methods were used for finding optimal parameters values, a grid search and a pattern search. The grid search tries to find values of each parameter across the specified search range using geometric steps. The pattern search starts at the centre of the search range and makes trial steps in each direction for each parameter.

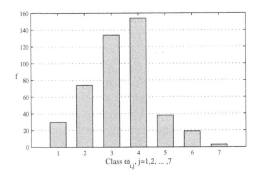

Fig. 1. Frequencies f of municipalities in classes $\omega_{i,j} \in \Omega$, j=1,2,...,7

Table 3. Optimal values of SVMs parameters obtained by grid search (pattern search) where the ranges for the parameters are as follows: $0.01 \le \alpha_i \le 10000$, $0.01 \le r_i \le 20$, $0 \le c \le 100$

	Kernel function			
Parameters	Linear	Polynomial	Sigmoidal	RBF
α_i	16.86(48.72)	0.36(28.27)	2782.56(133.24)	16.68(93.24)
Number of support vectors	168(141)	151(159)	140(174)	184(143)
Radius r_i of RBF	$-(-)$	2.21(0.29)	0.03(0.14)	2.21(0.38)
Shifting parameter c	$-(-)$	1.67(0.62)	0.60(0.00)	$-(-)$
Polynomial degree	$-(-)$	3(3)	$-(-)$	$-(-)$
$\xi[\%]$	88.05(88.50)	90.04(90.27)	88.50(88.27)	89.83(91.15)

To avoid over fitting, 10-fold cross-validation is used to evaluate the fitting provided by each parameter value set during the grid or pattern search process. The optimal values of SVMs parameters together with resulting classification accuracy $\xi[\%]$ are presented in Table 3. Lagrange multipliers α_i control misclassification error in the SVM algorithm. If the high value of parameter α_i is set, misclassification error is limited. However, if this value is low, certain misclassification error is allowed on training data O_{train}. Moreover, the algorithms are computationally demanding for large training datasets O_{train}.

The SVMs with supervised learning represent excellent results with the maximum classification accuracy $\xi_{max}=91.15[\%]$, the average classification accuracy $\xi_a=89.76[\%]$, and the standard deviation $\sigma=1.83[\%]$. The fact results from the analysis of designed SVMs that if training data O_{train} are linearly separable, and training set O_{train} is not large, then the generalization ability of SVMs does not deteriorate because the process of learning is realized with the view of margin maximization. Since the process of learning is formulated as the problem of quadratic programming, its result represents a globally optimal solution (there is no problem with trapping in a local extreme).

4.2 Modelling by Kernel-Based Approaches with Semi-supervised Learning

As stated by [13], from the practical point of view, the main motivation for semi-supervised learning algorithms is, of course, their capability to work in cases where there are much more unlabelled than labelled examples. Previous experiments have been mostly realized for benchmark datasets [13] as there is lack of complex real-world problems. The field of municipal creditworthiness modelling is suitable for the use of semi-supervised learning as the process of creditworthiness evaluation (credit rating process) is costly and time-consuming. Therefore, only low proportion of municipalities has been labelled by classes giving information about their creditworthiness so far.

Harmonic Gaussian Model (HGM) and Global Consistency Model (GCM) defined by [9] have been used as representatives of kernel-based approaches with semi-supervised learning. Again extensive number of experiments on municipal

Table 4. Classification accuracy ξ_{test}[%] and standard deviation σ[%] for s[%]

s[%]	ξ_{test}[%]	σ[%]	ξ_{test}[%]	σ[%]	ξ_{test}[%]	σ[%]	ξ_{test}[%]	σ[%]	ξ_{test}[%]	σ[%]
	2		3		4		5		10	
HGM(c)	47.82	13.96	50.90	11.79	61.77	8.20	66.33	7.71	77.64	4.10
HGM(e)	44.05	11.69	54.12	11.65	57.15	11.67	66.60	6.86	73.90	4.42
GCM(c)	61.00	7.93	64.14	6.58	70.21	5.54	72.48	4.28	77.53	2.59
GCM(e)	60.93	7.18	67.94	5.80	68.58	4.74	73.22	4.18	77.69	2.74
s[%]	15		20		25		30		35	
HGM(c)	81.75	2.72	83.59	2.08	83.72	2.08	84.65	2.13	85.62	1.81
HGM(e)	76.07	3.75	79.35	3.26	81.06	2.75	81.95	2.00	83.22	2.17
GCM(c)	79.49	2.88	81.26	2.63	82.71	1.99	82.82	2.12	83.12	2.16
GCM(e)	79.90	2.28	81.34	2.30	81.71	2.79	81.91	2.83	82.37	2.30
s[%]	40		45		50		55		60	
HGM(c)	86.40	1.59	86.79	2.04	86.95	1.78	86.89	2.00	87.15	2.28
HGM(e)	83.59	2.19	84.60	2.41	85.41	2.16	86.25	2.18	85.23	2.51
GCM(c)	83.00	2.04	82.24	2.87	82.53	2.73	79.99	3.80	76.87	3.09
GCM(e)	82.82	2.58	81.61	2.36	80.63	2.70	78.19	3.40	73.27	4.51

creditworthiness data set has been realized for different values of input parameters. As a result, HGM and GCM have been trained with the following values of parameters: radius of RBF r_i=10, Epochs=50, degree of the graph h=10, Distance=Cosine(c)/Euclidean(e). The experiments were designed for different number s[%] of the labelled data O_{trains}. Classification accuracy ξ_{test}[%] and standard deviation σ[%] have been observed for both kernel-based approaches with semi-supervised methods (Table 4).

The results are in line with the results observed by [9] which means that both HGM and GCM behave similarly, i.e. with an increase in the number of s[%] of

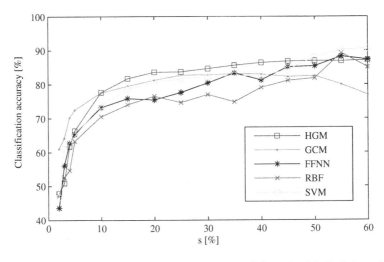

Fig. 2. Classification accuracy ξ_{test}[%] for different s[%] of the labelled data O_{trains}

labelled data the overall models' classification accuracies $\xi_{test}[\%]$ improve too, Fig. 2. Moreover, the results show that higher classification accuracy $\xi_{test}[\%]$ has been obtained by GCM for $s<15\%$ while HGM gives better results as the number $s[\%]$ of labelled data increases. This effect was expected as the proportion $s[\%]$ of the labelled data is unbalanced (the number of labelled data in each class $\omega_{i,j} \in \Omega$ differs very much), see [9]. The kernel-based approach with semi-supervised learning HGM and GCM have been compared to supervised methods feed-forward neural networks (FFNNs), RBF neural networks and SVMs. In all models, according to the expectations, with an increase in the number $s[\%]$ of labelled data O_{trainS}, the classification accuracies $\xi_{test}[\%]$ improve too. Global Consistency Model performed much better for extremely small number $s<10\%$ of the labelled data O_{trainS}. However, the HGM gives good results for higher number $s[\%]$ of the labelled data. Other methods (FFNN, RBF neural networks, kernel-based approaches) outperform the semi-supervised methods as the number $s[\%]$ of the labelled data goes over 50%. In this case SVMs with supervised learning gives best results.

5   Conclusion

The paper presents the design of parameters for municipal creditworthiness evaluation. The evaluation process is presented as a classification problem. The model is proposed where data matrix $\mathbf{X}$ is used as the input of the SVMs with supervised learning and kernel-based approaches with semi-supervised learning.

The SVMs with supervised learning were designed and studied for the classification of municipalities $o_i \in O$ into classes $\omega_{i,j} \in \Omega$ due to its high classification accuracy $\xi_{test}[\%]$ on testing data with a low standard deviation $\sigma[\%]$. The best results have been obtained for kernel RBF as optimal values of input parameters were gained by a pattern search method.

In economic praxis it is difficult to realize creditworthiness evaluation for each of the municipalities. This process is usually costly and time-consuming. Therefore, the kernel-based approaches with semi-supervised learning HGM and GCM for different number $s[\%]$ of the labelled data O_{trainS} were studied. The results of semi-supervised methods HGM and GCM are compared to FFNNs, RBF neural networks and SVMs with supervised learning considering classification accuracy $\xi_{test}[\%]$ for different number $s[\%]$ of the labelled data O_{trainS}. The results of the designed model for classification of municipalities $o_i \in O$ into classes $\omega_{i,j} \in \Omega$ show the possibility of evaluating municipal creditworthiness even if only low proportion of municipalities $o_i \in O$ are labelled with classes $\omega_{i,j} \in \Omega$. Thereby, the model presents an easier conception of the municipal creditworthiness evaluation for the public administration managers, banks, investors or rating agencies.

Acknowledgements

This work was supported by the scientific research of Czech Science Foundation, under Grant No: 402/09/P090 with title Modelling of Municipal Finance by Computational Intelligence Methods, and Grant No: 402/08/0849 with title Model of Sustainable Regional Development Management.

References

[1] Olej, V., Hajek, P.: Hierarchical Structure of Fuzzy Inference Systems Design for Municipal Creditworthiness Modelling. WSEAS Transactions on Systems and Control 2, 162–169 (2007)

[2] Olej, V., Hajek, P.: Modelling of Municipal Rating by Unsupervised Methods. WSEAS Transactions on Systems 7, 1679–1686 (2006)

[3] Hajek, P., Olej, V.: Municipal Creditworthiness Modelling by Kohonen's Self-organizing Feature Maps and LVQ Neural Networks. In: Rutkowski, L., Tadeusiewicz, R., Zadeh, L.A., Zurada, J.M. (eds.) ICAISC 2008. LNCS (LNAI), vol. 5097, pp. 52–61. Springer, Heidelberg (2008)

[4] Hajek, P., Olej, V.: Municipal Creditworthiness Modelling by Kohonen's Self-organizing Feature Maps and Fuzzy Logic Neural Networks. In: Kůrková, V., Neruda, R., Koutník, J. (eds.) ICANN 2008, Part I. LNCS (LNAI), vol. 5163, pp. 533–542. Springer, Heidelberg (2008)

[5] Cristianini, N., Shawe-Taylor, J.: An Introduction to Support Vector Machines and other Kernel-based Learning Methods. Cambridge University Press, Cambridge (2000)

[6] Abe, S.: Support Vector Machines for Pattern Classification. Springer, London (2005)

[7] Haykin, S.: Neural Networks: A Comprehensive Foundation. Prentice-Hall Inc., New Jersey (1999)

[8] Vapnik, V.N.: The Nature of Statistical Learning Theory. Springer, New York (1995)

[9] Huang, T.M., Kecman, V., Kopriva, I.: Kernel Based Algorithms for Mining Huge Data Sets. In: Supervised, Semi-supervised, and Unsupervised Learning. Studies in Computational Intelligence. Springer, Heidelberg (2006)

[10] Bennett, K.P., Demiriz, A.: Semi-supervised Support Vector Machines. In: Int. Conf. on Advances in Neural Information Processing Systems, vol. 2. MIT Press, Cambridge (1999)

[11] Chapelle, O., Scholkopf, B., Zien, A.: Semi-Supervised Learning. MIT Press, Cambridge (2006)

[12] Zhu, X.: Semi-Supervised Learning. Literature Survey, http://www.cs.wisc.edu/~jerryzhu/pub/ssl_survey (2005)

[13] Klose, A.: Extracting Fuzzy Classification Rules from Partially Labelled Data. Soft Computing 8, 417–427 (2004)

Clustering of Pressure Fluctuation Data Using Self-Organizing Map

Masaaki Ogihara, Hideyuki Matsumoto, Tamaki Marumo, and Chiaki Kuroda

Department of Chemical Engineering, Tokyo Institute of Technology,
Tokyo 152-8550, Japan
ogihara.m.ab@m.titech.ac.jp,
{hmatsumo,tmarumo,ckuroda}@chemeng.titech.ac.jp

Abstract. The batch Self-Organizing Map (SOM) is applied to clustering of pressure fluctuation in liquid-liquid flow inside a microchannel. When time-series data of the static pressure are computed by the SOM, several clusters of pressure fluctuation with different amplitudes are extracted in the visible way. Since the signal composition of the fluctuation is considered to change with flow rates of the water and the organic solvent, the ratio to the each cluster, which is estimated by the recalling, is classified by using the SOM. Consequently, the operating condition of flow rate is classified to three groups, which indicate characteristic behavior of interface between two flows in the microchannel. Furthermore, predictive performance for behavior of the interface is demonstrated to be good by the recalling.

Keywords: Self-Organizing Map, Pressure Fluctuation Data, Clustering.

1 Introduction

Many process engineers in recent years are interested in compact, safe, energy-efficient and environment-friendly sustainable process. Spread of concept of "process intensification" leads to innovation of new chemical device and various methods [1]. Microreactor consists of microfabricated channels (microchannels), which is one of the new chemical devices. The microfabrication technology has developed devices for microanalysis. Microchannel of size of several hundred micrometers has demonstrated to show rapid and precise control of temperature in an exothermic reaction process, because of large ratio of surface to reaction volume [2]. Therefore, the micro chemical technology has advantages of the reduction of solvent and the safe operation of dangerous substance.

Process based on multiphase reactions in microchemical systems, especially liquid-liquid two-phase reactions, occur in a broad range of application areas, such as nitration, extraction, emulsification, and so on [3]. For example, a three-phase flow, water/n-heptane/water, was constructed in a microchannel (100-μm width, 25-μm depth) and was used for separation of metal ions (Y^{3+}, Zn^{2+}) [4]. In the previous studies for extraction system on a microchip, difficulty in the system operation is how to stabilize the interface of two fluids with a high specific interface area. Hibara *et al.* proposed a method of modification of the wall inside microchannel by using octadecyl-silane group [5]. When the surface of the channel was chemically modified to be

D. Palmer-Brown et al. (Eds.): EANN 2009, CCIS 43, pp. 45–54, 2009.

hydrophobic, organic solvent and water flowed stably in the confluence. It was reported that two-phase flow could be operated stably by applying a microchip with patterned surface featuring deep and shallow microchannel areas [6].

While many methods for stabilization of two-phase flow have been investigated from the viewpoint of design of microchannel, there are few studies for online monitoring of the flow pattern in the start-up or long-term operation of microchemical system. It is not possible that we monitor the flow pattern inside microchannel by optical methods, if the transparent device like the glass microchip [5] cannot be used. Pressure sensors are considered to provide useful information about the two-phase flow inside the opaque device, and thus the integration system of microchannels and pressure sensors has been developed in recent years [7].

In our previous studies, the Y-shaped microchannel that pressure sensors were installed into was developed. When the pressure fluctuation data were processed by the spectrum analysis, it was seen that dynamic behaviors of interface between aqueous phase and organic phase could be distinguished by the spectrogram. In order to control dynamic behavior of the interface based on the spectrogram, it is necessary to analyze nonlinear relationships between waveform of the pressure fluctuation and operational conditions. Therefore, in the present paper, waveform of the pressure fluctuation is classified by using the self-organizing map (SOM), and the clustering by the SOM is related to classification of operational conditions.

A purpose of the present paper is to propose an application method of the batch SOM for classification of operational conditions based on the dynamic behavior of interface between immiscible liquid-liquid flows. A method of combining the clustering of pressure fluctuation data with the clustering of operational conditions will be discussed below.

2   Acquisition of Pressure Fluctuation Data

In order to acquire pressure fluctuation data, we carried out experiments using a Y-shaped microchannel as shown in Fig. 1. CH1 in Fig. 1 represents the pressure measurement point in the upstream of organic-phase flow. CH2 is the measurement point in the upstream of aqueous-phase flow. CH3 and CH4 represent the pressure measurement points in the downstream of organic-phase and aqueous-phase flows. Pressure data were sampled 100 times per second. Hexane was used as the test fluid of organic solvent.

The following data preprocessing is implemented to extract pressure fluctuation component from the sampled time-series data. First, the size of data window is set at five seconds as shown in Fig. 2(a). Fig. 2(b) shows pressure waveform that is extracted between time $T(X)$ and $T(X+249)$ in Fig. 2(a). Next, linear approximation of the extracted data is implemented based on the least square errors in order to disregard the effect of pressure value itself, and to focus on high frequency and minute pressure fluctuation. The pressure fluctuation component for five seconds in Fig. 2(c) is difference between pressure data in Fig. 2(b) and estimated data by the linear model. By sampling data of fluctuation component at intervals of 0.02 seconds, a 250-dimensional data vector of pressure fluctuation is prepared as an input to the SOM. When the above-mentioned data preprocessing is iterated by moving the data window at 0.05 seconds, about 1500 vectors of input data are prepared for a certain operational condition.

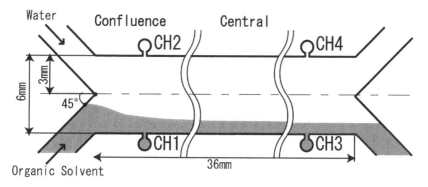

Fig. 1. Schematic diagram of microchannel for acquisition of pressure fluctuation data

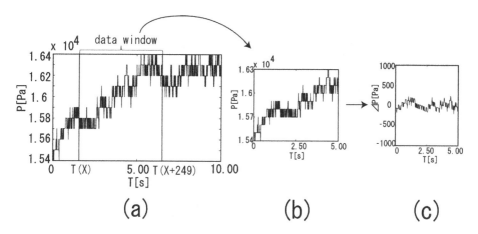

Fig. 2. Data preprocessing for extraction of pressure fluctuation

3 Methodology for Clustering by the SOM

3.1 Batch Self-Organizing Map

In the present paper, we used SOMine ver. 4.0 (Viscovery Software GmbH), which was based on the concept and algorithm of the "batch" SOM introduced by T. Kohonen [8]. Application of SOMine was demonstrated in our previous studies to be effective in monitoring of small displacement of an impeller shaft inside an agitation vessel [9]. Although two-dimensional Kohonen nets are used in the SOMine, its training mode is not the sequential mode that is used conventionally.

In the training mode, the batch SOM algorithm first processes all data vectors, and then updates the map once. The batch SOM algorithm is faster and more robust than the original Kohonen algorithm. According to user's manual of SOMine, the initial data vectors are decided by the principal component analysis (PCA) of the input data that are prepared for training. That is to say, in the present paper, training starts from a map representing linearized space for the multi-dimensional input data. Topological connection between two arbitrary nodes is defined by the Gaussian function. The radius of the Gaussian function is referred to as tension in the SOMine, which is used for determining the degree of smoothing of the map.

In the training of the map, the SOMine updates a node vector by setting it to the mean value of all weighted data vectors that match that node and its neighboring nodes, in a manner similar to the K-means method. During the training process, the number of nodes in a map is not fixed but grows from a fairly small number to the desired number of nodes, in order to implement more efficient training. This scheme is shown in Fig.3. Each map is trained for a certain number of batches using decreasing tension. When the number of nodes is increased, the growth of the map is compensated by a corresponding decrease in the tension.

In the clustering of a generated map after the training, the SOM-Ward clusters were used. This clustering method combines the local ordering information of the map with the classical hierarchical cluster algorithm for Ward.

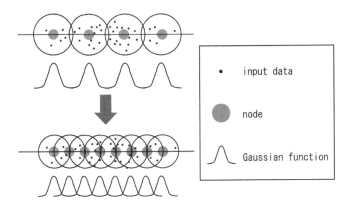

Fig. 3. A schematic diagram of increase of nodes in the training

3.2 Procedures of Clustering for Classification of Pressure Fluctuation Data and Operational Conditions

Clustering map of waveforms of pressure fluctuation would be generated by inputting the preprocessed pressure data, which was described in the section 2, into the batch SOM. In the previous study, it was seen that composition of waveforms of the fluctuation, which was contained in the time-series pressure data, changed with flow rates of the water and the organic solvent. The composition of waveforms of fluctuation was considered to be closely related to dynamic behavior of the interface between the water and the organic solvent. Therefore, it is necessary to prepare different data vectors for clustering of operational conditions, which are the ratio of flow rate of organic phase to aqueous phase (R) and the total flow rate of the two phases (Q). In the present paper, a new procedure of clustering that consists of four steps is proposed as shown in Fig.4.

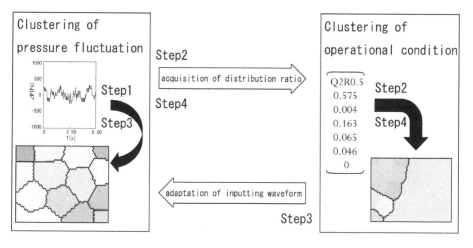

Fig. 4. A framework of clustering method

— **Step 1:** Multiple waveforms, features of which seem characteristic, are extracted arbitrarily from many waveforms of pressure fluctuation collected under various operational conditions. Next, a clustering map is computed by inputting the selected waveforms to the SOMine. The number of clusters on the generated map is usually equal to the number of inputted waveforms.

— **Step 2:** All waveforms of the fluctuation, which are extracted from the time-series pressure data acquired under a certain operational condition, are recalled on the clustering map that is generated in the step 1. Next, the ratio of distribution of recalled waveform to each cluster is estimated, and a set of the distribution ratios is an input data vector for the next clustering of operational conditions. Provisional clustering map of the operational conditions is computed by inputting the different sets of distribution ratios of waveform.

— **Step 3:** Some clusters, which seem to represent groups of waveforms with large amplitude, are selected from the provisional clustering map that are generated in the step 2. A new clustering map of the pressure fluctuation is computed again by inputting all waveforms classified to the selected clusters.

— **Step 4:** The same procedure as the step 2 is implemented by using the clustering map that is generated in the step 3. By inputting a new data vector of the distribution ratio to clusters, the last clustering map of operational conditions is computed, which could be easily analyzed.

4   Results and Discussions

4.1   Simulations of Clustering of Operational Conditions

As the first step of the clustering, fifteen characteristic waveforms were selected from time-series pressure data that were acquired under seventeen operational conditions. Fig. 5 showed four examples of the selected waveforms of pressure fluctuation.

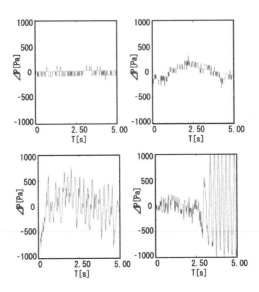

Fig. 5. Four examples of the waveforms of pressure fluctuation

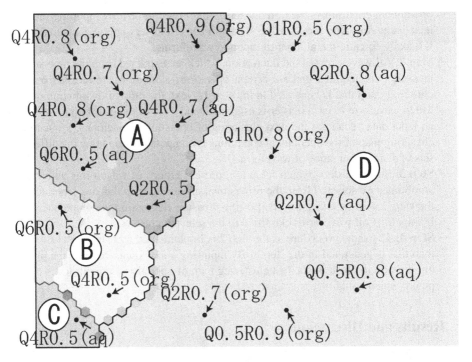

Fig. 6. Provisional clustering map of operational conditions

In the next step (step 2), the provisional clustering map of operational conditions was got as shown in Fig. 6. The generated map was seen to be divided into four clusters (A-D). "Q2R0.5 (org)" in Fig. 6, which is indicated by a bullet, represents a location of node recalled for the operational condition that the total flow rate Q is 2.00 ml/min and the ratio of flow rate R is 0.50. Then, "org" and "aq" in the parentheses represent measurement points of the organic phase side and the aqueous phase side, which are CH1 and CH2 in Fig. 1.

In the third step of clustering, we selected three clusters of A, B and C in Fig. 6, which were considered to be groups of operational conditions that pressure fluctuation with large amplitude were seen frequently. By inputting all waveforms classified to the three clusters, a new clustering map of the pressure fluctuation, which was separated into six clusters (a-f), was computed as shown in Fig. 7. Six waveforms around the map in Fig. 7 show the representative data of pressure fluctuation in the clusters. The cluster "a" was considered to be a group of the waveforms of small fluctuation, whereas the cluster "f" was a group of the waveforms with high frequency and large amplitude. The other clusters were considered to be groups of the waveforms, frequency of which was 0.2 – 0.5 Hz.

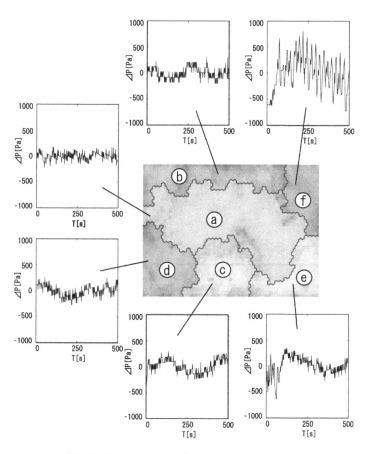

Fig. 7. Clustering map of the pressure fluctuation

The clustering map of operational conditions (Fig. 8) was computed in the last step (step 4). It was seen that the map could be separated into three clusters (A-C). As shown in Fig. 8, the operational conditions that Q was high and 4.0 – 6.0 ml/min were classified into the cluster C. On the other hand, the operating conditions that Q was 2.0 ml/min or less were classified into the clusters A and B. It was considered that the cluster A could be distinguished from the cluster B by value of R.

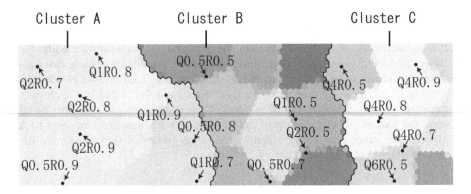

Fig. 8. Clustering map of the operational conditions

4.2   Prediction of Dynamic Behavior of Interface Based on Clustering Map

In the previous study [10], we analyzed dynamic behavior of interface between aqueous and organic phases by using two indices of W^* and α. As shown in Fig. 9, the angle which the segment OM and the base line l made was defined as the index of α. The point M was the contact of interface to the wall in the confluence of microchannels. Another index of W^* was ratio of the width of organic phase flow (W') to the width of channel (W), which was measured in the position that separated from the confluence at 18 mm.

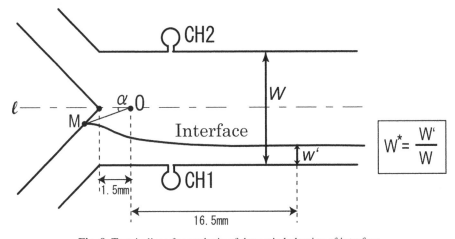

Fig. 9. Two indices for analysis of dynamic behavior of interface

As a result of the visualization analysis, dynamic behaviors of the interface were considered to be classified into three groups. Three groups (i) – (iii) could be distinguished by ranges of W^* and α, as shown in Fig. 10. Three outlines of the interface in Fig. 10 show examples of the representative form of interface in the groups. As making visual observations, interfaces in the groups (i) and (iii) were not seen to wave. On the other hand, small oscillations of interfaces in the group (ii) were observed.

When the above-mentioned result of clustering (Fig. 8) was compared with the result of visualization analysis, operational conditions in the cluster A, e.g. $[Q, R]$ = [2.0, 0.8], were considered to be classified by the group (ii). The cluster B in the map was considered to correspond to the group (i). However, for the case when the total flow rate Q was high, it was difficult to relate the cluster C to one of the three groups phenomenologically. It seemed necessary for classification of cases that Q was high to change the size of data window as shown in Fig. 2.

Finally, we carried out prediction of dynamic behavior of the interface for a new operational condition that Q was low, by using the clustering map shown in the subsection 3.2. First, a set of pressure fluctuation data was recalled on the clustering map of the waveforms (Fig. 7), which was got in the step 3. The ratio of distribution of recalled waveform to each cluster was estimated, and the set of distribution ratios was inputted to the clustering map of operational conditions (Fig. 8), which was got in the last step 4. In the case when $[Q, R]$ = [1.0, 0.6], this operational condition was seen to be recalled as a node in the cluster B in the map (Fig. 8). Because the cluster B could be related to the group (i), we estimated that the dynamic behavior of interface would not show wavy motion and that the two-phase flow could be stable flow. The prediction performance was demonstrated to be good by the experimental result of visualization of interface under the same operational condition.

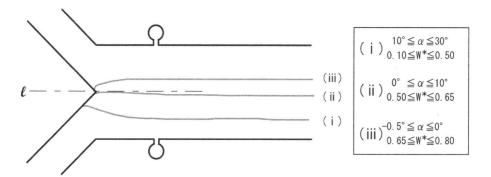

Fig. 10. Three examples of the representative form of interface in two-phase flow

5 Conclusions

The batch SOM was applied to the classification of operational conditions based on the dynamic behavior of interface between immiscible liquid-liquid flows in the microchannel. Pressure fluctuation data were used for monitoring the dynamic behavior of interface. We came up with the method of combining the clustering of pressure fluctuation data with the clustering of operational conditions in the present paper.

The clustering of pressure fluctuation and the clustering of operational conditions were iterated in turn. As a result of computation of the clustering, which consisted of four steps, a map with three clusters was seen. Two clusters A and B were considered to correspond respectively to the two groups (ii) and (i) that were classified by the previous analysis of visualization.

Prediction of dynamic behavior of the interface was carried out by recalling of a new operational condition that Q was low. Because the operational condition, $[Q, R] = [1.0, 0.6]$, was seen to be recalled as a node in the cluster B on the trained map, we estimated that the dynamic behavior of interface would not show wavy motion and that the two-phase flow could be stable flow. Therefore, the proposed clustering method by the batch SOM was demonstrated to show good predictive performance for estimation of the dynamic behavior of interface. In the future work, it will be necessary for the case when Q is high to discuss more appropriate method of the clustering.

References

1. Stankiewicz, A.I., Moulijn, A.: Process Intensification: Transforming Chemical Engineering. Chem. Eng. Progress. 96, 22–34 (2000)
2. Hessel, V., Hardt, S., Lowe, H.: Chemical Micro Process Engineering. Wiley-VCH, Weinheim (2004)
3. Zhao, Y., Chen, G., Yuan, Q.: Liquid-Liquid Two-Phase Flow Patterns in a Rectangular Microchannel. AIChE J. 52, 4052–4060 (2006)
4. Maruyama, T., Matsushita, H., Uchida, J., Kubota, F., Goto, M.: Liquid Membrane Operations in a Microfluidic Device for Selective Separation of Metal Ions. Anal. Chem. 76, 4495–4500 (2004)
5. Hibara, A., Nonaka, M., Hisamoto, H., Uchiyama, K., Kikutani, Y., Tokeshi, M., Kitamori, T.: Stabilization of Liquid Interface and Control of Two-Phase Confluence and Seaparation in Glass Microchips by Utilizing Octadecylsilane Modification of Microchannels. Anal. Chem. 74, 1724–1728 (2002)
6. Aota, A., Hibara, A., Kitamori, T.: Pressure Balance at the Liquid-Liquid Interface of Micro Countercurrent Flows in Microchips. Anal. Chem. 79, 3919–3924 (2007)
7. Kohl, M.J., Abdel-Khalik, S.I., Jetter, S.M., Sadowski, D.L.: An Experimental Investigation of Microchannel Flow with Internal Pressure Measurements. Int. J. Heat Mass Transfer 48, 1518–1533 (2005)
8. Kohonen, T.: Self-Organizing Maps (the Japanese language edition). Springer, Tokyo (2001)
9. Matsumoto, H., Masumoto, R., Kuroda, C.: Feature Extraction of Time-Series Process Images in an Aerated Agitation Vessel using Self Organizing Map. Neurocomputing (in press)
10. Marumo, T., Matsumoto, H., Kuroda, C.: Measurement of Pressure Fluctuation in Liquid-Liquid Two-Phase Flow in a Microchannel and Analysis of Stabilization of Liquid Interface. In: Proceeding of the 39th Autumn Meeting of the SCEJ(Japanese). S119 (2007)

Intelligent Fuzzy Reasoning for Flood Risk Estimation in River Evros

Lazaros S. Iliadis[1] and Stephanos Spartalis[2]

[1] Department of Forestry and Environmental Management, Pandazidou 193, str.,
PC 68200, Democritus University of Thrace, Greece
liliadis@fmenr.duth.gr
[2] Department of Production and Management Engineering, Xanthi,
Democritus University of Thrace, Greece

Abstract. This paper presents the design of a fuzzy algebra model and the implementation of its corresponding Intelligent System (IS). The System is capable of estimating the risk due to extreme disaster phenomena and especially due to natural hazards. Based on the considered risk parameters, an equal number of fuzzy sets are defined. For all of the defined fuzzy sets trapezoidal membership functions are used for the production of the partial risk indices. The fuzzy sets are aggregated to a single one that encapsulates the overall degree of risk. The aggregation operation is performed in several different ways by using various Fuzzy Relations. The degree of membership of each case to an aggregated fuzzy set is the final overall degree of risk. The IS has been applied in the problem of torrential risk estimation, with data from river Evros. The compatibility of the system to existing models has been tested and also the results obtained by two distinct fuzzy approaches have been compared.

Keywords: Fuzzy sets, trapezoidal membership function, fuzzy relations, T-Norms, natural disasters risk indices, torrential risk indices.

1 Introduction

Risk evaluation is a very important, complex and crucial task for scientists in our days. Especially, vulnerability estimation caused by natural disasters and extreme weather phenomena is of main concern. The term natural disasters includes many very serious problems of our times and it becomes worse every year, due to the climate changes and due to the deforestation.

This paper deals with the development of a Risk Approximation Model (RAM) incorporated by a prototype Intelligent System. The IS has been applied as a pilot approach in a real world environmental risk assessment problem, using actual data.

According to Turban and Aronson [1] decision making comprises of three phases: a) The intelligence (searching for conditions that call for decisions) b) The design (inventing, developing and analyzing possible courses of action) c) The choice (Selecting a course of action from the available ones). An unstructured problem is the one in which none of the three above phases has a predefined and routine solution. Risk estimation is exactly the case of an unstructured problem. It is a fact that fuzzy reasoning is widely used as a modern tool for decision making and it is often met in the literature for environmental risk problems [2].

D. Palmer-Brown et al. (Eds.): EANN 2009, CCIS 43, pp. 55–66, 2009.

There is no widely accepted definition of risk. Four items are found for risk in Webster Dictionary: (1) Possibility of loss or injury; (2) Someone or something that creates or suggests a hazard. According to Huang et al, [3] *Risk* is a scene in the future associated with some adverse incident. *Scene* means something seen by a viewer; a view or prospect. *Adverse* is contrary to one's interests or welfare; harmful or unfavorable [3]. A scene must be described with a system consisting of time, site and objects [3]. The association would be measured with a metric space whereas incident would be scaled with a magnitude [3].

Several successful efforts for risk estimation using fuzzy sets have also been made [3], [4], [5], [6]. The developed RAM uses specific fuzzy algebra concepts that include membership functions (MF) and distinct fuzzy conjunction operators (FCO). This is the second approach that has been performed on river Evros by our research team. The initial one [7] had encouraging results, whereas compatibility to the existing methods of Gavrilovic and Stiny [8], [9] was as high as 75%. It used triangular membership functions and four FCO [7] and all of the parameters had equal contribution in the determination of overall risk. Except from the different membership function and the additional FCO this paper presents a comparative study between the initial and the late approach in an effort to determine which MF fits better in this specific case. Also it enhances the initial model significantly by adding the option of scenarios performance with the use of uneven contribution weights to the parameters involved. So this research effort offers a significant step ahead.

The developed IS has been tested and validated on the actual problem of floods and torrential disasters in a specific area of Southern Balkans. Floods and erosion cause immense problems on settlements and infrastructure and intense feelings of insecurity among the citizens world wide. As a consequence, mitigation measures must be taken based on the design of an effective protection and prevention policy. The design of such a policy requires the correct estimation of the torrential risk for each river or stream watershed in the wider area of interest. Every year, the most significant phenomena of floods and erosion in the southern Balkans are observed in the mountainous part of the watershed of river Evros that belongs to Greece, Turkey and Bulgaria. Due to the lack of data from Turkey and Bulgaria the testing has been performed only in the watersheds that belong to Greece. This is obviously a limitation of the testing process. However its solution requires years of interstate cooperation. This is one of the reasons that make this effort a pilot one.

1.1 Necessity for a New Approach

The most widely used existing approaches are the ones of Gavrilovic and Stiny [8], [9]. The equation of Gavrilovic considers the average annual production of sediments, the average annual temperature, the average annual rain height of the watershed, its area, the kind of its geodeposition, the vegetation, the erosion of the watershed, and the calculation of the special degradation is really important. The method of Stiny determines the torrential risk based on sediment production with a periodicity of 100 years with the equation of Stiny-Herheulidze [9].

The main drawback of the existing modeling approaches is that they consider only partial risk indices (due to independent parameters) and that they apply crisp sets methods with specific boundaries, not capable of describing proper risk linguistics.

The basic innovative characteristic of the RAM is the production of an integrated risk index that has an overall consideration of all involved factors. The natural disasters risk estimation problem is a composite one and it should be seen under different perspectives before the design of a prevention and protection policy. For example other areas will be at highest risk under extreme weather phenomena and other areas will be at highest risk in average situations. The older methods do not make such distinctions. The IS performs both partial risk index estimation (for each factor affecting the problem) and it also calculates a unified risk index (URI) which can be considered as an estimation of the overall risk for each area. Finally a very important advantage of the proposed IS is its ability to assign different weights to the attributes involved. In this way it can either perform scenarios or it can consider the attributes as having equal importance. The Decision Support System (DSS) has been named TORRISDESSYS (Torrential Risk Decision Support System) and it can be applied to any area of the world as long as sufficient amount of data is present. Its outcome has been compared to the outcome of other established methodologies and the results have proven a high degree of compatibility between them. The results of this study can become a basis of the long-term planning of mitigation measures against floods and erosion.

1.2 Necessity for Applying Flexible Models

Two different algebraic approaches can be considered for the estimation of risk generally. The one applies crisp sets on existing data and the other uses fuzzy sets on data and on meta-data [10]. In this modelling approach the initial primitive data come from actual measurements. Then as the system learns this piece of information is transformed to degrees of membership (DOM) in the closed interval [0,1]. The DOMs are used as new sources of data for the production of the overall degree of membership which is the final risk measurement. If it is required by the case study, this overall risk index can also be transformed from a pure dimensionless number to a parameter with a physical meaning by applying known de-fuzzification equations and processes. From this point of view, the term meta-data is used here to indicate this continuous transformation in the nature of data. Fuzzy sets (FS) can be used to produce the rational and sensible representation of real world concepts [11]. For each FS there exists a degree of membership function $\mu_s(X)$ that is mapped on the closed interval of values [0,1] [12]. Figure 1 graphically presents the fuzzy sets *Tiny* and *Small* using a semi-trapezoidal and a triangular membership functions respectively.

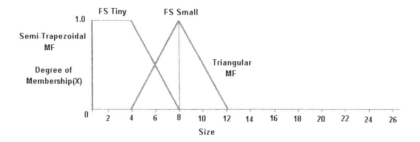

Fig. 1. The FS "tiny" and "small" with different MF

2 The Fuzzy Algebra Model

Most researchers in risk discipline accept fuzzy set theory for the risk screening and prioritization at plant and regional levels [3]. Fuzzy algebraic models constitute the modeling tools of soft computing and fuzzy logic is a tool for embedding structured human knowledge into workable algorithms [13]. This project aims in estimating an innovative and flexible means of risk estimation where the degree of risk can be calculated under different perspectives by applying various FCO from a pessimistic or an optimistic point of view. This will be clarified in the following section where various FCO will be described.

For this purpose the factors that influence the degree of risk have to be identified. Of course these factors vary according to the nature of the considered problem. For example in the specific application of the RAM that has been presented here, we are concerned with the early torrential risk estimation of mountainous watersheds (TRMW) using fuzzy membership functions and fuzzy relations.

The TRMW depends on various risk factors each one giving a partial degree of risk. Initially a set of indices representing the partial degrees of TRMW for each area of river Evros has been calculated. The torrential factors taken into consideration in the construction of the fuzzy logic model are the following:

1. The average altitude of each watershed. 2. The average slope of each watershed. 3. The average annual rain height in the average altitude of each watershed. 4. The percentage of forest-cover of each watershed. 5. The percentage of the compact geological forms of each watershed.

The above characteristics are divided in dynamic (factor number 4) and structural ones (factors 1,2,3,5). According to [14], [9], [15], [16], the most important morphometric characteristics of the watersheds that influence the torrential risk are the following: The area, the perimeter, the shape of the watershed, the maximum altitude, the minimum altitude, the average altitude, the average slope of the watershed, and its maximum altitude. Of course meteorological data play an important role. The selection of the considered factors has been performed based on the opinions of experts mentioned above and also on the availability of actual data. In this research not all of the parameters involved in the problem are covered due to insufficient amount of data. This is another reason for which the testing of the DSS can be seen as a pilot approach that introduces new modeling techniques indicating their potential performance. However both the designed model and the information tool are completely dynamic and capable of incorporating as many risk parameters as possible. From this point of view the TORRISDESSYS can be used in the future with a full data set in an integrated effort to estimate torrential risk.

Five FS have been constructed each one corresponding to a different torrential factor. The five FS used by the RAM model are the following: $\tilde{F}_1$ = "watersheds with High Average Altitude" $\tilde{F}_2$ "watersheds with High slope" $\tilde{F}_3$ "watersheds with high Average Annual Rain" $\tilde{F}_4$ "watersheds with high forest cover" $\tilde{F}_5$ "watersheds with Low Percentage of compact geological forms"

Of course several other FS can be included in other problems. The degree of membership of each area of to each one of the above FS is its partial degree of risk (PDR) for the corresponding torrential factor. This means that $PDR_i = \mu_i(FS_i)$ where μ_i is the function estimating the degree of membership (DOM) of area i to the FS_i. trapezoidal membership functions (TRAMF) have been applied on *Primitive data vectors* (PDV) for the estimation of the DOM of each area to the corresponding fuzzy set. The following equation 1 describes the TRAMF [13],[17],[18] [19].

$$\mu_s(X) = \begin{cases} 0, & \text{if } X \le a \\ (X-a)/(m-a), & \text{if } X \in (a,m) \\ 1, & \text{if } X \in [m,n] \\ (b-X)/(b-n), & \text{if } X \in (n,b) \\ 0, & \text{if } X \ge b \end{cases} \tag{1}$$

The TRAMF has a tolerance interval for the watersheds with the highest degree of risk. Also other membership functions (MF) like the triangular (having only one peak) are common in fuzzy algebra. In this research effort the TRAMF has been applied on PDV due to the importance of the tolerance interval. The following equation 2 presents a triangular membership function [13],[17] [19].

$$\mu_s(X) = \begin{cases} 0 \text{ if } X < a \\ (X-a)/(c-a) \text{ if } X \in [a,c) \\ (b-X)/(b-c) \text{ if } X \in [c,b) \\ 0 \text{ if } X >= b \end{cases} \tag{2}$$

The final target is the aggregation of all the partial DOM and the estimation of the degree of membership of each area to the final fuzzy set *"Torrentially risky Area"*. Various special types of fuzzy relations (the fuzzy T-Norms) have been applied in order to unify the partial risk indices. In this way *meta-data vectors* (MDV) that correspond to the unified risk indices are extracted from the PDV. Of course this is automatically performed by an intelligent prototype Information system. The following equations 3 and 4 offer a good definition of a fuzzy relation.

Let A be an input fuzzy region with element x using a membership function μ_A and let B be an output fuzzy region with element **y** and membership function μ_B. The fuzzy relation on the fuzzy product A X B is a mapping of the following type:

$$\mu_R : A \times B \rightarrow [0,1], \text{ where } \mu_R(X,Y) = \mu_A(X) \wedge \mu_B(Y) \tag{3}$$

In this case the fuzzy relation set is defined to be:

$$P = \{((X,Y),\mu_R(X,Y))/(X,Y) \in A \times B\} \tag{4}$$

A T-Norm performs the logical AND operation between the fuzzy sets. The result of each T-Norm is presented in the following equation 5.

$$T\_Norm_i = PDR_1 \wedge PDR_2 \wedge \dots \wedge PDR_i = \mu_1(FS_1) \wedge \mu_2(FS_2) \wedge \dots \wedge \mu_i(FS_i) \tag{5}$$

It should be defined that i is the number of independent parameters, and consequently the number of FS used. The following Table 1 contains the five different cases of T-Norms that have been used for the production of the unified risk index (URI) [20],[13],[21].

Table 1. Applied T-Norms

Algebraic Product URI= $\mu_A(X) * \mu_B(X)$	(6)
DrasticProduct URI = $MIN(\mu_A(X), \mu_B(X))..if..MAX(\mu_A(X), \mu_B(X)) = 1$ otherwise URI = 0	(7)
Einstein product URI = $\mu_A(X) * \mu_B(X) / (2-(\mu_A(X) + \mu_B(X) - \mu_A(X) * \mu_B(X)))$	(8)
Hamacher product URI = $\mu_A(X) * \mu_B(X) / (\mu_A(X) + \mu_B(X) - \mu_A(X) * \mu_B(X))$	(9)

After the application of the decision support system, each area is assigned a MDV which is a vector of unified risk indices due to the various T-Norms applied. Each element of each meta-data vector corresponds to the overall degree of risk of the examined case under a different angle. Consequently each element of the MDV has a different perspective of the risk problem. The interpretation of all the unified risk indices has equal importance towards risk estimation.

Since the T-Norms perform conjunction they are aggregation functions Agg(x). In this case we can have multiple attribute decision making with unequal weights assigned to the attributes [22], [23].

$$\mu_S(x_i) = Agg\left(f\left(\mu_A(x_i), w_1 \right), f\left(\mu_A(x_i), w_2 \right), \ldots f\left(\mu_A(x_i), w_n \right) \right)$$ (10)

where i=1,2…..k and k is the number of cases and n is the number of the attributes. The following equation 11 represents function f(μ,w).

$$f\left(\mu_A, w \right) = \mu_A^{\frac{1}{w}}$$ (11)

The above equations 10 and 11, offer a very significant approach in the consideration of the potential uneven contribution of each parameter to the final torrential risk. The TORRISDESSYS offers the option of uneven weights however in this paper we have performed the scenario of even contribution. This was done in order to have a similar and comparable approach to the one of [7]. According to [24] the T-Norms can be considered as optimistic conjunction operators. There are also other types of aggregation operators called S-Norms that assign higher degrees of memberships to risky areas and they view the problem exactly from a reverse angle, compared to the T-Norms. This research faces the problem from an optimistic point of view. Of course a future research effort should involve the application of S-Norms as well. The universal nature of the model is due to the fact that in a different application, the only thing that has to change is the nature of the parameters involved in the estimation of risk. The whole algebraic framework remains exactly the same.

2.1 Implementation of the IS

The IS has been developed in MS-Access. It uses a visual environment and it applies a modern graphical user interface. It has been designed and implemented to store the data in a relational access database. The retrieval mechanism of the DSS reads the required data from the database. Data is stored in several Tables according to the relational philosophy. The main Table contains six fields, the primary key and the five fields corresponding to the torrential risk factors. It has been designed to follow the first and second Normal form [25]. The DOM to the five fuzzy sets (partial risks) and the URI were calculated by performing SQL (Structured Query Language) operations on the database. The following statement is an example of a DOM estimation using SQL.

SELECT fuzzy_help.perioxi, fuzzy_help!help_drastic AS drastic_product FROM fuzzy_help ORDER BY fuzzy_help!help_drastic DESC;

3 Application in the Case of the Flood Risk Problem

The developed Fuzzy reasoning System has been applied in the case of flood Risk estimation in an area where the problem of torrential phenomena occurs quite frequently. The application area of the IS is located in the Northeastern part of Greece near the border of Greece Bulgaria and Turkey. It is a fact that this area contains the most important torrential streams of the Greek watershed of river Evros [7]. This area has been chosen due to its serious torrential phenomena that have consequences in the life of the local people.

The research area has been divided in three main sub-areas based on hydrological and geological homogeneity criteria and according to the use of land. The first part is called "Northern Evros" the second part is the one of "Central Evros" and third is the part of "Southern Evros". Finally the three parts have been divided in numerous sub-areas, each one containing an important stream or river watershed. These areas are encoded using integer numbers. Initially the limits of the research areas have been estimated, using maps of the Geographical Army Service (GAS) with a scale of 1:250.000. Then the torrential streams of each sub-area and their limits have been located from the maps (scale 1:100.000) of the GAS. The upper and lower limits of the watershed areas are 300 και 2 km^2 respectively [9]. For every research area and for each torrential stream the morphometric characteristics have been specified and they have been produced after the process of maps (scale 1:100.000) of the GAS. Meteorological data have been gathered from all of the sources that have meteorological stations in the area. They concern the monthly rain height and the average temperatures. The torrential streams of the Evros area have been determined from the 1:500.000, and 1:50.000 scale maps based on the clustering of Kotoulas [9]. The testing of the IS has been performed using actual data (several PDVs) concerning all of the watersheds of Northern, Central and Southern Evros area. The following Table 2 shows clearly the results of the unified risk indices estimation (URI) for each watershed of Northern Evros area. Trapezoidal membership functions are applied for the estimation of the five PDR. Finally the Einstein product, the Algebraic product and the Hamacher product T-Norms are used for the performance of the fuzzy conjunction operation and thus for the estimation of the final URI that constitutes the TRMW estimation.

Table 2. Torrential risk of Northern Evros with the TRAMF

Area	TRMW Einstein product	Area	TRMW Algebraic product	Area	TRMW Hamacher product	Area	TRMW Drastic product
16	0.61474	16	0.61475	16	0.15368	16	0.61475
1	0.36612	1	0.36612	1	0.09153	2	0.57072
2	0.17795	2	0.24282	2	0.06679	4	0.41556
5	0.12817	5	0.15889	5	0.04225	1	0.36612
4	0.07001	4	0.11956	4	0.03631	5	0.23077
17	0.00871	17	0.01909	17	0.00679	17	0.22225
11	0.00722	11	0.01586	11	0.00565	11	0.17578
12	0.00333	12	0.00805	12	0.00311	12	0.13706
3	0.00039	3	0.00104	3	0.00044	3	0.05188
7	0.00022	7	0.00059	7	0.00024	7	0.03639

Table 3 shows the influence of extreme values to the produced risk index. Al areas that are characterized by the complete lack of compact geological forms have a significant overall degree of risk. The drastic product T-Norm offers a good approach for the characterization of areas that have *extreme values* affecting positively the torrential phenomena in a watershed area.

Table 3. Torrential risk of Northern Evros using the Drastic product. The influence of the low level of compact geological forms to the TRMW.

Area	TRMW Drastic product	Compact geological forms
16	0.61475	≈0.00%
2	0.57072	≈0.00%
4	0.41556	≈0.00%
1	0.36612	≈0.00%
5	0.23077	≈0.00%
17	0.22225	≈0.00%

In the above case the overall DOM was estimated by co-evaluating the partial DOM to the FS $\tilde{F}$ ={Area with low percentage of Compact Geological Forms}. Due to the fact that for all areas presented in the above table, the DOM to the FS $\tilde{F}$ equals to 1, the Drastic Product produces an overall DOM equal to the minimum of all. The following Tables 4 and 5, show clearly the most risky areas for the area of Central Evros and Southern Evros respectively.

Table 4. Torrential risk of Central Evros by the TRAMF (10 most risky areas)

Area	Algebraic product	Area	Drastic product	Area	Einstein product	Area	Hamacher product
33	0.02706	33	0.24504	33	0.01259	33	0.00949
32	0.01955	32	0.21285	32	0.00843	32	0.00728
31	0.01048	31	0.20646	31	0.00420	31	0.00418
23	0.00867	26	0.20324	23	0.00352	23	0.00341
26	0.00425	23	0.18433	21	0.00163	26	0.00208
21	0.00417	21	0.14705	26	0.00143	21	0.00171
27	0.00287	20	0.13437	20	0.00104	27	0.00142
20	0.00280	22	0.11751	27	0.00096	20	0.00121
30	0.00211	30	0.10784	30	0.00077	30	0.00092
22	0.00189	18	0.08881	22	0.00070	29	0.00086

In both cases the TRAMF has been applied for the estimation of the partial degrees of risk and the Einstein, the algebraic product the Hamacher product and the Drastic product T-Norms for the final TRMW calculation.

Table 5. Torrential risk of Southern Evros by the TRAMF (10 most risky areas)

Area	Algebraic product	Area	Drastic product	Area	Einstein product
10	0.07390	7	0.34057	10	0.03902
7	0.05597	10	0.32789	7	0.02752
11	0.00624	13	0.06110	11	0.00240
19	0.00209	2	0.00010	19	0.00062
13	0.00058	22	0.00010	13	0.00019
5	0.00054	4	0.00010	5	0.00015
3	0.00024	6	0.00010	3	0.00006
6	0.000E-2	8	0.00010	6	0.00054E-2
8	0.00045E-2	17	0.00010	8	0.00029E-2
9	0.00045E-2	16	0.00010	9	0.00028E-2

4 Discussion – Comparison to Existing Approaches

Fuzzy logic is non-monotonic in the sense that if a new fuzzy fact is added to a database it may contradict conclusions previously derived [26]. Obviously, the major challenge in fuzzy risk analysis is to find a scientific approach to approximately show risk as objectively as possible. "Objectively" differs from "accurately" [3]. The former means "uninfluenced by emotions or personal prejudices;" the latter means "deviating only slightly or within acceptable limits from a standard." With fuzzy information to show risk, it is impossible to seek "accuracy," but it is possible to seek "objectiveness" meaning *do not ignore any available information; do not add any personal assumption*" [3].

Using the TRAMF the most risky areas are similar in the cases of the Algebraic product, Drastic product, Einstein product and Hamacher Product. This means that the most risky areas are more or less the same but they have small differences in their order. Only in the case of the drastic product that considers mainly the areas with extreme values there are some changes.

The five most risky areas for all types of T-Norms are identical whereas small variations in the rankings are met. More specifically for Northern Evros the watersheds 16,1,2,4 and 5 appear to be the most risky ones regardless the T-Norm applied. For central Evros the watersheds 33,32,31,23,26 are the five most risky. Finally for Southern Evros the streams 10,7,11 and 19 are the most risky, whereas the Drastic Product assigns high risk to the streams 2 and 22. These streams are classified as low risky by the other T-Norms. This is due to the fact that the watersheds 2 and 22 in Southern Evros have extremely low percentage of compact geological forms and values below average for the other fields of the PDV. Thus areas 2 and 22 are characterized as very risky by the Drastic Product only. This is not a contradiction but a view to the problem under a different perspective. Except from the drastic product, obviously the torrential risk estimation is not mainly T-Norm dependent. This means that when the TRAMF is used the most risky areas can be found regardless the fuzzy relation used Also another list of areas having extreme values for some factors (and thus are very risky) can be determined.

The TRAMF normalizes the differences between the areas of maximum risk. This is not the case when the triangular membership function is used because it has one peak point [7] [4]. This is mainly due to the nature of the TRAMF which has a whole tolerance interval of values corresponding to the maximum degree of membership and partly due to the nature of river Evros data. On the other hand our previous study [7] has shown that in the case of the triangular membership function (TRIMF) the situation is more or less the same. Again there is not a significant differentiation between the T-Norms in the areas characterized as risky. For example in the case of Southern Evros the five most risky streams are number 10,16,7,9 and 2.

Comparing the results of the torrential risk estimation between the triangular and the TRAMF, the first similarity that we have discovered is that they both have a high level of agreement for the watersheds at highest risk. On the other hand, although the order of these watersheds in terms of their absolute ranking position and their degree of risk is slightly differentiated there exists a perfect agreement in the first positions. More specifically areas with code number 10 and 7 are the most risky ones regardless the membership function but from the next ranking position and further there are differences. For example area number 19 is fourth based on the TRAMF when algebraic or Einstein Product is used and sixteenth when triangular membership function is used with the same T-Norms. This is due to the tolerance interval of the TRAMF which has a significant effect in the produced degree of membership. Finally we have performed a compatibility test between the results of the TRAMF and the output of other existing methods. In a previous research effort of ours [7] we had done the same test between the TRIMF and the same commonly used methods for the area of Evros river. As it will be shown in the next paragraph the TRAMF has a much higher level of compatibility and this is a serious indication towards its suitability to the torrential risk estimation. As it was mentioned in the early sections of this paper, this research is a pilot one. The results of the system have been compared to the outcome given by the method of Gavrilovic1 (G1), Gavrilovic2 (G2), [8] and the method of Stiny (St) [9] which were described briefly in the previous sections and they are widely used. The comparative study between the system and the other existing methods has proven that there exists a very high level of compatibility in most of the cases. In the case of Central Evros for the TRAMF the compatibility between the TORRISDESSYS and the Gavrilovic2 method is stable and it equals to 80% in the cases of the algebraic product, the min, the drastic product and the Hamacher product. Only in the case of Einstein product it is equal to 75%. The use of the Einstein T-Norm gives an average compatibility of 62.5% in Northern Evros, 63.3% in Central Evros and 52.77% in Southern Evros. The use of the algebraic product T-Norm gives an average compatibility of 62.5% in Northern Evros, 61.6% in Central Evros and 52.77% in Southern Evros.

The use of the Minimum T-Norm gives an average compatibility of 62.5% in Northern Evros, 61.66% in Central Evros and 52.77% in Southern Evros. The use of the Hamacher product T-Norm gives an average compatibility of 62.5% in Northern Evros, 61,66% in Central Evros and 52.77% in Southern Evros. In the case of the Drastic Product the compatibility between the TORRISDESSYS and the Gavrilovic2 method for Central Evros is 80% (agreement in 16 out of 20 cases). Generally the testing has proven that the DSS estimates the most risky torrential streams in the area of river Evros having an agreement higher than 50% with the methods of G1, G2 and Stiny that can reach up to 80%. According to the above results and to [7] the TRAMF has a much higher percentages of agreement with the established common methods than the TRIMF. An achievement of this research effort is the development of an effective and flexible risk estimation model that can be used in various risk situations and problems. The application of the system has been quite successful. A drawback of the system's testing process is the lack of torrential data from the Turkish and Bulgarian side of river Evros. It would be essential for the Greek side to cooperate with the other two sides and to input torrential data from both countries into the IS due to the fact that the natural disasters do not have any borders. This should be materialized in the future in a much wider effort.

References

1. Turban, E., Aronson, J.: Decision support systems and Intelligent systems, 5th edn. Prentice Hall, New Jersey (1998)
2. Carlsson, C., Fuller, R.: Fuzzy Reasoning in Decision-Making and Optimization, 1st edn. Physica-Verlag, Heidelberg (2001) (Studies in Fuzziness and soft computing)
3. Huang, C.F., Moraga, C.: A fuzzy risk model and its matrix algorithm. International Journal of Uncertainty, Fuzziness and Knowledge –based systems 10(4), 347–362 (2002)
4. Iliadis, L., Spartalis, S.: Fundamental fuzzy Relation Concepts of a D.S.S. for the estimation of Natural Disasters risk (The case of a trapezoidal membership function). Journal of Mathematical and Computer modelling 42, 747–758 (2005)
5. Kaloudis, S., Tocatlidou, A., Lorentzos, N., Sideridis, A., Karteris, M.: Assessing Wildfire Destruction Danger: a Decision Support System incorporating uncertainty. Journal Ecological Modelling 181(1), 25–38 (2005)

6. Loboda, T.V., Csiszar, I.: University of Maryland USA. Assessing the risk of ignition in the russian far east within a modeling framework of fire threat. Ecological Applications 17(3), 791–805 (2007)
7. Iliadis, L., Maris, F., Marinos, D.: A decision support system using fuzzy relations for the estimation of long-term torrential risk of mountainous watersheds: The case of river Evros. In: Proceedings of the 5th International Symposium on Eastern Mediterranean Geology, Thessaloniki, Greece (2004)
8. Gavrilovic, S.: Inzenjering o bujicnim tovoklima i eroziji. Beograd (1972)
9. Kotoulas, D.: Management of Torrents I. Publications of the University of Thessaloniki, Greece (1997)
10. Leondes, C.T.: fuzzy logic and Expert systems Applications. Academic Press, California (1998)
11. Kandel, A.: Fuzzy Expert systems. CRC Press, USA (1992)
12. Zadeh, L.A.: Fuzzy logic Computing with words. IEEE Trans. fuzzy systems 4(2), 103–111 (1996)
13. Kecman, V.: Learning and soft computing. MIT Press. London (2001)
14. Kotoulas, D.: Research on the characteristics of torrential streams in Greece, as a causal factor for the decline of mountainous watersheds and flooding, Thessaloniki, Greece (1987)
15. Stefanidis, P.: The torrent problems in Mediterranean Areas (example from Greece). In: Proc. XXIUFRO Congress, Finland (1995)
16. Viessman, J.W., Levis, G.L., Knappt, J.W.: Introduction to Hydrology. Harper and Raw Publishers, New York (1989)
17. Cox, E.: The fuzzy systems Handbook, 2nd edn. Academic Press, New York (1999)
18. Dubois, D., Prade, H., Yager, R.: Fuzzy Information Engineering. John Wiley and sons, New York (1996)
19. Nguyen, H.E., Walker, E.: A First Course in fuzzy logic. Chapman and Hall, Library of the Congress, USA (2000)
20. Cox, E.: Fuzzy modeling and Genetic Algorithms for data Mining and Exploration. Elsevier, USA (2005)
21. De Cock, M.: Representing the Adverb Very in fuzzy set Theory. In: Proceedings of the ESSLLI Student Session, Ch.19 (1999)
22. Calvo, T., Mayor, G., Mesira, R.: Aggragation Operators: New Trends and Applications (Studies in Fuzziness and soft computing). Physica-Verlag, Heidelberg (2002)
23. Fan, Z., Ma, P., Zhang, J.Q.: An approach to multiple attribute decision-making based on fuzzy preference information on alternatives. Fuzzy sets and systems 131(1), 101–106 (2002)
24. Iliadis, L.: Intelligent Systems and applications in risk estimation (Book in Greek). Stamoulh publishing co, Thessaloniki (2007)
25. Date, C.J.: An Introduction to database systems. Addison-Wesley, New York (2007)
26. Coppin, B.: Artificial Intelligence Illuminated Jones and Bartlett Publishers USA (2004)

Fuzzy Logic and Artificial Neural Networks for Advanced Authentication Using Soft Biometric Data

Mario Malcangi

Università degli Studi di Milano,
DICo – Dipartimento di Informatica e Comunicazione
Via Comelico 39
20135 Milan, Italy
malcangi@dico.unimi.it

Abstract. Authentication is becoming ever more important in computer-based applications because the amount of sensitive data stored in such systems is growing. However, in embedded computer-system applications, authentication is difficult to implement because resources are scarce. Using fuzzy logic and artificial neural networks to process biometric data can yield improvements in authentication performance by limiting memory and processing-power requirements. A multibiometric platform that combines voiceprint and fingerprint authentication has been developed. It uses traditional pattern-matching algorithms to match hard-biometric features. An artificial neural network was trained to match soft-biometric features. A fuzzy logic inference engine performs smart decision fusion and authentication. Finally, a digital signal processor is used to embed the entire identification system. The embedded implementation demonstrates that improvement in performance is attainable, despite limited system resources.

Keywords: Artificial neural networks, fuzzy logic engine, soft-biometric data, multibiometrics, embedded personal authentication systems, digital signal processor.

1 Introduction

Biometrics uses physiological and behavior characteristics possessed exclusively by one individual to attest her/his identity. These claim to be better than current and established authentication methods, such as personal identification numbers (PINs), passwords, smart cards, etc., because they offer key advantages such as:

- Availability (always)
- Uniqueness (to each person)
- Not transferable (to other parties)
- Not forgettable
- Not subject to theft
- Not guessable

As a result of these advantages, biometrics offers very high-level application security compared to the security level offered by traditional identification methods based on

D. Palmer-Brown et al. (Eds.): EANN 2009, CCIS 43, pp. 67–78, 2009.

possession of a token (e.g. an identification card) or of personal knowledge (e.g. a password). Although biometrics, because of these advantages, is a superior personal authentication method and its use in microelectronics has indeed spread rapidly, its adoption in mass-market applications has lagged. The main reason is that biometrics is not a sure-fire authentication method like a password, which can always perform.

In their original form, biometric features are analog information. Therefore they are subject to variability when captured by biometric scanning devices (fingerprint sensors, microphones, cameras, and so forth). Such features can be digitalized, but the data will nevertheless have been characterized fuzzily.

Classical pattern-matching methods may achieve very high scores in false acceptance rate (FAR) and false rejection rate (FRR) but are not enough good to match a 100-percent correct acceptance rate.

The intrinsic fuzziness of biometric features suggests that a soft-computing approach to designing processes that pattern-match biometric features [1], [2], [3], [4], [5], [9], [13] could lead to nearly 100-percent authentication. For example, humans never fail to recognize a well-known person, even in prohibitive conditions. They process biometric features fuzzily and neurally.

Soft-computing data are variations and uncertainties, so they work well for biometric measurement and matching because biometric features are always variable, difficult to describe analytically, and intrinsically not 100% true as belonging to their owner.

Humans also use multiple biometric data to gain nearly 100-percent authentication rates. For example, they combine speech and facial biometric features in a matching action to identify an individual. In other words human beings are neurally and fuzzily processing combined biometric features.

The adoption of traditional biometric authentication solutions has been especially slow in embedded systems. Embedded systems leave scarce resources (memory and processing power) available to applications, and thus cannot host such authentication technology. As a result, PIN codes or analogous authentication methods are currently being used.

The soft-biometrics [10] approach is potentially less computing-intensive than hard biometrics and needs less data to map a person's identity. In addition, multibiometric solutions [7], [11], [12] can be adapted to the input capacities of embedded systems. Combining both solutions makes it possible to optimize biometric-authentication functionality for implementation in embedded systems.

To demonstrate that this approach can lead to a biometric authentication system that overcomes the primary weaknesses in biometrics-based applications, two biometrics-based authentication processes (voiceprint and fingerprint) were integrated with a fuzzy logic-based processing model for individual authentication on an embedded-computing platform.

2   System Architecture

The biometric authentication system (fig. 1) consists of three processing layers, the feature-extraction layer, the matching layer, and the fuzzy logic-based decision layer. The feature-extraction layer uses signal processing-based algorithms for hard and soft feature extraction. The matching layer uses hard computing to identify hard features and soft computing (ANN) to identify soft features. The decision layer uses a fuzzy logic engine to fuse the identification scores and the crisp soft features.

A floating-point digital signal processor (DSP) executes the feature-extraction algorithms, the artificial neural network, and the fuzzy logic engine. This kind of processing device is the best choice to efficiently run both the hard computing and the soft computing engines.

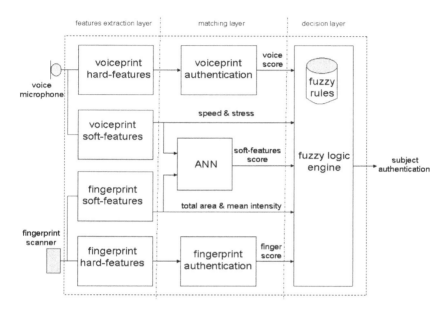

Fig. 1. System architecture

Speech is captured by a voice-type microphone, which is analog-conditioned (linearized, amplified, and filtered), 16-kHz sampled, and 16-bit quantized.

A Hamming window is then employed to extract ten-millisecond frames from the speech-data stream, using a 50% overlap between adjacent frames to avoid data loss at feature-extraction time. Each frame is processed to extract the speech features to be fuzzified at the fuzzy logic engine inputs.

The fingerprint image is captured using a solid-state fingerprint sensor as a bitmap, gray-scale image. A set of image-processing algorithms are applied to the gray-scale image to transform it into two colors, black for ridges and white for valleys. This image is then processed to extract fingerprint features to be fuzzified for input to the fuzzy logic engine.

3   Audio Hard Feature Extraction and Authentication

The success of the authentication system depends on the features extracted from speech. Several features can be extracted from speech [6]. The following features were selected as being most representative:

Root mean square (RMS), calculated as

$$RMS_j = \sqrt{\frac{1}{N}\sum_{m=0}^{N-1} s_j^{\,2}(m)}$$
(1)

Zero-crossing rate (ZCR), calculated as

$$ZCR_j = \sum_{m=0}^{N-1} 0.5\left|sign(s_j(m)) - sign(s_j(m-1))\right|$$
(2)

Autocorrelation (AC), calculated as

$$AC_j = \sum_{i=1}^{N}\sum_{j=1}^{N+1-i} s_j(j)s_j(i+j-1)$$
(3)

Cepstral linear prediction coefficients (CLPC), calculated as

$$CLPC_j = a_m + \sum_{k=1}^{m-1}\left(\frac{k}{m}\right)c_k a_{m-k}$$
(4)

with $c_0 = r(0)$, the first autocorrelation coefficient, a_m, the prediction coefficients.

The above measurements are executed on short speech frames $s_j(n)$, N samples-wide, weighted by a window $w(n)$ of the same length. A Hamming window is used to pre-process each frame. *RMS* is the short-time, root mean square of the windowed speech segment. Such measurement helps identify phonetic unit end-points. *ZCR* is a simple measurement of the spectral information used to determine whether or not the speech segment processed is voiced or unvoiced. It is also a useful indicator of the frequency with major energy concentration. AC is a good measurement of speech-pitch frequency that works better than Fourier transform. It preserves information about pitch-frequency amplitude while ignoring phase. Phase is perceptually and in-formationally unimportant for speech identification purpose.

Two methods are applied to score the person's identity. One is based on measuring Mahalanobis distance, the other on measuring the distance of dynamic time warp-ing—k-nearest neighbor (DTW-KNN).

The Mahalanobis distance measurement is

$$D_i(x) = (x - \bar{x})^T W^{-1}(x - \bar{x})$$
(5)

where W is the covariance array computed using the average and the standard deviation features of the utterance. The input pattern x is processed with reference to the utterance-averaged feature vector $\bar{x}$ that represents the person to be identified. The distance $D_i(x)$ is a score for the authorized user.

The DTW-KNN method combines the dynamic-time-warping measurement with k-nearest neighbor decision algorithm. The DTW first clusters similar elements that refer to a feature into classes. The cost function is computed using Euclidean distance, with a granularity of one frame. The KNN algorithm is then applied to select k minimal distance matching and to choose the most recurring person in k minimal distance matches. This results in lower false-positive and false-negative rates during identification, compared to the original DTW algorithm.

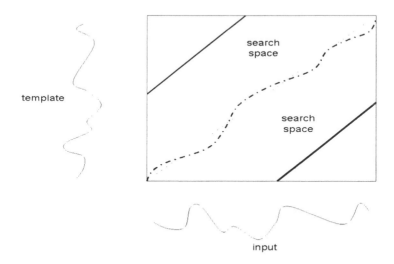

Fig. 2. DTW boundaries for voiceprint matching

4   Fingerprint Hard Feature Extraction and Authentication

Fingerprint pattern matching is based on minutiae to keep the amount of memory allocated to template storage to a minimum. Several preprocessing steps are executed on the captured gray-scale fingerprint image from the stage of bitmap to its minutiae-based representation.

The criteria we use for minutiae extraction is simple:

- If a ridge pixel has two or more 8-nearest, then it is a bifurcation
- If a ridge pixel has only one 8-nearest, then it a termination

This is a very fast algorithm for mapping fingerprint minutiae. The most critical step in this procedure is to avoid computing false minutiae caused by noise in the scanned fingerprint image. To overcome this problem, a backtrack control is executed on each feature pattern before it is validated, to check that each of the three branches of the bifurcation are significantly long.

The scanned fingerprint image is transformed into a set of coordinates x, y and the direction. Pattern matching consists of a procedure that first tries to align the template pattern and the input pattern, then computes an overlapping score. This score is a measurement of the authenticity of the person who created the input.

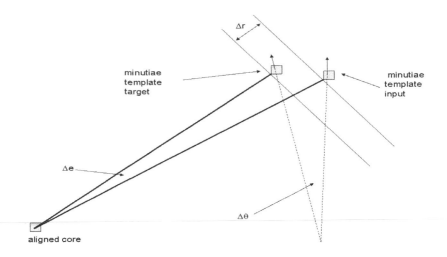

Fig. 3. Boundaries for fingerprint matching

5 Soft-Biometric Feature Extraction

Soft features are extracted from both speech and fingerprints. Speech offers much more soft-biometric data than fingerprints because it carries a lot of behavioral information. Only a limited amount of soft-biometric data can be extracted from fingerprints, but these data can also contribute to improving the whole authentication process.

The following soft features are extracted from speech:

- speed
- stress

Speed is measured as the total duration of the speech utterance. Stress is measured as the ratio between the peak amplitude of the stressed vowel and the average amplitude of the whole utterance. Both these voiceprint features are related to the way the person is used to speaking a requested word.

The following soft features are extracted from fingerprints:

- total area
- mean intensity

Total area is measured as the ratio between the total pixels available on the fingerprint-scanning device and the total pixels of the captured fingerprint image that have a value higher than the estimated peak noise level. Mean intensity is measured as the sum of the intensity of all the pixels with a value higher than the estimated peak level. Both these fingerprint features are related to the way the person approaches contact with the fingerprint sensor.

6 Artificial Neural Network-Based Soft-Biometric Feature Scoring

The artificial neural network used to authenticate soft-biometric features has three-layer, feed-forward, backpropagation architecture (FFBP-ANN). Its input nodes are fully connected to all the nodes in the hidden layer, and the hidden layer is fully connected to the output nodes.

Soft-biometric features are fed into units in the input layer. The authentication score comes out of the nodes at the output layer.

Inputs and outputs layer have a linear activation function that controls the connection. A non-linear (sigmoid) activation function connects hidden-layer nodes to output-layer nodes according to the following formula:

$$s_i = \frac{1}{1+e^{-I_i}}$$
$$I_i = \sum_j w_{ij} s_j$$
(6)

This FFBP-ANN is trained using data collected during the tuning and testing of the hard-biometric features of the person authorized for access. This data is then mixed with the biometric data of unauthorized persons and used to build the training patterns. The purpose is to train the ANN so it can learn to recognize behavior specific to the authorized individual.

7 Fuzzy Logic-Based Fusion and Authentication

The fuzzy logic engine evaluates three kinds of data input:

- voiceprint score
- fingerprint score
- soft-biometric measurements
- soft-biometric score

All these input data are fuzzified using optimally tuned membership functions (fig. 4a, fig. 4b, and fig. 4b).

A set of rules is then tuned to combine the fuzzified input as follows (only the most significant are reported):

1. IF voiceprint_score IS high AND fingerprint_score IS high AND soft_score IS high AND soft_score IS high THEN authentication IS very high

2. IF voiceprint_score IS medium AND fingerprint_score IS medium AND soft_score IS high THEN authentication IS high

3. IF voiceprint_score IS medium AND speech_speed IS high AND soft_score IS medium THEN authentication IS high

4. IF voiceprint_score IS low AND speech_speed IS high AND speech_stress IS high THEN authentication IS average

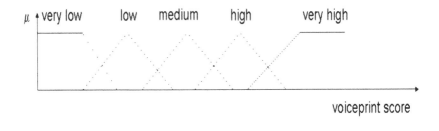

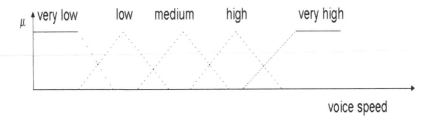

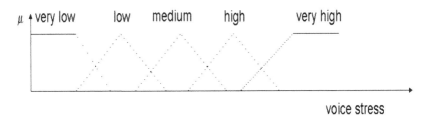

Fig. 4a. Membership function to fuzzify input (voiceprint)

5. *IF fingerprint_score IS medium AND fingerprint_total_area IS high THEN authentication IS high*

6. *IF finger_print_score IS low AND fingerprint_total_area IS high AND fingerprint_mean_intensity IS high THEN authentication IS average*

7. *IF voiceprint_score IS medium and   fingerprint_score IS medium AND soft_score IS low THEN authentication IS low*

8. *IF voiceprint_score IS high and fingerprint_score IS low AND soft_score IS low AND fingerprint_total_area_match IS low AND fingerprint_mean_intensity_match IS low AND speech_speed_match IS low AND speech_stress_match IS low THEN authentication IS very low*

9. *IF voiceprint_score IS low and fingerprint_score IS high AND soft_score IS low AND fingerprint_total_area_match IS low AND fingerprint_mean_intensity_match IS low AND speech_speed_match IS low AND speech_stress_match IS low THEN authentication IS very low*

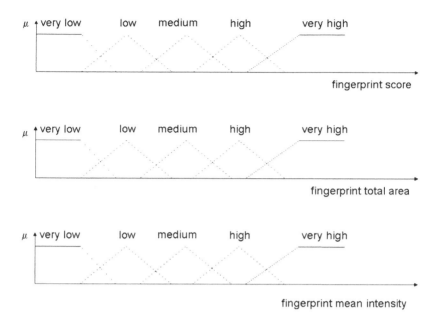

Fig. 4b. Membership function to fuzzify input (fingerprint)

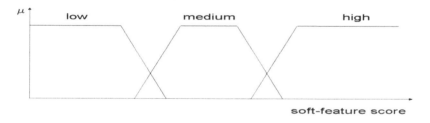

Fig. 4c. Membership function to fuzzify input (soft-features score)

Rules were derived from feature distribution. Each rule was manually tuned using a fuzzy-logic development environment specially adapted to this purpose. The primary purpose of this tuning action is to integrate single biometric matchers into a single smart-decision system.

Rule 1 is a reinforcement of voiceprint and fingerprint matchers.

Rule 2 combines a voiceprint matcher and a fingerprint matcher when both scores are too close to the decision threshold.

Rules 3, 4, 5, and 6 act as recovery rules when the voiceprint or fingerprint matchers generate a false rejection.

Rules 7, 8, and 9 protect against false acceptance.

For fine tuning, many other rules can be generated to take additional soft-biometric measurements into account. Using more rules leads monotonically toward greater reliability in the authentication process.

Triangular membership functions are used to process inputs. The inference rule set is then applied. The result of all the rules is evaluated using the centroid method [8], so a crisp output value can be calculated by finding the variable value of the center of gravity. A singleton membership function was used to defuzzify the final decision.

8 Performance Evaluation

Performance evaluation aims to measure the performance of previously implemented biometric identification based on hard-computing [13] against the improved implementation integrated with soft-computing methods. Voiceprint and fingerprint authentication were first implemented and tested separately, then jointly, and finally combined through the fuzzy logic inference engine and the artificial neural network applied to soft-biometric features. An "all-against-all" test strategy was applied to obtain match and mismatch scores.

Evaluation of joint voiceprint and fingerprint authentication consisted of taking as good the best of the two matches (OR). Single-user authentication was performed, so this test had minimal system requirements. The following results were produced:

- Voiceprint alone: 90.5% correctly accepted;
- Fingerprint alone: 85.7 % correctly accepted;
- Voiceprint OR fingerprint: 92.3% correctly accepted;
- Fuzzy logic decision fusion of voiceprint, fingerprint, and artificial neural network evaluated soft features: 95.8 correctly accepted.

The OR test confirmed that multibiometrics can improve performance compared to single-biometric authentication. System performance can be significantly improved, while keeping complexity to a minimum, using fuzzy logic as decision fusion and reinforcing it with an artificial neural network applied to soft-biometric features.

9 Embedded Implementation

This multibiometric mixed-method authentication system was embedded to demonstrate that such a model can be scaled to an embedded computing platform while keeping authentication performance high. A single-chip DSP, a limited memory bank, a fingerprint sensor, and a voice CODEC were assembled as a prototype board to run the soft-multibiometric authentication process.

Response time was less than 1 second when the DSP runs at 100 MHz, while memory requirements were less than 64 kwords. These hardware requirements are a minimal portion of current system-on-chip (SoC) processing devices for embedded applications.

Fig. 5. Prototype of the embedded system

10 Conclusions

Preliminary results of this research demonstrate that a soft-multibiometric approach to personal authentication on embedded systems can require fewer system resources if smart logic is used to implement the decision process. Artificial neural networks can be used to make inferences about soft-biometric features. Fuzzy logic proves a very practical solution for implementing smart decision logic.

The integration of multiple biometric identification systems through a fuzzy logic inference engine is a good strategy for keeping system complexity low while increasing performance. A positive side effect of this approach is that embedded information such as behavior and emotion can be included in final decision. Artificial neural networks prove to work optimally in mapping such information, enabling an effective multimodal approach to personal biometric authentication.

References

1. Bishop, C.: Neural Networks for Pattern Recognition. Oxford University Press, Oxford (1995)
2. Ciota, Z.: Improvement of Speech Processing Using Fuzzy Logic Approach. In: Proceedings of IFSAWorld Congress and 20th NAFIPS International Conference (2001)
3. Bosteels, R.T.K., Kerre, E.E.: Fuzzy Audio Similarity Measures Based on Spectrum Histogram and Fluctuation Patterns. In: Proceedings of the International Conference Multimedia and Ubiquitous Engineering 2007, Seul, Korea, April 27–28 (2007)
4. Malcangi, M.: Soft-computing Approach to Fit a Speech Recognition System on a Single-chip. In: Proceedings of 2002 International Workshop System-On-Chip For Real-Time Applications, Banff, Canada, July 6-7 (2002)
5. Malcangi, M.: Improving Speech Endpoint Detection Using Fuzzy Logic-based Methodologies. In: Proceedings of the Thirteenth Turkish Symposium on Artificial Intelligence and Neural Networks, Izmir, Turkey, June 10-11 (2004)
6. O'Shaugnessy, D.: Speech Communication – Human and Machine. Addison-Wesley, Reading (1987)
7. Pak-Sum Hui, H., Meng, H.M., Mak, M.: Adaptive Weight Estimation in Multi-biometric Verification Using Fuzzy Logic Decision Fusion. In: Proceedings of IEEE International Conference on Acoustic, Speech, and Signal Processing (2007)
8. Runkler, T.A.: Selection of Appropriate Defuzzification Methods Using Application Specific Properties. IEEE Transactions on Fuzzy Systems 5(1) (Feburary 2007)

 9. Wahab, A., Ng, G.S., Dickiyanto, R.: Speaker Authentication System Using Soft Comput-
    ing Aprroaches. Neurocomputing 68, 13–17 (2005)
10. Jain, A.K., Dass, S.C., Nandakumar, K.: Soft Biometric Traits for Personal Recognition
    Systems. In: Zhang, D., Jain, A.K. (eds.) ICBA 2004. LNCS, vol. 3072, pp. 731–738.
    Springer, Heidelberg (2004)
11. Jain, A.K., Nandakumar, K., Lu, X., Park, U.: Integrating faces, fingerprints, and soft bio-
    metric traits for user recognition. In: Maltoni, D., Jain, A.K. (eds.) BioAW 2004. LNCS,
    vol. 3087, pp. 259–269. Springer, Heidelberg (2004)
12. Hong, A., Jain, S., Pankanti, S.: Can Multibiometrics improve performance? In: Proceed-
    ings of IEEE Workshop on Automatic Identification of Advanced Tecnologies (1999)
13. Malcangi, M.: Robust Speaker Authentication Based on Combined Speech and Voiceprint
    Recognition, proposed to IECCS (2008)

Study of Alpha Peak Fitting by Techniques Based on Neural Networks

Javier Miranda[1], Rosa Pérez[2], Antonio Baeza[1], and Javier Guillén[1]

[1] Laboratorio de Radiactividad Ambiental de la Universidad de Extremadura
Dpto. Física Aplicada, Universidad de Extremadura
Avda. de la Universidad, s/n 10071 Cáceres
[2] Dpto. de Tecnología de los Computadores y de las Comunicaciones
Escuela Politécnica, Universidad de Extremadura
Avda. de la Universidad, s/n 10071 Cáceres
`{jmircar,rosapere,abaeza,fguillen}@unex.es`

Abstract. There have been many studies which analyze complex alpha spectra based on numerically fitting the peaks to calculate the activity level of the sample. In the present work we propose a different approach – the application of neural network techniques to fit the peaks in alpha spectra. Instead of using a mathematical function to fit the peak, the fitting is done by a neural network trained with experimental data corresponding to peaks of different characteristics. We have designed a feed-forward (FF) multi-layer perceptron (MLP) artificial neural network (ANN), with supervised training based on a back-propagation (BP) algorithm, trained on the peaks of Polonium, extracted from many spectra of real samples analyzed in the laboratory. With this method, we have achieved a fitting procedure that does not introduce any error greater than the error of measurement, evaluated to be 10%.

1 Introduction

In nature, there are unstable atomic nuclei which disintegrate emitting ionizing radiation, commonly known as radioactivity. One of these types of radiation is the emission of alpha particles. It is observed in elements heavier than lead in the periodic table, including uranium, thorium, and polonium. This emission consists of helium nuclei (two protons and two neutrons) expelled from the parent nuclei with high kinetic energies, in the order of MeV (mega-electron-volt). These energies are discrete and characteristic of the emitting radionuclide. The emission consists therefore basically of charged particles with a significantly greater mass than electrons. This implies that they can only travel a very short distance until they completely deplete their energy. Indeed, even a simple sheet of paper can stop them. Even though they are short-range particles, their determination is frequently necessary to characterize a material radiologically.

The alpha particle detection process is usually carried out using PIPS (Passivated Implanted Planar Silicon) detectors in which the energy of the alpha particle is deposited. Due to the strong interaction of alpha particles with matter, to avoid losing energy in the detection the distance between sample and detector must be as short as possible,

D. Palmer-Brown et al. (Eds.): EANN 2009, CCIS 43, pp. 79–85, 2009.

typically a few millimeters, and a certain degree of vacuum is also needed in the detection chamber. Figure 1 shows a typical polonium alpha spectrum. One observes that each alpha-emitting radionuclide has a discrete and characteristic emission energy, which allows its complete identification. However, despite the aforementioned precautions, it is not possible for all the alpha particles to deposit their complete energy in the detector, and low-energy tails are commonly observed in the spectrum of an alpha analysis (see Fig. 1). An alpha spectrum consists of a plot of the number of alpha particles detected, called counts, at defined energy intervals, called channels.

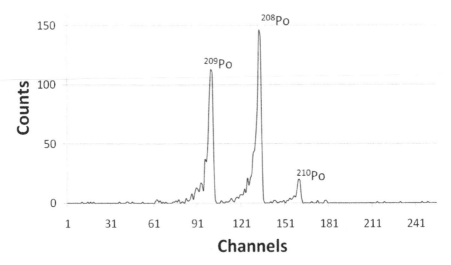

Fig. 1. Polonium alpha spectrum

The resolution of the detectors used is in the order of 25 keV. So, the low-energy tails usually pose no problem if the alpha peaks are well separated, i.e., their energies are so different that they do not overlap. Unfortunately, that is not always possible. The content of an alpha-emitting radionuclide in a sample is measured by analysing the spectrum, specifically by determining the total number of counts (integral) in several Regions Of Interest (ROIs), defined as intervals of energy in which an alpha particle peak is detected. These ROIs should be wide enough to include the aforementioned low-energy tail.

1.1 Existing Solutions

The analysis of alpha spectra is usually carried out using models that represent the shape of a single-energy alpha peak. In this sense, there are numerous works dealing with the generation of a function that takes the shape of a theoretical peak and then fits a combination of them to the real peaks present in the alpha spectrum. Bortels and Collaers [2] proposed a function based on the convolution of several components, as also did Westemeier and Van Aarle [7]. Most of these models depend on a number of parameters that define the function. The fitting procedure optimizes these parameters to best approximate the real spectrum. The more parameters the function contains, the better the results obtained, although the optimization process is more complicated and several types of ambiguities appear.

There are many programs that cover all types of needs in the analysis of alpha spectra from low to high statistics. Most focus on guaranteeing software performance that is very stable and reliable (Blaauw et al. [1]). There are also non-commercial programs whose focus is more on high performance optimization algorithms. Some examples are FITBOR (Martín Sánchez et al. [4]), ALPHA (Pomme and Sibbens [6]), and ALFIT (Lozano and Fernandez [3]).

1.2 Proposed Solution

The present communication describes a different approach to the problem. We developed a neural network in software and trained it with a set of peaks selected as models for the network to reproduce the shape of an isolated alpha peak from the knowledge acquired. The difference between this and previous solutions is that, instead of generating certain parameterized model functions to fit the peaks, a generalization based on the shape of peaks taken from real alpha spectra is used. The purpose of this work is thus the application of neural network techniques for the peak-fitting analysis of alpha spectra obtained from the measurement of real samples. The fitting procedure is based on the experimental observables that are characteristic of standard alpha peaks. A further advantage of the proposed solution is that the goodness of the results is compared directly with the real integral of the alpha peaks, calculated as the simple sum of the counts of each of the channels of the ROI.

2 Method

2.1 Training Data

First, we selected the set of isolated peaks to form part of the overall training of the network. These singlet peaks are obtained from the measurement of different real samples. In a pre-selection process, alpha peaks with too low integrals or a visually unrepresentative shape, such as double peaks, were discarded. If this information were included, it would impede the convergence of the network in minimizing the fitting error, and lead to poor generalization of the shape of an alpha peak.

The spectra selected were obtained from sources of polonium. These spectra contain alpha peaks of different polonium isotopes – ^{208}Po, ^{209}Po, and ^{210}Po (see Fig. 1). This type of source was chosen because the peaks observed in their alpha spectra are sufficiently separated as not to overlap, and consequently each one is due to a single energy emission. They can thus be considered as representative of the shape of a singlet alpha peak.

We grouped these peaks based on their integrals. The total number of counts of a peak was therefore the unit of measurement to establish the working range of the neural network. This range was set between 50 and 5000 counts. The criteria for establishing the limits of this range were based on experience with the analysis of alpha spectra. A peak with an integral below 50 counts is considered to have too low statistics for the analysis to be reliable because it implies an uncertainty of about 14% in

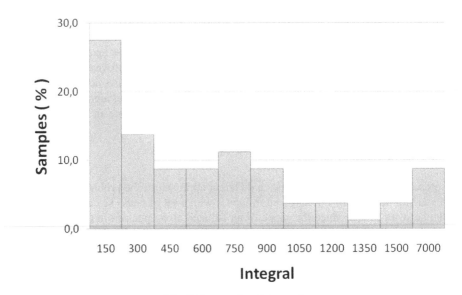

Fig. 2. Range of peak integrals

the determination of the activity of the radionuclide. Alpha peaks with integrals above 5000 counts are rarely obtained in the measurement of environmental radioactive samples. The training set must cover the entire working range. Figure 2 shows the percentage of peaks within each integral interval used to train the network.

In order to extract these peaks from the data files of spectra of the samples as they were obtained in the laboratory, we modified the software we used for alpha spectra analysis (AlphaSpec [5]) to allow the export of the data corresponding to selected peaks to a file in a format compatible with the tool constructed to test and experiment with the neural networks of this work.

2.2 Network Design

The purpose of the network design was to reproduce alpha peaks registered in the spectra. Since this function is performed during the operation of the network and not in training, the time required to conduct comprehensive training of the network is not a limiting factor in the design of the system. Therefore, the main objective was not optimization of the network's training time, but faithful reproduction of the peak shape. The quality criterion of the procedure was that the reproduction of the peak shape does not introduce an error greater than that produced in the process of measurement of the spectrum, estimated to be less than 10%. Therefore, the difference between the integral re-created by the system and the expected integral for one of the given training set of peaks should not exceed 10%.

The neural network we designed for this work was a multilayer perceptron (MLP) with supervised training based on back-propagation, and sigmoid activation functions for the hidden and output layers and a linear activation function for the input layer. The training algorithm uses the mean-square-error (MSE) cost function to determine when the network has been trained sufficiently.

2.3 Inputs and Output

The inputs to the network are a set of numerical values which identify the shape of the alpha peak obtained from real samples, and are sufficiently descriptive to allow the network to determine the shape on their basis (see Fig. 3). In particular, we considered the following 10 characteristics of a peak:

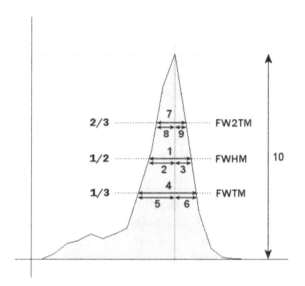

Fig. 3. Peak characteristics

1.   FWHM, full-width at half-maximum
2.   FWHM-a, first half of FWHM
3.   FWHM-b, second half of FWHM
4.   FWTM, full-width at third-maximum
5.   FWTM-a, first half of FWTM
6.   FWTM-b, second half of FWTM
7.   FW2TM, full-width at two-thirds-maximum
8.   FW2TM-a, first half of FW2TM
9.   FW2TM-b, second half of FW2TM
            (All are expressed in "keV" energy units)
10.   Max, maximum counts of the peak in a channel

The first 9 inputs coincide with the first 9 characteristics, and the tenth input is the inverse of characteristic number 10 (1/Max). An eleventh input is the ratio between the maximum and the FWHM (Max/FWHM). With respect to the output, we observed with the spectra used to train the network that the standard width of the ROIs of an isolated alpha peak, including its low-energy tail, was less than 21 channels. Therefore, the output of the network was designed to consist of 21 neurons in the last layer. With the sigmoid activation function of the output layer, the network output values will be in the range 0 to 1. The output values of the training set peaks had therefore to be scaled to this range for the correct calculation of the MSE.

3 Results

Multiple training cycles were tested by varying the parameters of the network, using a tool specifically designed to experiment with different configurations. The network parameters used were: Learning-rule (0.009), Momentum (0.8), Error (0.001). The network was trained with these parameters using a cross-validation method on 102 peaks of different integrals. The algorithm takes 60% of peaks for the training set, 20% for the validation set, and the rest for the test set.

Table 1 presents the detailed results of the test set, with a 22-neuron hidden layer configuration, 8395 epochs, and 2 minutes of training (Intel Core 2 Duo 2,6 GHz).

Table 1. MAX is the maximum counts in a channel, $I_{EXPECTED}$ is the expected integral, I_{ANN} is the ANN output, and ERROR is the difference between expected integral and the output, expressed in %

PEAK TYPE	MAX	$I_{EXPECTED}$	I_{ANN}	ERROR (%)
PO208	183	656	600	8,495
PO208	46	133	129	2,450
PO209	95	323	340	5,301
PO208	63	220	198	9,617
PO209	162	621	559	9,941
PO210	11	45	44	1,775
PO209	177	482	530	9,885
PO209	189	710	723	1,950
PO209	85	338	337	0,199
PO209	245	838	886	5,714
PO210	183	560	606	8,275
PO210	49	193	182	5,516
PO208	142	581	561	3,374
PO210	382	1189	1271	6,907
PO208	455	1524	1526	0,169
PO210	36	125	127	2,214
PO210	155	569	536	5,777
PO210	64	164	181	10,761
PO209	27	79	83	5,091
PO208	1373	1373	1483	8,070
PO210	118	308	352	14,276

4   Conclusions

The neural network that we have described here is capable of reproducing the shape of a single alpha peak from real samples without needing the generation of a theoretical function. The goodness of the result provided by the neural network used was compared directly with the real value of the integral of these alpha peaks. The errors for the 90.4% of the test set of 21 peaks different from the training and validation sets were below 10%, with a mean error of 5.98%.

The technique is therefore currently being extended, studying its application to the analysis of multi-emission peaks, and to radionuclides other than polonium.

Acknowledgment

This work has been financed by the Spanish Ministry of Science and Education under the project number CTM2006-11105/TECNO, entitled "Characterization of the time evolution of radioactivity in aerosols in a location exempt of a source term". Also we are grateful to the Autonomous Government of Extremadura for the "Studentship for the pre-doctoral formation for researchers (resolved in D.O.E. 130/2007)", and for the financial support to the LARUEX research group (GRU09041).

References

1. Blaauw, M., García-Toraño, E., Woods, S., et al.: The 1997 IAEA Intercomparison of Commercially Available PC-Based Software for Alpha-Particle Spectrometry. Nuclear Instruments and Methods in Physics Research Section A: Accelerators, Spectrometers, Detectors and Associated Equipment 428, 317–329 (1999)
2. Bortels, G., Collaers, P.: Analytical Function for Fitting Peaks in Alpha-Particle Spectra from Si Detectors. International Journal of Radiation Applications and Instrumentation. Part A. Applied Radiation and Isotopes 38, 831–837 (1987)
3. Lozano, J.C., Fernández, F.: ALFIT: A Code for the Analysis of Low Statistic Alpha-Particle Spectra from Silicon Semiconductor Detectors. Nuclear Instruments and Methods in Physics Research Section A: Accelerators, Spectrometers, Detectors and Associated Equipment 413, 357–366 (1998)
4. Martín Sánchez, A., Rubio Montero, P., Vera Tomé, F.: FITBOR: A New Program for the Analysis of Complex Alpha Spectra. Nuclear Instruments and Methods in Physics Research, Section A: Accelerators, Spectrometers, Detectors and Associated Equipment 369, 593–596 (1996)
5. Miranda Carpintero, J., Pérez Utrero, R.: Universidad de Extremadura. Escuela Politécnica: Desarrollo del software básico de análisis en espectrometría alfa (2006)
6. Pommé, S., Sibbens, G.: A New Off-Line Gain Stabilisation Method Applied to Alpha-Particle Spectrometry. Advanced Mathematical and Computational Tools in Metrology VI, 327–329 (2004)
7. Westmeier, W., Van Aarle, J.: PC-Based High-Precision Nuclear Spectrometry. Nuclear Instruments and Methods in Physics Research Section A: Accelerators, Spectrometers, Detectors and Associated Equipment 286, 439–442 (1990)

Information Enhancement Learning: Local Enhanced Information to Detect the Importance of Input Variables in Competitive Learning

Ryotaro Kamimura

IT Education Center, Tokai University,
1117 Kitakaname Hiratsuka Kanagawa 259-1292, Japan
ryo@cc.u-tokai.ac.jp

Abstract. In this paper, we propose a new information-theoretic method called "information enhancement learning" to realize competitive learning and self-organizing maps. In addition, we propose a computational method to detect the importance of input variables and to find the optimal input variables. In our information enhancement learning, there are three types of information, namely, self-enhancement, collective enhancement and local enhancement. With self-enhancement and collective enhancement, we can realize self-organizing maps. In addition, we use local enhanced information to detect the importance of input units or input variables. Then, the variance of local information is used to determine the optimal values of the enhanced information. We applied the method to an artificial data. In the problem, information enhancement learning was able to produce self-organizing maps close to those produced by the conventional SOM. In addition, the importance of input variables detected by local enhanced information corresponded to the importance obtained by directly computing errors.

1 Introduction

In this paper, we propose a new information-theoretic method in which a network is enhanced to respond more explicitly to input patterns [1]. We propose here three types of enhancement: self-enhancement, collective enhancement and local enhancement. With self-enhancement, a state of a network is self-divided into an enhanced and relaxed state, and the relaxed state tries to reach the enhanced state. With collective enhancement, all neurons in a network are collectively enhanced and respond to input patterns. With self- and collective enhancement, we can realize self-organizing maps similar to those obtained by the conventional SOM [2], [3]. Finally, local information is used to detect the importance of some components, such as input units, variables, competitive units and so on. In this paper, we focus in particular upon the local enhanced information to detect important input variables.

Input variable selection or the detection of the importance of variables has been considered to be one of the most important problems in neural networks as well as machine learning. In particular, study on feature and variable selection

D. Palmer-Brown et al. (Eds.): EANN 2009, CCIS 43, pp. 86–97, 2009.
© Springer-Verlag Berlin Heidelberg 2009

has revived, because it is necessary to develop efficient computational methods to classify the massive amount of data accumulated daily. Guyon [4], in her excellent introduction to variable and feature selection, pointed out that feature selection had many benefits in machine learning, that is, *facilitating data visualization, data interpretation, reducing storage requirement, reducing training time, improving generalization performance, understanding underlying processes that generate the data* and so on. Though many methods have been proposed, the majority of are restricted to supervised learning [5], [6], [7], [8]. In this paper, our method is concerned mainly with unsupervised competitive learning.

2   Theory and Computational Methods

2.1   Enhancement and Relaxation

We have shown that the state of networks can be seen from different points of view. The simplest case is a binary one, where the state is divided into an enhanced and a relaxed state. Figure 1 shows a process of enhancement and relaxation for a network itself. Figure 1(a) shows an original situation obtained by competitive learning, in which three neurons are differently activated for input units. With enhancement, as shown in Figure 1(b), the characteristics of competitive unit activations are enhanced, and only one competitive unit is strongly activated. On the other hand, Figure 1 (c) shows a state obtained by relaxation,

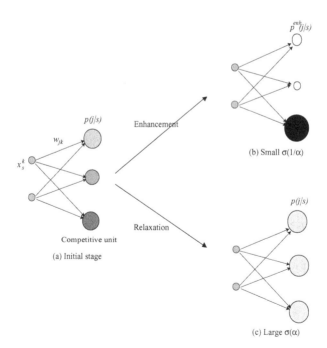

Fig. 1. Enhancement (b) and relaxation (c) of competitive units for a network itself

in which all competitive units respond uniformly to input units. Because competitive units cannot differentiate between input patterns, no information on input patterns is stored.

As shown in Figure 1(a), a network is composed of connection weights w_{jk} from the kth input unit to the jth competitive unit. The kth element of the sth input pattern is represented by x_k^s. A neuron output can be defined by the Gaussian-like function:

$$v_j^s \propto \exp\left\{-\frac{1}{2}(\mathbf{x}^s - \mathbf{w}_j)^T \mathbf{\Sigma}^{-1}(\mathbf{x}^s - \mathbf{w}_j)\right\}, \qquad (1)$$

where $\mathbf{\Sigma}$ denotes the scaling matrix[1]. The enhanced scaling matrix can be defined by

$$\begin{aligned} \mathbf{\Sigma} &= \mathbf{\Sigma}_{enh}^{(\alpha)} \\ &= (1/\alpha)^2 \mathbf{I}, \end{aligned} \qquad (2)$$

where $\mathbf{I}$ is an identity matrix. When the relaxation is applied, we have

$$\begin{aligned} \mathbf{\Sigma} &= \mathbf{\Sigma}_{rel}^{(\alpha)} \\ &= \alpha^2 \mathbf{I}. \end{aligned} \qquad (3)$$

2.2   Self-enhancing

We have shown that an initial state can be split into an enhanced and a relaxed state. Then, we must decrease the gap between the two states as much as possible. In the self-enhancing process, the difference between two probabilities should be as small as possible.

We now present update rules for self-enhancement learning for a general case. At an enhanced state, competitive units can be computed by

$$v_{j,enh}^s \propto \exp\left\{-\frac{1}{2}(\mathbf{x}^s - \mathbf{w}_j)^T (\mathbf{\Sigma}_{enh}^{(\alpha)})^{-1}(\mathbf{x}^s - \mathbf{w}_j)\right\}. \qquad (4)$$

Normalizing this output, we have enhanced firing probabilities

$$p^{enh}(j \mid s) = \frac{\exp\left\{-\frac{1}{2}(\mathbf{x}^s - \mathbf{w}_j)^T (\mathbf{\Sigma}_{enh}^{(\alpha)})^{-1}(\mathbf{x}^s - \mathbf{w}_j)\right\}}{\sum_{j=1}^{M} \exp\left\{-\frac{1}{2}(\mathbf{x}^s - \mathbf{w}_j)^T (\mathbf{\Sigma}_{enh}^{(\alpha)})^{-1}(\mathbf{x}^s - \mathbf{w}_j)\right\}}. \qquad (5)$$

Suppose that $p(j|s)$ denotes the probability of the jth neuron's firing in a relaxed state. Then, we should make these probabilities as close as possible to the probabilities of enhanced neurons' firing. Thus, we have an objective cross entropy function defined by

$$I = \sum_{s=1}^{S} p(s) \sum_{j=1}^{M} p(j \mid s) \log \frac{p(j \mid s)}{p^{enh}(j \mid s)}, \qquad (6)$$

---

[1] We used the scaling matrix instead of the ordinary covariance matrix, because the output does not exactly follow the Gaussian function.

where M and S denote the number of competitive units and input patterns, respectively. In addition, we should decrease quantization errors, defined by

$$E = \sum_{s=1}^{S} p(s) \sum_{j=1}^{M} p(j \mid s)\|\mathbf{x}^s - \mathbf{w}_j\|^2. \tag{7}$$

It is possible to differentiate this cross entropy and the quantization error function to obtain update rules [9], [10]. However, the update rules become complicated, with heavy computation required for computing conditional probabilities.

Fortunately, we can skip the complicated computation of conditional entropy by introducing the free energy used in statistical mechanics [11], [12], [13], [14], [15], [16], [17]. Borrowing the concept from statistical mechanics, let us introduce free energy, or a free energy-like function, defined by

$$F = -2\alpha^2 \sum_{s=1}^{S} p(s) \log \sum_{j=1}^{M} p^{enh}(j \mid s)$$

$$\times \exp\left\{ -\frac{1}{2}(\mathbf{x}^s - \mathbf{w}_j)^T (\mathbf{\Sigma}_{rel}^{(\alpha)})^{-1} (\mathbf{x}^s - \mathbf{w}_j) \right\}. \tag{8}$$

In an optimal state, we have

$$F = \sum_{s=1}^{S} p(s) \sum_{j=1}^{M} p(j \mid s)\|\mathbf{x}^s - \mathbf{w}_j\|^2$$

$$+2\alpha^2 \sum_{s=1}^{S} p(s) \sum_{j=1}^{M} p(j \mid s) \log \frac{p(j \mid s)}{p^{enh}(j \mid s)}. \tag{9}$$

Thus, to decease the free energy, we must decrease the cross entropy and the corresponding error function. By differentiating the free energy, we have

$$w_{jk} = \frac{\sum_{s=1}^{S} p(j \mid s) x_k^s}{\sum_{s=1}^{S} p(j \mid s)}, \tag{10}$$

where $p(s)$ is set to $1/S$. We should note that, thanks to the excellent work of Heskes [18], the free energy can be interpreted in the framework of an EM algorithm.

In the above formulation, we have dealt with a general case of self-enhancement learning. However, this is a self-enhancement learning version of competitive learning. In application to competitive learning, an enhanced state is one where a winner takes all in the extreme case. This is realized by the enhancement parameter $1/\alpha$. On the other hand, a relaxed state is one where competitive units respond to input patterns almost equally, which is realized by setting the enhancement parameter to α. The self-enhancement learning, in terms of competitive learning, tries to attain a state where the winner-take-all is predominant.

2.3   Collective Enhancement

To demonstrate the performance of our method clearly, we use self-organizing maps, because it is easy to interpret final results intuitively. In the previous section, we applied self-enhancement learning to competitive learning. Thus, it is possible to apply it to self-organizing maps just by introducing lateral interactions in an enhanced state. Instead of lateral interactions, we introduce the concept of collectiveness in an enhanced state, as shown in Figure 2. Collectiveness means that all competitive units are related to each other.

Now, let us explain collective activations. Figure 2 shows a network architecture that is composed of a competitive and a collective layer. Figure 2 shows that the jth neuron cooperates with all the other neurons on the map and responds to input patterns. We realize this cooperation by summing all the neighboring units' activities. Then, we have collected activations

$$V_j^s \propto \sum_{m=1}^{M} W_{jm} \exp\left\{-\frac{1}{2}(\mathbf{x}^s - \mathbf{w}_m)^T (\boldsymbol{\Sigma}_{enh}^{(\alpha)})^{-1}(\mathbf{x}^s - \mathbf{w}_m)\right\}, \qquad (11)$$

where W_{jm} denotes connection weights from the mth competitive unit to the jth competitive unit, and M is the number of competitive units. We can imagine many kinds of collectiveness on the map; however, we usually use distance between competitive units for expressing collectiveness. For example, when competitive units are closer to their neighbors, they should be linked to them more intensely. A distance function between two neurons can be defined by

$$W_{jm} = \exp\left(-\frac{1}{2}\|\mathbf{r}_j - \mathbf{r}_m\|^2\right), \qquad (12)$$

where $\mathbf{r}_j$ denotes a position for the jth unit, and $\mathbf{r}_m$ denotes a position for the mth neighboring neuron. Thus, we have

$$V_j^s \propto \sum_{m=1}^{M} \exp\left(-\frac{1}{2}\|\mathbf{r}_j - \mathbf{r}_m\|^2\right)$$
$$\exp\left\{-\frac{1}{2}(\mathbf{x}^s - \mathbf{w}_m)^T (\boldsymbol{\Sigma}_{enh}^{(\alpha)})^{-1}(\mathbf{x}^s - \mathbf{w}_m)\right\}. \qquad (13)$$

We can compute a normalized activity

$$p^{coll}(j \mid s) = \frac{V_j^s}{\sum_{m=1}^{M} V_m^s}. \qquad (14)$$

This normalized activity is considered to represent collective firing rates. Then, we have the free energy

$$F = -2\alpha^2 \sum_{s=1}^{S} p(s) \log \sum_{j=1}^{M} p^{coll}(j \mid s)$$
$$\times \exp\left\{-\frac{1}{2}(\mathbf{x}^s - \mathbf{w}_j)^T (\boldsymbol{\Sigma}_{rel}^{(\alpha)})^{-1}(\mathbf{x}^s - \mathbf{w}_j)\right\}. \qquad (15)$$

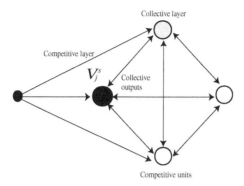

Fig. 2. A concept of collective activations

2.4   Local Enhancement

For the local enhancement, we focus upon input units for enhancement. For example, we use input units for enhancement, and we can then estimate the importance of input variables. Now, suppose that the tth input unit is a target for enhancement. For this, we use a scaling matrix $\mathbf{\Sigma}^{(t,\alpha)}$, meaning that the tth unit is a target with the enhancement parameter α. The scaling parameter $\mathbf{\Sigma}^{(t,\alpha)}$ is defined by

$$\mathbf{\Sigma}^{(t,\alpha)} = \mathbf{I}_t \mathbf{\Sigma}_{enh}^{(\alpha)} + (\mathbf{I} - \mathbf{I}_t)\mathbf{\Sigma}_{rel}^{(\alpha)}, \tag{16}$$

where the kk'th element of the matrix $\mathbf{I}_t$ is defined by $(\mathbf{I}_t)_{kk'}$,

$$(\mathbf{I}_t)_{kk'} = \delta_{kk'}\delta_{kt}. \tag{17}$$

Thus, when the tth input unit is a target for enhancement, we have competitive unit outputs v_{jt}^s computed by

$$v_{jt}^s \propto \exp\left(-\frac{1}{2}(\mathbf{x}^s - \mathbf{w}_j)^T(\mathbf{\Sigma}^{(t,\alpha)})^{-1}(\mathbf{x}^s - \mathbf{w}_j)\right). \tag{18}$$

This means that, when the tth input unit is a target for enhancement, the tth element of the diagonal scaling matrix is set to $(1/\alpha)^2$, while all the other elements in the diagonal are set to α^2. In this case, we can detect the importance of input variables. We can normalize these activations for probabilities,

$$p^t(j \mid s) = \frac{\exp\left\{-\frac{1}{2}(\mathbf{x}^s - \mathbf{w}_j)^T(\mathbf{\Sigma}^{(t,\alpha)})^{-1}(\mathbf{x}^s - \mathbf{w}_j)\right\}}{\sum_{m=1}^{M} \exp\left\{-\frac{1}{2}(\mathbf{x}^s - \mathbf{w}_m)^T(\mathbf{\Sigma}^{(t,\alpha)})^{-1}(\mathbf{x}^s - \mathbf{w}_m)\right\}}. \tag{19}$$

And we have

$$p^t(j) = \sum_{s=1}^{S} p(s)p^t(j \mid s). \tag{20}$$

By using these probabilities, we have local enhanced information for the tth input unit

$$I_t = \sum_{s=1}^{S} \sum_{j=1}^{M} p(s)p^t(j \mid s) \log \frac{p^t(j \mid s)}{p^t(j)}. \tag{21}$$

We can also define the free energy, and we have

$$F_t = -2\alpha^2 \sum_{s=1}^{S} p(s) \log \sum_{j=1}^{M} p^t(j \mid s)$$

$$\times \exp\left\{ -\frac{1}{2}(\mathbf{x}^s - \mathbf{w}_j)^T (\mathbf{\Sigma}_{rel}^{(\alpha)})^{-1}(\mathbf{x}^s - \mathbf{w}_j) \right\}. \tag{22}$$

3   Results and Discussion

In this section, we present the experimental results of using information enhancement learning. For easy interpretation, comparison and reproduction, we used the SOM toolbox with default parameter values [19] where no special options were used. All data in this experiment were normalized, with a range between zero and one. The number of competitive units was automatically determined by the software package. For information enhancement learning, no special techniques, for example, to accelerate learning, were used. Thus, experimental results presented here can easily be reproduced. We computed the quantitative measures for performance comparison, namely, quantization errors. The quantization error is simply the average distance from each data vector to its BMU [19].

3.1   Artificial Data

First, we used artificial data to demonstrate how to compute enhanced information. Figure 3(a) show data in which input patterns are symmetrical to input units and patterns. In the data, black and white squares represent one and zero, respectively. Thus, the number of black squares is increased gradually and then decreased to one in the latter part. Figure 3(b) shows quantization errors obtained by information enhancement and SOM. As can be seen in the figure, the quantization error becomes 0.428 when the enhancement parameter α is 13. Thus, a slightly better performance could be obtained by information enhancement in terms of quantization errors. Figures 4(a) and (b) show U-matrices and maps with labels obtained by SOM and by information enhancement, respectively. We can see that almost the same type of maps and labels can be obtained by both methods. However, when we more closely examine labels in Figure 4(a2) and (b2), labels obtained by information enhancement show the more natural ordering of input patterns on the map. In addition, warmer-colored boundaries in the middle are more explicit with information enhancement.

Then, we computed local enhanced information for each input variable. One of the main problems in computing the enhanced information is that enhanced

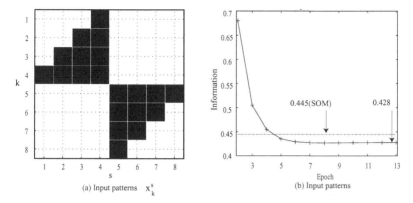

Fig. 3. Input patterns (a) and quantization errors (b) by enhancement learning in blue and by SOM in red. Black and white squares represent one and zero, respectively.

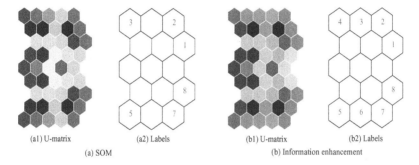

Fig. 4. U-matrix and labels obtained by SOM (a) and enhancement learning (b)

information varies greatly depending upon the chosen enhancement parameter α. Among many possibilities of enhanced information, we must choose one according to a certain criterion. When the variance of the enhanced information is larger, the difference between different enhanced information is easily detected. Thus, we compute the ratio between the variance and the means of enhanced information. The ratio is computed by the variance of enhanced information by the mean of the enhanced information and computed by

$$R(\alpha) = \frac{Var(I_t)}{\bar{I}_t}, \tag{23}$$

where Var and $\bar{I}$ mean the variance and the average of the enhanced information. Figure 5 shows this ratio for the artificial problem. As can be seen in the figure, the ratio reaches its peak for $\alpha = 3$, and then the ratio gradually decreases. This means that the local information is optimal when the parameter α is three. Figures 6(a) to (f) show the ratio of the variance to the mean of the enhanced information as the enhancement parameter α is increased from two (a) to seven (f). When the enhancement parameter α is increased from two (a) to three (b),

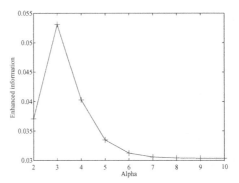

Fig. 5. Ratio of the variance to the mean of enhanced information

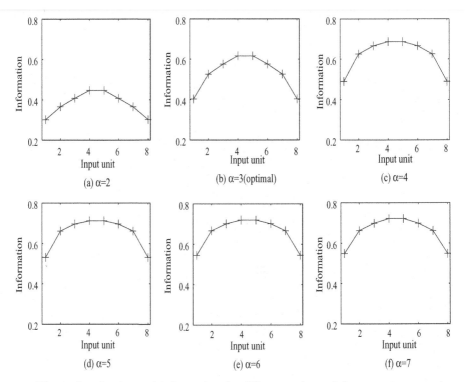

Fig. 6. Local enhanced information for different values of the parameter α

the magnitude of the ratio is significantly increased. On the other hand, even if the enhancement parameter α is increased from four to seven, little difference can be seen, and all we can see is the increase of the magnitude of enhanced information.

Figures 7(a) and (b) show enhanced information and quantization errors, respectively. As the input unit moves to the center of the input pattern, the enhanced information is gradually increased. Then, Figure 7(b) shows quantization

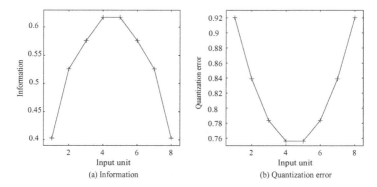

(a) Information   (b) Quantization error

Fig. 7. Local enhanced information and quantization errors. The correlation coefficient is -0.999.

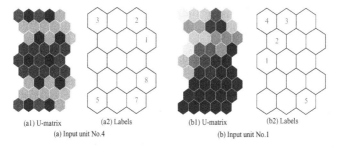

(a1) U-matrix   (a2) Labels   (b1) U-matrix   (b2) Labels

(a) Input unit No.4   (b) Input unit No.1

Fig. 8. U-matrices and labels for connection weights with only a unit with maximum information (No. 4) (a) and with minimum information (No. 1) (b)

errors; we can see that errors are decreased as the input units move to the center. These results are natural, because the input patterns are symmetrical to the center. The correlation between enhanced information and quantization errors becomes -0.999, and this means that enhanced information is closely correlated with quantization errors.

Figures 8 (a) and (b) show examples of U-matrices and labels for input No. 4 (maximum variance) and for input unit No. 1 (minimum variance). As can be seen in Figure 8(a), input patterns in the U-matrix are clearly divided into two parts by the strongly warmer-colored boundaries in the middle. On the other hand, for the input unit with the minimum variance, the U-matrix is significantly different from the U-matrix obtained by SOM and information enhancement. In addition, the labels in Figure 8(b2) show a map with a smaller number of labels, meaning that the number of empty cells is significantly increased.

4   Conclusion

In this paper, we have proposed information enhancement learning to realize self-organizing maps, detect the importance of input variables and find the optimal input variables. In our information enhancement learning, there are three types of information, namely, self-enhancement, collective enhancement and local enhancement. With self-enhancement and collective enhancement, we can realize self-organizing maps. In addition, we use local enhanced information to detect the importance of input units or input variables. An input variable is considered to be optimal when the variance of local information is maximized. We applied the method to an artificial data problem. In the problem, information enhancement learning was able to produce self-organizing maps close to those produced by the conventional SOM in terms of quantization errors and U-matrices. In addition, the importance of input variables detected by local enhanced information corresponded to the importance obtained by directly computing errors.

One of the problems of this method is that of how to find optimal local information. We have proposed the ratio of the variance to the average as a criterion. However, we have found that this ratio is not necessarily valid for all cases. Thus, a more general approach to this optimality problem is necessary. In addition, to evaluate the performance of the method, we should apply it to larger and more practical problems.

Acknowledgment

The author is very grateful to two reviewers and Mitali Das for her valuable comments.

References

1. Kamimura, R.: Feature discovery by enhancement and relaxation of competitive units. In: Fyfe, C., Kim, D., Lee, S.-Y., Yin, H. (eds.) IDEAL 2008. LNCS, vol. 5326, pp. 148–155. Springer, Heidelberg (2008)
2. Kohonen, T.: The self-organizing maps. Proceedings of the IEEE 78(9), 1464–1480 (1990)
3. Kohonen, T.: Self-Organizing Maps. Springer, Heidelberg (1995)
4. Guyon, I., Elisseeff, A.: An introduction of variable and feature selection. Journal of Machine Learning Research 3, 1157–1182 (2003)
5. Rakotomamonjy, A.: Variable selection using svm-based criteria. Journal of Machine Learning Research 3, 1357–1370 (2003)
6. Perkins, S., Lacker, K., Theiler, J.: Grafting: Fast, incremental feature selection by gradient descent in function space. Journal of Machine Learning Research 3, 1333–1356 (2003)
7. Reunanen, J.: Overfitting in making comparison between variable selection methods. Journal of Machine Learning Research 3, 1371–1382 (2003)
8. Castellano, G., Fanelli, A.M.: Variable selection using neural-network models. Neurocomputing 31, 1–13 (1999)
9. Kamimura, R.: Cooperative information maximization with gauissian activation functions for self-organizing maps. IEEE Transactions on Neural Networks 17(4), 909–919 (2006)
10. Linsker, R.: How to generate ordered maps by maximizing the mutual information between input and output. Neural Computation 1, 402–411 (1989)

11. Ueda, N., Nakano, R.: Deterministic annealing variant of the em algorithm. In: Advances in Neural Information Processing Systems, pp. 545–552 (1995)
12. Rose, K., Gurewitz, E., Fox, G.C.: Statistical mechanics and phase transition in clustering. Physical review letters 65(8), 945–948 (1990)
13. Martinez, T.M., Berkovich, S.G., Schulten, K.J.: Neural-gas network for vector quanitization and its application to time-series prediction. IEEE transactions on neural networks 4(4), 558–569 (1993)
14. Erdogmus, D., Principe, J.: Lower and upper bounds for misclassification probability based on renyi's information. Journal of VLSI signal processing systems 37(2/3), 305–317 (2004)
15. Torkkola, K.: Feature extraction by non-parametric mutual information maximization. Journal of Machine Learning Research 3, 1415–1438 (2003)
16. Kamimura, R.: Free energy-based competitive learning for mutual information maximization. In: Proceedings of IEEE Conference on Systems, Man, and Cybernetics, pp. 223–227 (2008)
17. Kamimura, R.: Free energy-based competitive learning for self-organizing maps. In: Proceedings of Artificial Intelligence and Applications, pp. 414–419 (2008)
18. Heskes, T.: Self-organizing maps, vector quantization, and mixture modeling. IEEE Transactions on Neural Networks 12(6), 1299–1305 (2001)
19. Versanto, J., Himberg, J., ALhoniemi, E., Parhankagas, J.: Som toolbox for matlab 5. Tech. Rep. A57, Helsinki University of Technology (2000)

Flash Flood Forecasting by Statistical Learning in the Absence of Rainfall Forecast: A Case Study

Mohamed Samir Toukourou[1], Anne Johannet[1], and Gérard Dreyfus[2]

[1] Ecole des Mines d'Alès, CMGD, 6 av. de Clavières
30319 Alès cedex, France
[2] ESPCI-Paristech, Laboratoire d'Electronique,
10, rue Vauquelin75005 Paris, France

Abstract. The feasibility of flash flood forecasting without making use of rainfall predictions is investigated. After a presentation of the "cevenol flash floods", which caused 1.2 billion Euros of economical damages and 22 fatalities in 2002, the difficulties incurred in the forecasting of such events are analyzed, with emphasis on the nature of the database and the origins of measurement noise. The high level of noise in water level measurements raises a real challenge. For this reason, two regularization methods have been investigated and compared: early stopping and weight decay. It appears that regularization by early stopping provides networks with lower complexity and more accurate predicted hydrographs than regularization by weight decay. Satisfactory results can thus be obtained up to a forecasting horizon of three hours, thereby allowing an early warning of the populations.

Keywords: Forecasting, identification, neural network, machine learning, generalization, weight decay, early stopping.

1 Introduction

The need for accurate predictions of flash floods has been highlighted by the recent occurrences of catastrophic floods such as in *Vaison-la-Romaine* (1991), *Nîmes* (1988), *Gardons* (2002), *Arles* (2003), to name only a few, located in the south of France. These disasters result from intense rainfalls on small (some hundreds of km^2), high-slope watersheds, resulting in flows of thousands of m^3/s with times of concentration of a few hours only. The death toll (over 100) in these circumstances in the southeast of France, and the cost of more than 1.2 billion Euros in 2002, showed that the design of a reliable tool to forecast such phenomena is mandatory.

Faced with this major risk, the French Ministry in charge of Sustainable Development (currently MEEDADT) created in 2003 the national center for flood forecasting and warning SCHAPI (*Service Central d'Hydrométéorologie et d'Appui à la Prévision des Inondations*), which is in charge of the "*vigicrue*" surveillance service. The *Gardon d'Anduze*, in the South-East of France, has been chosen by this Center as a pilot site to compare the flash floods (concentration time of 2h-4h) forecasting models. In this context, this paper describes the study of neural network models to build an efficient real time flash flood forecaster.

D. Palmer-Brown et al. (Eds.): EANN 2009, CCIS 43, pp. 98–107, 2009.
© Springer-Verlag Berlin Heidelberg 2009

Real time flash flood forecasting is usually addressed by coupling complex atmospheric and hydrologic models. The complexity generated by this coupling is huge, and the performances of the present models are limited by several factors: the observations may not be accurate enough for these models to produce useful predictions, the models may be biased by a lack of observations on the ground at an appropriate scale, and the models themselves do not take into account the whole complexity of the phenomena.

An alternative approach consists of capitalizing on the available data in order to build models by statistical machine learning. This will reduce the computational burden and free the model designers from the limitations of physical modeling when the phenomena are too complex, or when the estimation of physical parameters is difficult.

Due (i) to the lack of accurate estimations of rainfalls, and (ii) to the high noise level in water level measurements, and in order to guarantee the best possible generalization capabilities, complexity control is a particularly critical issue. Two traditional regularization methods have been investigated: early stopping and weight decay. After careful variable and model selection, the ability of models, obtained by either regularization method, to predict the most dramatic event of the database (September 2002) is assessed. Hydrographs are displayed and comparisons between the results of both methods are performed. Finally, we conclude that, in the present case, training with early stopping provides networks with lower complexity, longer training but more satisfactory predictions.

2 Problem Statement

2.1 Flash Flood Forecasting

Real time flood forecasting is currently addressed on a disciplinary basis by coupling atmospheric and hydrologic models, hydrologic and geographic models, sometimes in conjunction with a risk management system or an expert system. The difficult issue is the modeling of flash floods in mountainous areas. That problem is usually considered from the hydraulic or hydrologic viewpoints, and, in sharp contrast to the present study, the forecast of rainfall or climate is an important and necessary part of current projects (European projects Flood Forecasting System [1] and PREVIEW [2]). In PREVIEW, Le Lay [3] derives a space-dependent rainfall-runoff relation, whose inputs are the rainfall radar observations and/or the rainfall forecasts provided by the weather models, and the output is the discharge for the *Gardon d'Anduze* watershed. This work shows a fundamental limitation of hydrologic models: they need rainfall forecasts because they can only propagate the rainfall over the watershed, in agreement with the physical behavior. However, in the case of small basins subject to intense storms, no rainfall forecast is available with suitable time scale and accuracy. Traditionally, two hypotheses are postulated: either null rainfall, or persistency of past observed rainfalls. Obviously these assumptions are inappropriate, and, as a consequence, the forecasts of the model are not satisfactory. A probabilistic approach is possible in order to downscale the rainfall predictions [4]; nevertheless the time scale is not appropriate given the rising time of the *Gardon d'Anduze* flood. The coupling of atmospheric and geographical data can also be performed with remote sensing data [5]. As a consequence, the huge quantity of data to be processed leads also to investigate parallel simulation as in the CrossGrid project, where a prototype of flood forecasting operates on Grid technologies [6]. From the viewpoint of the end users, the very short computation times involved in the execution of neural network algorithms once training has been performed makes them very attractive as components of a warning system, without having to resort to grid computing. Another advantage is that any nonlinear, dynamical behavior may be modeled by neural networks, particularly the relation between the rainfall up to time t and the discharge at time $t+f$. Forecast is thus possible without estimating future rainfalls. Although neural networks were applied previously to the forecasting of outflows at several forecasting horizons [7] [8], or for water supply management in mountainous areas [9], they were never applied to events of such speed and intensity.

2.2 *Gardon d'Anduze* Flash Floods

The *Gardon* catchment is emblematic of flash flood behavior: first, its floods are very irregular and may rise up to several meters in a few hours; in addition, the basin is populated, which explains the huge damage costs and loss of human lives.

In a few words, the *Gardon d'Anduze* catchment, sub-catchment of the *Gardon* catchment (*Rhône* river tributary) is located in the southeast of France, in the *Cévennes* mountainous area. The basin area is 546 km^2, the catchment is mountainous and has a large mean slope of 40%, which explains the velocity of the floods. The basin contains three main geological units: schist (60 %), granite (30%) and limestone (10%), which results in heterogeneous soil moisture and permeability.

The *Anduze* catchment is subject to very intense storms delivering huge amounts of water: for example, a 500 mm rainfall was recorded in the *Anduze* rain gauge in less than nine hours in 2002. These storms occur most often in autumn, when the Mediterranean Sea is almost warm. They are called "*épisodes cévenols*".

Fifteen flash flood events are available in the database, whose characteristics are shown in Table 1 (1700 records sampled every 30 min). Five very intense events are indicated.

Table 1. Characteristics of events of the database

Date	Duration (hours)	Maximum level (m)	Discharge Peak (m^3/s)	Very intense
September, 21-24, 1994	35	3,71	181	
October, 4-5, 1995	54	5.34	975	y
October, 13-14, 1995	92	5	864	
November, 10-12, 1996	82	2,71	268	
December, 18-19, 1997	104	5.37	985	y
October, 20-21 1997	34	3,64	473	
November, 5-7 1997	74	4,20	624	
November, 26-27 1997	66	2,58	244	
December, 18-19 1997	104	5,37	985	y
September, 28-29, 2000	46	4.80	800	
September, 8-9, 2002	29	9.71	2742	y
September, 24-25 2006	23	2,24	186	
October, 19-20, 2006	55	6.61	1436	y
November, 17 2006	34	2,75	275	
November, 20-23 2007	70	2,69	264	

In the present paper, we show that, despite the difficulty of the task, the evolution of the water level at *Anduze* can be forecast up to 3 hours ahead of time, without any assumption about the evolution of future rainfall, for the catastrophic, most intense event of the database, namely the event of September 2002.

2.3 Noise and Accuracy

In the present section, we focus on the nature and quality of the information available in the database.

Rainfall measurements are performed with rain gauges. These are very accurate sensors, which broadcast the water level every five minutes; however, they provide very local information, so that the heterogeneity of the rainfalls is an important source of inaccuracy: for example, for the event of September 2002, the cumulated rainfalls were 3 times as large in *Anduze* as in *Saint-Jean-du-Gard*, which is only ten kilometers away. Therefore, the most important rainfall may be located between rain gauges, thereby causing inaccurate estimates due to the too large mesh of the rain gauge network. For this reason, radar acquisition of rainfalls with a definition of 1 km^2 has been performed since 2002, but complete, homogeneous sequences are not yet available for all events.

Water level measurements are available with several sampling periods: 1 hour from 1994 to 2002, and 5 minutes after this date. However, because of real time constraints, the sampling period used in this work is $T = 30$ mn, although variance analysis has shown that 15 minutes would be more appropriate. Thus for events recorded before 2002, the peak value is probably underestimated, possibly by 10% to 30%. For the event of 2002, the error results from an accident: instrumentation was damaged during the event, and the water level was estimated *a posteriori*.

Therefore, the unreliability of the available data makes the forecasting of such catastrophic events a challenging task.

3 Model Design

3.1 Definition of the Model

Given a forecasting horizon f, the model is intended to forecast, at discrete time kT, ($k \in N^+$) the water level at *Anduze* at time $(k + f) T$ $(f \in N^+)$.

The available information for the *Anduze* catchment is the water level at the *Anduze* station, the rainfalls at 6 rain gauges delivering the cumulated rainfalls over the sampling period (30 min), and the soil moisture (Soil Water Index, given by the ISBA (Interactions between Soil, Biosphere, and Atmosphere) model [10]).

The 6 rain gauges: *Barre-des-Cévennes, Saint-Roman de Tousques, Saumane, Mialet, Soudorgues* and *Anduze* are spatially well distributed and one can consider that each of them is important. The information about rainfalls is conveyed to the network as sliding windows. All sliding windows have equal width w, whose optimal value is chosen as described in section 3.2. Different values were found, depending on the forecasting horizon (Table 2). Similarly, the information about past water levels is conveyed as sliding windows, whose optimal width was found to be $r = 2$, irrespective of the forecasting horizon.

Table 2. Sliding window width for rainfalls

Forecasting horizon (f)	0.5 hour	1 hour	2 hours	3 hours	4 hours	5 hours
w	2.5	3	3	2	0.5	0.5

Because of the non-linearity of the physical process, we take advantage of the universal approximation property of a neural network with one hidden layer of sigmoid neurons and a linear output neuron [11]. The water level at time $t+f$ is forecast from (i) the measured rainfalls in a sliding window of width w, and (ii) from the measured water levels in a sliding window of width r. The training data is chosen (see section 3.2) in the set of flood sequences recorded over several years (1994-2007), described in Table 1.

Since the model takes into account measured past values of the water level, during the same flood, the available information about soil moisture is not explicitly conveyed to the model since it is implicitly present in the input data.

3.2 Model Selection

One of the events was set apart for use as a test set (see section 3.3); another event was selected for use either as an early stopping set when the latter regularization technique was used, or as an additional test set when regularization was performed by weight decay (see section 4 for more details on regularization). In the latter case, it was also set apart and used neither for training nor for model selection.

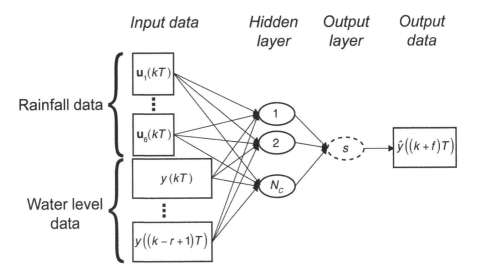

Fig. 1. The model is fed by cumulated rainfall measurements provided by the 6 rain gauges over a temporal window of width w: ui(kT) is the vector of the w rainfall level measurements provided by rain gauge i (i = 1... 6) at times kT, (k−1)T, ..., (k−w+1)T. Water level measurements, over a sliding window of width r, are also input to the model. The output is the forecast water level, f sampling periods ahead.

Model complexity selection was performed by partial K-fold cross-validation on the remaining $N–2$ events of the database. K models were trained from the $N–3$ remaining floods; therefore, $N–K–2$ events (the least intense ones) were present in the training set of all K models.

The generalization ability of the model was assessed by the cross-validation score (see e.g. [12]):

$$S = \sqrt{\frac{1}{K} \cdot \sum_{i=1}^{K} MSE_i} \qquad (1)$$

where MSE_i is the mean squared forecasting error of the model, for the time sequence recorded during event i of the validation set.

In the present case, $K = 4$ events were chosen: the 1995, 1997, and 2006 very intense events reported in Table 1. The 2002 event was selected as a test set because it is typical of events whose forecasting is crucial for early warning. In addition, this event is also more intense than those used for training and validation: it is a difficult test for the models. The other nine floods were always present in the training set.

The above procedure was used for complexity selection, spanning the space of rainfall window width w, water level window width r and number of hidden neurons N_C. For each model, 100 different parameter initializations were performed. Complexity selection was performed separately for each forecasting horizon.

After selecting the appropriate complexity (i.e. after selecting the appropriate values of w, r and N_C) for a given forecasting horizon f, a final model was trained for that horizon, from thirteen sequences: all floods except the test sequence and the early stopping sequence, or all floods except the two test sequences when weight decay regularization was performed. Its performance was assessed on the test sequence(s).

3.3 Training

The usual squared error cost function was minimized by the Levenberg-Marquardt algorithm [13] during training, after computation of the gradient of the cost function by backpropagation.

4 Regularization

In addition to performing input and model selection by cross-validation, regularization methods were applied during training. Two different methods were assessed in this study: weight decay and early stopping.

4.1 Weight Decay

Weight decay prevents parameters from taking excessive values (resulting in overfitting), by introducing a term in the cost function that penalizes large parameter values; this idea is implemented in a systematic fashion in Support Vector Machines. In the present case, the new cost function is expressed as:

$$J = \gamma MSE + (1 - \gamma)\|\theta\|^2 . \tag{2}$$

where MSE is the usual mean squared prediction error, θ is the vector of parameters and γ is the hyperparameter that controls the balance between the terms of the cost function.

Similarly to model selection, the hyperparameter γ was selected by cross validation, for each forecasting horizon, for γ varying from 0.5 to 0.95 with an increment step of 0.05. Table 3 shows the optimal value of γ obtained for each forecasting horizon.

Table 3. Optimal hyperparameter values for each forecasting horizon

Forecasting horizon (f)	0.5 hour	1 hour	2 hours	3 hours	4 hours	5 hours
γ	0.6	0.9	0.7	0.55	0.75	0.75

4.2 Early Stopping

As an alternative regularization technique, early stopping was used in the present investigation. Early stopping consists of terminating training when the prediction error, assessed on a stopping set, different from the training set, starts increasing. It has been shown [14] that it is theoretically equivalent to weight decay. However, due to finite sample size, the results may vary widely depending on the choice of the stopping set. In the present investigation, the stopping set was the event of September 2000 of the database, which was well learnt when it was in the training set, and well predicted when it was in the test set. Therefore, it appeared as a "prototype" of the behavior of the flood process.

5   Results and Discussion

The quality of the forecast can be estimated by various criteria, each of which focuses on a particular desired feature of the model: accuracy of the prediction of the water level at the peak of the flood, accuracy of the prediction of the time of occurrence of the peak, absence of spurious water level peak, etc. The most widely used criterion is the coefficient of determination of the regression

$$R^2 = 1 - \frac{MSE}{\sigma^2}$$

where σ^2 is the variance of the observations. If the model simply predicts the mean of the observations, $R^2 = 0$; conversely, if the model predicts the observations with perfect accuracy, $R^2 = 1$. In the hydrology literature, the coefficient of determination is known as the "Nash-Sutcliffe criterion".

The "persistency criterion" is more specific to forecast models [15]; it is defined as:

$$P = 1 - \frac{\displaystyle\sum_{test\ sequence} (y(t+f) - \hat{y}(t+f))^2}{\displaystyle\sum_{test\ sequence} (y(t) - y(t+f))^2} . \tag{3}$$

where y is the observed water level and $\hat{y}$ is the estimated water level. P is equal to 0 if the predictor is perfectly dumb, i.e. it always predicts that the future value is equal to the present one, and it is equal to 1 if the predictor provides perfectly accurate forecasts.

Tables 4 and 5 describe the models and the accuracy of their predictions on the 2002 flood.

Table 4. Models obtained with regularization by weight decay

Forecasting horizon (f)	0.5h	1 h	2 h	3 h	4 h	5 h	mean
N_C	7	7	5	7	5	3	5.6
Persistency criterion	0.62	0.75	0.62	0.63	0.70	0.68	0.66
R^2 (Nash-Sutcliffe criterion)	0.96	0.91	0.81	0.78	0.72	0.49	0.78
Estimated/Observed peak values	0.86	0.81	0.77	0.69	0.69	0.57	0.73

Table 5. Models obtained with regularization by early stopping

Forecasting horizon (f)	0.5 h	1 h	2 h	3 h	4 h	5 h	mean
N_C	2	2	5	3	3	3	3
Persistency criterion	0.45	0.65	0.32	0.28	0.23	0.59	0.42
R^2 (Nash-Sutcliffe criterion)	0.98	0.93	0.87	0.93	0.84	0.58	0.85
Estimated/Observed peak values	0.90	0.84	0.73	0.82	0.79	0.60	0.78

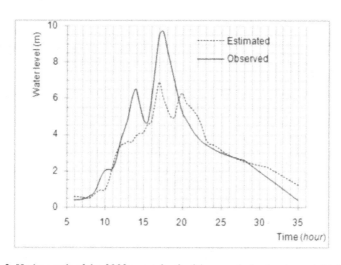

Fig. 2. Hydrograph of the 2002 event for $f = 3$ h – regularization by weight decay

In the present case, early stopping provides consistently more parsimonious models than weight decay, with consistently higher values of the determination criterion. However, the persistency criterion is larger for models obtained with weight decay.

Figures 2 and 3 show the predicted and observed curves for the test sequence (2002 event). Given the difficulty of the task, these results are extremely encouraging, since they show that the model would have allowed the public services to issue early warnings to the population if it had been available during that event.

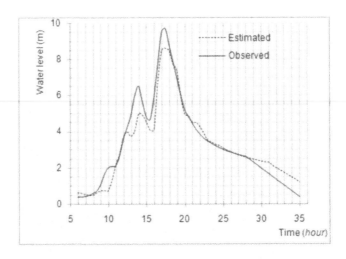

Fig. 3. Hydrograph of the 2002 event for $f = 3$ h – regularization by early stopping

6 Conclusion

Flash flood forecasting is a very challenging task due to high variability and noise in the data, especially when no rainfall forecast is available. In the present study, we have shown the feasibility of forecasting the catastrophic event of September 2002 in *Anduze* with an accuracy and forecasting horizon that are compatible with an early warning of the populations.

This requires a careful methodology for model selection and regularization; it is shown that early stopping and weight decay result in different generalization capabilities, and that, in this specific case, early stopping provides more satisfactory results on the test set. This is not claimed to be a general result, but it shows that a variety of methods must be used in order to solve such difficult problems satisfactorily.

From the viewpoint of hydrology, this methodology should easily be applied to small (less than 1000 km^2), fast (concentration time less than 10 h) basins providing only rainfalls and water level. Because no exogenous data is necessary, the method should be applicable to many European mountainous watersheds.

Acknowledgments. The authors acknowledge the SCHAPI and its regional service SPC of Gard for their collaboration on this exciting application.

References

1. European Flood Forecasting System (2003), `http://effs.wldelft.nl/`
2. PREVIEW (2008), `http://www.preview-risk.com`
3. Le Lay, M., Saulnier, G.-M.: Exploring the Signature of Climate and Landscape Spatial Variabilities in Flash Flood Events: Case of the 8–9 September 2002 Cévennes-Vivarais Catastrophic Event. Geophysical Research Letters 34, 5 page (2007)
4. Taramasso, A.C., Gabellani, S., Marsigli, C., Montani, A., Paccagnella, T., Parodi, A.: Operational flash-flood forecasting chain: an application to the Hydroptimet test cases. Geophysical Research Abstracts 7, 9–14 (2005)
5. Jasper, K., Gurtz, J., Lang, H.: Advanced flood forecasting in Alpine watersheds by coupling meteorological observations and forecasts with a distributed hydrological model. Journal of Hydrology 267, 40–52 (2002)
6. CrossGrid (2005), `http://www.eu-crossgrid.org`
7. Zealand, C.M., Burn, D.H., Simonovic, S.P.: Short term streamflow forecasting using artificial neural networks. Journal of Hydrology 214, 32–48 (1999)
8. Schmitz, G.H., Cullmann, J.: PAI-OFF: A new proposal for online flood forecasting in flash flood prone catchments. Journal of Hydrology 1, 1–14 (2008)
9. Iliadis, S.L., Maris, F.: An artificial neural networks model for mountainous water-resources management: the case of Cyprus mountainous watersheds. Environmental Modelling & Software 22, 1066–1072 (2007)
10. Noilhan, J., Mahfouf, J.F.: The ISBA land surface parameterization scheme. Global and Planetary Change 13, 145–159 (1996)
11. Hornik, K., Stinchcombe, M., White, H.: Multilayer Feedforward Networks Are Universal Approximators. Neural Networks 2, 359–366 (1989)
12. Dreyfus, G.: Neural Networks, Methodology and Applications. Springer, Heidelberg (2005)
13. Hagan, M.-T., Menhaj, M.-B.: Training feedforward networks with the Marquardt Algorithm. IEEE Transaction on Neural Networks 5(6), 989–993 (1994)
14. Sjöberg, J., Ljung, L.: Overtraining, regularization, and searching for a minimum, with application to neural networks. International Journal of Control 62(6), 1391–1407 (1995)
15. Kitadinis, P.K., Bras, R.L.: Real-time forecasting with a conceptual hydrologic model: 2 applications and results. Water Resour. Res. 16, 1034–1044 (1980)

An Improved Algorithm for SVMs Classification of Imbalanced Data Sets

Cristiano Leite Castro, Mateus Araujo Carvalho, and Antônio Padua Braga

Federal University of Minas Gerais
Department of Electronics Engineering
Av. Antônio Carlos, 6.627
Campus UFMG - Pampulha, 30.161-970
Belo Horizonte, MG, Brasil
{crislcastro,mateus.carvalho,apbraga}@ufmg.br

Abstract. Support Vector Machines (SVMs) have strong theoretical foundations and excellent empirical success in many pattern recognition and data mining applications. However, when induced by imbalanced training sets, where the examples of the target class (minority) are outnumbered by the examples of the non-target class (majority), the performance of SVM classifier is not so successful. In medical diagnosis and text classification, for instance, small and heavily imbalanced data sets are common. In this paper, we propose the Boundary Elimination and Domination algorithm (BED) to enhance SVM class-prediction accuracy on applications with imbalanced class distributions. BED is an informative resampling strategy in input space. In order to balance the class distributions, our algorithm considers density information in training sets to remove noisy examples of the majority class and generate new synthetic examples of the minority class. In our experiments, we compared BED with original SVM and Synthetic Minority Oversampling Technique (SMOTE), a popular resampling strategy in the literature. Our results demonstrate that this new approach improves SVM classifier performance on several real world imbalanced problems.

Keywords: Support Vector Machines, supervised learning, imbalanced data sets, resampling strategy, ROC Curves, pattern recognition applications.

1 Introduction

Since their introduction by V. Vapnik and coworkers [1], [2], [3], Support Vector Machines (SVMs) have been successfully applied to many pattern recognition real world problems. SVMs are based on Vapnik-Chervonenkis' theory and the structural risk minimization principle (SRM) [2], [4] which aims to obtain a classifier with high generalization performance through minimization of the global training error and the complexity of the learning machine. However, it is well established that in applications with imbalanced data sets [5], where the training

D. Palmer-Brown et al. (Eds.): EANN 2009, CCIS 43, pp. 108–118, 2009.

examples of the target class (minority) are outnumbered by the training examples of the non-target class (majority), the SVM classifier performance becomes limited. This probably occurs because the global training error considers different errors as equally important assuming that the class prior distributions are relatively balanced [6]. In the case of most real world problems, when the imbalanced class ratio is huge, one can observe that the separation surface learned by SVMs is skewed toward the minority class [5]. Consequently, test examples belonging to the small class are more often misclassified than those belonging to the prevalent class [7].

In order to improve SVMs performance on applications with imbalanced class distributions, we designed the Boundary Elimination and Domination Algorithm (BED). It eliminates outliers and noisy examples of the majority class in lower density areas and generates new synthetic examples of the minority class by considering both the analyzed example, and its k minority nearest neighbors. By increasing the number of minority examples in areas near the decision boundary, the representativity of support vectors of this class is improved. Moreover, the BED has parameters that allow the control of the resampling process intensity and can be adjusted according to the level of imbalance and overlapping of the problem.

In our experiments, we used several real world data sets extracted from UCI Repository [8] with different degrees of class imbalance. We compared the BED algorithm with Synthetic Minority Oversampling Technique (SMOTE) [9], a popular oversampling strategy. In both algorithms, we used SVMs as base classifiers. The performances were evaluated using appropriate metrics for imbalanced classification such as F-measure [10], G-mean [11] and ROC (Receiver Operating Characteristics) curves [12].

This paper is organized as follows: Section 2 reviews the SVMs learning algorithm and previous solutions to the learning problem with imbalanced data sets, specially in the context of SVMs. Section 3, presents our approach to the problem and describes the BED algorithm. Section 4 describes how the experiments were performed and the results obtained. Finally, Section 5 is the conclusion.

2   Background

2.1   Support Vector Machines

In their original formulation [1], [2], SVMs were designed to estimate a linear function $f(\mathbf{x}) = \text{sgn}\left(\langle \mathbf{w} \cdot \mathbf{x} \rangle + b\right)$ of parameters $\mathbf{w} \in \Re^d$ and $b \in \Re$, using only a training set drawn i.i.d. according to an unknown probability distribution $P(\mathbf{x}, y)$. This training set is a finite set of samples,

$$(\mathbf{x}_1, y_1, \cdots, \mathbf{x}_n, y_n) \ . \tag{1}$$

where $\mathbf{x}_i \in \Re^d$ and $y_i \in \{-1, 1\}$. The SVMs learning aims to find the hyperplane which gives the largest separating margin between the two classes. For a linearly separable training set, the margin ρ is defined as euclidean distance between the separating hyperplane and the closest training examples. Thus, the learning problem can be stated as follows: find $\mathbf{w}$ and b that maximize the margin while ensuring that all the training samples are correctly classified,

$$min_{(\mathbf{w},b)} \quad \frac{1}{2}\|\mathbf{w}\|^2 \quad . \tag{2}$$

$$s.t. \quad y_i\left(\langle \mathbf{w} \cdot \mathbf{x}_i \rangle + b\right) \geq 1, \quad i = 1,\ldots,n \quad . \tag{3}$$

For the non-linearly separable case, slack variables ε_i are introduced to allow for some classification errors (soft-margin hyperplane) [3]. If a training example is located inside the margins or the wrong side of the hyperplane, its corresponding ε_i is greater than 0. The $\sum_{i=1}^{n} \varepsilon_i$ corresponds to an upper bound of the number of training errors. Thus, the optimal hyperplane is obtained by solving the following constrained (primal) optimization problem,

$$min_{(\mathbf{w},b,\varepsilon_i)} \quad \frac{1}{2}\|\mathbf{w}\|^2 + C\sum_{i=1}^{n}\varepsilon_i \quad . \tag{4}$$

$$s.t. \quad y_i\left(\langle \mathbf{w} \cdot \mathbf{x}_i \rangle + b\right) \geq 1 - \varepsilon_i, \quad i = 1,\ldots,n \quad . \tag{5}$$

$$\varepsilon_i \geq 0, \quad i = 1,\ldots,n \quad . \tag{6}$$

where the constant $C > 0$, controls the trade-off between the margin size and the misclassified examples. Instead of solving the primal problem directly, one considers the following dual formulation,

$$max_{(\alpha)} \quad \sum_{i=1}^{n}\alpha_i - \frac{1}{2}\sum_{i,j=1}^{n} y_i y_j \alpha_i \alpha_j \langle \mathbf{x}_i \cdot \mathbf{x}_j \rangle \quad . \tag{7}$$

$$s.t. \quad 0 \leq \alpha_i \leq C, \quad i = 1,\ldots,n \quad . \tag{8}$$

$$\sum_{i=1}^{n}\alpha_i y_i = 0 \quad . \tag{9}$$

Solving this dual problem the Lagrange multipliers α_i are obtained whose sizes are limited by the *box constraints* $(\alpha_i \leq C)$; the parameter b can be obtained from some training example (support vector) with non-zero corresponding α_i. This leads to the following decision function,

$$f(\mathbf{x}_j) = \text{sgn}\left(\sum_{i=1}^{n} y_i \alpha_i \langle \mathbf{x}_i \cdot \mathbf{x}_j \rangle + b\right) \quad . \tag{10}$$

The SVM formulation presented so far is limited to linear decision surfaces in input space, which are definitely not appropriate for many classification tasks. The extension to more complex decision surfaces is conceptually quite simple and, is done by mapping the data into a higher dimensional feature space F, where the problem becomes linear. More precisely, a non-linear SVM first maps the input vectors by $\Phi : \mathbf{x} \rightarrow \Phi(\mathbf{x})$, and then estimates a separating hyperplane in F,

$$f(\mathbf{x}) = \text{sgn}\left(\langle \Phi(\mathbf{x}) \cdot \mathbf{w} \rangle + b\right) \quad . \tag{11}$$

It can be observed, in (7) and (10), that the input vectors are only involved through their inner product $\langle \mathbf{x}_i \cdot \mathbf{x}_j \rangle$. Thus, to map the data is not necessary to consider the non-linear function Φ in explicit form. The inner products can only be calculated in the feature space F. In this context, a *kernel* is defined as a way to directly compute this product [4]. A *kernel* is a function K, such that for all pair $\mathbf{x}, \mathbf{x}'$ in input space,

$$K(\mathbf{x}, \mathbf{x}') = \langle \Phi(\mathbf{x}) \cdot \Phi(\mathbf{x}') \rangle \; . \tag{12}$$

Therefore, a non-linear SVM is obtained by only replacing the inner product $\langle \mathbf{x}_i \cdot \mathbf{x}_j \rangle$ in equations (7) and (10) by the *kernel*[1] function $K(\mathbf{x}_i, \mathbf{x}_j)$ that corresponds to that inner product in the feature space F.

2.2 Related Works

Two main approaches have been designed to address the learning problem with imbalanced data sets: algorithmic and data preprocessing [13]. In the algorithmic approach, learning algorithms are adapted to improve performance of the minority (positive) class. In the context of SVMs, [14] and [15] proposed techniques to modify the threshold (parameter b) in the decision function given by the equation (10). In [16] and [17], the error for positive examples was distinguished from the error for negative examples by using different constants C^+ and C^-. The ratio C^+/C^- is used to control the trade-off between the number of false negatives and false positives. This technique is known by Asymmetric Misclassification Costs SVMs [17].

Another algorithmic approach to improve SVMs on imbalanced classification is to modify the employed kernel. Thus, based on kernel-target alignment algorithms [18], Kandola and Taylor [19] assigned different alignment targets to positive and negative examples. In the same direction, [5] proposed the kernel boundary alignment algorithm (KBA) which adapts the decision function toward the majority class by modifying the kernel matrix.

In the data preprocessing approach, the objective is to balance the class distribution by resampling the data in input space, including oversampling examples of the minority class and undersampling examples of the majority class. Oversampling works by duplicating pre-existing examples (oversampling with replacement) or generating new synthetic data which is usually obtained by interpolating. For instance, in the SMOTE algorithm [9], for each minority example, its nearest minority neighbors are identified and new minority examples are created and placed randomly between the example and the neighbors.

Undersampling involves the elimination of class majority examples. The examples to be eliminated can be selected randomly (random undersampling) or through some prior information (informative oversampling). The one-sided

---

[1] *Kernel* functions used in this work:
  Linear *kernel*: $K(\mathbf{x}, \mathbf{x}') = \mathbf{x} \cdot \mathbf{x}'$
  RBF kernel: $K(\mathbf{x}, \mathbf{x}') = \exp\left(\frac{-(\mathbf{x}-\mathbf{x}')^2}{2r^2}\right)$.

selection proposed by Kubat and Matwin [11], for instance, is a informative undersampling approach which removes noisy, borderline, and redundant majority examples.

Previous data preprocessing strategies that aim to improve the SVMs learning on imbalanced data sets include the following: [20] combined the SMOTE algorithm with Asymmetric Misclassification Costs SVMs mentioned earlier. [21] used SMOTE and also random undersampling for SVM learning on intestinal-contraction-detection task. Recently, [22] proposed the Granular SVM - Repetitive Undersampling algorithm (GSVM-RU) whose objective is to minimize the negative effect of information loss in the undersampling process.

3   Boundary Elimination and Domination Algorithm

In this Section, we describe the Boundary Elimination and Domination algorithm (BED) to improve SVMs performance on imbalanced applications. The basic idea behind BED is to increase the representativity of the minority class on the training set through data preprocessing in input space.

The data preprocessing is done by removing (undersampling) noisy examples of the majority class and generating (oversampling) new examples of the minority class. The whole process of resampling is guided by density information around the training examples. This information makes it possible for the BED algorithm to identify both isolated examples or examples that belong to class overlapping area. The process of elimination and synthesis of examples in these regions improves the representativity of support vectors so that a SVM classifier will estimate a better surface separation.

To obtain the density information, BED defines a credibility score for each example of the training set. This score is calculated from the k nearest neighbors (based on Euclidean distance) of the evaluated example according to the following equation,

$$cs(\mathbf{x}_i^c) = \frac{nC^c}{nC^+ + nC^-} \ . \tag{13}$$

where, $\mathbf{x}_i^c$ corresponds to the ith training example of the C^c class. The symbol $c = \{+, -\}$ indicates if the example belongs to the positive (minority) class or the negative (majority) class, respectively. The nC^+ value corresponds to the number of positive neighbors of the $\mathbf{x}_i^c$. For a given positive example, $cs(\mathbf{x}_i^+)$ evaluates the proportion of positive examples between the k nearest neighbors of the $\mathbf{x}_i^+$. Therefore, if $cs(\mathbf{x}_i^+) \approx 1$, one can state that $\mathbf{x}_i^+$ is on a region with high density of the positive examples. Equivalent definitions hold for the negative class.

For majority class examples $(\mathbf{x}_i^-)$, the credibility score $cs(\mathbf{x}_i^-)$ is used to find noisy examples that occupy class overlapping areas and also isolated examples belonging to the minority class regions. Thus, BED establishes a rule to detect and eliminate these examples. This rule depends on the *maxtol* parameter and is given by,

$$\text{if } cs(\mathbf{x}_i^-) < maxtol \rightarrow \mathbf{x}_i^- \text{ is eliminated,}$$

$$\text{if } cs(\mathbf{x}_i^-) \geq maxtol \rightarrow \mathbf{x}_i^- \text{ is not eliminated.}$$

The $maxtol$ parameter value ($0 \leq maxtol \leq 1$) should be defined by the user and will determine the intensity of undersampling stage. The higher the $maxtol$ value the more examples belonging to C^- are eliminated. In the case of high levels of class imbalance and overlapping, values of $maxtol$ close to 1 are suggested given that it is likely that a large number of the negative examples in areas belonging to the C^+ exist. When the the imbalance is softer smaller values for $maxtol$ are suggested.

For minority class examples ($\mathbf{x}_i^+$), the credibility score $cs(\mathbf{x}_i^+)$ is also used to synthesize new examples. If the $cs(\mathbf{x}_i^+)$ is considered valid by BED, a new example $\hat{\mathbf{x}}^+$ is created as follows:

1. for each continuous attribute (feature) m, its value is calculated based on the mean of m-values of the nC^+ nearest neighbors and the example $\mathbf{x}_i^+$,

$$\hat{\mathbf{x}}^+(m) = \frac{\sum_{j=1}^{nC^+} \mathbf{x}_j^+(m) + \mathbf{x}_i^+(m)}{nC^+ + 1} . \tag{14}$$

2. each nominal attribute p assumes the value most frequently observed between the p-values of the nC^+ nearest neighbors and the example $\mathbf{x}_i^+$.

Otherwise, if $cs(\mathbf{x}_i^+)$ is not valid, $\mathbf{x}_i^+$ is considered an isolated or a noisy example and, a new example should not be created around it. The rule which evaluates the positive example is based on the $mintol$ parameter and is given by,

$$\text{if } cs(\mathbf{x}_i^+) < mintol \rightarrow \mathbf{x}_i^+ \text{ is not eliminated,}$$

$$\text{if } cs(\mathbf{x}_i^-) \geq mintol \rightarrow \hat{\mathbf{x}}^+ \text{ is created.}$$

The parameter $mintol$ defines the validity degree of positive examples in the input space and similar to the $maxtol$ parameter can have values from 0 to 1. The lower the $mintol$ value the higher the probability of a positive example on the class overlapping area to be considered valid and used to generate a new synthetic example $\hat{\mathbf{x}}^+$. $mintol$ values however, should not be very close to 0 in order to ensure that isolated examples belonging to the C^+ do not generate new examples. The $mintol$ adjustment should also be done by the user according to the level of class imbalance and overlapping.

At each iteration of BED algorithm, a new training example is analyzed. The algorithm ends when the classes become balanced. The effect caused by BED algorithm on an imbalanced training set is illustrated in Fig. 1: majority examples in regions of prevalence of minority examples are eliminated. Meanwhile, new minority examples are better represented in these regions.

It is expected, therefore, that our resampling strategy allows that a SVM classifier obtains a separation surface that maximize the number of correct positive classifications. Moreover, it is important to notice that BED can be used with any classification algorithm.

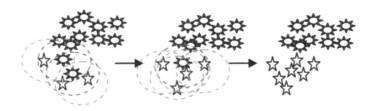

Fig. 1. Illustration of resampling process of the BED algorithm

4   Experiments and Results

4.1   Experiment Methodology

Six real world datasets from the UCI Repository [8] with different levels of class imbalance were used in our experiments (Table 1). In order to have the same negative to positive ratio, stratified 7-fold crossvalidation was used to obtain training and test subsets (ratio 7:3) for each data set.

To improve the performance of SVM classifiers in imbalanced applications, original SVM was used as base classifier in all experiments conducted. Furthermore, we compared our algorithm BED with a resampling strategy well known in the literature, called SMOTE [9]. For all the SVM classifiers, we employed linear and RBF *kernels*. The parameters C (box-contraints) and r (radius of Gaussian function) were kept equal in all runs for each data set. Thus, the algorithms could be compared without the influence of SVM parameters.

For the SMOTE algorithm the percentage of minority class oversampling for each data set is shown in Table 1. The parameters of BED algorithm (k, *maxtol* and *mintol*) were set empirically from average results obtained in several runs. The optimal choices of these parameters are also in Table 1.

Table 1. Characteristics of the six data sets used in experiments: number of attributes and number of positive and negative examples. For some data sets, the class label in the parentheses indicates the target class we chose. Moreover, this table shows the optimal choice of parameters for SMOTE (% oversampling) and BED (k, *maxtol* and *mintol*).

Data Set	#Attrib	#POS	#NEG	%overs.	k	$maxtol$	$mintol$
Diabetes	08	268	500	200%	4.0	0.5	0.5
Breast	33	47	151	200%	10.0	0.5	0.5
Heart	44	55	212	300%	5.0	0.5	0.5
Car(3)	06	69	1659	1000%	3.0	0.5	0.5
Yeast(5)	08	51	1433	1000%	10.0	0.5	0.5
Abalone(19)	08	32	4145	1000%	10.0	0.5	0.5

After setting of the parameters, algorithm performances were evaluated using appropriate metrics for imbalanced classification. For each metric, the mean and standard deviation were calculated from 7 runs with different training and test subsets obtained from stratified 7-fold crossvalidation. The metrics used in evaluation process and the average results achieved for the test set are presented in details in Section 4.2 below.

4.2   Results

Table 2 illustrates the results using the *G-mean* metric, defined as $\sqrt{tpr \cdot fpr}$ [11], which corresponds to the geometric mean between the correct classification rates for positive (*sensitivity*) and negative (*specificity*) examples, respectively. Note that, in five out of the six data sets evaluated, BED achieved better results than SMOTE and original SVM (the best results are marked in bold). The *G-mean* values in Table 2 indicate that BED achieved a better balance between *sensitivity* and *specificity*. It is worth noting that in the case of Abalone data set, characterized by a huge imbalance degree, when both original SVM and SMOTE were unable to give satisfactory values for *G-mean*, the BED algorithm worked well.

In Table 3, we evaluated the algorithms using the metric *F-measure* [10] that considers only the performance for the positive class. *F-measure* is calculated through two important measures: *Recall* and *Precision*. *Recall* (R) is equivalent to *sensitivity* and denotes the ratio between the number of positive examples correctly classified and the total number of original positive examples. *Precision* (P), in turn, corresponds to the ratio between the number of positive examples correctly classified and the total number of examples identified as positives by the classifier. Thus, *F-measure* is defined as $\frac{2 \cdot R \cdot P}{R+P}$ and represents the harmonic mean between *Recall* and *Precision*. As shown in Table 3, the BED algorithm produced better results. Compared to the original SVM and SMOTE algorithms,

Table 2. This table compares *G-mean* values on UCI data sets. The first column lists the data sets used. The following columns show the results achieved by the algorithms: Support Vector Machines (SVMs) (column 2), Synthetic Minority Oversampling Technique (SMOTE) (column 3) and Boundary Elimination and Domination (BED) (column 4). Mean and standard deviation values for each data set were calculated for 7 runs with different test subsets obtained from stratified 7-fold crossvalidation.

Data Set	SVMs	SMOTE	BED
Diabetes	0.70 ± 0.04	0.71 ± 0.02	**0.75 ± 0.06**
Breast	0.58 ± 0.07	0.66 ± 0.08	**0.71 ± 0.04**
Heart	0.63 ± 0.05	**0.76 ± 0.04**	**0.76 ± 0.02**
Car	0.91 ± 0.08	0.94 ± 0.02	**0.95 ± 0.02**
Yeast	0.15 ± 0.25	0.54 ± 0.10	**0.68 ± 0.04**
Abalone	0.00 ± 0.00	0.00 ± 0.00	**0.66 ± 0.12**

Table 3. This table compares *F-measure* values on UCI data sets. The first column lists the data sets used. The following columns show the results achieved by the algorithms: Support Vector Machines (SVMs) (column 2), Synthetic Minority Oversampling Technique (SMOTE) (column 3) and Boundary Elimination and Domination (BED) (column 4). Mean and standard deviation values for each data set were calculated for 7 runs with different test subsets obtained from stratified 7-fold crossvalidation.

Data Set	SVMs	SMOTE	BED
Diabetes	0.63 ± 0.06	0.65 ± 0.01	$\mathbf{0.68 \pm 0.04}$
Breast	0.48 ± 0.23	$\mathbf{0.56 \pm 0.10}$	0.55 ± 0.05
Heart	0.46 ± 0.07	0.52 ± 0.08	$\mathbf{0.56 \pm 0.05}$
Car	0.86 ± 0.08	0.87 ± 0.03	$\mathbf{0.97 \pm 0.01}$
Yeast	0.12 ± 0.21	0.12 ± 0.04	$\mathbf{0.38 \pm 0.07}$
Abalone	0.00 ± 0.00	0.00 ± 0.00	$\mathbf{0.33 \pm 0.01}$

BED performance was superior especially for data sets with higher imbalance degree which is the case of most of the real world problems. The *F-measure* values, described in Table 3, shows that BED improves the classifier performance for the positive class.

Average ROC curves (Receiver Operating Characteristics) [12] were plotted for all data sets and gave similar results. The ROC curve for a binary classifier shows graphically the true positive rate as a function of the false positive rate when the decision threshold varies. To illustrate, Fig. 2 shows the example for the Diabetes data set. Note that BED generates a better ROC curve than SMOTE and original SVM.

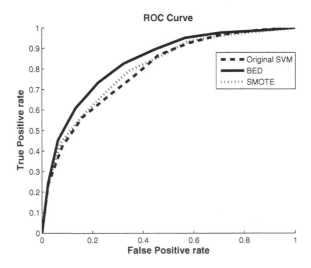

Fig. 2. Average ROC curves for test set obtained from Diabetes data set

5   Conclusions

It is known that representativeness in a training set is the most important feature to achieve classifiers with high generalization performance. However in most classification problems, representativeness is not only expensive but often a very difficult task. In general, data sets available are small, sparse, with missing values and with heavily imbalanced prior probabilities.

SVMs classifiers try to solve the representativeness problem in data sets by controlling complexity of decision function through maximizing separation margin. When applied to small and imbalanced data sets problems, SVMs tend to smooth the response. One of the possible explanations is the asymptotic bounds imposed by the reduced size of the training and validation data sets. In addition, SVMs do not take into consideration the differences (costs) between the class distributions during the learning process.

Here, we propose the Boundary Elimination and Domination algorithm (BED) to tackle the representativeness problem in training sets. The results obtained with real world applications demonstrated that BED is an efficient resampling strategy when used in small and imbalanced data sets leading to a better surface separation.

References

1. Boser, B.E., Guyon, I.M., Vapnik, V.: A training algorithm for optimal margin classifiers. In: Proceedings of the fifth annual workshop on Computational learning theory, pp. 144–152. ACM Press, New York (1992)
2. Vapnik, V.N.: The nature of statistical learning theory. Springer, New York (1995)
3. Cortes, C., Vapnik, V.: Support-Vector Networks. Mach. Learn. 20, 273–297 (1995)
4. Cristianini, N., Shawe-Taylor, J.: An introduction to support vector machines and other kernel-based learning methods. Cambridge University Press, London (2000)
5. Wu, G., Chang, E.Y.: KBA: kernel boundary alignment considering imbalanced data distribution. IEEE Trans. Knowl. Data Eng. 17, 786–795 (2005)
6. Provost, F., Fawcett, T.: Robust classification for imprecise environments. Mach. Learn. 42, 203–231 (2001)
7. Sun, Y., Kamel, M.S., Wong, A.K.C., Wang, Y.: Cost-sensitive boosting for classification of imbalanced data. Pattern Recognition 40, 3358–3378 (2007)
8. Asuncion, A., Newman, D.J.: UCI Machine Learning Repository. University of California, Irvine, School of Information and Computer Sciences,
   http://www.ics.uci.edu/mlearn/MLRepository.html
9. Chawla, N.V., Bowyer, K.W., Hall, L.O., Kegelmeyer, W.P.: SMOTE: synthetic minority over-sampling technique. J. Artif. Intell. Res. 16, 321–357 (2002)
10. Tan, P., Steinbach, M.: Introduction to Data Mining. Addison Wesley, Reading (2006)
11. Kubat, M., Matwin, S.: Addressing the curse of imbalanced training sets: one-sided selection. In: Proceedings of 14th International Conference on Machine Learning, pp. 179–186. Morgan Kaufmann, San Francisco (1997)
12. Egan, J.P.: Signal detection theory and ROC analysis. Academic Press, London (1975)

13. Weiss, G.M.: Mining with rarity: a unifying framework. SIGKDD Explor. Newsl. 6, 7–19 (2004)
14. Karakoulas, G., Shawe-Taylor, J.: Optimizing classifiers for imbalanced training sets. In: Proceedings of Conference on Advances in Neural Information Processing Systems II, pp. 253–259. MIT Press, Cambridge (1999)
15. Li, Y., Shawe-Taylor, J.: The SVM with uneven margins and Chinese document categorization. In: Proceedings of the 17th Pacific Asia Conference on Language, Information and Computation, pp. 216–227 (2003)
16. Veropoulos, K., Campbell, C., Cristianini, N.: Controlling the sensitivity of support vector machines. In: Proceedings of the International Joint Conference on Artificial Intelligence, pp. 55–60 (1999)
17. Joachims, T.: Learning to classify text using support vector machines: methods, theory and algorithms. Kluwer Academic Publishers, Norwell (2002)
18. Cristianini, N., Shawe-Taylor, J., Kandola, J.: On kernel target aligment. In: Proceedings of the Neural Information Processing Systems NIPS 2001, pp. 367–373. MIT Press, Cambridge (2002)
19. Kandola, J., Shawe-Taylor, J.: Refining kernels for regression and uneven classification problems. In: Proceedings of International Conference on Artificial Intelligence and Statistics. Springer, Heidelberg (2003)
20. Akbani, R., Kwek, S., Japkowicz, N.: Applying support vector machines to imbalanced datasets. In: Proceedings of European Conference on Machine Learning, pp. 39–50 (2004)
21. Vilariño, F., Spyridonos, P., Vitri, J., Radeva, P.: Experiments with SVM and stratified sampling with an imbalanced problem: detection of intestinal contractions. In: Proceedings of International Workshop on Pattern Recognition for Crime Prevention, Security and Surveillance, pp. 783–791 (2005)
22. Tang, Y., Zhang, Y.Q., Chawla, N.V., Krasser, S.: SVMs modeling for highly imbalanced classification. IEEE Trans. Syst., Man, Cybern. B 39, 281–288 (2009)

Visualization of MIMO Process Dynamics Using Local Dynamic Modelling with Self Organizing Maps

Ignacio Díaz[1], Abel A. Cuadrado[1], Alberto B. Diez[1], Juan J. Fuertes[2],
Manuel Domínguez[2], and Miguel A. Prada[2]

[1] Ingeniería de Sistemas y Automática
Universidad de Oviedo
[2] Instituto de Automática y Fabricación
Universidad de León

Abstract. In this paper we propose a visual approach for the analysis of nonlinear multivariable systems whose dynamic behaviour can be defined in terms of locally linear MIMO (Multiple Input, Multiple Output) behaviours that change depending on a given set of variables (such as e.g. the working point). The proposed approach is carried out in two steps: 1) building a smooth 2-D map of such set of variables using Self-Organizing Maps (SOM) and 2) obtaining a local MIMO ARX (Auto-Regressive with eXogenous input) model for each SOM unit. The resulting structure allows to estimate the process data with an accuracy comparable to other state-of-the-art nonlinear estimation techniques but, in addition, it allows to visualize the MIMO dynamics information stored in the SOM using component planes as done in the SOM literature, bringing the power of visualization to acquire insight useful for process understanding and for control system design. The proposed approach is applied to an industrial-scale version of the well known 4-tank plant, showing a comparison in terms of estimation accuracy with a global linear estimator and with a NARX (Nonlinear Auto-Regressive with eXogenous input) estimator based on a Multi-Layer Perceptron (MLP), as well as, visualizations of MIMO dynamic features such as directionality, RGA (Relative Gain Array), and singular frequency gains for the aforementioned plant.

1 Introduction

Obtaining a good model of the process dynamics is a key part in the control system design workflow, as well as for process understanding towards its optimization and supervision. Many real industrial processes, however, are tightly coupled nonlinear MIMO systems whose dynamics depend heavily on the working point. Moreover, we often lack prior knowledge about their behaviour, such as theoretical (first principles) models, having only large amounts of input-output data, what makes us rely on system identification techniques. There exists a large amount of works in the literature to identify nonlinear systems –see [1,2] for comprehensive overviews and [3] for an exhaustive survey on the field– and

D. Palmer-Brown et al. (Eds.): EANN 2009, CCIS 43, pp. 119–130, 2009.

particularly for MIMO systems [4,5,6]. However, most of them produce black box models which are able to accurately reproduce the system behaviour, but do not leave room for interpretation of the dynamic nature of the process.

In this paper, we propose a MIMO system identification method inspired on the SOM-based local dynamic modelling approach described in [7,8] that, in addition allows to visualize multivariable dynamic features such as singular gains and directions of transfer function matrices. To show the possibilities of the proposed method, we apply it to an industrial-scale version of the well known 4-tank process [9]. The paper is organized as follows. In Sec. 2 a generalization to the MIMO case of the SOM local dynamic modelling approach is described, showing also how this technique can be further exploited as a powerful way to explore the system dynamics. In Sec. 3 experimental results using an industrial-scale 4-tank model are described, comparing the accuracy of the estimator with other approaches, and showing some of its possibilities for visual exploration of the process dynamics. Finally, Sec. 4 provides a general discussion and concludes the paper.

2   Local Linear Modelling of Dynamics

Let's consider the following MIMO parametric Nonlinear AutoRegresive with eXogenous inputs (p-NARX) system

$$\mathbf{y}(k) = f(\varphi(k), \mathbf{p}(k)) \tag{1}$$

where $\varphi(k) = [\mathbf{y}(k-1), \cdots, \mathbf{y}(k-n_y), \mathbf{u}(k), \cdots, \mathbf{u}(k-n_u)]^T$ is the *data vector* that contains past inputs and outputs which are known at sample k, $\mathbf{p}(k) = [p_1(k), \cdots, p_p(k)]^T$ is a vector of parameters, and $f(\cdot, \cdot)$ is a given functional relationship that may be linear or nonlinear, being $\mathbf{u}(k) = [u_1(k), \cdots, u_M(k)]^T$ and $\mathbf{y}(k) = [y_1(k), \cdots, y_L(k)]^T$ the input and output vectors at sample k. Model (1) expresses the present output as a function of past outputs and inputs, as well as of a set of *model parameters* $\mathbf{p}(k)$ that determine a dynamic relationship between the sequences $\{\mathbf{u}(k)\}$ and $\{\mathbf{y}(k)\}$.

2.1   Clustering Dynamics

Let's consider a set of available process variables $\mathbf{s}(k) = [s_1(k), \cdots, s_s(k)]^T \in \mathcal{S}$ that are known (or supposed) to discriminate different dynamical process behaviours. We call these variables *selectors of dynamics* [10].

In a first stage, a SOM with N units is trained in the space of selectors $\mathcal{S}$ to cluster the process dynamics. Each unit i is associated to a s-dimensional prototype vector $\mathbf{m}_i$ in $\mathcal{S}$ and a position vector in a low dimensional regular grid, $\mathbf{g}_i$, in a 2D visualization space for representation purposes. For each vector $\mathbf{s}(k)$ the best matching unit is computed as

$$c(k) = \arg\min_i \{\mathbf{s}(k) - \mathbf{m}_i(t)\} \ . \tag{2}$$

Then, an adaptation stage is performed for every unit i according to

$$\mathbf{m}_i(t+1) = \mathbf{m}_i(t) + \alpha(t)h_{c(k),i}\left[\mathbf{s}(k) - \mathbf{m}_i(t)\right] \qquad (3)$$

where $\alpha(t)$ is the learning rate and $h_{ij} = \exp\left(\frac{\|\mathbf{g}_i - \mathbf{g}_j\|}{2\sigma^2(t)}\right)$ is the neighborhood function, being $\sigma(t)$ the width of the kernel. After convergence, the result is a set of prototype vectors that divide $\mathcal{S}$ into a finite set of Voronoi regions, each defining a different dynamic behaviour.

2.2   Local Model Estimation

Once a SOM has been trained with the $\{\mathbf{s}(k)\}$, a model i may be estimated for each prototype $\mathbf{m}_i$ using all the pairs $\{\mathbf{y}(k), \varphi(k)\}$ such that

$$\|\mathbf{g}_{c(k)} - \mathbf{g}_i\| \leq \sigma_{\text{loc}} \qquad (4)$$

where

$$c(k) = \arg\min_j\{\mathbf{s}(k) - \mathbf{m}_j\} \qquad (5)$$

that is, those pairs whose corresponding selectors $\mathbf{s}(k)$ are mapped onto a neighborhood of unit i of width σ_{loc}. For each SOM unit i, a particular case of the parametric model (1) can be the following local MIMO ARX(n_y, n_u) model

$$y_l(k) = \sum_{j=1}^{n_y} a_{lj}^i y_l(k-j) +$$

$$\sum_{m=1}^{M}\sum_{j=0}^{n_u} b_{lmj}^i u_m(k-j) + B_l^i + \varepsilon(k) \qquad (6)$$

for $l = 1, \cdots, L$, whose parameters $\mathbf{p}_i = \{a_{lj}^i, b_{lmj}^i, B_l^i\}$ can be obtained using standard least squares regression to fit the aforementioned pairs $\{\mathbf{y}(k), \varphi(k)\}$. The set of equations (6) are equivalent to a transfer function matrix relationship

$$\mathbf{Y}(z) = \mathbf{G}^i(z) \cdot \mathbf{U}(z) \qquad (7)$$

where $\mathbf{Y}(z) = \mathcal{Z}\{\mathbf{y}(k)\}$, $\mathbf{U}(z) = \mathcal{Z}\{\mathbf{u}(k)\}$ and $\mathbf{G}^i(z) = \left[G_{lm}^i(z)\right]$ being

$$G_{lm}^i(z) = \frac{b_{lm0}^i + b_{lm1}^i z^{-1} + \cdots + b_{lmn_u}^i z^{-n_u}}{1 - a_{l1}^i z^{-1} - \cdots - a_{ln_y}^i z^{-n_y}} \qquad (8)$$

the transfer function between input u_m and output y_l.

2.3   Retrieval

Once the model is trained, a local transfer function matrix model $\mathbf{G}^i(z)$ is assigned to each neuron i. The problem of retrieval, at sample k, is stated as to get

$\mathbf{y}(k)$ given the data vector $\varphi(k)$ and the dynamic selectors $\mathbf{s}(k)$. This is accomplished in two steps: 1) obtain the best matching unit, $c = \arg\min_i \{\mathbf{s}(k) - \mathbf{m}_i\}$ and 2) apply the local model $\mathbf{G}^c(z)$, equivalent to (6), whose parameters $\mathbf{p}_c = \{a_{lj}^c, b_{lmj}^c, B_l^c\}$ were previouly estimated. Note that depending on the problem (one- or multiple-step ahead prediction) data vector $\varphi(k)$ may contain real or estimated past outputs.

2.4   Visualization of Dynamics

The proposed technique, as shown for SISO (Single Input, Single Output) models by the authors in [10] and [11] is also a powerful technique to visualize the different dynamic behaviours of the system, allowing to obtain insight. Since the resulting model is a SOM map trained on the space of selectors $\mathcal{S}$, it defines a map of the space spanned by the selector samples $\{\mathbf{s}(k)\}$, that can be visually explored using the component planes of each selector as described in the SOM literature [12,13]. The important issue here is that, provided our prior assumption that different selectors actually define different dynamics is true (a common case would be the working point in nonlinear processes), we can explore *how* these dynamics change with the selectors. In effect, since every SOM unit i is associated to a MIMO ARX(n_y, n_u) model $\mathbf{G}^i(z)$, any scalar feature of this MIMO model can be visualized using a new SOM component plane. Thus, some useful component planes would be

- *Frequency maps* defined by the singular values $\sigma_i(e^{j\theta})$ of $\mathbf{G}^i(e^{j\theta})$ at normalized frequency θ. These elements describe the system gains at the singular directions.
- *Direction maps* defined by the elements of singular vectors $\mathcal{U}_i(e^{j\theta}), \mathcal{V}_i(e^{j\theta})$ of $\mathbf{G}^i(e^{j\theta})$ at normalized frequency θ. These vectors represent the output and input directions at which the system attains the singular gains $\sigma_i(e^{j\theta})$ at normalized frequency θ.
- *Relative gain array (RGA) maps* defined by the elements λ_{ij} of the RGA matrix $\Lambda = \mathbf{G}^i(1) * \mathbf{G}^i(1)^{-T}$, where the Schur product '$*$' denotes elementwise multiplication.

3   Experimental Results

3.1   Industrial-Scale 4-Tank Model

The proposed analysis of the dynamic behaviour of multivariable systems has been tested using a quadruple-tank industrial-scale model. This equipment was developed by the group of Automatic Control of the University of León [14]. Figure 1 shows a picture of the industrial-scale model. It keeps the original structure of the quadruple-tank process proposed by [9] but it is built using common industrial instrumentation:

- *Grundfos UPE 25-40* flow pumps equipped with expansion modules *Grudfos MC 40/60* that control them by means of an analog signal.

- *Samson 3226-3760* three-way valves with positioners *Samson 3760* to allow an external signal to control the opening degree of the valves.
- *Endress&Hauer PMC 731* pressure transmitters, for acquisition of the tank level.
- *SMC* digital electrovalves to simulate perturbations on the level of the tank.
- *Opto 22 SNAP Ultimate I/O* data acquisition system, used as an interface between the applications and the industrial-scale model.

A block diagram of the 4-tank, including the variables and conventions used in this paper is shown in Fig. 2. As seen in Fig. 3, the system is modelled as a MIMO process with two inputs, two outputs having external parameters γ_1, γ_2.

Fig. 1. Industrial-scale model of the quadruple-tank process used to test the proposed method

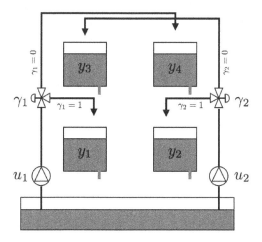

Fig. 2. Schematic diagram of the 4-tank process

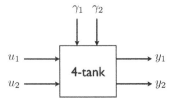

Fig. 3. Functional block diagram of the 4-tank process

3.2   Experiment Description

Let's denote $\mathbf{y} = [y_1, y_2]^T$, the vector of levels of tanks 1 and 2, $\mathbf{u} = [u_1, u_2]^T$ the power of pumps 1 and 2 and $\mathbf{s} = [\gamma_1, \gamma_2]^T$, the positions of valves 1 and 2 (mixing valves) at sample k. In the experiment, control loops are included to control the levels y_1 and y_2 using u_1 and u_2 as control actions, and the system is excited by random steps of variable duration in the reference values for the levels. Along the experiment, the position of both mixing valves were smoothly changed and all the variables were recorded at a rate of 8 samples per second. Finally, since the process is much slower, the data were decimated with a ratio 10:1 resulting in a final sample rate of $f_s = 0.8$ Hz.

3.3   Model Training and Validation

The whole data set was divided into training, validation and test subsets. The space of dynamic selectors (valve positions γ_1 and γ_2) was mapped by training a 10×10 SOM taking $\mathbf{s} = (\gamma_1, \gamma_2)^T$ as input data using the batch algorithm during 100 epochs, varying monothonically the neighborhood kernel width $\sigma(t)$ from 3.33 down to 0.5. In a second stage, a battery of estimations of local MIMO ARX(n_y, n_u) models for all SOM units was carried out ranging all combinations of $n_y = \{2, 3, 4\}$, $n_u = \{2, 3, 4\}$ and $\sigma_{loc} = \{2, 3\}$, to yield the best AIC (Akaike's Information Criterium) model structure –see Table 1.

The best AIC result is achieved for model #18, that corresponds to an ARX(4,4) model, with a $\sigma_{loc} = 3$. The result was a transfer function matrix for each SOM unit i,

$$\mathbf{G}^i(z) = \begin{pmatrix} G_{11}(z) & G_{12}(z) \\ G_{21}(z) & G_{22}(z) \end{pmatrix} \tag{9}$$

whose elements are ARX$(4, 4)$ SISO models. Using the retrieval procedure described above, the resulting local models can estimate the outputs y_1, y_2. Figure 4 shows the estimation performance of the local model approach.

For comparison purposes, an alternative MLP-based NARX model was obtained as a particular case of (1), where

$$\varphi(k) = [\mathbf{y}(k-1), \cdots, \mathbf{y}(k - n_y),$$
$$\mathbf{u}(k), \cdots, \mathbf{u}(k - n_u), \gamma_1, \gamma_2]^T \tag{10}$$

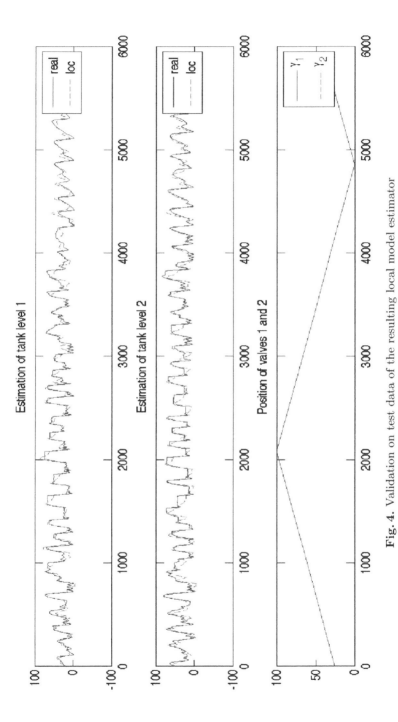

Fig. 4. Validation on test data of the resulting local model estimator

Table 1. Comparison of different model orders

	n_u	n_y	σ_{loc}	RMSE (loc)	RMSE (lin)	AIC (loc)
1	2	2	2	14.1	23	5.29
2	2	2	3	13.7	23	5.24
3	2	3	2	14.2	25.7	5.32
4	2	3	3	14.1	25.7	5.3
5	2	4	2	13.9	23.1	5.26
6	2	4	3	13.5	23.1	5.22
7	3	2	2	13.9	21.9	5.27
8	3	2	3	13.6	21.9	5.23
9	3	3	2	13.7	21.8	5.24
10	3	3	3	13.5	21.8	5.21
11	3	4	2	14.2	24.3	5.31
12	3	4	3	13.9	24.3	5.27
13	4	2	2	14	22.1	5.28
14	4	2	3	13.6	22.1	5.23
15	4	3	2	13.8	21.9	5.25
16	4	3	3	13.5	21.9	5.21
17	4	4	2	13.7	21.5	5.24
* 18	4	4	3	13.3	21.5	5.19

and using an MLP nonlinear mapping to approximate function $f(\cdot, \cdot)$. A battery of trainings using an MLP with two hidden layers was carried out, making three trainings for different combinations of the n_y, n_u orders and number of neurons in both hidden layers, resulting the best structure for a NARX(4,4) model with 8 neurons in both hidden layers.

The accuracy of the three estimators was measured for multiple-step ahead predictions, feeding the estimators only with the input data, that is, the power of both pumps u_1, u_2 and the positions of the valves γ_1, γ_2, while past values of the outputs needed to build the data vector φ were obtained from previous estimations, except for the initial value. The results are summarized in Table 2.

Table 2. Comparison of training, validation and test, errors for local models

	Global (linear) model	Local models	MLP-NARX
Training	19.2	9.43	7.87
Validation	21.5	13.3	–
Test	19.2	10.3	8.12

3.4 Visualization of MIMO Dynamic Features

It can be seen that the local model approach shows a superior performance with respect to the global (linear) MIMO model (that is, a single transfer function matrix for all the data). On the other hand, while its performance is slightly below the MLP-NARX, the SOM local model approach brings all the power of SOM visualization to explore the system dynamics in an insightful way, useful to understand the underlying nonlinear phenomena as well as for control system design.

One interesting kind of visualization are *frequency response maps* [11]. These maps show how the system pumps-to-levels gains at a given frequency vary with respect to γ_1 and γ_2. Since, after training, a transfer function matrix is defined for every unit, two new component planes can be defined for any normalized frequency $\theta \in (0, \pi)$ as the absolute values of the singular values of $\mathbf{G}^i(z)$

$$|\sigma_1(e^{j\theta})|, \quad |\sigma_2(e^{j\theta})| \ . \tag{11}$$

These maps were obtained for $\theta = 0.5$ in Fig. 5. It can be seen how the principal system gains are strongly correlated to $\gamma_1 + \gamma_2$, showing that the system dynamics is strongly influenced by changes in the valves positions, producing larger variations in the levels for large values of the γ's when the powers of the pumps vary at frequency θ.

Also shown in these maps are the singular input and output vectors $\mathcal{V}_1, \mathcal{U}_1$ for singular value σ_1 and the singular input and output vectors $\mathcal{V}_2, \mathcal{U}_2$ for singular value σ_2. Looking at the maps of $\gamma_1 + \gamma_2$ of Fig. 5, it can be observed that for low values of $\gamma_1 + \gamma_2$, where pump 1 feeds mostly tank 4 and pump 2 feeds mostly tank 3, the singular input vector points in the direction of input u_1 and its corresponding output vector points in the direction of output y_2 (tank 2, which is just below tank 4). Conversely, as expected, when the second singular input vector points to input u_2 the second output singular vector points to output y_1. On the other hand, for high values of $\gamma_1 + \gamma_2$, where pumps 1 and 2 feed mostly tanks 1 and 2 respectively, the singular input vector has the same components for u_1 and u_2, and points to the same direction as the corresponding output vector, showing that the highest gain is achieved if the powers of both pumps are equal, producing also equal changes in the levels of both tanks.

Another parameter representing a measure of interaction in multivariable control systems is the relative gain array (RGA) [9]. The RGA can be defined for discrete systems as $\Lambda = \mathbf{G}(1) * \mathbf{G}^{-T}(1)$. Since the elements of each row and

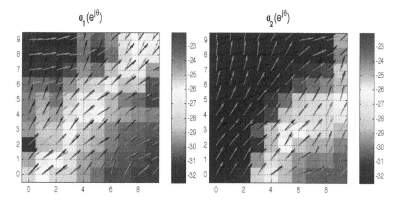

Fig. 5. Frequency maps showing the singular values (gains) $|\sigma_1|$ and $|\sigma_2|$ as well as the singular input and output directions at frequency $\theta = 0.5$. The gains (in dB) are represented by colors and the directions are represented by two arrows (input in grey and output in black) at each SOM unit.

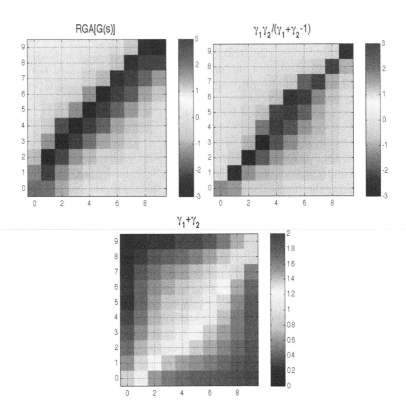

Fig. 6. Empirical RGA obtained from the data (left), and theoretical RGA (center) obtained from relationship (12). Also the component plane of $\gamma_1 + \gamma_2$ is shown (right) for comparison.

column sum up to one, the element $\lambda = \Lambda_{11}$ uniquely determines the RGA. In [9], the author shows that for the 4-tank process this parameter is given by

$$\lambda = \frac{\gamma_1 \gamma_2}{\gamma_1 + \gamma_2 - 1} \ . \tag{12}$$

To show the consistency of the local model approach, we obtained two RGA component planes: 1) computing the RGA for every $\mathbf{G}^i(z)$ obtained *empirically* from the data after the training process, and 2) using the *theoretical* relationship (12) using the γ_1 and γ_2 for every SOM unit i. Both maps along with the map of $\gamma_1 + \gamma_2$ are shown in Fig. 6.

The region on the right map of Fig. 6 where $\gamma_1 + \gamma_2$ is approximately 2 is in agreement with the same region on the left and center maps, where the theoretical and empirical values of the RGA are about 1. That reveals a high correlation between input u_1 - output y_1 and between input u_2 - output y_2. This conclusion admits a highly intuitive physical explanation, because pump 1 sends most of the flow to tank 1 and pump 2 sends most of the flow to tank 2 in these operation points. Likewise, the region where $\gamma_1 + \gamma_2$ adds up to 0, i.e. most of the flow in pumps 1 and 2 is sent to the upper tanks, matches the region where RGA is approximately 0. In this case, it denotes a strong correlation between input u_1 - output y_2 and between input u_2 - output y_1.

4 Conclusion

In this work we have proposed an extension to MIMO systems of a SOM-based local linear modelling algorithm and shown its performance on an industrial-scale prototype of the well known 4-tank plant system. While its performance in terms of estimation accuracy is slightly lower than other state of the art nonlinear identification methods, such as MLP-NARX models, it is much superior to any single global linear MIMO model (i.e. transfer function matrix). Moreover, unlike the MLP-NARX model, the nature of the proposed approach, based on local transfer function matrix models organized in an ordered fashion by the SOM algorithm, makes it possible to explore in a visual way the dynamic features using maps, allowing to discover subtle relationships about its behaviour.

While the results presented in this paper are still in a preliminary stage, they are encouraging and reveal the potencial of the proposed approach to improve supervision, control and optimization of industrial processes, based on a deeper understanding of their dynamic behaviour.

Acknowledgements

This work was supported by the spanish Ministerio de Educación y Ciencia and FEDER Funds, under grants DPI-2006-13477-C01 and DPI-2006-13477-C02.

References

1. Sjoberg, J., Zhang, Q., Ljung, L., Benveniste, A., Delyon, B., Glorennec, P., Hjal-marsson, H., Juditsky, A.: Nonlinear black-box modeling in system identification: a unified overview. Automatica 31(12), 1691–1724 (1995)
2. Chen, S., Billings, S., Grant, P.: Non-linear system identification using neural networks. International Journal of Control 51(6), 1191–1214 (1990)
3. Giannakis, G., Serpedin, E.: A bibliography on nonlinear system identification. Signal Processing 81(3), 533–580 (2001)
4. Dayal, B., MacGregor, J.: Multi-output process identification. Journal of Process Control 7, 269–282 (1997)
5. Peng, H., Wu, J., Inoussa, G., Deng, Q., Nakano, K.: Nonlinear system modeling and predictive control using the RBF nets-based quasi-linear ARX model. Control Engineering Practice (2008)
6. Yu, D., Gomm, J., Williams, D.: Neural model input selection for a MIMO chemical process. Engineering Applications of Artificial Intelligence 13(1), 15–23 (2000)
7. Principe, J.C., Wang, L., Motter, M.A.: Local dynamic modeling with self-organizing maps and applications to nonlinear system identification and control. Proceedings of the IEEE 86(11), 2240–2258 (1998)

8. Cho, J., Principe, J.C., Erdogmus, D., Motter, M.A.: Modeling and inverse controller design for an unmanned aerial vehicle based on the self-organizing map. IEEE Transactions on Neural Networks 17(2), 445–460 (2006)
9. Johansson, K.: The quadruple tank process: A multivariable laboratory process with an adjustable zero. IEEE Transactions on Control Systems Technology 8(3), 456–465 (2000)
10. Díaz Blanco, I., Cuadrado Vega, A.A., Diez González, A.B., Fuertes Martínez, J.J., Domínguez González, M., Reguera, P.: Visualization of dynamics using local dynamic modelling with self organizing maps. In: de Sá, J.M., Alexandre, L.A., Duch, W., Mandic, D.P. (eds.) ICANN 2007. LNCS, vol. 4668, pp. 609–617. Springer, Heidelberg (2007)
11. Díaz Blanco, I., Domínguez González, M., Cuadrado, A.A., Fuertes Martínez, J.J.: A new approach to exploratory analysis of system dynamics using som. applications to industrial processes. Expert Systems With Applications 34(4), 2953–2965 (2008)
12. Kohonen, T., Oja, E., Simula, O., Visa, A., Kangas, J.: Engineering applications of the self-organizing map. Proceedings of the IEEE 84(10), 1358–1384 (1996)
13. Kohonen, T.: Self-Organizing Maps. Springer, Heidelberg (1995)
14. Domínguez González, M., Reguera, P., Fuertes, J.: Laboratorio Remoto para la Enseñanza de la Automática en la Universidad de León (España). Revista Iberoamericana de Automática e Informática Industrial (RIAI) 2(2), 36–45 (2005)

Data Visualisation and Exploration with Prior Knowledge

Martin Schroeder, Dan Cornford, and Ian T. Nabney

Aston University, NCRG, Aston Triangle, Birmingham, B4 7ET, UK
shroderm@aston.ac.uk

Abstract. Visualising data for exploratory analysis is a major challenge in many applications. Visualisation allows scientists to gain insight into the structure and distribution of the data, for example finding common patterns and relationships between samples as well as variables. Typically, visualisation methods like principal component analysis and multi-dimensional scaling are employed. These methods are favoured because of their simplicity, but they cannot cope with missing data and it is difficult to incorporate prior knowledge about properties of the variable space into the analysis; this is particularly important in the high-dimensional, sparse datasets typical in geochemistry. In this paper we show how to utilise a block-structured correlation matrix using a modification of a well known non-linear probabilistic visualisation model, the Generative Topographic Mapping (GTM), which can cope with missing data. The block structure supports direct modelling of strongly correlated variables. We show that including prior structural information it is possible to improve both the data visualisation and the model fit. These benefits are demonstrated on artificial data as well as a real geochemical dataset used for oil exploration, where the proposed modifications improved the missing data imputation results by 3 to 13%.

1 Introduction

Data visualisation is widely recognised as a key task in exploring and understanding data sets. Including prior knowledge from experts into probabilistic models for data exploration is important since it constrains models, which usually leads to more interpretable results and greater accuracy. As measurement becomes cheaper, datasets are becoming steadily higher dimensional. These high-dimensional data sets pose a great challenge when using probabilistic models since the training time and the generalisation performance of these models depends on the number of free parameters.

A common fix for Gaussian models is to reduce the number of parameters and to ensure sparsity in the model by constraining the covariance matrix to be either diagonal, or spherical in the most restricted case. These constraints exclude valuable information about the data structure, especially in cases where there is some understanding of the structure of the covariance matrix.

A good example of this is data from chemical analysis like Gas Chromatography-Mass Spectrometry (GC-MS). When one examines the results of GC-MS runs over

D. Palmer-Brown et al. (Eds.): EANN 2009, CCIS 43, pp. 131–142, 2009.

different samples, one knows that certain compounds are highly correlated with each other. This information can be incorporated into the model with a block-matrix covariance structure. This will help to reduce the number of free parameters without losing too much valuable information. In this paper we will look at a common probabilistic model for data exploration called the Generative Topographic Mapping (GTM) [2]. The standard GTM uses a spherical covariance matrix and we will modify this algorithm to work with an informative block covariance matrix.

The paper has the following structure. First we shortly review models for data visualisation and put them into context. Then we introduce the standard GTM model, extend it to the case of a block covariance matrix and describe how GTM can deal with missing data. Then we present some experiments on artificial data and real data, where we compare the block version of GTM against spherical and full covariance versions and show where the advantages of the models lie. Finally we conclude the paper and point out further areas of research.

2   Data Exploration

A fundamental requirement for visualisation of high-dimensional data is to be able to map, or project, the high-dimensional data onto a low-dimensional representation (a 'latent'space) while preserving as much information about the original structure in the high-dimensional space as possible.

There are many possible ways to obtain such a low-dimensional representation. Context will often guide the approach, together with the manner in which the latent space representation will be employed. Some methods such as PCA and factor analysis [5] linearly transform the data space and project the data onto the lower-dimensional space while retaining the maximum information. Other methods like the Kohonen, or Self Organising, Maps [11] and the related Generative Topographic Mapping (GTM) [2,1] try to capture the topology of the data. Another recent topology-preserving method, the Gaussian Process Latent Variable Model (GP-LVM) [12] utilises a Gaussian Process prior over the mapping function. Instead of optimising the mapping function one considers a space of functions given by the Gaussian Process. One then fits the model by directly optimising the positions of the point in the latent space. Geometry-preserving methods like multi-dimensional scaling [3] and Neuroscale [13] try to find a representation in latent space which preserves the geometric distances between the data points. The later approach can even be extended through a technique called Locally Linear Embedding [15,9], which defines another metric to calculate the geometric distances, one used to optimise the mapping function.

In this paper we will focus on the classical Generative Topographic Mapping (GTM) and an extension which we will call Block Generative Topographic Mapping (B-GTM).

2.1   Standard GTM

The essence of GTM is to try to fit a *density model*, which is constrained to lie on a 2-dimensional manifold, to the data in order to capture the structure

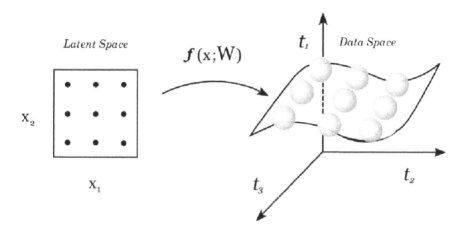

Fig. 1. The non-linear function $f(\mathbf{x}, \mathbf{W})$ defines a manifold S embedded in the data space given by the image of the latent variable space under the mapping $\mathbf{x} \to \mathbf{t}$

in the high-dimensional data space. This manifold consists of a grid of points in the latent space which are connected via a non-linear mapping function to a distorted grid of Gaussian centres in the data space. Thus the GTM may be described as a mixture of Gaussians whose centres are constrained to lie on a manifold. To fit the intrinsic structure in the data, the non-linear mapping function is learned using an Expectation Maximisation (EM) algorithm [6] to maximise the data likelihood.

First one considers a function $\mathbf{t} = f(\mathbf{x}, \mathbf{W})$ which maps points $\mathbf{x}$ in the L-dimensional latent space into an L-dimensional non-Euclidean manifold S embedded within the D-dimensional data space onto the points $\mathbf{t}$, shown for the case $L = 2$ and $D = 3$ in Figure 1.

Defining a probability distribution $p(\mathbf{x})$ for the data points in the latent space will induce a corresponding distribution $p(\mathbf{t}|\mathbf{W})$ in the data space considering a mapping function $\mathbf{t} = f(\mathbf{x}, \mathbf{W})$. The conditional distribution of $\mathbf{t}$ is chosen to be a radially-symmetric Gaussian centred on $f(\mathbf{x}, \mathbf{W})$ with variance β^{-1}. For a given value of $\mathbf{W}$, the distribution $p(\mathbf{t}|\mathbf{W})$ is obtained by integration over the distribution $p(\mathbf{x})$

$$p(\mathbf{t}|\mathbf{W}, \beta) = \int p(\mathbf{t}|\mathbf{x}, \mathbf{W}, \beta)p(\mathbf{t}) \, dx \ . \tag{1}$$

A specific form of $p(\mathbf{x})$ is considered, where $p(\mathbf{x})$ is given by a sum of delta functions centred on the nodes of a regular grid in latent space

$$p(\mathbf{x}) = \frac{1}{K} \sum_{i=1}^{K} \delta(\mathbf{x} - \mathbf{x}_i) \tag{2}$$

in which case the integral in (1) can be evaluated and and the corresponding log likelihood becomes

$$L(\mathbf{W}, \beta) = \sum_{n=1}^{N} \ln \left\{ \frac{1}{K} \sum_{i=1}^{K} p(\mathbf{t}_n | \mathbf{x}_i, \mathbf{W}, \beta) \right\} . \tag{3}$$

To derive the EM algorithm for the GTM model, $f(\mathbf{x}, \mathbf{W})$ is chosen to be a linear-in-parameters regression model of the form $f(\mathbf{x}, \mathbf{W}) = \mathbf{W}\Phi(\mathbf{x})$ with the elements of $\Phi(\mathbf{x})$ consisting of M fixed radial basis functions [4] and $\mathbf{W}$ being a $D \times M$ matrix. Formalising the EM algorithm one computes the responsibility, that each Gaussian generated the given data point in the E-Step:

$$R_{in}(\mathbf{W}_{old}, \beta_{old}) = p(\mathbf{x}_i | \mathbf{t}_n, \mathbf{W}_{old}, \beta_{old}) \tag{4}$$

$$= \frac{p(\mathbf{t}_n | \mathbf{x}_i, \mathbf{W}_{old}, \beta_{old})}{\sum_{j=1}^{K} p(\mathbf{t}_n | \mathbf{x}_j, \mathbf{W}_{old}, \beta_{old})} . \tag{5}$$

Maximising the log likelihood one gets the updates for $\mathbf{W}$ and β in the M-step:

$$\Phi^T \mathbf{G}_{old} \Phi \mathbf{W}_{new}^T = \Phi^T \mathbf{R} \mathbf{T} , \tag{6}$$

with Φ being a $K \times M$ matrix with elements $\Phi_{ij} = \Phi_j(x_i)$, $\mathbf{T}$ being a $N \times D$ matrix with elements t_{nk}, $\mathbf{R}$ being a $K \times N$ matrix with elements R_{in} and $\mathbf{G}$ being a $K \times K$ diagonal matrix with elements

$$G_{ii} = \sum_{n=1}^{N} R_{in}(\mathbf{W}_{old}, \beta_{old}) . \tag{7}$$

and for β:

$$\frac{1}{\beta_{new}} = \frac{1}{ND} \sum_{n=1}^{N} \sum_{i=1}^{K} R_{in}(\mathbf{W}_{old}, \beta) \| \mathbf{W}_{new} \Phi(\mathbf{x}_i) - \mathbf{t}_n \|^2 . \tag{8}$$

2.2   Extension to Block GTM

A novel approach to include prior information about the correlations of variables into the model is to use a full covariance matrix in the GTM noise model and to enforce a block structure onto it. This still results in a relatively sparse covariance matrix, keeping the number of unknown parameters at an acceptable level, while helping the model to fit the data including prior information about its structure. We assume that the groups of highly correlated variables are known a priori. After ordering the variables by their known groups, the covariance matrix will have the following structure:

$$\Sigma = \begin{bmatrix} \Sigma_1 & 0 & \cdots & 0 \\ 0 & \Sigma_2 & \ddots & \vdots \\ \vdots & \ddots & \ddots & 0 \\ 0 & \cdots & 0 & \Sigma_p \end{bmatrix} \tag{9}$$

with Σ_1 to Σ_p being the submatrices of correlated group of variables. We further assume that there is no correlation between variables in distinct groups. The extension of the learning algorithm is straightforward since the only changes occur in the computation of R in the E-step and of Σ in the M-step, where the calculation of the former is straightforward. For the M-step we have to derive the update for the full block covariance matrices Σ_b, $b = 1, \ldots, p$. Taking the derivative of the negative log likelihood with respect to Σ_b we get:

$$\Sigma_b = \frac{1}{ND} \sum_{n=1}^{N} \sum_{k=1}^{K} R_{in} \mathbf{a}_{kn}^{b} \mathbf{a}_{kn}^{T,b} , \tag{10}$$

where $\mathbf{a}_{kn}^{b} = (f(\mathbf{x}_k, \mathbf{W})^b - \mathbf{t}_n^b)$ with $\mathbf{t}_n^b$ being the point $\mathbf{t}_n$ only at the dimensions for block b.

2.3 Extension of GTM for Missing Data Using EM

The EM algorithm can be used to estimate mixture model parameters in the presence of missing data[7]. This work has been extended to the GTM model [18] and it has been shown that the GTM performs quite well as an imputation method [16]. The extension for the block version of GTM is straightforward and will not be discussed here, but further details can be found in [17].

2.4 Stabilising the EM Algorithm

A problem we encountered during the usage of the EM algorithm in conjunction with the block and full GTM on high dimensional data was a collapse of the variance to very small values. This collapse was due to numerical errors when calculating the activation given by (4). The calculation of the activation involves a negative exponential term from the Gaussian, which in very high dimensions becomes very small and thus gets truncated and substituted by 0 due to the limited precision of the floating point representation. This will ultimately result in points where the responsibility R_{ij} is 0 for all Gaussians. The result is a collapse of the variance down to very small values, which in return leads to meaningless results in the projection. To prevent this from happening we use heuristics while calculating the activation as well as the covariance. One simple working heuristic in regards to the responsibilities is to substitute the 0 with a uniform distribution for the k nearest points to the Gaussian centre in question. Another simple working heuristic in regards to the covariance matrix is to add positive noise to the diagonal if the covariance matrix becomes non positive definite.

2.5 Assessing Unsupervised Learning

The dimensionality reduction methods discussed in this report are all examples of unsupervised learning. Thus we cannot tell *a priori* what is the expected or desired outcome. This makes it very difficult to judge which method is the best in the sense of telling us the most about a certain dataset. In the simple case of artificial data one can use prior knowledge about the structure of the data in the original space to quantify the error on the projection. For the more complex case of real data there are various approaches to this problem ranging from different resampling methods [19] to a Bayesian approach using the GP-LVM [8].

In this paper we are going to focus mainly on the following measures of the quality of a projection:

Nearest-Neighbour Label Error (NNLE): The nearest-neighbour label error can only be computed on labelled data, where we know the class of each data point. The idea is to consider the projected data and calculate for each point how many of the k nearest points are in the same class. Then we average the fraction of k-nearest neighbours in the same class over all the points. Finally we average over all the classes as well.

Missing Data or Data Resampling (RMSE): Another alternative approach we developed is driven by the capabilities of the models to estimate missing data. Re-estimating missing data can be seen as a resampling approach [14,19] when the missing data patterns are created artificially and one retains the original value for comparison. Most probabilistic methods can be be modified to incorporate assumptions about missing data. In the case where one introduces missing data into the experiment after the model fit and where the model allows a back projection from the latent into the data space one can then use the model for data imputation. This way once can use missing data as benchmark method for these class of models. To benchmark the different methods against each other we are going to iteratively and piecewise delete every dimension $d = 1, \ldots, D$ from every point and see which estimates the model produce for the missing data. We then calculate the average root mean square error (RMSE) over all the points $n = 1, \ldots, N$, where $\mathbf{t}$ are the original values and $\hat{\mathbf{t}}$ are the estimates:

$$RMSE = \frac{1}{N} \sum_{i=1}^{N} \sqrt{\sum_{d=1}^{D} (\mathbf{t}_{id} - \hat{\mathbf{t}}_{id})^2}.$$

3   Experiments on Artificial Data

To evaluate the effectiveness of block GTM (B-GTM) we carried out comparative experiments with spherical (i.e. standard) GTM (S-GTM), full GTM (F-GTM), and PCA. The data were sampled from a GTM with an 8×8 grid in the latent space. The grid was projected into a higher dimensional space using a 2×2 RBF. network. The weights were randomly sampled from a normal distribution with zero mean and unit standard deviation. Since the RBF was chosen with random weights the restriction to a 2×2 RBF ensured a non-linear but smooth and realistic mapping. The GTM used to generate the data had a block diagonal

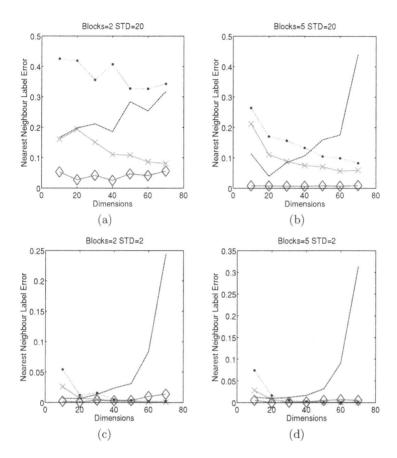

Fig. 2. The nearest neighbour label error on the artificial test data with **high (STD=20) and low(STD=2) structure** for the GTM model with different co-variance structures. PCA=(blue, dotted line with big dot), S-GTM=(green, constant line with X), B-GTM=(red, slashed line with diamond), F-GTM=(black, slashed and dotted line).

covariance matrix and experiments were conducted with a range of levels of variance and correlation. The overall variance of the data varied from 6.45 to 7.55, with covariances around the single Gaussians varying from 2 to 20, denoted by STD, in Figures 2 and 3. The amount of STD is in general controls the amount of structure in the data. A low value for STD means no structure, while a high value means a lot of structure. In each experiment 100 data points were sampled from this GTM and each experiment was conducted 20 times, with a different randomly generated GTM each time.

To calculate the NNLE the 8 × 8 grid was split into 4 classes with the 16 Gaussians in one corner of the grid being defined as one class. The results for this experiment, shown in Figure 2, indicate that in the case of little or no

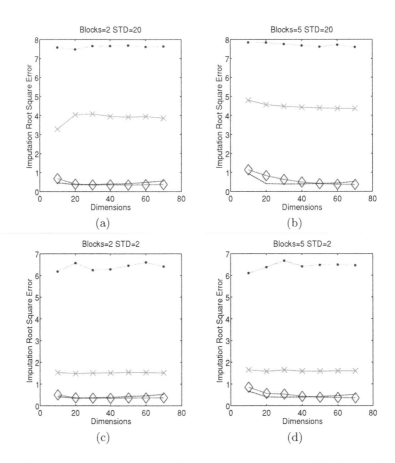

Fig. 3. The root mean square error for imputation on the artificial test data with **high (STD=20) and low(STD=2) structure** for the GTM model with different covariance structures. PCA=(blue, dotted line with big dot), S-GTM=(green, constant line with X), B-GTM=(red, slashed line with diamond), F-GTM=(black, slashed and dotted line).

structure in the data the B-GTM performs as well as, or only slightly worse than, S-GTM, while F-GTM is clearly struggling with increasing dimensionality. However once more structure is present B-GTM clearly outperforms S-GTM, albeit once dimensionality increases the performance difference narrows. The difference in number of blocks is significant as well since more blocks mean fewer parameters for the B-GTM model. This results in improved performance as well.

The RMSE was calculated on the same 20 projections. The results for this experiment, shown in Figure 3, indicate that the block as well as full version of GTM always outperform the S-GTM regardless of the amount of block structure in the data. However the amount by which the spherical GTM is outperformed depends on the amount of block structure. Further there is no significant difference between the performance of block or the full version of GTM. This can be explained due to the nature of imputation as a model validation technique, which only assess the fit of the model in the data space. If the GTM for example is warped on itself and thus gives poor results in the projection space, it will still give good imputation results if it properly covers the data cloud.

4   Experiments on Geochemical Data

To measure the performance of the projection methods for a real-world application domain we used a geochemical data set. The data set consists of 133 different oil samples from the North Sea. The variables are peak heights from gas chromatograph-mass spectrometry featuring up to 61 different alkanes, steranes and hopanes. The data in the data sets have come from a variety of sources. It is normal in petroleum geochemistry, however, for the compound peaks of interest to be identified first, and then measured in height from top to bottom. The bottom is normally given by a realistic background signal (baseline). After the peak identification this is done by the software on the GC-MS system of the source laboratory. As is standard in machine learning we pre-processed all variables to have zero mean and unit variance. Since missing data are a common occurrence in geochemistry, we first excluded very sparse samples. Afterwards we excluded all non complete variables to obtain our complete data set with 61 remaining variables. Reasons for missing data include but are not exclusive to non performance of certain analysis due to cost savings, contaminated samples as well as errors in the peak detection.

Using the data set, 20 random replications of missing data patterns for a given $0 < p_i < 1$ proportion of values missing were generated across all the variables. The patterns are chosen to be completely random, since we do not understand real missing data patterns sufficiently to reproduce them. Multiple Regression Imputation (MRI) [16] as well as mean imputation were chosen as benchmarks to evaluate how the models perform in general.

The apparently missing values are imputed using the GTM model and the Root Mean Square Error (RMSE) is used to assess the performance. The GTM models are all the same (3x3 RBF and 25x25 for the latent space) except for the covariance structure. In the case of the B-GTM the covariance block structure was chosen according to the input from geochemical experts.

The results from the experiment indicate that the B-GTM with the informed block structure is superior to the GTM with a spherical and full covariance matrix as can be seen in Figure 5. The B-GTM consistently, and over a variety of missing data patterns, performs as well as or better than the F-GTM, while

% Missing	5	10	20	30	40	50	60	70
RMSE Improvement in %	0.04	0.13	0.03	0.04	0.13	0.10	0.10	0.04

Fig. 4. The overall improvement of performance in the RMSE from B-GTM to S-GTM

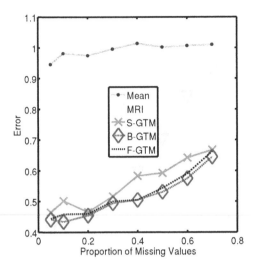

Fig. 5. The imputation root mean square error on North Sea Oil data. S=spherical, B=Block, F=Full.

always outperforming the normally-used spherical version of GTM. The gains in RMSE performance of B-GTM over S-GTM ranged on average between 3 and 13 % as can be see in the table in figure 4. The spread clearly shows that more random replications will be needed for smoother results, however these kinds of benchmarks are highly computational intensive and not the focus of this paper.

5 Conclusions

In this work we have introduced an extension to the GTM algorithm, which allows the user to specify a block structure for the covariance matrix. This block structure can either be the result from extensive data analysis, the inputs from experts in the field and preferably a combination of both. The experiments show that the block extension is a beneficial to the GTM model if the data exhibit a certain amount of structure in the covariance matrix. When structure is present and correctly specified block GTM tends to perform at least as well as spherical or full GTM. In the case of the real geochemical data the gain from B-GTM over S-GTM ranged from 3 to 13 %. The RMSE also proved to be a helpful performance indicator, which however has to be treated with caution since one has to be aware that is only assess the fit of the model in the data space. The experiments further show that the assessment of unsupervised learning is a delicate task. One must understand the model behaviour as well as the restrictions of the model as well as the used performance indicators.

6 Future Work

Future work in this area will be aimed at assessing the possibility of including methods like Bayesian Correlation Estimation [10] into the algorithm in order to learn the correlation structure rather than rely on it being imposed *a priori*. Another approach might be the variational formulation of the GTM algorithm which includes the estimate of the correlation structure as well.

Acknowledgements

MS would like to thank the EPSRC and IGI Ltd. for funding his studentship under the CASE scheme. As well as the geochemical experts from IGI Ltd. (Chris Cornford, Paul Farrimond, Andy Mort, Matthias Keym) for their help and patience.

References

1. Bishop, C.M., Svensen, M., Williams, C.K.I.: Gtm: a principled alternative to the self-organizing map. In: Vorbrüggen, J.C., von Seelen, W., Sendhoff, B. (eds.) ICANN 1996. LNCS, vol. 1112, pp. 165–170. Springer, Heidelberg (1996)
2. Bishop, C.M., Svensen, M., Williams, C.K.I.: Developments of the generative topographic mapping. Neurocomputing 21, 203–224 (1998)
3. Borg, I., Groenen, P.: Modern Multidimensional Scaling: theory and applications. Springer, Heidelberg (2005)
4. Broomhead, D., Lowe, D.: Feed-forward neural networks and topographic mappings for exploratory data analysis. Complex Systems 2, 321–355 (1988)
5. Chatfield, C., Collins, A.J.: Introduction to Multivariate Analysis. Chapman and Hall, Boca Raton (1980)
6. Dempster, A., Laird, N., Rubin., D.: Maximum likelihood from incomplete data via the em algorithm. Journal of the Royal Statistical Society 39, 1–38 (1977)
7. Ghahramani, Z., Jordan, M.I.: Learning from incomplete data. Technical Report AIM-1509 (1994)
8. Harmeling, S.: Exploring model selection techniques for nonlinear dimensionality reduction. Technical report, Edinburgh University, Scotland (2007)
9. de Silva, V., Tenenbaum, J.B., Langford, J.C.: A global geometric framework for nonlinear dimensionality reduction. Science 290, 2319–2323 (2000)
10. Liechty, M.W., Liechty, J.C., Müller, P.: Bayesian correlation estimation. Biometrika 91, 1–14 (2004)
11. Kohonen, T.: Self-Organizing Maps. Springer, Heidelberg (1995)
12. Lawrence, N.D.: A scaled conjugate gradient algorithm for fast supervised learning. Journal of Machine Learning Research 6, 1783–1816 (2005)
13. Lowe, D., Tipping, M.E.: Feed-forward neural networks and topographic mappings for exploratory data analysis. Neural Computing and Applications 4, 84–95 (1996)
14. Moeller, U., Radke, D.: Performance of data resampling methods for robust class discovery based on clustering. Intelligent Data Analysis 10, 139–162 (2006)
15. Roweis, S.T., Saul, L.K.: Locally linear embedding. Science 290, 2323–2326 (2000)

16. Schroeder, M., Cornford, D., Farrimond, P., Cornford, C.: Addressing missing data in geochemistry: A non-linear approach. Organic Geochemistry 39, 1162–1169 (2008)
17. Schroeder, M., Nabney, I.T., Cornford, D.: Block gtm: Incorporating prior knowledge of covariance structure in data visualisation. Technical report, NCRG, Aston University, Birmingham (2008)
18. Sun, Y.: Non-linear Hierarchical Visualisation. PhD thesis, Aston University (2002)
19. Yu, C.H.: Resampling methods: concepts, applications, and justification. Practical Assessment, Research and Evaluation 8 (2003)

Reducing Urban Concentration Using a Neural Network Model

Leandro Tortosa[1], José F. Vicent[1], Antonio Zamora[1], and José L. Oliver[2]

[1] Departamento de Ciencia de la Computación e Inteligencia Artificial
Universidad de Alicante
Ap. Correos 99, E–03080, Alicante, Spain
[2] Departamento de Expresión Grafica y Cartografia
Universidad de Alicante
Ap. Correos 99, E–03080, Alicante, Spain

Abstract. We present a 2D triangle mesh simplification model, which is able to produce high quality approximations of any original planar mesh, regardless of the shape of the original mesh. This model is applied to reduce the urban concentration of a real geographical area, with the property to maintain the original shape of the urban area. We consider the representation of an urbanized area as a 2D triangle mesh, where each node represents a house. In this context, the neural network model can be applied to simplify the network, what represents a reduction of the urban concentration. A real example is detailed with the purpose to demonstrate the ability of the model to perform the task to simplify an urban network.

1 Introduction

The urban design concerns primarily with the design and management of public space in towns and cities, and the way public places are experienced and used. We can see [2] and [10] as general references to introduce general ideas and concepts related to urban development.

One of the more relevant topics that urban design considers is the urban topology, density and sustainability. We want to remark the aspect of density. The geographical area were the authors of this paper are living presents a very high level of urban density. We are convinced that a change in the pattern of urban development is absolutely desirable. In this sense what is proposed is a reduction in the concentration of buildings using a neural network model.

See, for example, [6] for a detailed description on surface simplification methods. Several geometric problems for sets of n points require to triangulate planar points sets [8]. This motivates an increase attention in problems related to the triangulation of planar point sets and 2D triangle meshes, as we can see in recent works [13], [4], [3], [5], and [12].

A recent topic in mesh simplification is the preservation of topological characteristics of the mesh and of data attached to the mesh [14].

D. Palmer-Brown et al. (Eds.): EANN 2009, CCIS 43, pp. 143–152, 2009.

Self-organizing networks are able to generate interesting low-dimensional representations of high-dimensional input data. The most well known of these models is the Kohonen Feature Map [9]. It has been used in the last decades to study a great variety of problems such as vector quantization, biological modeling, combinatorial optimization and so on.

In this context, the model we present here is applied to reduce urban concentration of a real geographical area. More exactly, we propose a 2D triangle mesh simplification model, taking as a basis the fact that we consider the representation of an urbanized area as a 2D triangle mesh, where each node represents a house and we establish links between nodes in order to create networks or two-dimensional meshes. Then, in fact, we are going to work with planar meshes or networks and the nodes in these networks will represent any type of buildings (houses, cottages, and so on).

2   The Neural Network Model

The artificial neural network that we present in this paper is able to simplify any 2D triangle mesh, regardless of the irregular shape of the original mesh. The primary objective of the model is to obtain a planar point set that will constitute the vertices of the simplified mesh and, finally, a triangulation process based on a comparison between the original triangle mesh and the reduced set of vertices is performed. After this process, a new simplified two-dimensional mesh is generated with a similar shape as the original one. The model is appropriate to approximate or simplify any two-dimensional triangle mesh, even presenting irregular shapes.

The model may be subdivided in two different phases. The first one consists on getting the set of vertices to simplify the original mesh while the second phase consists on the triangulation of the vertices obtained in the previous step with the aim to generate the final simplified mesh.

The main idea of the algorithm coincides with the model GNG3D used to reduce three-dimensional objects (see [1] for a detailed description of the model). However, some deep modifications and simplifications have been introduced in the optimization algorithm of the GNG3D model, in order to adapt it to the case of working with point sets in the plane. It has also been reduced the number of parameters that control the training process.

We consider, as a starting point, a 2D triangle mesh consisting of

- a set $A = \{n_1, n_2, \ldots, n_N\}$ of vertices or nodes,
- a set $T = \{t_1, t_2, \ldots, t_L\}$ of triangles among node pairs.

We can say that the set T constitutes the set of triangles or faces that make up the original 2D mesh. Consequently, each element of the set T is given by

$$t_i = \{n_j, n_k, n_l\}, \quad n_j, n_k, n_l \in A.$$

It is important to remark that we are going to fix the number of vertices of the simplified mesh; therefore, if we set the number of vertices of the simplified mesh as M, then the objective is to find a set of vertices

$$K = \{k_1, k_2, \ldots, k_M\},$$

where k_i, for $i = 1, 2, \ldots, M$ are the new nodes in the simplified triangle mesh. Note that always $M < N$ and M is fixed. The set of vertices K must be computed by means of the following algorithm.

Phase 1: the self-organizing algorithm.
INIT: Start with M points k_1, k_2, k_M at random positions $w_{k_1}, w_{k_2}, \ldots, w_{k_M}$ in R^2. Initialize a local counter to zero for every point.

1. Generate an input signal ξ that will be a random point $n_i \in A$ from the original mesh.
2. Find the nearest node s_1.
3. Find the second and third nearest nodes, s_2 and s_2 to the input signal.
4. Increment the local counter of the winner node s_1.
5. Move s_1 towards ξ by fractions ϵ_{win} respect to the total distance

$$\Delta w_{s_1} = \epsilon_{win}(\xi - w_{s_1}).$$

6. Move s_2 and s_3 towards ξ by fractions ϵ_n respect to the total distance:

$$\Delta w_{s_2} = \epsilon_n(\xi - w_{s_2}),$$

$$\Delta w_{s_3} = \epsilon_n(\xi - w_{s_3}).$$

7. Repeat steps 1 to 6 λ times, with λ an integer.
   - If the local counter for any $k_i, i = 1, 2, \ldots, M$ is zero, then remove this node and add a new node between the node with higher local counter value and any node adjacent to it.
   - If the local counter for every $k_i, i = 1, 2, \ldots, M$ is greater than zero, then continue and repeat steps 1 to 6.
8. Stop when the maximum number of iterations has been reached.

Phase 2: the triangulation process.
INIT: Consider the set A of the original nodes, T the triangles of the original 2D triangle mesh, and K the set of the nodes obtained by the above self-organizing algorithm.

1. Associate each node of the original mesh with a node of the set K. For every n_i, for $i = 1, 2, \ldots, N$, find $j \in \{1, 2, \ldots, M\}$ such that

$$|w_{n_i} - w_{k_j}| < |w_{n_i} - w_{k_l}|, \quad l \in \{1, 2, \ldots, M\},$$

where w_{n_i} represents the position of the node n_i. Save (n_i, k_j). We say that k_j is the node associated to n_i.

2. Change the nodes of the original triangles by their associated nodes. For every $t_i = \{n_{i_1}, n_{i_2}, n_{i_3}\} \in T$, substitute

$$\{n_{i_1}, n_{i_2}, n_{i_3}\} \longrightarrow \{k_{j_1}, k_{j_2}, k_{j_3}\},$$

where $k_{j_1}, k_{j_2}, k_{j_3}$ are the associated nodes of $n_{i_1}, n_{i_2}, n_{i_3}$, with $j_1, j_2, j_3 \in \{1, 2, \ldots, M\}$.

- If $k_{j_1} \neq k_{j_2} \neq k_{j_3}$, then save $t_i = \{n_{i_1}, n_{i_2}, n_{i_3}\}$.
- If $k_{j_1} = k_{j_2}$, or $k_{j_1} = k_{j_3}$, or $k_{j_2} = k_{j_3}$, then continue.

3. Graph the set

$$C = \{t_i = \{n_{i_1}, n_{i_2}, n_{i_3}\}, \quad k_{j_1} \neq k_{j_2} \neq k_{j_3}\}.$$

4. If some node is isolated we add a new triangle, linking this node with their adyacent nodes.

3   Some Highlights of the Model

This first phase of the model can be seen as a training process, where a set of nodes representing the new vertices of a planar mesh is obtained. So far, nothing about the triangles or faces of the original planar mesh has been mentioned. The second phase of the model is a triangulation process developed with the aim to reconstruct the new mesh. Our proposal constitutes a post-process which uses the information provided by the self-organizing algorithm and the information about the nodes and triangles of the original mesh.

Respect to the self-organizing algorithm presented in the Phase 1 of the model, we want to consider, in broad terms, that although the central idea of the self-organizing models to develop an unsupervised incremental clustering algorithm is shared, some basic considerations must be remarked from this algorithm.

- Initially we take M nodes randomly as a starting point, what represents a quite different initial approach to the proposed by the GNG3D model, where the number of nodes of the final mesh was totally controlled by the user throughout the training process of the construction of the mesh. This is a critical aspect because it eliminates the entire process of adding and removing nodes. Therefore, we focus on the process of self-learning the shape of the original mesh.
- Throughout this process we do not take into account the triangles or faces that form the initial mesh.
- The parameters involved in the training process are ϵ_{win}, ϵ_n and λ. The parameter ϵ_{win} is related to the displacement of the winner node, while ϵ_n is related to the displacement of the neighbor nodes in the plane. The parameter λ is introduced in order to be sure that any node of the initial set of random points will stay isolated during all the execution of the algorithm.

Some clarification is necessary regarding the parameter λ and the point 7 of the self-organizing algorithm. As we choose in a random way the initial geometric positions of the initial M nodes (point 1), it is possible that some of these initial points are generated in positions far away from the area covered by the original mesh. Accordingly, no signal will be near these points and, therefore, these nodes will never be *winners* in the training process. Consequently, its local counter will remain zero as the number of iterations increase. So, this is the way we have to detect when a node is outside the area of the initial triangle mesh. This is the reason to introduce the condition in the point 7 of the self-organizing algorithm.

In the self-organizing algorithm the positions of the initial vertices are modified with the purpose to learn the shape of the original mesh. After this process we have no information about the triangles that make up the simplified mesh. We only have the geometric position of the new vertices; therefore, an efficient algorithm may be implemented to reconstruct the 2D mesh preserving the original shape.

The most popular triangulation method is the Delaunay triangulation [7], which basically finds the triangulation that maximizes the minimum angle of all triangles, among all triangulations of a given point set. We follow, as it has been described in the algorithm, a rather different approach. The idea underlying our approach is to establish a special equivalence relation in the original set of vertices. The equivalence relation consists on assigning to every node of the set A a node of the set K which is nearest to it. After this association, we will have some equivalence sets because some nodes of the original set A will have as a representative the same node of the set K (remember that the cardinal of K is much less that the cardinal of A). After this concordance process, we only have to determine the triangles with different representatives to determine the nodes of the set K that must be linked to graph the final simplified mesh.

We have added a condition at the end of the triangulation algorithm. The primary objective of this condition is to avoid the existence of holes or unconnected regions in the final mesh. When the shapes of the original meshes are not too complicated, we have no problem in the final result with holes or isolated regions; however, as it has been observed in the examples performed, when the shape of the original mesh is really complex, we can have problems with the appearance of unconnected regions or isolated vertices, specially when $M \ll N$.

4   A Real Example

To assess the performance of the model proposed in this paper, several experiments were conducted using some original 2D triangle meshes. In this section we want to analyze in detail the urban reduction and simplification of an area representing a real residential area with geographical coordinates $(38.25, -0.7)$ (latitude and longitude). The residential area to study is shown in Figure 1.

In the residential area shown in Figure 1, we identify each of the houses or buildings with a node in the mesh and perform a triangulation process with these nodes, obtaining a two-dimensional grid made up of planar triangles, as we can see in Figure 2. This initial mesh has 293 vertices or nodes (houses) and 352 triangles.

Note that our goal is to achieve, while maintaining the same topology of the mesh, an optimal distribution of the new nodes (houses) with the aim to create more space between them to establish new services or enhance green spaces. So, we consider, as an initial parameter, that the final mesh will be composed by 125 nodes, that is, we have that $N = 273$ and $M = 125$. Therefore, the new

Fig. 1. Residential area to be simplified

Fig. 2. Two-dimensional mesh generated from Figure 1

design of the residential area will have 125 houses, with the same shape as the original one.

To reach this objective we apply our simplification model to the mesh in Figure 2.

The starting point of the self-organizing algorithm is to place K points, $k_1, k_2, \ldots, k_K$, at random positions in the graphic. We take, for this example, $K = 125$ points because it is the number of nodes (houses) that we want for the simplified network. After generating, at random, the initial positions of the 125 nodes of the simplified mesh, we continue running the training process in the self-organizing algorithm. In this case, after 5000 iterations, we check the local counters of the $n_1, \ldots, n_{293}$ nodes of the original network; then, we proceed to remove the vertices that have not been winners in any iteration. After this step, we follow with the iterations until the stopping criteria is reached.

Figure 3 shows us the positions of the initial nodes, after the process of removing isolated nodes is completed. It may be remarked that thanks to the process of removing isolated vertices as it has been described above, we get that all the initial nodes are close to the area covered by the original network. This is very important in the further development of the algorithm.

Now, we continue the training process with the rest of the iterations. After performing 750000 iterations, we stop the process, obtaining the final positions of the vertices for the simplified network. The experimental result obtained is shown in Figure 4.

At this moment it is possible to carry out the triangulation process from the information provided by the self-organizing algorithm. The basis of triangulation process is the comparison between the original nodes of the network and the new nodes. The reconstruction of the simplified mesh is shown in Fig. 5.

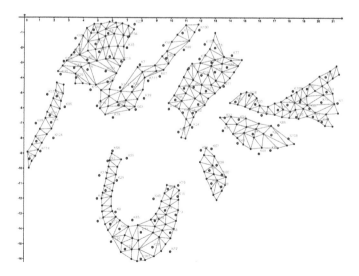

Fig. 3. The original 2D triangle mesh with the initial 125 nodes placed at random

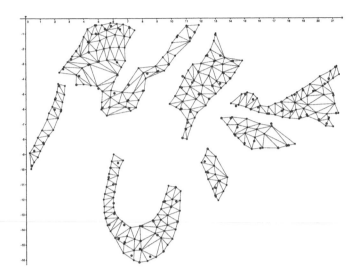

Fig. 4. The original 2D triangle mesh with the final 125 nodes placed after running the self-organizing algorithm

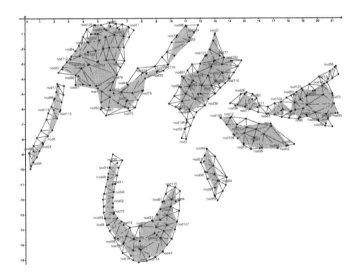

Fig. 5. The reconstructed 2D triangle mesh

The simplified mesh of Fig. 5 has been drawn to show how the mesh conforms to the final topology of the original network. In this example we see the efficiency of the model to approximate the shape of the original mesh. But, what is more relevant, Fig. 5 shows the new design that could have the residential area studied. By means of this figure we visualize a new distribution of the nodes (houses) with the property that the nodes are now located in the best places to approximate the shape of the original area, maximizing the space between them, with the advantages that this gives us regarding the creation of green spaces, bicycle tours, sporting facilities, ...

To carry out this example, it has been necessary to specify a set of parameters to run the self-organizing algorithm. The parameters involved in it are the following:

- ϵ_{win} is related to the displacement of the winner node.
- ϵ_n is related to the displacement of the neighbor nodes.
- λ was introduced in order to be sure that any node of the initial set of random points will stay isolated during all the execution of the algorithm.

There is no theoretical result that provides us a parameter set that can always produce the best results for the simplified mesh. Consequently, the determination of the set of parameters to run the algorithm is obtained experimentally. The values of these parameters used in our examples have been $\epsilon_{win} = 0.8$, while $\epsilon_n = 0.1$. The parameter λ is not fixed, as it depends on the number of nodes N; we used $\lambda = 13 \cdot N$.

In all the original 2D meshes studied with different sizes and shapes, the results have been equally successful in the sense that we simplify the original mesh, preserving the topology of the initial network. No matter how irregular it is the shape of the original mesh, as the self-organizing algorithm always *learns* how the shape is and place the initial nodes in the original mesh.

5   Conclusion

A new 2D triangle mesh simplification model has been introduced with the central property of preserving the shape of the original mesh. The model presented consists of a self-organizing algorithm which objective is to generate the positions of the nodes of the simplified mesh; afterwards, a triangulation algorithm is carried out to reconstruct the triangles of the new simplified mesh.

Among the various applications of the model we have chosen one related to urban density. An example using a real geographical area is shown to demonstrate that the model is able to reduce urban concentration, keeping the exact topology of the terrain and creating more space between buildings.

Some experimental results using original meshes with irregular shapes show that the final simplified meshes are generated very fast and with a high level of approximation respect to the original ones, no matter how irregular shapes they present.

References

1. Alvarez, R., Noguera, J., Tortosa, L., Zamora, A.: A mesh optimization algorithm based on neural networks. Information Sciences 177, 5347–5364 (2007)
2. Barnett, J.: An introduction to urban design. Harper and Row, New York (1982)

3. Bereg, S.: Transforming pseudo-triangulations. Information Processing Letters 90(3), 141–145 (2004)
4. Bose, P., Hurtado, F.: Flips in planar graphs. Computational Geometry: Theory and Applications 42(1), 60–80 (2009)
5. Castelló, P., Sbert, M., Chover, M., Feixas, M.: Viewpoint-based simplification using f-divergences. Information Sciences 178(11), 2375–2388 (2008)
6. Cignoni, P., Montani, C., Scopigno, R.: A comparison of mesh simplification algorithms. Computer Graphics 22(1), 37–54 (1998)
7. Delaunay, B.: Sur la sphere vide. Bull. Acad. Sci. USSR VII:Class. Scil, Mat. Nat., 793–800 (1934)
8. Held, M., Mitchell, J.S.B.: Triangulating input-constrained planar point sets. Information Processing Letters 109, 54–56 (2008)
9. Kohonen, T.: Self-Organizing formation of topologically correct feature maps. Biological Cybernetics 43, 59–69 (1982)
10. Larice, M., MacDonald, E. (eds.): The Urban Design Reader, Routledge, New York (2007)
11. Luebke, D.P.: A developer's survey of polygonal simplification algorithms. IEEE Computer Graphics and Applications 21(3), 24–35 (2001)
12. Nguyen, H., Bunkardt, J., Gunzburger, M., Ju, L., Saka, Y.: Constrained CVT meshes and a comparison of triangular mesh generators. Comput. Geom. 42(1), 1–19 (2009)
13. Sharir, M., Welzl, E.: Random triangulations of planar point sets. In: Proceedings of the twenty-second annual symposium on Computational geometry, pp. 273–281 (2006)
14. Vivodtzev, F., Bonneau, G.P., Le Texier, P.: Topology Preserving Simplification of 2D Non-Manifold Meshes with Embedded Structures. Visual Comput. 21(8), 679–688 (2005)

Dissimilarity-Based Classification of Multidimensional Signals by Conjoint Elastic Matching: Application to Phytoplanktonic Species Recognition

Émilie Caillault, Pierre-Alexandre Hébert, and Guillaume Wacquet

Université Lille Nord de France,
Laboratoire d'Analyse des Systèmes du Littoral, EA 2600, ULCO,
Maison de la Recherche Blaise Pascal,
50 rue Ferdinand Buisson,
B.P. 699,
F-62228 Calais Cedex, France
name@lasl.univ-littoral.fr

Abstract. The paper describes a classification method of multidimensional signals, based upon a dissimilarity measure between signals. Each new signal is compared to some reference signals through a conjoint dynamic time warping algorithm of their time features series, of which proposed cost function gives out a normalized dissimilarity degree. The classification then consists in presenting these degrees to a classifier, like k-NN, MLP or SVM. This recognition scheme is applied to the automatic estimation of the Phytoplanktonic composition of a marine sample from cytometric curves. At present, biologists are used to a manual classification of signals, that consists in a visual comparison of Phytoplanktonic profiles. The proposed method consequently provides an automatic process, as well as a similar comparison of the signal shapes. We show the relevance of the proposed dissimilarity-based classifier in this environmental application, and compare it with classifiers based on the classical DTW cost-function and also with features-based classifiers.

1   Introduction

The survey of marine ecosystem is a major current concern in our society since its impact is essential for many domains: ecology (biodiversity, production, survey), climate, economy (tourism, resources control, transport). In 2000, Directive DCE adopted by the European Parliament [1] defines the Phytoplankton as a biologic factor for marine quality assessment. In this context, we propose an automatic dissimilarity-based classification method designed to assess the marine water quality by Phytoplankton species classification and counting.

The available signals are fluorescence and scatter parameter scans of each particle detected by a flow cytometer in a marine sample. So our problem comes down to the classification of multidimensional signals, whose shape profile is class-specific. Up to date, this classification is made by visual comparison of the obtained profiles and references ones, or by the inverted microscope method [2][3]. The major difficulty of discrimination task is that Phytoplankton is a live vegetal species. So its internal structure (pigments, size, nucleus position) varies according to its belonging group but also

D. Palmer-Brown et al. (Eds.): EANN 2009, CCIS 43, pp. 153–164, 2009.

and above all according to its physiological condition (life cycle, cell or colony) and its environment [4][5]. To make a Phytoplankton classifier system robust to these variabilities, it appears relevant to use an elastic measure to compare two signal profiles. So our approach is based on a classical elastic matching from Sakoe and Chiba (Dynamic Time Warping DTW [6]). This method was largely tested, initially for speech recognition (comparison of 1D time-frequency amplitude patterns) or for handwritten pattern recognition (1D spatial matching) [7].

In order to get a more understandable qualitative measure, simple and comprehensible for any biologist, we adapt the matching cost of this algorithm so as to get a [0, 1]-normalized dissimilarity degree that deals with multidimensional time signals.

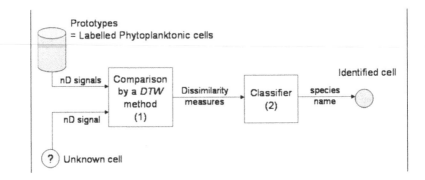

Fig. 1. Scheme of the recognition system for Phytoplanctonic cells

Figure 1 presents the global scheme of the recognition system based on a dissimilarity measure applied to Phytoplankton characterization. It is composed of two classical parts:

1. a feature-like extraction module which computes for the unknown cell a dissimilarity vector in relation to some reference cells;
2. a standard classifier which takes in entry this vector and gives in output the recognized species name.

Next section describes the proposed accomodation of Sakoe and Chiba's algorithm to get a [0,1]-dissimilarity degree between nD signals. Section 3 presents the experimentation protocol and results that show the efficiency of the system with different classifiers. Different variants are tested and compared first with features-based classifiers, then with classical DTW algorithm, in term of recognition rate.

2 Dissimilarity Measure for Multidimensional Signals by Conjoint Elastic Matching

2.1 Comparison of Two 1D Signals by the Classical Method *Dynamic Time Warping*

Dynamic Time Warping (DTW) method proposed by Sakoe and Chiba [6] gives a distance measure for two time series, whose lengths are not necessarily equal. This measure represents the quantity of geometrical distortion needed to match both curves, regardless of some time distortions.

More precisely, the method matches the points of both signals, and defines their matching cost as the mean-distance between the paired points. For example, in the ideal case where all paired points are identical, the matching is perfect, and the cost is consequently zero. The softness of the algorithm comes from its ability to pair some points time shifted, with a cost equal to zero.

Let $X = \{(x_i), i = 1, \ldots, n_x\}$ and $Y = \{(y_j), j = 1, \ldots, n_y\}$ be the two signals to be compared, with i and j their time index, and n_x and n_y their respective length. We first consider monodimensional signals: each value x_i or y_j belongs to $\Re$.

The algorithm builds a matching $P = \{(i_k, j_k), k = 1, \ldots, n_k\}$ between the points of signals X and Y, according to some time conditions. The resulting matching is defined as the one minimizing the following weighted mean distance C between paired points, based upon some distance d and a weight vector W:

$$C(X,Y,P,W) = \frac{\sum_{k=1}^{n_k} d(x_{i_k}, y_{j_k}).w(k)}{\sum_{k=1}^{n_k} w(k)} = \frac{Dist(X,Y,P,W)}{\sum_{k=1}^{n_k} w(k)}. \tag{1}$$

Selected conditions of the pairing in this DTW variant are the following:

1. End-Points conditions: first (and last) points of both signals are paired: $(1,1) \in P$ (and $(n_x, n_y) \in P$);
2. Continuity conditions: all points are matched;
3. Monotonicity conditions: pairs are time ordered: $i_{k-1} \le i_k$ and $j_{k-1} \le j_k$.

According to these conditions, each possible pairing may be represented as a path in the bidimensional space of the pairs $\{(i,j), i = 1, \ldots n_x, j = 1, \ldots n_y\}$, i.e. the full set of the possible pairs of points of X and Y. The algorithm then backtracks the optimal matching, going from initial pair $(1,1)$ to final pair (n_x, n_y) (condition 1) that minimizes the cost C.

Sakoe and Chiba showed that cost C can only be optimized by dynamic programming if denominator $\sum_{k=1}^{n_k} w(k)$ does not depend from matching P. This may be obtained by letting the weigth sum equal to $n_x + n_y$, n_x or n_y, for example.

Considering that the weights verify this condition, then an optimal path is a path which minimizes the accumulated distance $Dist$.

If the optimal paths leading to pairs $(i-1, j)$, $(i, j-1)$ and $(i-1, j-1)$ are supposed to be known, then the optimal path leading to pair (i, j) may easily be defined as the one of the three previous paths whose cost is minimal, followed by pair (i, j). Its cost $Dist$ may be computed in the same way. Now, the optimal paths to the pairs $(1, \ldots)$ and $(\ldots, 1)$ are known thanks to conditions 2 and 3. Then, a recursive optimization is used to compute the path corresponding to the best matching between signals X and Y, as well as its cost $Dist$. In practice, it consists in sequentially computing the (n_x, n_y)-sized matrix of costs $Dist(i, j)$, which measures the minimal cost of the path leading to pair (i, j). The final cost is consequently and directly obtained in the last element (n_x, n_y) of the matrix.

Here is given the part of the algorithm DTW (cf. Algorithm 1) allowing a quick computation of cost C of the optimal path. The path itself may then be retrieved following the way the least cost was computed, from last pair (n_x, n_y) to first pair $(1,1)$. This is generally used to normalize final cost $Dist(n_x, n_y)$, so as to get the mean distortion measure by matched pair.

In order to get the accumulated distance $Dist$ minimization equivalent to the cost C minimization, the weigths are defined according to the *symetric* solution proposed by Sako and Chiba. Let (i_k, j_k) be the k-th pair of matching P:

- $w(k) = 2$, if $k = 1$ or if $(i_{(k-1)}, j_{(k-1)}) = (i_k - 1, j_k - 1)$;
- $w(k) = 1$, otherwise.

Then $\sum_{k=1}^{n_k} w(k) = n_x + n_y$ does not depend on P, and optimizing $Dist(n_x, n_y)$ makes C optimized too.

Algorithm 1. DTW algorithm computing the accumulated distance of the best matching

$Dist(1,1) = 2.d(x_i, y_j)$
for all $i = 2, \ldots, n_x$ **do**
$\quad Dist(i,1) = Dist(i-1,1) + d(x_i, y_1)$
end for
for all $j = 2, \ldots, n_y$ **do**
$\quad Dist(1,j) = Dist(1, j-1) + d(x_1, y_j)$
end for
for all $i = 2, \ldots, n_x$ **do**
$\quad$ **for all** $j = 2, \ldots, n_y$ **do**
$\quad\quad Dist(i,j) = \min\{Dist(i-1,1) + d(x_i, y_j), Dist(i, j-1) + d(x_i, y_j), \ldots$
$\quad\quad \ldots, Dist(i-1, j-1) + 2.d(x_i, y_j)\}$
$\quad$ **end for**
end for
return $Dist(n_x, n_y)$

2.2 Neighborhood Restrictions of DTW Algorithm

The previous version of DTW algorithm allows extremely soft matchings without penalty: the first point of a signal may indeed be matched to the last point of the other one. This is an extreme distortion, that may be avoided by narrowing the possible matched points thanks to a limited time window.

The strictest restriction consists in making each point of the longest signal - denoted X - possibly paired to a unique point of the other signal - denoted Y: the time nearest, once the length of the curves are fitted. This is a "linear" DTW:

$$P = \left\{ (i, j_i^*); i \in \{1, \ldots, n_x\}, j_i^* = E\left(1 + \frac{(i-1)}{(n_x - 1)}(n_y - 1)\right) \right\}, \tag{2}$$

where E denotes the round function.

In this particular case, the distortion is global: the one which fits the duration of both signals. No optimization is required to get the distortion cost.

In order to allow local time distortion, a less restrictive version could be prefered, which limits the matchings through a time window. Its size may be defined as a ratio p of the whole duration:

$$\forall (i,j) \in P, \ (i,j) \in \{(i,j_i); i \in \{1,\ldots,n_x\}, j_i \in [j_i^* - p.n_y, j_i^* + p.n_y]\}, \qquad (3)$$

where j_i^* denotes the indice defined in the previous linear matching.

2.3 Dissimilarity Measure of Positive Signals

The cost function provided by DTW is a relative measure, which can not be easily interpreted by itself: it is a mean distance, which depends on the intensities of both signals. In order to make the response similar to the one of a human expert, we prefer a bounded measure of dissimilarity, between 0 and 1. We then propose to replace the distance d with a dissimilarity s, built as a ratio of distances:

$$s(x_{i_k}, y_{j_k}) = \frac{d(x_{i_k}, y_{j_k})}{\max\{d(x_{i_k},0), d(y_{j_k},0)\}}. \qquad (4)$$

In order to make this ratio consistent, we suppose the signals to be positive, otherwise the dissimilarity degree could exceed 1.

Using this measure, the cost function C becomes a mean dissimilarity between paired points, and consequently a global dissimilarity measure for signals.

2.4 Conjoint Elastic Matching of nD Signals

We now consider nD signals $\bar{X} = \{(\bar{x}_i), i = 1,\ldots,n_x\}$ and $\bar{Y} = \{(\bar{y}_j), j = 1,\ldots,n_y\}$, whose all time measures $\bar{x}_i = \{(x_{ic}), c = 1,\ldots,n_c\}$ and $\bar{y}_j = \{(y_{ic}), c = 1,\ldots,n_c\}$ belong to $(\Re^+)^{n_c}$. To sum up, each nD signal consists of n_c monodimensional positive signals, identically sampled.

In the classical DTW, we simply consider a Manhattan distance measure:

$$d(\bar{x}_i, \bar{y}_j) = \sum_{c=1}^{n_c} d_{L_1}(x_{ic}, y_{ic}), \qquad (5)$$

with d_{L_1} the L_1-distance. The choice consists in accumulating the distortion measures over the n_c curves.

In our dissimilarity version of DTW, we accumulate dissimilarity measures instead of L_1-distances for the n_c positive curves, and we then normalize the result:

$$s(\bar{x}_i, \bar{y}_j) = \frac{1}{n_c} \sum_{c=1}^{n_c} s(x_{ic}, y_{ic}). \qquad (6)$$

2.5 Matching Visualizations

In order to visualize the quality of DTW matching for two 1D signals, both signals are usually plotted on the same figure, while paired points are connected with a segment. The weakness of this technique is that it does not help to assess the dissimilarity between the signals.

Then, we propose an other way to visualize both signals X and Y. Each point of pair $(i_k, j_k), k \in \{1,\ldots,n_k\}$ of the optimal matching P is represented by a bidimensional point $\hat{x}_k$ (or $\hat{y}_k$). The set of pairs is totally ordered, then X-axis values of points $\hat{x}_k$ and $\hat{y}_k$ are set to k, following this time order. Y-axis values simply are the 1D measures x_{i_k} and y_{j_k}, then:

$$\hat{x}_k = (k, x_{i_k}) \; ; \; \hat{y}_k = (k, y_{j_k}). \tag{7}$$

Plotting points this way makes easier the perception of the distances or dissimilarities accumulated over the pairs of P, because they are directly measured along the Y-axis. Moreover time distortions are visualized. Note that we used dotted lines to differentiate the time distortions from any initial constant part of the curves.

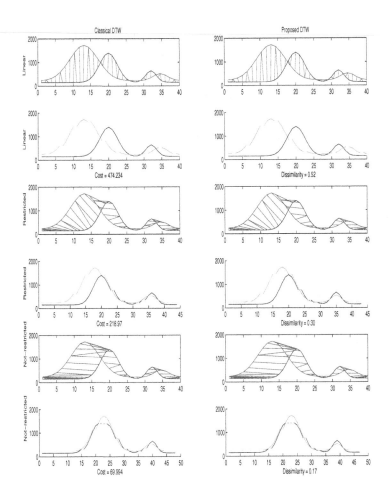

Fig. 2. Different DTW variants applied to two artificial 1D signals

Figure 2 shows the comparison of two artificial 1D curves according to the variants of DTW algorithm previously described; results are presented as follows:

- two columns: left, Sakoe and Chibas's original algorithm; right, our variant with a bounded dissimilarity degree;
- three couples of rows, for the different ways of limiting the neighborhood: first the "linear", then the "p-restricted", and finally the "no-restricted" variant. Results are presented in the classical way, then in the way we propose.

Matchings visualized in Figure 2 show how similar both types of algorithms are, either distance or dissimilarity oriented: the only significative difference appears in the costs, normalized in [0,1]. Furthermore this example attests the role of the neighborhood restriction, that allows limitation of the time distortions.

3 Application to the Phytoplanktonic Species Identification

3.1 Data Presentation

ND signals acquisition. In this study, nD signals were gathered in the LOG laboratory[1] from different phytoplanktonic species living in Eastern Channel, with a CytoSense flow cytometer (CytoBuoy[2]), and labelled by biologists [3], once having them isolated from the natural environment.

Flow cytometry is a technique used to characterize individual particles (cells or bacteria) drived by a liquid flow at high speed in front of a laser light (cf. Figure 3). Different signals either optical or physical are provided: forward scatter (which reflects the particle length), sideward scatter (which is more dependant on the particle internal structure) and several wavelengths of fluorescence (which depend upon the type of its photosynthetic pigments) measures.

More precisely, in the used signals library, each detected particle is described by 8 monodimensional raw signals issued from the flow cytometer in identical experimental conditions (same sampling rates, same detection threshold, etc.):

- a signal on forward scatter (FWS), corresponding to the cell length;
- two signals on sideward scatter (SWS), corresponding to the internal structure, in high and low sensitivity levels (SWS_HS, SWS_LS);
- two signals on red fluorescence (FLR), $\lambda em > 620nm$, in high and low sensitivity (FLR_HS, FLR_LS), which characterize chlorophyll pigments;
- a signal on orange fluorescence (FLO), $565nm < \lambda em < 592nm$, in low sentitivity(FLO_HS);
- two signals on yellow fluorescence (FLY), $545nm < \lambda em < 570nm$, in high and low sensitivity (FLY_HS, FLY_LS).

These signals are composed of voltage measures (mV), and their sampling period was here chosen to correspond to 0.5μ-meter displacement of the water flow. Consequently,

---

[1] Laboratoire d'Océanologie et de Géosciences, UMR 8187: http://log.univ-littoral.fr
[2] Cytobuoy system: http://www.cytobuoy.com

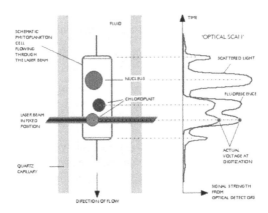

Fig. 3. Signals acquisition with a flow cytometer, image extracted from CytoBuoy's site

the longer the cell is, the higher the number of sampled measures is, and the time axis can be interpreted as a spatial length axis.

Phytoplanktonic species identification is a hard task, that is the reason why all these signals are used to make the particles characterization as complete as possible. Each particle of our experiment is consequently characterized by a 8D signal.

Derived features. Classification process requires an efficient characterization of the particles. This may be obtained either directly from the raw nD signals, or from some features which synthesize information of these signals. 4 attributes per signal are then extracted: length, height, integral, and number of peaks. Each Phytoplankton cell may then be described by 32 derived features.

Description of the studied Phytoplankton cells. The dataset is issued from a unique culture cells sample, whose particles belong to 7 distinct Phytoplanktonic species: *Chaetoceros socialis, Emiliania Huxleyi, Lauderia annulata, Leptocylindrus minimus, Phaeocystis globosa, Skeletonema costatum* and *Thalassiosira rotula*.

Each species is equally represented by 100 Phytoplanktonic cells, which were labelled by biologists using a microscope [3].

Figures in Table 1 show some signal samples of species *Lauderia annulata* and *Emiliania huxleyi*. For the first species, three individuals are selected: two very close, and an outlier.

Despite a high similarity between the profiles, intra-species variability can be quite important. In particular rising and falling edges of *Lauderia annulata* signals are not exactly synchronous. The curves SWS_HS (the highest ones) of *L. annulata* species show a size variability (*L. annulata* 10: 45μm, *L. annulata* 11: 55μm and *L. annulata* 5: 90μm), as well as a variability of the nucleus position (at the center of the cell for *L. annulata* 10-11, but clearly left shifted for *L. annulata* 5). In the case of FL_RED_HS signals (the second highest ones), we can also see differences in spatial shifts and in intensity levels between *L. annulata* 5 and the two others: this is due to different positions and different numbers of chloroplasts in cells (cf. Fig. 3).

Table 1. nD signals describing two species

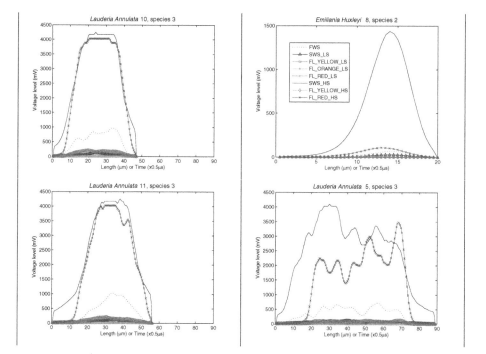

Last example *E. huxleyi* is an extreme case showing how similar cytometric curves of distinct species can be. However the length of this particle is clearly smaller in this particular case.

3.2   Applied Classification Methods

Two main classification approaches are experimented:

1. a features-based or "absolute" approach, which consists in applying classical classifiers on the 32 features extracted from the signals;
2. a dissimilarity-based or "relative" approach, which consists in comparing each new Phytoplanktonic cell to a set of labelled cells, directly from its 8 signals, thanks to a DTW variant; the distance or dissimilarity vector is then used as a feature vector, and processed by a classical classifier.

The classifiers are selected among the commonly used:

- k-nearest-neighbor with $k = 1$ (1-NN);
- multi-layer perceptron (MLP), with 1 hidden layer, and the sigmoid transfer function;
- support-vector machine (SVM1), with a first order polynomial kernel;
- support-vector machine (SVM2), with a fourst order polynomial kernel.

DTW-MLP structure used is 175/91/7 neurons (input/hidden/output-layer), features-MLP structure is 32/19/7 neurons. DTW-SVM and features-SVM method used around 300 support vectors in each training fold.

Distance and dissimilarity measures are issued from all the DTW variants previously described:

- first, classical DTW (a mean-distance) versus the proposed DTW variant (dissimilarity-based);
- then, three neighborhood restrictions were compared: "linear", "p-restricted" (with percentage $p = 10\%$ then $p = 20\%$) and the "no-restricted" variants.

In order to better estimate the variability of the recognition scores, 4-fold cross-validation is used: the dataset of 100×7 Phytoplanktonic cells is divided into 4 subsets of 25×7 cells, which are successively used as training fold while the union of the three other subsets is used as a test set.

3.3 Classification Results

Features-based classifiers. Table 2 shows the recognition scores of the features-based classification methods. The multi-layer perceptron obtains the best scores (mean score is 95.6%) as well as the least standard deviation (1.1%).

Table 2. Recognition rates (%) of the features-based classifiers

Training fold	Fold 1	Fold 2	Fold 3	Fold 4	Mean	Std
1-NN	93.7	90.2	93.7	92.5	**92.5**	1.7
MLP	96.9	94.8	96	94.8	**95.6**	1.1
SVM1	90	87.4	91	92.5	**90.2**	2.2
SVM2	95	91.2	90	93.9	**92.5**	2.4

Distance-based and dissimilarity-based classifiers. We now focus on the second approach, based upon the distance and dissimilarity measures. First, classifier 1-NN is used to measure the impact of the neighborhood restriction and the impact of the DTW measure, either classical, or the one proposed. Table 3 shows that the dissimilarity measure reaches higher score than the classical distance measure. This may be explained by the fact that matched pairs can have extremely high distance, penalizing the final mean-distance cost; but their dissimilarity degree is necessarily bounded by 1: a single badly matched pair can not extremely affect the final mean-dissimilarity cost.

Then, it appears that features-based classifiers scores are surpassed by the dissimilarity-based approaches. This tends to prove that the choosen DTW approach is relevant for this application.

Then, as expected, the best neighborhood restriction appears to be obtained with a moderate window: $p = 10\%$. Consequently, following comparisons between different classical classifiers are conducted using the 10%-restricted DTW algorithms (cf. Table 4).

Table 4 finally shows that the multi-layer perceptron is able to reach the highest mean score (97.3%) , with a very low standard deviation (0.7%).

Table 3. Recognition rates (%) of the dissimilarity-based 1-NN classifiers

Training folds	Fold 1	Fold 2	Fold 3	Fold 4	Mean	Std
Classical distance-based DTW						
linear	93.3	90.8	94.2	92.1	**92.6**	1.4
10%-restricted	94.8	92.5	94.8	93.7	**94.0**	1.0
20%-restricted	96.3	92.9	94.6	93.1	**94.2**	1.5
no-restricted	96.1	90.2	93.5	91.8	**92.9**	2.5
Proposed dissimilarity-based DTW						
linear	97.7	94.8	95.0	96.1	**95.9**	1.3
10%-restricted	97.9	94.6	96.0	96.1	**96.1**	1.3
20%-restricted	98.2	95.4	96.1	97.1	**96.7**	1.2
no-restricted	97.3	95.6	96.0	96.9	**96.4**	0.8

Table 4. Recognition rates (%) of the 10%-restricted dissimilarity-based classifiers

Training fold	Fold 1	Fold 2	Fold 3	Fold 4	Mean	Std
1-NN	98.2	95.4	96.1	97.1	**96.7**	1.3
MLP	98.2	97.3	97.3	96.7	**97.3**	0.7
SVM1	98.8	95.6	95.6	96.1	**96.5**	1.6
SVM2	92.3	93.5	93.3	92.9	**93**	0.6

4   Conclusion

In this paper, we proposed a conjoint dissimilarity [0,1]-measure for signals, based upon their shape. Such a bounded measure makes the interpretation by human users easier, and it can also be more relevant than a simple distance in some applications like the one presented. This dissimilarity measure was adapted to multidimensional signals, by equally weighting each dimension.

The proposed measure was applied to the automatic classification of Phytoplanktonic cells, which appears to be an innovative method: only few automatic species recognitions have yet been proposed. The experiment was performed on a labelled set of 700 Phytoplankton cells, with 100 cells per species. The quality of the obtained rates (which reach 97.2%) tends to show the relevance of the proposed dissimilarity measure, first in comparison with more classical distortion measures, then in comparison with a feature-based characterization.

These promising results encourage some future works, like the use of other distances (for instance in order to weight the distinct signal dimensions), or like the fusion of this distortion dissimilarity with some other dissimilarity measures (for instance, a duration dissimilarity).

Acknowledgements. The authors thank all the members of the LOG laboratory who took part in the data acquisition of the nD signals data, in particular L. Felipe Artigas, Natacha Guiselin, Elsa Breton, and Xavier Mériaux. The data were collected thanks to project CPER "Phaeocystis Bloom" funded by Région Nord-Pas-de-Calais, Europe (FEDER) and Université du Littoral Côte d'Opale.

They also thank Pr. Denis Hamad and our institute ULCO, for coordinating and earning BQR project "PhytoClas".

References

1. The European Parliament, the European Council: Directive 2000/60/ec of the european parliament and of the council of 23 october 2000 establishing a framework for community action in the field of water policy. Official Journal of the European Communities EN 2000/60/EC (2000)
2. Lund, J., Kipling, G., Cren, E.L.: The inverted microscope method of estimating algal numbers and the statistical basis of estimation by counting. Hydrobiologia 11, 143–170 (1958)
3. Guiselin, N., Courcot, L., Artigas, L.F., Jéloux, A.L., Brylinski, J.M.: An optimised protocol to prepare phaeocystis globosa morphotypes for scanning electron microscopy observation. Journal of Microbiological Methods 77(1), 119–123 (2009)
4. Cloern, J.E.: Phytoplankton bloom dynamics in coastal ecosystems: a review with some general lessons from sustained investigation of san francisco bay, california. Reviews of Geophysics 34, 127–168 (1996)
5. Takabayashi, M., Lew, K., Johnson, A., Marchi, A., Dugdale, R., Wilkerson, F.P.: The effect of nutrient availability and temperature on chain length of the diatom, skeletonema costatum. Journal of Plankton Research 28(9), 831–840 (2006)
6. Sakoe, H., Chiba, S.: Dynamic programming algorithm optimization for spoken word recognition. IEEE Transactions on Acoustics, Speech, and Signal Processing ASSP-26(1), 43–49 (1978)
7. Niels, R., Vuurpijl, L.: Introducing trigraph - trimodal writer identification. In: Proc. European Network of Forensic Handwr. Experts (2005)

Revealing the Structure of Childhood Abdominal Pain Data and Supporting Diagnostic Decision Making

Adam Adamopoulos[1,2], Mirto Ntasi[2], Seferina Mavroudi[2], Spiros Likothanassis[2], Lazaros Iliadis[3], and George Anastassopoulos[4]

[1] Medical Physics Laboratory, Democritus University of Thrace, GR-68100, Alexandroupolis, Hellas
[2] Pattern Recognition Laboratory, Department of Computer Engineering and Informatics, University of Patras, GR-26504, Patras, Hellas
[3] Democritus University of Thrace, 193 Pandazidou str, GR 68200, Orestias, Hellas
[4] Medical Informatics Laboratory, Democritus University of Thrace, GR-68100, Alexandroupolis, Hellas
adam@med.duth.gr, {ntasi,mavroudi,likothan}@ceid.upatras.gr, liliadis@fmenr.duth.gr, anasta@med.duth.gr

Abstract. Abdominal pain in childhood is a common cause of emergency admission in hospital. Its assessment and diagnosis, especially the decision about performing a surgical operation of the abdomen, continues to be a clinical challenge. This study investigates the possibilities of applying state of the art computational intelligence methods for the analysis of abdominal pain data. Specifically, the application of a Genetic Clustering Algorithm and of the Random Forests algorithm (RF) is explored. Clinical *appendicitis prediction* involves the estimation of at least 15 clinical and laboratory factors (features). The contribution of each factor to the prediction is not known. Thus, the goal of abdominal pain data analysis is not restricted to the classification of the data, but includes the exploration of the underlying data structure. In this study a genetic clustering algorithm is employed for the later task and its performance is compared to a classical K-means clustering approach. For classification purposes, tree methods are frequently used in medical applications since they often reveal simple relationships between variables that can be used to interpret the data. They are however very prone to overfitting problems. Random Forests, applied in this study, is a novel ensemble classifier which builds a number of decision trees to improve the single tree classifier generalization ability. The application of the above mentioned algorithms to real data resulted in very low error rates, (less than 5%), indicating the usefulness of the respective approach. The most informative diagnostic features as proposed by the algorithms are in accordance with known medical expert knowledge. The experimental results furthermore confirmed both, the greater ability of the genetic clustering algorithm to reveal the underlying data patterns as compared to the K-means approach and the effectiveness of RF-based diagnosis as compared to a single decision tree algorithm.

Keywords: Random Forest, decision trees, clustering, genetic algorithms, K-means, abdominal pain, appendicitis.

D. Palmer-Brown et al. (Eds.): EANN 2009, CCIS 43, pp. 165–177, 2009.
© Springer-Verlag Berlin Heidelberg 2009

1 Introduction

Physicians' medical diagnosis using medical protocols and observation of the patient often lurks risks, which can be effectively moderated by intelligent computational techniques.

Intelligent computational techniques are able to make a reliable suggestion about a particular instance of the problem [20] thereby facilitating accurate diagnosis. It is proven that the use of such methods during the medical diagnosis not only increases the overall diagnostic accuracy but improves markedly the physician's diagnostic performance [13,14,11,12,10], helping focus his effort on the selection of relevant knowledge and the proper use of the selected facts [21].

Machine learning is one of major branches of artificial intelligence and indeed machine learning algorithms were from the very beginning designed and used to analyse medical data sets [18]. Therefore, an abundance of corresponding applications have been developed for solving medical diagnosis problems: Artificial Neural Networks (ANN), Genetic Algorithms (GA) and Symbolic Learning are some of these approaches. A very incomplete and only indicative list of applications of machine learning in medical diagnosis includes applications in oncology [39], rheumatology [37], neuropsychology [38], gynaecology [40], urology and cardiology [18].

One of the most common chronic pain syndromes in children is recurrent abdominal pain, affecting 10-30% of all school aged children [2,3]. Although many cases of acute abdominal pain are benign, some require rapid diagnosis and herald a surgical or medical emergency. Rapid and certain diagnosis in these cases is essential to avoid undesirable complications, such as peritonitis, hemorrhage, infertility, unnecessarily performed laparotomy and chronic abdominal pain or anxiety disorders in adult life [6,7].

Abdominal pain often presents as a diagnostic problem to the clinician, because the disease is too protean to admit to a certain methodology or to be diagnosed by clinical instinct. Hospitalization, followed by active clinical observation, seems to be the most widely used method of clinical management of abdominal pain, for many years. However, this method entails negative consequences to young patients: children miss, on average, 21 more days of school per year [29], as the median length of stay in hospital for appendicitis ranges from 2 days (non-ruptured appendicitis) to 11 days (ruptured appendicitis) and days on antibiotic vary from 4.6 to 7.9 days [1]. Furthermore, the predictive value of clinical diagnosis reached with this method has been estimated between 68%-92% [8,9], while generally, the diagnostic inaccuracy rates from 15%-20% the last years [11]. This raises the need to adopt computer-aided methods on medical diagnosis, in order to improve the clinical outcomes.

In this study the application of a genetic clustering algorithm and a random forest classifier for appendicitis diagnosis is explored, using 15 clinical and laboratorial factors, which are usually used in clinical practice for the diagnosis of appendicitis in children [22,23].

The paper is organized as follows: Initially, in Section 2 the Random Forests classifier is introduced. Then, Section 2.1 presents the Experiment Setup and Section 2.2 discusses the results of the application of the RF classifier to real data from the Pediatric Surgery Department of the University Hospital of Alexandroupolis, Hellas. Section 3 deals with the Genetic Clustering and Feature Selection Approach. Section 3.1 presents the Experiment Setup and Experimental Results are given in Section 3.2. Finally, Section 4 summarizes and concludes the paper.

2 Random Forest Implementations

One of the most promising approaches for medical data analysis relies on tree based classifiers which achieve robustness and learning efficiency over the data set. The tree representation is the most widely used logic method for efficiently producing classifiers from data. Thus there exist a large number of tree algorithms which are primarily described in the machine-learning and applied-statistics literature [19]. Leo Breiman's Random Forests (RF) [25] is a recent development that has quickly proven to be one of the most important algorithms. It has shown robust and improved results of classification on data sets.

Random Forests combine two powerful ideas: resampling and random feature selection. Resampling is done by sampling multiple times with replacement from the original training data set. Thus in the resulting samples, about 2/3rd of the training data are used for each bootstrap sample and the remaining one-third of the cases are left out of the sample, creating trees that generalize independently. The essence of random feature selection is to build multiple trees in randomly selected subspaces of the feature space. Hence, in each tree growing process, at each branch, only a random subset of the possible features is available to use [24]. This has the effect of washing out the typical training instabilities of decision trees which are fast classifiers but liable to overfitting. This way, random forests generate multiple independent decision trees, for a given training set and use a weighted average of the trees as the final decision metric. As a result, RFs attain an extremely good classification accuracy compared to decision trees, which form the base for constructing RF [19].

Although RFs do not provide information on individual trees separately, they do provide several metrics useful in interpretation [41]. They incorporate for example a variable importance (VI) index to measure the relative importance of variables. The index is defined in terms of prediction performance and can be used to filter out unnecessary attributes.

In addition, RFs are very user-friendly in the sense that they have only two parameters (the number of variables in the random subset at each node and the number of trees in the forest) and are usually not very sensitive to their values [26].

2.1 Experiment Setup

The data was collected from a sufficient sample of patient records of the Pediatric Surgery Department of the University Hospital of Alexandroupolis, Hellas. The data set as presented in Table 1 consists of 516 cases, whereof 437 (84.69%) were normal (discharge, observation, no-findings cases) and 79 (15.31%) had different stages of appendicitis (focal, phlegmonous, gangrenous appendicitis and peritonitis). The data set was divided in a training set consisted of 416 cases and an independent testing set of 100 cases.

The diagnosis was based on the following 15 features (clinical and laboratorial factors) presented in Table 4: sex, age, religion, demographic data, duration of pain, vomitus, diarrhea, anorexia, tenderness, rebound, leucocytosis, neutrophilia, urinalysis, temperature and constipation.

According to the subset of features used during the classification we distinguished three different inputs presented in Table 5. Input Subset1 included all 15 features. Input Subset 2 included only a subset of 6 features (i.e., sex, age, duration of pain, tenderness, leucocytosis, netrofilia). Input Subset 3 was formed in accordance with the feature scoring of the RF output and included a subset of 5 features (tenderness, rebound, leucocytocis, netrofilia, urinalysis).

Table 1. Diagnostic categories Coding – Diagnosis corresponding to 7 categories

Diagnosis		Coding	Number of cases in training set	Number of cases in testing set
Normal	discharge	-2	157	17
	observation	-1	150	29
	no findings	0	42	34
Operative treatment (pathological)	focal appendicitis	1	31	11
	Phlegmonous, or supurative appendicitis	2	23	4
	gangrenous appendicitis	3	7	2
	peritonitis	4	6	3

Table 2. Diagnostic categories Coding – Diagnosis corresponding to 3 categories

Diagnosis		Coding	Number of cases in training set	Number of cases in testing set
Normal	discharge, observation, no findings	-2, -1, 0	349	80
Operative treatment	focal appendicitis, phlegmonous, or supurative appendicitis	1, 2	54	15
Emergency operative treatment	gangrenous appendicitis, peritonitis	3, 4	13	5

Table 3. Diagnostic categories Coding – Diagnosis corresponding to 2 categories

Diagnosis		Coding	Number of cases in training set	Number of cases in testing set
Normal	Discharge, observation, no findings	-2, -1, 0	349	88
Operative treatment (pathological)	Focal appendicitis Phlegmonous or supurative appendicitis, gangrenous appendicitis, peritonitis	1, 2, 3, 4	67	12

The samples were classified, as presented in Tables 1-3, in either two categories (normal vs operative treatment category), or in three categories (normal category, operative treatment category or emergency operative treatment category) or in seven categories (discharge, observation, no-findings, focal, phlegmonous, gangrenous, appendicitis or peritonitis category).

RF experiments were performed using the Matlab interface of Ting Wang on top of the Random Forest implementation (version 3.3) written by Leo Breiman and Adele Cutler. During RF experiments we varied two parameters: the number N of trees that formed the forest and the number m of variables which are selected at random out of the totality of variables, in order to be used to split the internal nodes of each tree. Specifically, we increased the number of trees from 50 to 200 with a step of 50 and the number of variables m from 2 to 8 with a step of 1.

Table 4. Clinical and laboratorial parameters (features) coding

Feature	Diagnostic factor	Coding
1	sex	1 (Boy) - 2 (Girl)
2	age	2 – 14
3	religion	1 (Christian), 2 (Muslim)
4	demographic data	1 (Alexandroupolis), 2 (Komotini), 3 (Xanthi)
5	duration of pain	0 (> 24 hours), 1 (< 24 hours)
6	vomitus	0 (absence), 1 (existence)
7	diarrhea	0 (absence), 1 (existence)
8	anorexia	0 (absence), 1 (existence)
9	tenderness	0 (diffuse), 1 (right abdomen), 2 (left abdomen)
10	rebound	0 (absence), 1 (existence)
11	leucocytosis	3.500 – 10.800 K/µl
12	neutrophilia	40 – 75%
13	urinalysis	-1 (-), 0 (non-abnormal), 1 (microhematuria), 2 (pyuria)
14	temperature	37 ^{O}C – 38 ^{O}C
15	constipation	0 (absence) - 1 (existence)

Table 5. Input subsets

Subset of features	Diagnostic features	Coding	Selected by
1	all 15 diagnostic features	1 – 15	
2	sex, age,, duration of pain, tenderness, leucocytosis, neutrofillia	1, 2, 5, 9, 11, 12	medical research
3	Tenderness, rebound, leucocytosis, neutrofillia, urinalysis	9, 10, 11, 12, 13	feature scoring

In order to compare with the RF application, CART Decision trees were implemented using the Statistics Toolbox, Version 2, of MATLAB. Throughout the experiments two parameters were adjusted using an internal crossvalidation procedure: the criterion for choosing a split and the number of observations per node needed for a split to be done.

Table 6. RF error rate (%) for 3 categories of classification

RF parameters	% error in training set			% error in testing set		
[N, m]	Subset of features			Subset of features		
	1	2	3	1	2	3
[50, 2]	3,05	4.51	4.04	4.35	3.00	4.00
[50, 3]	3.19	4.38	4.28	4.00	3.00	4.00
[50, 4]	3.11	4.95	5.04	4.00	3.00	4.00
[50, 5]	3.53	5.09	4.57	3.75	3.00	4.30
[50, 6]	3.64	5.04	4.59	4.00	3.00	3.95
[50, 7]	3.26	5.12	-	4.00	3.00	-
[50, 8]	3.91	-	-	4.00	-	-
[100,2]	3.23	4.62	4.57	4.00	3.00	4.00
[100,3]	2.90	4.78	4.40	4.00	3.00	4.00
[100,4]	3.07	5.06	4.54	4.00	3.00	4.03
[100,5]	3.37	5.39	4.81	4.00	3.00	4.00
[100,6]	3.50	5.69	5.13	4.00	3.00	4.15
[100,7]	3.68	5.52	-	4.00	3.00	-
[100,8]	3.72	-	-	4.90	-	-
[150,2]	3.25	4.51	3.93	4.00	3.00	4.00
[150,3]	3.06	4.51	4.38	4.00	3.00	4.00
[150,4]	3.23	4.93	4.46	4.00	3.00	4.20
[150,5]	3.48	5.52	4.99	4.00	3.00	4.00
[150,6]	3.43	5.60	5.14	4.00	3.00	4.00
[150,7]	3.58	5.65	-	4.00	3.00	-
[150,8]	3.77	-	-	4.60	-	4.00
[200,2]	3.35	4.44	4.04	3.85	3.00	4.00
[200,3]	2.87	4.71	4.25	4.00	3.00	4.00
[200,4]	3.05	5.05	4.87	4.00	3.00	4.00
[200,5]	3.20	5.38	4.95	4.00	3.00	4.00
[200,6]	3.53	5.73	5.24	4.00	3.00	4.00
[200,7]	3.55	5.88	-	4.00	3.00	-
[200,8]	3.73	-	-	4.00	-	-

2.2 Experimental Results and Discussion

Due to the limited amount of data cases, the 7-categories RF classification resulted in poor results and was not further considered. The random forest classifier performed well over the data using a 3-categories classification. Interestingly, the 2-categories classification also resulted in a very high error rate suggesting that the Operative Treatment and the Emergency Operative Treatment cases could have different diagnostic patterns which should not be grouped together. To further investigate on this we employed a genetic clustering method described in Section 3.

The results of the Random Forest application for the 3-categories classification averaged for several experiments and for different pairs of parameters are shown in Table 6. The test set error rate ranged from 3%-4.90 % using all the features (input with Subset 1) and limited to 3%-4.30 % when using part of the features (input with Subset 2 and 3). As expected, the performance was relatively insensitive to the parameter values.

Observing indicative confusion matrices in Table 7 we can see in each column the way that the samples portioned out in different categories according to RF simulation. The main diagonal denotes the performance. As it is shown the higher percentage of misclassification is observed in the operative treatment category, while there is no divergence in the emergency operative treatment category. This could suggest that only the emergency operative treatment cases have a diagnostic pattern that differs significantly from the normal cases, explaining the difficulties in the diagnosis of abdominal pain symptoms.

In Fig. 1 the average Variable Importance Graphs are displayed, as outputted from Breimans' RF implementation. The important features as depicted in the graph in Fig. 1(a) are related to the diagnostic factors of tenderness, rebound, leucocytosis, neutrofilia and urinalysis (diagnostic features #9, #10, #11, #12 and #13, according to Table 4). This result motivated the formation of the input subset 3, as presented in Table 5, which consisted of only these factors. Experimenting only with subset 3 inputs resulted in graphs as shown in Fig 1(b). Again the diagnostic factors relating to leucocytosis and neutrophilia (diagnostic features #11 and #12) achieved a very high scoring, which appears very rational from a medical point of view.

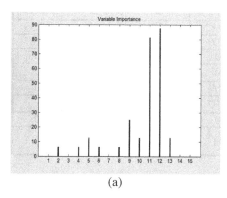
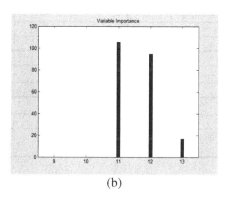

(a)   (b)

Fig. 1. Average Variable Importance Graphs for (a) subset 1 (all diagnostic features), and (b) subset 3 (diagnostic features #9, #10, #11, #12 and #13)

In general we did not observe significant differences while experimenting with the 3 different input subsets confirming that for Random Forests we need not to consider the dimensionality of the data since the algorithm automatically selects only the important features for classification. In contrary, Decision Trees classifiers performances verified their tendency to overfit the data. A 31% error rate was obtained using a 3-categories classification and 84%-86% error rate using 7-categories classification. The decision-trees disability is ascertained through the confusion matrices presented in Tables 8 - 9 whose main diagonal assembles low percentages. The results were not improved using a subset of features, therefore only the average results are presented.

Table 7. Confusion matrices of RF experiments for 3 categories of classification. Results on the testing set

Diagnosis	Normal (%)			Operative treatment (%)			Emergency operative treatment (%)		
	Subset of features			Subset of features			Subset of features		
	1	2	3	1	2	3	1	2	3
Normal	98.86	97.73	98.86	33.33	11.11	33.33	0	0	0
Operative treatment	1.14	2.27	1.14	66.67	88.89	66.67	0	0	0
Emergency operative treatment	0	0	0	0	0	0	100	100	100
Number of cases	88	88	88	9	9	9	3	3	3

Table 8. Average confusion matrices (%) for Decision Trees implementations for 7 categories of classification. Results on the testing set.

Coding	-2	-1	0	1	2	3	4
-2	23.53	58.62	55.88	81.82	75	50	33.33
-1	47.06	34.48	29.41	9.09	0	50	66.67
0	0	0	0	0	0	0	0
1	0	0	11.77	0	25	0	0
2	29.41	0	2.94	0	0	0	0
3	0	3.45	0	9.09	0	0	0
4	0	3.45	0	0	0	0	0
Number of cases	17	29	34	11	4	2	3

Table 9. Average confusion matrices (%) for Decision Trees implementations for 3 categories of classification. Results on the testing set.

Diagnosis	Normal	Operative Treatment	Emergency Operative Treatment
Normal	85	86.67	100
Operative treatment	12.5	6.67	0
Emergency operative treatment	2.5	6.67	0
Cases	80	15	5

3 Genetic Algorithm Clustering and Genetic Feature Selection

Recently, research has investigated the use of optimization techniques such as Genetic Algorithms (GA) in disease diagnosis. Genetic algorithms are a part of evolutionary computing based on an analogy to biological evolution. The algorithms focus on the application of selection, mutation, and recombination to a population of competing problem solutions [22], [23]. Thus, GA managed to pass the appropriate solutions from generation to generation in order to be combined and produced even better, while the inappropriate ones tend to be excluded. Genetic Algorithms can also be used for clustering purposes of clinical data. The very idea is to develop a GA to search for an appropriate classification of clinical data in clusters in such a way that the total number of misclassified cases is minimized [31].

3.1   Experiment Setup

Early works using GA for clustering, feature selection and classification can be found in [32, 33]. In [34] a centroid-based real encoding of the chromosomes was proposed. According to this technique, each chromosome is a sequence of real numbers representing the K clusters centers. For a data set considering N features, the length of each chromosome of the GA is $N \cdot K$. For each one of the individual of the GA population, the K cluster centers are initialized to randomly chosen points form the data set. If x_i, $i = 1, 2, \ldots n$, are the data points in the N-dimensional space, each point is assigned to the cluster C_j with center z_j, if:

$$\left\| x_i - z_j \right\| \le \left\| x_i - z_p \right\|, \quad p = 1, 2, \ldots, K . \tag{1}$$

After the clustering is done, the cluster centers are replaced by the mean points of the perspective clusters. The task for the GA is to minimize the distance of the data points from the corresponding cluster centers and therefore to search and find for the optimal classification of the data points to the corresponding clusters.

After the formation of the clusters, each cluster is assigned a class label according to a   majority vote of the corresponding examples that populate the cluster (i.e. each example votes for its known true class).   The classification performance depends on the number of the examples with class labels that differ from the corresponding cluster class label. The goal of using a clustering algorithm is not only to measure the classification performance, but more important, to study if the selected features are able to inherently discover the different classes and to thereby prove their effectiveness.

Moreover, in the present work, GA were used for the achievement of an additional task: that of genetically selection of features that will be considered for the classification of the clinical cases into clusters. Thus, the genome of the GA individuals is composed of two parts. The first part is a substring of 15 bits related to the corresponding 15 features of the clinical data set presented in Table 4. In this binary substring a value of 0 denotes that the corresponding feature is ignored and therefore it is not considered during the clustering procedure of the clinical cases. On the opposite, a value of 1 in this binary substring denotes that the corresponding feature is considered during the clustering procedure. In addition, the genome contains in binary form the coordinates of the centers of the clusters. In order to further investigate on the hypothesis outlined in Section 2.2, two clusters are considered: the normal cluster, and the pathological operative treatment cluster. Again, in accordance to Table 3, in the normal cluster are grouped the cases that have no need for additional medical treatment (coded by -2, -1 and 0, in the data set). All the rest cases (coded by 1, 2, 3 or 4 in the data set), are grouped in the pathological cluster. Therefore, the GA deals with two objectives: (a) to search and find the most suitable subset of features that will be considered for the classification of the cases into clusters, and (b) to find the coordinates of the clusters centers that minimize the distance of the points corresponding to the clinical cases from that cluster centers. Considering the issue of the fitness function of the GA, the number of correctly classified cases gives the fitness of the GA individuals. Due to the fact that the GA genome is coded in binary form all the well known genetic operators for binary crossover and binary mutation can be applied. For the experiments performed for the purposes of the present work, the GA algorithm was implemented in the Matlab® programming environment, using the *GA and Direct Search* tool incorporated there. Mostly, the *scattered crossover operator* was used for crossover and the *binary flip mutation operator* was used for the generation of mutants. For purpose of comparison, the genetic clustering results are compared to the corresponding ones obtained by using the K-means clustering algorithm [35, 36].

Table 10. Error rates (%) of the K-means clustering algorithm (testing set)

Number of clusters	Error (%)		
	Subset of features		
	1	2	3
2	4.07	48.26	4.26
3	45.74	49.42	21.90
7	83.72	82.95	87.98

Table 11. Error rates (%) of the Genetic Algorithm clustering algorithm and Genetic Feature Selection

Experiment Nr.	Error (%)	Features considered
1	1.98	2, 11, 14
2	2.13	2, 11, 14
3	2.52	11, 14
4	3.49	2, 11, 13, 14
5	3.68	2, 11, 12, 14
6	3.88	2, 11, 12, 14

3.2 Experimental Results and Discussion

The results obtained in several experiments of genetic clustering and feature selection are presented in Table 11. The implemented GA managed to perform clustering of all the 516 cases into two clusters (the "normal" and the "pathological" ones) in a very satisfactory fashion, resulting to a classification error ranging from 1.94% (10 misclassified cases out of 516 in total), to 3.88 (20 misclassified cases out of 516 in total). Most significant is that the features used by the GA are limited from 2 up to 4 features. According to Table 4, the features considered by the GA for clustering of the cases refer to the religion of the patients (feature #2), the leucocytosis and neutrophilia levels, (features #11 and #12), and the body temperature (feature #14). As far as for the clustering classification itself, the GA clustering outperforms the results obtained by K-means clustering algorithm (presented in Table 10).

4 Conclusions

In this paper, a Random Forest approach and a Genetic Clustering Algorithm were employed for the prediction of abdominal pain in children. To the best of our knowledge no similar analysis strategy for abdominal pain data has been reported in literature previously. Random Forests were able to efficiently tackle the diagnosis prediction problem of the 3-categories classification approach achieving performance up to 97%. Error rates fluctuated from 3%-4.9% using all the three types of input subsets. A crucial point is that the RF classifiers performed well over the emergency operative treatment cases (100% performance). However as far as the operative treatment cases are concerned the percentage of misclassification was quite high (up to 33% cases were misclassified to the normal category), which denoted that RF fail to recognize the seriousness of corresponding symptoms. Decision trees classifiers in contrast failed to even distinguish the necessity of operative treatment, misclassifying 86% of the dataset.

The Genetic clustering and feature selection approach revealed the underlying data structure and succeeded to cluster the 2-categories data for which the RF approach failed, with an error rate ranging from 1.94% to 3.38%. Additionally, it clearly outperformed the K-Means clustering approach, which resulted in an error rate above 4%.

The important features as automatically predicted for both approaches –RF and Genetic Clustering- were limited and in agreement with medical knowledge. Interestingly, the Genetic Algorithm selected the religion among the important features. Although at a first glance this selection seems inappropriate, it reflects the way the religion influences different dietary habits which ultimately are associated with abdominal pain symptoms.

References

1. Newman, K., Ponsky, T., Kittle, K., Dyk, L., Throop, C., Gieseker, K., Sills, M., Gilbert, J.: Appendicitis 2000: Variability in practice, outcomes, and resource utilization at thirty pediatric hospitals. In: 33rd Annual Meeting of the American Pediatric Surgical Association, Phoenix, Arizona, May 19-23 (2002)
2. Apley, J., Naish, N.: Recurrent abdominal pains: A field survey of 1000 school children. Archives of Disease in Childhood 33, 165–170 (1958)
3. Hyams, J.S., Burke, G., Davis, P.M., Rzepski, B., Andrulonis, P.A.: Abdominal pain and irritable bowel syndrome in adolescents: A community-based study. J Pediatr 129, 220–226 (1996)
4. Campo, J.V., DiLorenzo, C., Chiappetta, L., Bridge, J., Colborn, D.K., Gartner Jr., J.C., Gaffney, P., Kocoshis, S., Brent, D.: Adult outcomes of pediatric recurrent abdominal pain: do they grow out of it? Pediatrics 108(e1) (2001) doi: 10.1542/peds.108.1.e1
5. Berry Jr., J., Malt, R.A.: Appendicitis near its centenary. Massachussetts General Hospital and the Department of Surgery, Harvard Medical School
6. Paterson-Brown, S.: Emergency laparoscopy surgery. Br. J. Surg. 80, 279–283 (1993)

7. Olsen, J.B., Myrén, C.J., Haahr, P.E.: Randomized study of the value of laparoscopy before appendectomy. Br. J. Surg. 80, 922–923 (1993)
8. Raheja, S.K., McDonald, P., Taylor, I.: Non specific abdominal pain an expensive mystery. J. R. Soc. Med. 88, 10–11 (1990)
9. Hawthorn, I.E.: Abdominal pain as a cause of acute admission to hospital. J. R. Coll. Surg. Edinb. 37, 389–393 (1992)
10. McAdam, W.A., Brock, B.M., Armitage, T., Davenport, P., Chan, M., de Dombal, F.T.: Twelve years experience of computer-aided diagnosis in a district general hospital. Airedale District General Hospital, West Yorkshire
11. Sim, K.T., Picone, S., Crade, M., Sweeney, J.P.: Ultrasound with graded compression in the evaluation of acute appendicitis. J. Natl. Med. Assoc. 81(9), 954–957 (1989)
12. Graham, D.F.: Computer-aided prediction of gangrenous and perforating appendicitis. Br. Med. J. 26 2(6099), 1375–1377 (1977)
13. de Dombal, F.T., Leaper, D.J., Horrocks, J.C., Staniland, J.R., McCann, A.P.: Human and Computer-aided Diagnosis of Abdominal Pain: Further Report with Emphasis on Performance of Clinicians. Br. Med. J. 1(5904), 376–380 (1974)
14. de Dombal, F.T., Leaper, D.J., Staniland, J.R., McCann, A.P., Horrocks, J.C. Computer-aided Diagnosis of Acute Abdominal Pain. Br Med J. 1, 2(5804), 9–13 (1972)
15. Weydert, J.A., Shapiro, D.B., Acra, S.A., Monheim, C.J., Chambers, A.S., Ball, T.M.: Evaluation of guided imagery as treatment for recurrent abdominal pain in children: a randomized controlled trial. BMC Pediatr. 6, 29 (2006)
16. Gardikis, S., Touloupidis, S., Dimitriadis, G., Limas, C., Antypas, S., Dolatzas, T., Polychronidis, A., Simopoulos, C.: Urological Symptoms of Acute Appendicitis in Childhood and Early Adolescence. Int. Urol. Nephrol. 34, 189–192 (2002)
17. Mantzaris, D., Anastassopoulos, G., Adamopoulos, A., Gardikis, S.: A non-symbolic implementation for abdominal pain estimation in childhood. Information Sciences 178(20), 3860–3866 (2008)
18. Kononenko, I.: Machine Learning for Medical Diagnosis: History, State of the Art and Perspective. University of Ljubljana Faculty of Computer and Information Science
19. Boinee, P., de Angelis, A., Foresti, G.L.: Meta Random Forests. International Journal of Computational Intelligence 2, 3 (2006)
20. Tsirogiannis, G.L., Frossyniotis, D., Stoitsis, J., Golemati, S., Stafylopatis, A., Nikita, A.S.: Classification of Medical Data with a Robust Multi-Level Combination Scheme. School of Electrical and Computer Engineering National Technical University of Athens
21. Rubin, A.D.: Artificial Intelligence approaches to medical diagnosis. MIT Press, Cambridge
22. McCollough, M., Sharieff, G.Q.: Abdominal Pain in Children. University of California, San Diego, Pediatr Clin. N Am. 53, 107–137 (2006)
23. de Edelenyi, F.S., Goumidi, L., Bertrans, S., Phillips Ross McManus, C., Roche, H., Planells, R., Lairon, D. Springer, Heidelberg (2008)
24. Bailey, S.: Lawrence Berkeley National Laboratory (University of California), Paper LBNL-696E (2008)
25. Breiman, L.: Technical Report 670, Statistics Department University of California, Berkeley, September 9 (2004)
26. Liaw, A., Wiener, M.: Classification and Regression by Random Forest. The Newsletter of the R Project, vol. 2/3 (December 2002)
27. Holland, J.H.: Adaptation in Natural and Artificial Systems. Univ. Michigan Press, Ann Arbor (1975)

28. Goldberg, D.: Genetic Algorithms in Search, Optimization, and Machine Learning. Addison-Wesley, Reading (1989)
29. Li, B.U.: Recurrent abdominal pain in childhood: an approach to common disorders. Comprehensive Therapy 13, 46–53 (1987)
30. Gislason, P.A., Benediktsson, J.A., Sveinsson, J.R.: Random Forests for land cover classification. Pattern Recognition Letters 27, 294–300 (2006)
31. Bandyopadhyay, S., Pal, S.K.: Classification and learning using genetic algorithms. Springer, Heidelberg (2007)
32. Tseng, L., Yang, S.: Genetic Algorithms for clustering, feature selection, and classification. In: Proceedings of the IEEE International Conference on Neural Networks, Houston, pp. 1612–1616 (1997)
33. Bhuyan, N.J., Raghavan, V.V., Venkatesh, K.E.: Genetic algorithms for clustering with an ordered representation. In: Proceedings of the Fourth International Conference Genetic Algorithms, pp. 408–415 (1991)
34. Maulik, U., Bandyopadhyay, S.: Genetic algorithm based clustering technique. Pattern Recognition 33, 1455–1465 (2000)
35. Jain, A.K., Dubes, R.C.: Algorithms for clustering data. Prentice-Hall, Englewood Cliffs (1988)
36. Tou, J.T., Gonzalez, R.C.: Pattern Recognition Principles. Addison-Wesley, Reading (1974)
37. Karalic, A., Pirnat, V.: Significance Level Based Classification with Multiple Trees. Informatica 15(1), 54–58 (1991)
38. Muggleton, S.: Inductive Acquisition of Expert Knowledge. Turing Institute Press &Addison_Wesley (1990)
39. Bratko, I., Kononenko, I.: Learning Rules from Incomplete and Noisy Data. In: Phelps, B. (ed.) Interactions in Artificial Intelligence and Statistical Methods. Technical Press, Hampshire (1987)
40. Nunez, M.: Decision Tree Induction Using Domain Knowledge. In: Wielinga, B., et al. (eds.) Current Trends in Knowledge Acquisition. IOS Press, Amsterdam (1990)
41. Prasad, A.M., Iverson, L.R., Liaw, A.: Newer Classification and Regression Tree Techniques: Bagging and Random Forests for Ecological Prediction. Ecosystems 9, 181–199 (2006)

Relating Halftone Dot Quality to Paper Surface Topography

Pekka Kumpulainen, Marja Mettänen, Mikko Lauri, and Heimo Ihalainen

Tampere University of Technology, Department of Automation Science and Engineering,
P.O. Box 692, FI-33101 Tampere, Finland
{pekka.kumpulainen,marja.mettanen,mikko.lauri,
heimo.ihalainen}@tut.fi

Abstract. Most printed material is produced by printing halftone dot patterns. One of the key issues that determine the attainable print quality is the structure of the paper surface but the relation is non-deterministic in nature. We examine the halftone print quality and study the statistical dependence between the defects in printed dots and the topography measurement of the unprinted paper. The work concerns SC paper samples printed by an IGT gravure test printer. We have small-scale 2D measurements of the unprinted paper surface topography and the reflectance of the print result. The measurements before and after printing are aligned with subpixel resolution and individual printed dots are detected. First, the quality of the printed dots is studied using Self Organizing Map and clustering and the properties of the corresponding areas in the unprinted topography are examined. The printed dots are divided into high and low print quality. Features from the unprinted paper surface topography are then used to classify the corresponding paper areas using Support Vector Machine classification. The results show that the topography of the paper can explain some of the print defects. However, there are many other factors that affect the print quality and the topography alone is not adequate to predict the print quality.

Keywords: Print quality, surface topography, SOM, clustering, SVM classification.

1 Introduction

The structure of the paper surface is one of the key issues that determine the attainable print quality. The overall roughness and surface compressibility measurement have shown to be important in predicting the quality of gravure print [3, 9]. More recent studies have shown with aligned 2D measurements that topography can explain part of the missing ink in fulltone printing [13, 1].

The quality of halftone print is important since most of the printed material is produced by printing dot patterns (a.k.a raster dots). That introduces a new challenge for the image-based analysis of the small-scale quality properties of the print: the regular dot pattern area must be separated from the void areas between the dots to be able to focus the analysis of the 2D data to those points that were supposed to be covered by ink. This ensures the correct detection of missing ink and poor quality dots. A robust

D. Palmer-Brown et al. (Eds.): EANN 2009, CCIS 43, pp. 178–189, 2009.

and accurate method for detecting the raster dot grid is presented in [14]. We use it to extract each individual raster dot area from the 2D measurements of print reflectance and surface topography.

In this paper we analyze the individual raster dots from two paper samples. First, in the explorative analysis part, we verify the hypothesis that there is a relation between the print quality and the unprinted topography. The qualities of the printed dots are investigated and described using Self Organizing Map (SOM) [12]. We then calculate descriptive features of the printed dots to train another SOM, which is clustered [18] to further emphasize the most significant properties of the dots. We compare the results with properties of the corresponding areas in the unprinted surface topography. Finally, for the second part of the study, we select one paper sample and divide the printed dots into high and low print quality. We use the features from the unprinted paper surface topography to classify the corresponding print areas using Support Vector Machine classification [5].

In the following section we briefly introduce the paper samples under study and the process required to achieve the data for our analysis. In the next sections we describe the methods applied to the data and present the results of the analysis of the print quality and the classification of the dots based on the topography. Concluding remarks are given in the last section.

2 Data Acquisition

We examine supercalendered (SC) paper samples printed by an IGT gravure test printer [10]. The test set contains paper samples of three roughness levels, each one printed with three printing nip pressure levels. In this work we study the two extreme cases: the one with highest roughness and lowest force in the printing nip, and the lowest roughness and the highest force. We refer to these cases as sample A and sample B correspondingly. We focus on the conventional screening area of the Heliotest strip that consists of a dense regular raster dot pattern due to the engraved cells on the printing cylinder. Examples of printed paper strips with an arrow pointing the area under study are shown in Fig. 1. The upper corners are marked with pencil for the alignment of the measurements before and after printing.

Fig. 1. Example of printed paper strips

The selected areas have been imaged with a photometric stereo device that applies the principles described in [8]. It is based on photographic imaging with slanting illumination and it provides both reflectance and surface topography maps of the paper sample. The image size is 22.5 x 15 mm and contains 2268 x 1512 pixels, thus the pixel size is approximately 10 x 10 μm. The same area of each paper strip is imaged before and after printing, and these images are aligned at subpixel accuracy [15]. The photometric stereo images are in RGB colors but the topography map is computed from the mean of the color channels. In the analysis of the reflectance of the printed paper, we use only green channel which is the most sensitive to the red color used in printing.

The basic regular structure of the printed dots is detected using 2D FFT and the final positions of the dots are estimated at subpixel resolution by maximizing the local spatial correlation between the print reflectance and a raster pattern model [14]. The latter step is essential to allow the detected point grid to slightly deviate from the perfectly regular pattern. The raster dots are extracted from the print reflectance map by selecting a square from around each grid point and interpolating the pixel values inside each square so that the subpixel grid point becomes the geometrical centre of the square. Each of the final interpolated dots is presented by a matrix of 13 by 13 pixels. Finally, certain erroneous measurement values are eliminated from the data. They occur at the transparent fibers on the paper surface that appear as depressions in the topography map although they are really elevations. They are detected from the measurements of the unprinted paper by the principal component analysis [11] of the color channels. After removing the areas that contain fibers or other artifacts we have 12628 individual raster dots remaining in sample A and 10822 dots in sample B.

3   Analysis Methods

In section 4 we use Self Organizing Maps (SOM) to present the main characteristics of the print quality. We use clustering in section 5 to further emphasize the most significant properties of the printed dots. Support Vector Machines (SVM) are used in section 6 to study if the surface topography of unprinted paper is sufficient to classify the paper areas to high and low final print quality.

All the methods we use have several parameters which affect the results. They all are also influenced by the preprocessing of the data, especially scaling and weighting of the variables [6, 7]. We will show the computations for optimizing the parameters on SVM classification only. The details of other decisions are discussed in appropriate sections.

3.1   Self Organizing Maps

Self Organizing Map (SOM) is an unsupervised neural network which approximates and visualizes multi-dimensional data [12]. SOM usually consists of a two-dimensional regular grid of nodes and a model of observations in the high-dimensional data space, often called *code vector*, is associated with each node. During the training phase, the code vectors are computed to optimally describe the distribution of the data. We use the SOM toolbox for MATLAB [19] (available at: http://www.cis.hut.fi/projects/somtoolbox).

SOM can be used to visualize the properties of the high-dimensional data in several ways [17]. In this paper we visualize the properties of the raster dots on the SOM grid. On each node we display an image of the mean of all the dots that belong to the node. Belonging to a node means that the code vector of that node has the minimum distance of all nodes to the data point. The standard deviations of the data are presented in a similar manner. We also use the hit histogram, which shows the number of the data points that belong to each of the nodes.

3.2 Clustering

Clustering is a term for methods that discover groups of similar objects in multivariate data where no predefined classes exist and thus there are no known right or false results [6]. In this study we find clusters of the code vectors of the SOM [18], which adds another condensed view to the data. Using the SOM as an intermediate step reduces the computational load. Even though these data do not have evident clusters hierarchical clustering with complete linkage [11] reveals informative groups from the data.

3.3 Support Vector Machine Classification

Support Vector Machines (SVM) are learning systems that use hypothesis space of linear functions in a high-dimensional kernel-induced feature space [5]. In section 6 we use SVM classifiers on the features of the topography measurement before printing to classify the printed reflectance into good and poor quality.

In the first subsection we use C-Support Vector Classification (C-SVC) [2, 4]. Given l samples of training vectors $\mathbf{x}_i$ in two classes, defined by $y_i \in \{-1, 1\}$, C-SVC solves the primal problem

$$\min_{\mathbf{w}, b, \xi} \quad \frac{1}{2}\mathbf{w}^T\mathbf{w} + C\sum_{i=1}^{l}\xi_i$$
$$\text{subject to} \quad y_i(\mathbf{w}^T\phi(\mathbf{x}_i) + b) \geq 1 - \xi_i,$$
$$\xi_i \geq 0, i = 1, \dots, l. \tag{1}$$

In the second subsection we use v-SVM [16] to find out if there are more sparse solutions available. The training errors and the number of support vectors can be controlled by an additional parameter v. The primal form is

$$\min_{\mathbf{w}, b, \xi, \rho} \quad \frac{1}{2}\mathbf{w}^T\mathbf{w} - v\rho + \frac{1}{l}\sum_{i=1}^{l}\xi_i$$
$$\text{subject to} \quad y_i(\mathbf{w}^T\phi(\mathbf{x}_i) + b) \geq \rho - \xi_i,$$
$$\xi_i \geq 0, i = 1, \dots, l, \rho \geq 0. \tag{2}$$

The dual for the scaled version used in [4] is

$$\min_{\alpha} \quad \frac{1}{2}\alpha^T Q\alpha,$$
$$\text{subject to} \quad 0 \leq \alpha_i \leq 1, i = 1, \dots, l,$$
$$\mathbf{e}^T\alpha = vl, \mathbf{y}^T\alpha = 0. \tag{3}$$

where $\mathbf{Q}$ is a positive semidefinite matrix $Q_{ij} \equiv y_iy_jK(\mathbf{x}_i, \mathbf{x}_j)$ and the kernel $K(\mathbf{x}_i, \mathbf{x}_j) \equiv \phi(\mathbf{x}_i)^T\phi(\mathbf{x}_j)$.

The decision function with output α/ρ is

$$\mathrm{sgn}\left(\sum_{i=1}^{l} y_i \, (\alpha_i/\rho)(K(\boldsymbol{x}_i, \boldsymbol{x}) + b)\right). \tag{4}$$

In addition to the classification, the probabilities of the classes can be calculated for each $\boldsymbol{x}_i$ [20]. In the last subsection we use the probabilities of the classes to select the raster dots that are most likely classified correctly.

4 Analysis of Print Quality

The individual pixels of each raster dot in the printed reflectance measurement are used, converted to N x 169 matrix, where N is the number of raster dots. We use a 2-dimensional rectangular grid, which enables to show the mean images in each node.

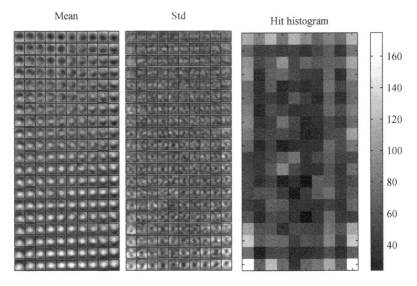

Fig. 2. Typical raster dots in sample A. Average of the raster dots in each node (*left*), standard deviation (*middle*) and the number of dots in each node (*right*).

The map size is 20 x 10 nodes and it is initialize along the first two principal components of the data space. Results of paper sample A are depicted in Fig. 2. Darker gray level in the figure presents more ink in print and the lighter levels and white depict points where ink is missing. There are a lot of poor quality dots. The highest values in the hit histogram are in the lower corners, which present poor quality.

The corresponding results of the paper sample B are shown in Fig. 3. Most of the map is covered with high quality dots. The highest values in the hit histogram are also at the upper side, which presents good quality.

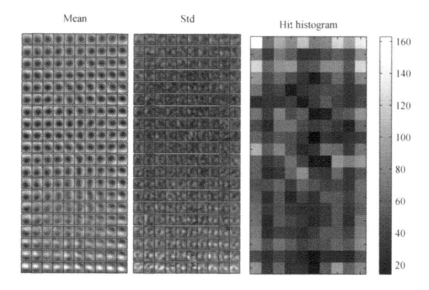

Fig. 3. Typical raster dots in sample B. Average of the raster dots in each node (*left*), standard deviation (*middle*) and the number of dots in each node (*right*).

These results give a general description of the variety of the printed dots. The fundamental properties of the dots seem to be related to the overall level of darkness and variance as well as the level and variance in the middle and at the edges of each dot. These observations are utilized in the following section.

5 Analysis of Topography and Print

For further analysis we calculate only few features that describe the main properties of the raster dots. For each raster dot we calculate the overall mean and the standard deviation of all the pixels in the reflectance measurement. The mean values are subtracted from each dot. Thereafter we divide the dots in two areas: the inner part, which contains a circle with a radius of 4 pixels, and the outer part, which contains the rest of the pixels towards the edges. The mean and the standard deviation are calculated in both areas. This yields a total of 6 features. The features are normalized to zero mean and unit variance and then multiplied by weight factors. The circular means and standard deviations are given weight 0.9 and 0.4 respectively. The overall mean and standard deviation are not weighted.

We train a 1-dimensional SOM of 200 nodes initialized along the first principal component. Using the 1-dimensional SOM line, instead of the often used 2-dimensional grid facilitates the interpretation of the results. The high and low quality dots tend to be divided into the opposite ends of the line. The component planes of the SOM are replaced by component lines shown in Fig. 4. The running number of the SOM nodes is on the x-axis.

The values of the overall mean, which represents the total amount of ink in the dots, rise from left to right in both samples. This clearly states that the good quality dots are represented by the nodes in the left end and the poor quality in the right. This is also proven by the mean values in the inner and outer circles.

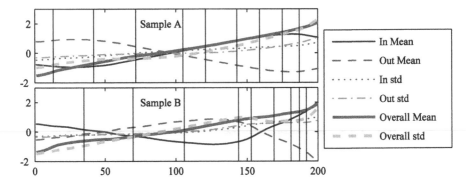

Fig. 4. Component lines of the 1-dimensional SOM in the 6-dimensional feature space

For more condensed presentation we divide the data into ten clusters. The cluster borders are marked with vertical lines in Fig. 4.

The code vectors present a lower dimensional feature space and they cannot be used to visualize the data. Therefore we calculate the pixel-wise mean and standard deviation in each cluster. We also calculate the mean and standard deviation of the topography at the locations determined by the clustering of the reflectance.

The results of the paper sample A are shown in Fig. 5. The numbers at the top refer to the percentage of the dots contained in each cluster. The following line tells the cumulative percentage.

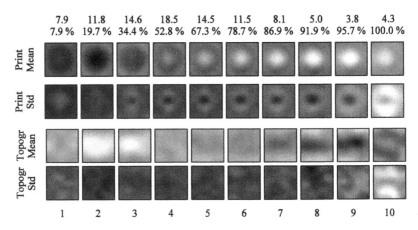

Fig. 5. Mean and standard deviation of the printed reflectance in paper sample A in clusters 1 to 10 (*top*). Results from the corresponding areas in unprinted topography (*bottom*).

In average the high and low quality dots are organized to the opposite ends of the line. Less than half of the dots are of proper quality. The rest towards the right end are more or less defective. The variation within the cluster also increases from left to right. The mean of the topography is lower in the clusters on the right, which means that there are deeper depressions in the paper in these clusters.

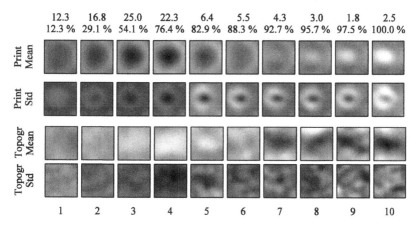

Fig. 6. Mean and standard deviation of the printed reflectance in paper sample B in clusters 1 to 10 (*top*). Results from the corresponding areas in unprinted topography (*bottom*).

The corresponding results of the paper sample B are illustrated in Fig. 6. Up to 90% of the dots seem to be of decent quality. Again the topography measurement shows the deeper average depressions at the poor print quality.

On the average the poor quality raster dots with missing ink seem to coincide with the depressions in the topography. However, there is a great amount of variance in both reflectance and topography especially at the low print quality end. This indicates that, in addition to the depressions in the surface topography, various other factors in the printing process cause missing printing ink.

6 SVM Classification

We select the highest and lowest quality raster dots from the paper sample A for the classification study. We form two classes using only the data in two clusters from both ends of the 1-D SOM line. Clusters 1 and 2 form one class, presenting the high print quality and clusters 9 and 10 form the class for the poor quality. We use LIBSVM, a library for support vector machines [4] for the classification.

We balance the data set so that each class has the same number of dots, 1023 in both classes. 10% of these data are dedicated as test data, and thus there are 1840 dots in the training data and 206 in the test data. The same six features that were used in SOM are now calculated from the measurement of the topography before printing. We scale the topography features to the range [0 1] and give the same weights that were used for the reflectance features: 0.9 and 0.4 for the circular means and standard deviations, respectively.

6.1 C-SVC Classification

We use the C-SVC classification with Radial Basis Function (RBF) kernels: $K(\mathbf{x}_i, \mathbf{x}_j) = \exp(-\gamma \|\mathbf{x}_i - \mathbf{x}_j\|^2)$, $\gamma > 0$. The classification accuracy from the 5-fold cross validation on the training data is depicted in Fig. 7 as function of the coefficient C in eq. (2) and the parameter γ of the RBF kernel.

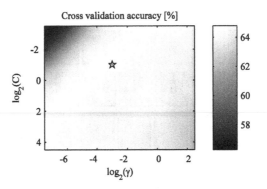

Fig. 7. Classification accuracy of cross validation in C-SVC

The maximum value 64.8%, highlighted by a star was achieved at values $C = 0.5$ and $\gamma = 2^{-3}$. The total number of support vectors in this model is 1530, thus it is not a very sparse solution.

6.2 ν-SVM Classification

In this subsection we study if it is possible to have more sparse solutions and reduce the number of support vectors by ν-SVM. The parameter C of the C-SVC is now replaced by ν, which acts as an upper bound of the fraction of training errors and a lower bound of the fraction of support vectors. Similar 5-fold cross validation procedure as above is conducted on the training data set.

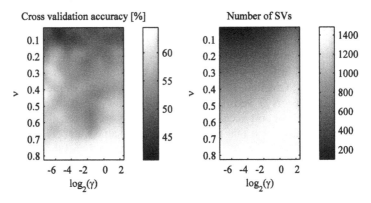

Fig. 8. Accuracy of the ν-SVM classification (*left*) and the number of support vectors (*right*) as functions of the parameters ν and γ

The results are presented in Fig. 8. The best classification accuracy 64.5% is achieved at values $\gamma = 2^2$ and $v = 0.8$. This model has 1485 support vectors, slightly less than in C-SVC.

6.3 Class Probabilities

In this final example we select $v = 0.8$ and the $\gamma = 2^{-4}$, from the previous experiments and calculate the probabilities for each data vector. Differences of the probabilities in the two classes are shown in Fig. 9. Negative value stands for class -1 and positive for 1. The data are arranged so that the negative class is on the left and the positive on the right, separated by a line.

Table 1 shows the proportion of the raster dots covered and the classification accuracy when only dots with high probability are considered. The probability threshold in this example is 0.7. Large variance in both reflectance and topography at the low print quality end as observed in Figs. 5 and 6 indicates that other factors than topography depressions also cause printing defects. Thus the accuracy of classification based only on the relation of reflectance and topography remains low.

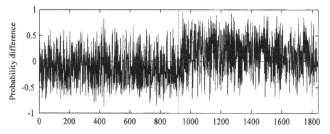

Fig. 9. Difference of the probabilities in classes -1 and 1

Table 1. Classification results including only the data where the probability p > 0.7

	Proportion of the data	Classification accuracy
Training data	29.7 %	79.7 %
Test data	30.1 %	72.6 %

7 Conclusion

We have examined the quality of raster dots printed by an IGT gravure test printer. Two samples were studied: sample A with high roughness level and low printing pressure, and sample B with low roughness level and high printing pressure. The overall goals were to describe the quality of the raster dots and to find out if the topography measurement before printing can be used to distinguish the high and low quality dots, thus predicting the final print quality.

Visualizing the dots by SOM reveals several kinds of defects in the dots, missing ink in the middle being the most severe. The overall quality in sample B is higher than in sample A as expected considering the smoothness of the paper surface and the printing pressure. The best SVM classification results were below 65%. Although topography partly determines the contact between the paper surface and the printing roll, it does not classify the print quality very well. This is in line with prior studies which also recognize that several other factors, such as surface compressibility and porosity, have a large impact on the printing result. However, the current work has shown the efficiency of SOM and SVC in the analysis and visualization of large data sets such as thousands of small images of printed dots, and the connections found between print quality and surface topography are plausible from the paper physical point of view.

In future work we will apply these methods to a wider range of print samples. Utilizing the classification probabilities seems promising in finding the common features of the surface topography that most likely affect the print quality. In this study we have only used the areas covered by the printed raster dots. The defects in print quality can be caused by properties of larger neighborhood in the unprinted topography, which will be further studied.

References

1. Barros, G.G.: Influence of substrate topography on ink distribution in flexography. Ph.D. thesis, Karlstad University (2006)
2. Boser, B.E., Guyon, I., Vapnik, V.: A training algorithm for optimal margin classifiers. In: Fifth Annual Workshop on Computational Learning Theory, pp. 144–152. ACM Press, New York (1992)
3. Bristow, J.A., Ekman, H.: Paper properties affecting gravure print quality. TAPPI Journal 64(10), 115–118 (1981)
4. Chang, C-C., Lin, C-J.: LIBSVM: a library for support vector machines, http://www.csie.ntu.edu.tw/~cjlin/papers/libsvm.pdf (Last updated February 27, 2009), Software available at, http://www.csie.ntu.edu.tw/~cjlin/libsvm (2001)
5. Cristianini, N., Shawe-Taylor, J.: An Introduction to Support Vector Machines and Other Kernel-based Learning Methods. Cambridge University Press, Cambridge (2000)
6. Everitt, B., Landau, S., Leese, M.: Cluster analysis, 4th edn., Arnold (2001)
7. Gnanadesikan, R., Kettenring, J., Tsao, S.: Weighting and selection of variables for cluster analysis. Journal of Classification 12(1), 113–136 (1995)
8. Hansson, P., Johansson, P.-Å.: Topography and reflectance analysis of paper surfaces using a photometric stereo method. Optical Engineering 39(9), 2555–2561 (2000)
9. Heintze, H.U., Gordon, R.W.: Tuning of the GRI proof press as a predictor of rotonews print quality in the pressroom. TAPPI Journal 62(11), 97–101 (1979)
10. IGT Testing Systems: IGT Information leaflet W41 (2003)
11. Johnson, R.A., Wichern, D.W.: Applied multivariate statistical analysis, 4th edn. Prentice-Hall, Englewood Cliffs (1998)
12. Kohonen, T.: Self-Organizing Maps. Series in Information Sciences, vol. 30. Springer, Heidelberg (1995)
13. Mettänen, M., Hirn, U., Lauri, M., Ritala, R.: Probabilistic analysis of small-scale print defects with aligned 2D measurements. In: Transactions of the 14th Fundamental Research Symposium, Oxford, UK (Accepted for publication, 2009)

14. Mettänen, M., Lauri, M., Ihalainen, H., Kumpulainen, P., Ritala, R.: Aligned analysis of surface topography and printed dot pattern maps. In: Proceedings of Papermaking Research Symposium 2009, Kuopio, Finland (2009)
15. Mettänen, M., Ihalainen, H., Ritala, R.: Alignment and statistical analysis of 2D small-scale paper property maps. Appita Journal 61(4), 323–330 (2008)
16. Schölkopf, B., Smola, A., Williamson, R.C., Bartlett, P.L.: New support vector algorithms. Neural Computation 12, 1207–1245 (2000)
17. Vesanto, J.: SOM-based data visualization methods. Intell. Data Anal. 3(2), 111–126 (1999)
18. Vesanto, J., Alhoniemi, E.: Clustering of the self-organizing map. IEEE Transactions on Neural Networks 11(3), 586–600 (2000)
19. Vesanto, J., Himberg, J., Alhoniemi, E., Parhankangas, J.: Self-organizing map in Matlab: the SOM toolbox. In: Proceedings of the Matlab DSP Conference 1999, Espoo, Finland, pp. 35–40 (1999)
20. Wu, T.-F., Lin, C.-J., Weng, R.C.: Probability estimates for multi-class classification by pairwise coupling. Journal of Machine Learning Research 5, 975–1005 (2004)

Combining GRN Modeling and Demonstration-Based Programming for Robot Control

Wei-Po Lee* and Tsung-Hsien Yang

Department of Information Management, National Sun Yat-sen
University Kaohsiung, Taiwan
wplee@mail.nsysu.edu.tw

Abstract. Gene regulatory networks dynamically orchestrate the level of expression for each gene in the genome. With such unique characteristics, they can be modeled as reliable and robust control mechanisms for robots. In this work we devise a recurrent neural network-based GRN model to control robots. To simulate the regulatory effects and make our model inferable from time-series data, we develop an enhanced learning algorithm, coupled with some heuristic techniques of data processing for performance improvement. We also establish a method of programming by demonstration to collect behavior sequence data of the robot as the expression profiles, and then employ our framework to infer controllers automatically. To verify the proposed approach, experiments have been conducted and the results show that our regulatory model can be inferred for robot control successfully.

Keywords: Recurrent neural network, gene regulation, time series data, bio-inspired robot control, learning by demonstration.

1 Introduction

Gene regulatory networks (GRNs) are essential in cellular metabolism during the development of living organism. They dynamically orchestrate the level of expression for each gene in the genome by controlling whether and how the gene will be transcribed into RNA [1]. In a GRN, the network structure is an abstraction of the chemical dynamics of a system, and the network nodes are genes that can be regarded as functions obtained by combining basic functions upon the inputs. These functions have been interpreted as performing a kind of information processing within the cell, which determines cellular behaviors. As can be observed, GRNs act as analog biochemical computers to specify the identity of and level of expression of groups of target genes. Such systems often include dynamic and interlock feedback loops for further regulation of network architecture and outputs. With the unique characteristics, GRNs can be modeled as reliable and robust control mechanisms for robots.

The first important step in applying GRNs to controlling robots is to develop a framework for GRN modeling. In the work of GRN modeling, many regulation models have been proposed [2][3]; they can range from very abstract models (involving Boolean values only) to very concrete ones (including fully biochemical interactions with

---

* Corresponding author.

D. Palmer-Brown et al. (Eds.): EANN 2009, CCIS 43, pp. 190–199, 2009.

stochastic kinetics). The former approach is mathematically tractable that provides the possibility of examining large systems; but it cannot infer networks with feedback loops. The latter approach is more suitable in simulating the biochemical reality and more realistic to experimental biologists. To construct a network from experimental data, an automated procedure (i.e., reversely engineering) is advocated, in which the GRN model is firstly determined, and then different computational methods are developed for the chosen model to reconstruct networks from the time-series data [2][3]. As can be observed from the literature, works in modeling GRNs shared similar ideas in principle. However, depending on the research motivations behind such works, different researchers have explored the same topic from different points of view; thus the implementation details of individual work are also different. This is especially apparent in the application of GRNs for robot control. For example, in the work by Quick et al. [4] and Stewart et al. [5], the authors have emphasized on describing the operational details of their artificial genes, enzymes, and proteins to show how their GRN-based control systems are close to the biological process, rather than how their models can be used to control robot practically. Different from their work, we focus on investigating whether the presented approach can be used to model GRNs in practice and on how to exploit this approach to construct robot controllers in particular.

As recurrent neural networks (RNNs) consider the feedback loops and take internal states into account, they are able to show the system dynamics of the network over time. With these characteristics, this kind of network is especially suitable for control systems. Therefore, in this work we develop a RNN-based regulatory model for robot control. To simulate the regulatory effects and make our model inferable from time-series expression data, we also implement an enhanced RNN learning algorithm, coupled with some heuristic techniques of data processing to improve the learning performance. After establishing a framework for GRN modeling, we develop a method of programming by demonstration to collect behavior sequence data of a robot as the time series profiles, and then employ our framework to infer behavior controllers automatically. To verify the presented approach, two series of experiments have been conducted to demonstrate how it operates and how it can be used to construct controllers for robots successfully.

2 Developing GRN Controllers

The most important issues in using GRNs for robot control are the development of GRN model and the computational method for constructing the model from available expression data. To address the relevant problems, this section describes how we take biological, computational, and engineering points of view to develop an inferable control model.

2.1 RNN-Based Regulatory Model

There are several recurrent neural network architectures, ranging from restricted classes of feedback to full interconnection between nodes [6][7]. Vohradsky and colleagues have proposed the use of fully recurrent neural network architecture in studying regulatory genetic networks such as those involved in the transcriptional and translational control of gene expression [8]. In this work, we also take a similar architecture to model GRN, but unlike their work that mainly simulates regulatory effects, our goal is to infer regulatory networks from measured expression data.

In a fully recurrent net, each node has a link to any node of the net, including itself. Using such a model to represent a GRN is based on the assumption that the regulatory effect on the expression of a particular gene can be expressed as a neural network in which each node represents a particular gene and the wiring between the nodes define regulatory interactions. When activity flows through the network in response to an input, each node influences the states of all nodes in the same net. Since all activity changes are determined by these influences, each input can be seen as setting constraints on the final state that the system can settle into. When the system operates, the activities of individual nodes will change in order to increase the number of constraints satisfied. The ideal final state would be a set of activities for the individual nodes where all the constraints are satisfied. The network would then be stable because no node would be trying to change the state of any of the nodes to which it is connected. Similarly, in a GRN, the level of expression of genes at time t can be measured from a gene node, and the output of a node at time $t+\Delta t$ can be derived from the expression levels and connection weights of all genes connected to the given gene at the time t. That is, the regulatory effect to a certain gene can be regarded as a weighted sum of all other genes that regulate this gene. Then the regulatory effect is transformed by a sigmoidal transfer function into a value between 0 and 1 for normalization.

The same set of the above transformation rules is applied to the system output in a cyclic fashion until the input does not change any further. As in [8], here we use the basic ingredient to increase the power of empirical correlations in signaling constitutive regulatory circuits. It is to generate a network with nodes and edges corresponding to the level of gene expression measured in microarray experiments, and to derive correlation coefficients between genes. To calculate the expression rate of a gene, the following transformation rules are used:

$$\frac{dy_i}{dt} = k_{1,i}G_i - k_{2,i}y_i$$

$$G_i = \{1 + e^{-(\sum_j w_{i,j}y_j + b_i)}\}^{-1}$$

where y_i is the actual concentration of the i-th gene product; $k_{1,i}$ and $k_{2,i}$ are the accumulation and degradation rate constants of gene product, respectively; G_i is the regulatory effect on each gene that is defined by a set of weights (i.e., $w_{i,j}$) estimating the regulatory influence of gene j on gene i, and an external input b_i representing the reaction delay parameter.

When the above GRN model is used to control a robot, each gene node now corresponds to an actuator of the robot in principle. Two extra nodes are added to serve as inter-genes and their roles are not specified in advance. The redundancy makes the controllers easier to be inferred from data. Fig. 1 illustrates the architecture of our GRN controller. In this architecture, the sensor information received from the environment is continuously sent to all nodes of the fully interconnected network, and the outputs of the actuator nodes (i.e., a_i) are interpreted as motor commands to control the robot. For control tasks in which the sensor information is not required, for example the locomotion task, the perception part in the figure is simply disabled.

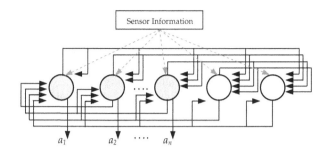

Fig. 1. The GRN control architecture

2.2 Learning Algorithm for Constructing GRN Controllers

After the network model is decided, the next phase is to find settings of the thresholds and time constants for each neuron as well as the weights of the connections between the neurons so that the network can produce the most approximate system behavior (that is, the measured expression data). By introducing a scoring function for network performance evaluation, the above task can be regarded as a parameter estimation problem with the goal of maximizing the network performance (or minimizing an equivalent error measure). To achieve this goal, here we use the backpropagation through time (BPTT) [9] learning algorithm to update the relevant parameters of recurrent networks in discrete-time steps.

Instead of mapping a static input to a static output as in a feedforward network, BPTT maps a series of inputs to a series of outputs. The central idea is the "unfolding" of the discrete-time recurrent neural network (DTRNN) into a multilayer feedforward neural network when a sequence is processed. Once a DTRNN has been transformed into an equivalent feedforward network, the resulting feedforward network can then be trained using the standard backpropagation algorithm.

The goal of BPTT is to compute the gradient over the trajectory and update network weights accordingly. As mentioned above, the gradient decomposes over time. It can be obtained by calculating the instantaneous gradients and accumulating the effect over time. In BPTT, weights can only be updated after a complete forward step during which the activation is sent through the network and each processing element stores its activation locally for the entire length of the trajectory. More details on BPTT are referred to Werbos' work [9].

In the above learning procedure, learning rate is an important parameter. Yet it is difficult to choose an appropriate value to achieve an efficient training, because the cost surface for multi-layer networks can be complicated and what works in one location of the cost surface may not work well in another location. Delta-bar-delta is a heuristic algorithm for modifying the learning rate in the training procedure [10]. It is inspired by the observation that the error surface may have a different gradient along each weight direction so that each weight should have its own learning rate. In our modeling work, to save the effort in choosing appropriate learning rate, this algorithm is implemented for automatic parameter adjustment.

2.3 Demonstration-Based Programming

After establishing a framework for GRN modeling, we then use it to construct behavior controllers for robots. In this work, we develop an imitation-based method to collect data of behavior sequence as gene expression profiles, and employ the presented methodology to infer behavior controllers automatically.

The imitation mechanism has two parts: one is an active process that is for acquiring new behaviors; the other is a passive process for imitating known behaviors. For a robot, the former is to try to employ a certain learning strategy to produce the behavior that is currently shown but not known previously. And the latter is to recognize what kind of behavior a demonstrator is performing and to retrieve the same kind of behavior that has been previously developed and recorded in the memory. As the passive process can be achieved by a straightforward way (i.e., building a mapping table to link the extracted behavior trajectory to the most similar one recorded previously), in this work we concentrate on the active process (i.e., learning new behaviors). For active imitation, we take an engineering point of view and consider imitation as a vehicle for learning new behaviors. It can be considered as a method of programming by demonstration [11][12]. In this method, the robot is firstly shown how to perform the desired behavior: it is driven manually to achieve the target task. In this stage, the robot can be regarded as a teacher showing the correct behavior. During the period of human-driven demonstration, at each time step the relevant information received from the robot's sensors and actuators are recorded to form a behavior data set for later training. In other words, it is to derive the time-series expression profiles of sensors and actuators from the qualitative behavior demonstrated by the robot.

After the behavior data is obtained, in the second stage the robot plays the role of a learner that is trained to achieve the target task. As described in the above sections, a RNN-based GRN model is adopted as the behavior controller here for the learner, and the corresponding learning algorithm is used to train the controller. To cope with different environment situations, the robot is operated to achieve the target behavior a few times so that a reliable and robust controller can be obtained. All expression data from different behavior trials are collected and arranged in a single training set. By minimizing the accumulated action error over the entire training data set, the robot can improve its behavior and finally achieve the task. If the robot cannot produce a similar behavior as in the demonstration, the user can modify the training set by driving the robot to repeat the sub-behaviors that it failed to achieve, and then adding the newly obtained patterns to the data set to start the re-learning procedure.

3 Experiments and Results

Following the proposed GRN model and the corresponding learning procedure, we conduct two series of experiments to evaluate our methodology. The first series is to examine whether the proposed model can be inferred from a given set of time series data. The second series of experiments is to investigate whether the proposed approach can construct behavior controllers for robots, in which the data sets of behavior sequences are collected from the demonstration procedure.

3.1 Modeling GRNs from Expression Data

In the experiments of GRN modeling, we firstly used the GRN simulation software Genexp (reported in [8]) to produce expression data. Different gene networks were defined in which the accumulation and degradation rate constants of gene product were chosen from preliminary test. Due to the space limitation, only one set of the experiments is reported here.

In this experiment, a four genes network was defined. The simulation was run for 30 time steps for data collection. Then the proposed approach was employed to infer the above network. The result is shown in Fig. 2. It compares the system behaviors of the original and reconstructed networks, in which the x-axis represents time step and y-axis, the concentrations of different gene components. As can be observed, the behaviors of the two systems are nearly identical and the accumulated error for the five nodes is very small. It shows that the network can be reconstructed from the expression data by the modeling framework presented.

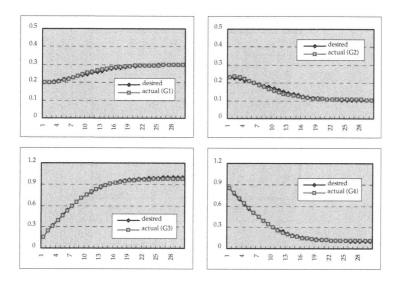

Fig. 2. Behaviors of the target (desired) and inferred (actual) systems

The second experiment for evaluating our GRN modeling method is to infer a S-system that has a power-law form and has been proposed as a precise model for regulatory networks. Here, we chose the same regulatory network reported in [13] as the target network. It consists of five nodes and their relationship can be described as the following:

$$\dot{X}_1 = 15.0X_3 X_5^{-0.1} - 10.0X_1^{2.0}$$
$$\dot{X}_2 = 10.0X_1^{2.0} - 10.0X_2^{2.0}$$
$$\dot{X}_3 = 10.0X_2^{-0.1} - 10.0X_2^{-0.1}X_3^{2.0}$$
$$\dot{X}_4 = 8.0X_1^{2.0}X_5^{-0.1} - 10.0X_4^{2.0}$$
$$\dot{X}_5 = 10.0X_4^{2.0} - 10.0X_5^{2.0}$$

Again, the proposed approach has been employed to infer the above network. The upper part of Fig. 3 shows the original (desired) time series data for the five nodes, and the lower part, the expressions of the synthesized system. As can be observed, the behaviors of the two systems are nearly identical and the accumulated error for the five nodes is very small. It shows that a S-system can also be modeled by our RNN-based network, and the network can be reconstructed from the expression data by the learning mechanism presented.

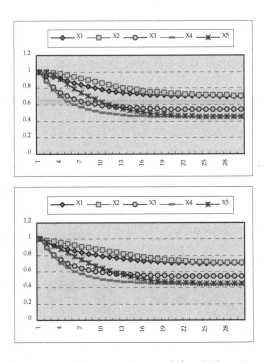

Fig. 3. Behaviors of the target (up) and inferred (down) systems

3.2 Learning GRNs for Robot Control

The second series of experiments is to examine whether the GRN controllers can also be inferred by the learning mechanism from the data set collected in the human-driven demonstration. Experiments have been conducted for two kinds of robots: the mobile robot and the walking robot. To save evaluation time in the procedure of learning, the experiments were performed in simulation. Many studies have shown that the controllers developed in realistic simulators can be transferred successfully to real robots [14].

In the first experiment the robot was expected to use its gripper (originally raised) to pick up the object located in the environment, and then put the object down outside the working area. The simulated robot has a gripper that can be lifted up and put down, and the two fingers of the gripper can be opened and closed to grab and release the object. In the simulation, a fine time-slice technique was used and each time step lasted for 100 ms. For this task, a GRN model with six gene nodes was used to control the robot: two nodes for wheel motors, two nodes for controlling the action of the gripper (to pick up/release the object), and two nodes as inter-genes. The perception information from infrared sensors mounted in the front side of the robot and the sensor equipped on the inner side of the gripper (to detect the object) was used as extra input for each nodes. The simulated robot was driven manually to perform the target task, and the relevant information was recorded to train the robot. With the quantitative behavior data set, the learning algorithm associated with the GRN model was employed to derive a controller that satisfied the collected data by minimizing the accumulated motor error. Fig. 4 illustrates the robot and the typical behavior produced by the controller that was successfully inferred by the proposed approach.

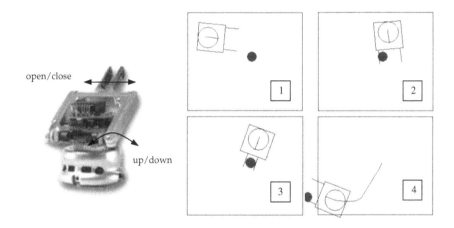

Fig. 4. The robot and its typical behavior

Different from the above reactive control task, the second task involves internal states: to develop a locomotion controller for a walking robot. It has been shown that animals walk in a stereotyped way by adopting rhythmic patterns of movement, and the neural control of these stereotyped movements is hierarchically organized. It has also become clear that these kinds of motion patterns are controlled by the rhythm generator mechanism called central pattern generator (CPG) that provides feedforward signals needed for locomotion even in the absence of sensory feedback and high-level control [15]. In this experiment, the proposed GRN model works as the role of CPG to control the motors of the robot legs.

As the main goal here is to examine whether the locomotion task can be achieved by the presented approach, we did not built a walking robot but simply took the activation data of the successful controller developed in [16] as the time-series data for the inference of a GRN model. The robot used in their work has six legs that may be either up or down, and the robot can swing the leg forward or backward when it is up. Each leg has a limited angular motion that can be used as sensory information of the leg controller. To simplify the control architecture for a six legs insect, they assumed that the locomotion controller exhibits left-right and front-back symmetries. Therefore, only one set of leg controller parameters needs to be determined, and then these parameters can be copied to each of the six legs. More details of the robot can be found in [16].

The left part of Fig. 5 shows the activation data produced by a typical leg controller reported in [16], in which the x-axis represents the time steps and y-axis, the normalized activations of the actuators (from top to down: foot, swing forward, and swing backward). To obtain a controller with the same behavior, in our experiments, a GRN model with five gene nodes was used to control a leg: one node for foot control, two nodes for forward and backward swing control, and two nodes as inter-genes. As in [21], two sets of experiments, with (a) and without (b) utilizing sensory information, have been conducted for the locomotion task here. The proposed approach was employed again to derive controllers by minimizing the accumulated motor error. The results are presented in the right part of Fig. 5, which show the controllers can be inferred successfully.

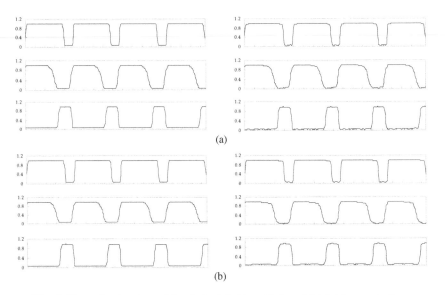

Fig. 5. The activation data of the desired (left) and inferred (right) leg controllers

4 Conclusions and Future Work

Gene regulatory networks have been shown to play important roles in the development of living organism. With many unique characteristics, they can be modeled to work as reliable and robust robot controllers. Therefore in this work we develop a methodology, including deriving a RNN-based GRN model, building a learning mechanism, and designing a procedure of programming by demonstration, to achieve GRN-based robot control. To verify the proposed methodology, different sets of experiments have been conducted. The results have shown that our approach can be successfully used to infer regulatory systems from measured data to control a mobile robot and a walking robot.

Currently we are extending our work in several directions. The first is to use the proposed approach to construct controllers for tasks involving more internal states. In those cases, the controllers can take the advantage of their feedback links to operate in the environment with and without sensory information. Meanwhile, it is worthwhile to investigate how to employ other types of learning algorithms to improve the modeling performance. Another direction is to deal with the scalability problem in which more network nodes and connections will be needed eventually for more complicated control tasks. We plan to develop and integrate an efficient clustering scheme into our framework to construct networks in a hierarchical way.

References

1. Davison, E.H., Rast, J.P., Oliveri, P., et al.: A genomic regulatory network for development. Science 295, 1669–1678 (2002)
2. deJong, H.: Modeling and simulation of genetic regulatory systems: a literature review. Journal of Computational Biology 9, 67–103 (2002)
3. Cho, K.H., Choo, S.M., Jung, S.H., et al.: Reverse engineering of gene regulatory networks. IET System Biology 1, 149–163 (2007)
4. Quick, T., Nehaniv, C.L., Dautenhahn, K., et al.: Evolving embodied genetic regulatory network-driven control systems. In: Proceeding of the Seventh European Conference on Artificial Life, pp. 266–277 (2003)
5. Stewart, F., Taylor, T., Konidaris, G.: METAMorph: Experimenting with genetic regulatory networks for artificial development. In: Capcarrère, M.S., Freitas, A.A., Bentley, P.J., Johnson, C.G., Timmis, J. (eds.) ECAL 2005. LNCS, vol. 3630, pp. 108–117. Springer, Heidelberg (2005)
6. Blasi, M.F., Casorelli, I., Colosimo, A., et al.: A recursive network approach can identify constitutive regulatory circuits in gene expression data. Physica A 348, 349–377 (2005)
7. Beer, R.D.: Dynamical approaches to cognitive science. Trends in Cognitive Sciences 4, 91–99 (2000)
8. Vohradsky, J.: Neural network model of gene expression. The FASEB Journal 15, 846–854 (2001)
9. Werbos, P.J.: Backpropagation through time: what it does and how to do it. Proceedings of the IEEE 78, 1550–1560 (1990)
10. Jacobs, R.A.: Increased rates of convergence through learning rate adaptation. Neural Networks 1, 295–307 (1988)
11. Dillmann, R.: Teaching and learning of robot tasks via observation of human performance. Robotics and Autonomous Systems 47, 109–116 (2004)
12. Nakaoka, S., Nakazawa, A., Kanehiro, F., et al.: Learning from observation paradigm: leg task models for enabling a biped humanoid robot to imitate human dance. International Journal of Robotics Research 26, 829–844 (2007)
13. Ando, S., Sakamoto, E., Iba, H.: Evolutionary modeling and inference of gene network. Information Sciences 145, 237–259 (2002)
14. Nolfi, S., Floreano, D.: Evolutionary Robotics: The Biology, Intelligence, and Technology of Self-Organizing Machine. MIT Press, MA (2000)
15. Ijspeert, A.J.: Central pattern generators for locomotion control in animals and robots : a review. Neural Networks 21, 642–653 (2008)
16. Beer, R.D., Gallagher, J.C.: Evolving dynamical neural networks for adaptive behavior. Adaptive Behavior 1, 91–122 (1992)

Discriminating Angry, Happy and Neutral Facial Expression: A Comparison of Computational Models

Aruna Shenoy[1], Sue Anthony[2], Ray Frank[1], and Neil Davey[1]

[1] Department of Computer Science, University of Hertfordshire, Hatfield, AL10 9AB, UK
[2] Department of Psychology, University of Hertfordshire, Hatfield, AL10 9AB, UK
{a.1.shenoy,s.1.anthony,r.j.frank,n.davey}@herts.ac.uk

Abstract. Recognizing expressions are a key part of human social interaction, and processing of facial expression information is largely automatic for humans, but it is a non-trivial task for a computational system. The purpose of this work is to develop computational models capable of differentiating between a range of human facial expressions. Raw face images are examples of high dimensional data, so here we use two dimensionality reduction techniques: Principal Component Analysis and Curvilinear Component Analysis. We also preprocess the images with a bank of Gabor filters, so that important features in the face images are identified. Subsequently the faces are classified using a Support Vector Machine. We show that it is possible to differentiate faces with a neutral expression from those with a happy expression and neutral expression from those of angry expressions and neutral expression with better accuracy. Moreover we can achieve this with data that has been massively reduced in size: in the best case the original images are reduced to just 5 components with happy faces and 5 components with angry faces.

Keywords: Facial Expressions, Image Analysis, Classification, Dimensionality Reduction.

1 Introduction

According to Ekman and Friesen [1] there are six easily discernible facial expressions: anger, happiness, fear, surprise, disgust and sadness, apart from neutral. Moreover these are readily and consistently recognized across different cultures [2]. In the work reported here we show how a computational model can identify facial expressions from simple facial images. In particular, we show how happy faces with neutral faces and angry faces with neutral faces can be differentiated.

Data presentation plays an important role in any type of recognition. High dimensional data is normally reduced to a manageable low dimensional data set. We perform dimensionality reduction using Principal Component Analysis (PCA) and Curvilinear Component Analysis (CCA). PCA is a linear projection technique and it may be more appropriate to use a non linear Curvilinear Component Analysis (CCA) [3]. The Intrinsic Dimension (ID) [4], which is the true dimension of the data, is often much less than the original dimension of the data. To use this efficiently, the actual dimension of the data must be estimated. We use the Correlation Dimension to estimate the Intrinsic

D. Palmer-Brown et al. (Eds.): EANN 2009, CCIS 43, pp. 200–209, 2009.

Dimension. We compare the classification results of these methods with raw face images and of Gabor Pre-processed images [5, 6]. The features of the face (or any object for that matter) may be aligned at any angle. Using a suitable Gabor filter at the required orientation, certain features can be given high importance and other features less importance. Usually, a bank of such filters is used with different parameters and later the resultant image is a *L2 max* (at every pixel the maximum of feature vector obtained from the filter bank) superposition of the outputs from the filter bank.

2 Background

We perform feature extraction with Gabor filters and then use dimensionality reduction techniques such as Principal Component Analysis (PCA) and Curvilinear Component Analysis (CCA) followed by a Support Vector Machine (SVM) [7] based classification technique and these are described below.

2.1 Gabor Filters

A Gabor filter can be applied to images to extract features aligned at particular orientations. Gabor filters possess the optimal localization properties in both spatial and frequency domains, and they have been successfully used in many applications [8]. A Gabor filter is a function obtained by modulating a sinusoidal with a Gaussian function. The useful parameters of a Gabor filter are orientation and frequency. The Gabor filter is thought to mimic the simple cells in the visual cortex. The various 2D receptive field profiles encountered in populations of simple cells in the visual cortex are well described by an optimal family of 2D filters [9]. In our case a Gabor filter bank is implemented on face images with 8 different orientations and 5 different frequencies.

Recent studies on modeling of visual cortical cells [10] suggest a tuned band pass filter bank structure. Formally, the Gabor filter is a Gaussian (with variances S_x and S_y along x and y-axes respectively) modulated by a complex sinusoid (with centre frequencies U and V along x and y-axes respectively) and is described by the following equation:-

$$g(x,y) = \frac{\exp\left[-\frac{1}{2}\left\{\left(\frac{x}{S_x}\right)^2 + \left(\frac{y}{S_y}\right)^2\right\} + 2\pi j(Ux+Vy)\right]}{2\pi S_x S_y} \quad (1)$$

The variance terms S_x and S_y dictates the spread of the band pass filter centered at the frequencies U and V in the frequency domain. This filter has real and imaginary part.

A Gabor filter can be described by the following parameters: The S_x and S_y of the Gaussian explain the shape of the base (circle or ellipse), frequency (f) of the sinusoid, orientation (Θ) of the applied sinusoid. Figure 1 shows examples of various Gabor filters. Figure 2 b) shows the effect of applying a variety of Gabor filters shown in Figure 1 to the sample image shown in Figure 2 a). Note how the features at particular orientations are exaggerated.

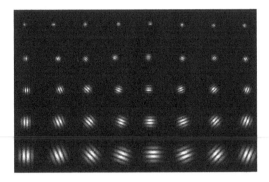

Fig. 1. Gabor filters - Real part of the Gabor kernels at five scales and eight orientations

An augmented Gabor feature vector is created of a size far greater than the original data for the image. Every pixel is then represented by a vector of size 40 and demands dimensionality reduction before further processing. So a 63 × 63 image is transformed to size 63 × 63 × 5 × 8. Thus, the feature vector consists of all useful information extracted from different frequencies, orientations and from all locations, and hence is very useful for expression recognition.

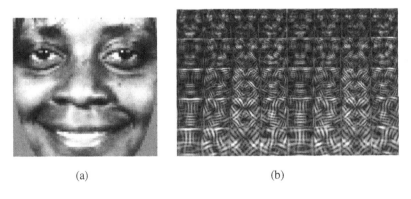

(a)   (b)

Fig. 2. a) Original face image 63 × 63 (3969 dimensions). b) Forty Convolution outputs of Gabor filters.

Once the feature vector is obtained, it can be handled in various ways. We simply take the *L2 max* norm for each pixel in the feature vector. So that the final value of a pixel is the maximum value found by any of the filters for that pixel.

The *L2 max* norm Superposition principle is used on the outputs of the filter bank and the Figure 3 b) shows the output for the original image of Figure 3 a).

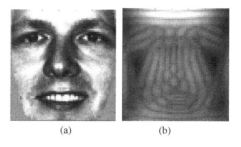

(a)   (b)

Fig. 3. a) Original Image used for the Filter bank b) Superposition output (*L2 max* norm)

2.2 Curvilinear Component Analysis

Curvilinear Component Analysis (CCA) is a non-linear projection method that preserves distance relationships in both input and output spaces. CCA is a useful method for redundant and non linear data structure representation and can be used in dimensionality reduction. CCA is useful with highly non-linear data, where PCA or any other linear method fails to give suitable information [3]. The D–dimensional input X should be mapped onto the output d-dimensional space Y. Their d-dimensional output vectors { y_i } should reflect the topology of the inputs { x_i }. In order to do that, Euclidean distances between the x_i's are considered. Corresponding distances in the output space y_i's is calculated such that the distance relationship between the data points is maintained.

(a)   (b)   (c)

Fig. 4. (a) 3D horse shoe dataset (b) 2D CCA projection (c) $dy - dx$ plot

CCA puts more emphasis on maintaining the short distances than the longer ones. Formally, this reasoning leads to the following error function:

$$E = \frac{1}{2} \sum_{i=1}^{N} \sum_{j=1}^{N} \left(d_{i,j}^{X} - d_{i,j}^{Y} \right)^2 F_\lambda \left(d_{i,j}^{Y} \right) \forall \; j \neq i \qquad (2)$$

where $d^x_{i,j}$ and $d^Y_{i,j}$ are the Euclidean distances between the points i and j in the input space X and the projected output space Y respectively and N is the number of data points. $F\!\left(d^Y_{i,j}\right)$ is the neighbourhood function, a monotonically decreasing function of distance. In order to check that the relationship is maintained a plot of the distances in the input space and the output space ($dy - dx$ plot) is produced. For a well maintained topology, dy should be proportional to the value of dx at least for small values of $dy's$. Figure 4 shows CCA projections for the 3D data horse shoe data. The $dy - dx$ plot shown is good in the sense that the smaller distances are very well matched [3].

2.3 Intrinsic Dimension

One problem with CCA is deciding how many dimensions the projected space should occupy, and one way of obtaining this is to use the intrinsic dimension of the data manifold. The Intrinsic Dimension (ID) can be defined as the minimum number of free variables required to define data without any significant information loss. Due to the possibility of correlations among the data, both linear and nonlinear, a D-dimensional dataset may actually lie on a d-dimensional manifold ($D \geq d$). The ID of such data is then said to be d. There are various methods of calculating the ID; here we use the correlation Dimension [8] to calculate the ID of face image dataset.

2.4 Classification Using Support Vector Machines

A number of classifiers can be used in the final stage for classification. We have concentrated on the Support Vector Machine. Support Vector Machines (SVM) are a set of related supervised learning methods used for classification and regression. SVM's are used extensively for many classification tasks such as: handwritten digit recognition [11] or Object Recognition [12]. A SVM implicitly transforms the data into a higher dimensional data space (determined by the kernel) which allows the classification to be accomplished more easily. We have used the LIBSVM tool [7] for SVM classification.

The SVM is trained in the following way:

```
1. Transform the data to a format required for using the
   SVM software package - LIBSVM -2.83 [7].
2. Perform simple scaling on the data so that all the
   features or attributes are in the range [-1, +1].
3. Choose a kernel. We used the RBF kernel,
```

$$k(x, y) = e^{-\gamma\left|x-y\right|^2}.$$

4. Perform fivefold cross validation with the specified kernel to find the best values of the cost parameter C and γ.

5. By using the best value of C and γ, train the model and finally evaluate the trained classifier using the test sets.

3 Experiments and Results

We experimented on 264 faces (132 female and 132 male) each with three classes, namely: *Neutral* and *Happy* and *Angry* (88 faces for each expression). Neutral and Happy are used in one experiment and Neutral and Angry faces are used in another. The images are from the BINGHAMTON dataset [13] and some examples are shown in Figure 5. Two training and test sets are used. One training set had 132 faces (with 66 female, 66 male and equal numbers of them with neutral and happy expression) and another training set had 132 faces (with 66 female and 66 male and equal numbers of them with neutral and angry expression). The original 128×128 image was reduced to 63×63. As we have two training sets, we have two test sets. Each consists of 44 faces (11 female, 11 male and equally balanced number of expression: neutral with angry and neutral with happy).

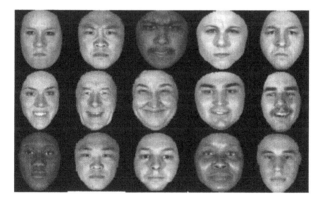

Fig. 5. Example BINGHAMTON images used in our experiments which are cropped to the size of 128×128 to extract the facial region and reduced to 63×63 for all experiments. The first row has examples of angry expression, middle row has happy expression and last row has images with neutral expression.

For PCA reduction we always use the first principal components which account for 95% of the total variance of the data, and project the data onto these principal components - we call this is our standard PCA reduction. With Neutral and Happy faces, this resulted in using 100 components of the raw dataset and 23 components in the Gabor pre-processed dataset. With Neutral and Angry faces, this resulted in using 97 components of the raw dataset and 22 components in the Gabor pre-processed dataset. As CCA is a highly non-linear dimensionality reduction technique, we use the intrinsic dimensionality technique and reduce the components to its Intrinsic Dimension. The Intrinsic Dimension of the raw faces with Neutral and Happy was approximated as 6 and that of Gabor pre- processed images was 5. Likewise, the Intrinsic Dimension of the raw faces with Neutral and Angry was approximated as 5 and that of Gabor pre-processed images was 6. Figure 6 shows the Eigenfaces obtained by the PCA technique with raw faces (Happy with Neutral set and Angry with Neutral set).

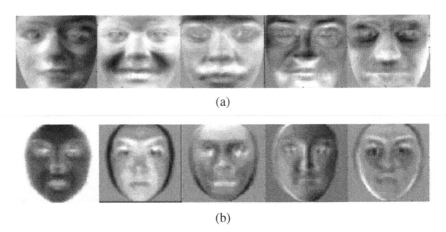

(a)

(b)

Fig. 6. a) The first 5 eigenfaces of the neutral and happy data set. b) The first 5 eigenfaces of the neutral and angry data set.

The results of the SVM classification for Neutral and Happy are as in Table 1 and for Neutral and Angry are as in Table 2. The PCA, being a linear dimensionality reduction technique, did not do quite as well as CCA with happy and neutral data set; however, there has been no difference with the angry and neutral dataset. With CCA there was good generalization, but the key point to be noted here is the number of components used for the classification. The CCA makes use of just 6 components with raw faces get good classification result and 5 components with the Gabor pre-processed images with the neutral and happy dataset. With the angry and neutral dataset, the CCA makes use of 5 components with raw faces and 6 components with Gabor pre-processed faces with results comparable with the raw faces.

Table 1. SVM Classification accuracy of raw faces and Gabor pre-processed images with PCA and CCA dimensionality reduction techniques for Neutral and Happy dataset

SVM% accuracy	Happy and Neutral (44 faces)
Raw faces	100 (44/44)
Raw with PCA100	88.64 (39/44)
Raw with CCA6	93.18 (41/44)
Gabor pre-processed faces	90.91(40/44)
Gabor with PCA23	95.45 (42/44)
Gabor with CCA5	68.18 (30/44)

Table 2. SVM Classification accuracy of raw faces and Gabor pre-processed images with PCA and CCA dimensionality reduction techniques for Neutral and Angry dataset

SVM% accuracy	Angry and Neutral (44 faces)
Raw faces	79.54 (35/44)
Raw with PCA97	68.18 (30/44)
Raw with CCA5	68.18 (30/44)
Gabor pre-processed faces	68.18 (30/44)
Gabor with PCA22	61.36 (27/44)
Gabor with CCA6	63.64 (28/44)

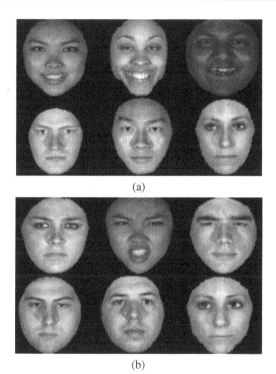

(a)

(b)

Fig. 7. Examples of the most often misclassified set of faces. (a) Top row shows happy faces wrongly classified as neutral. Bottom row shows neutral faces wrongly classified as happy. (b) Top row shows angry faces wrongly classified as neutral. Bottom row shows neutral faces wrongly classified as angry.

The classification results for the Neutral and Happy face images shown in Table 1, indicates best classification using raw faces. The intrinsic dimensionality of the raw images is found to be just 6 and the CCA projection therefore reduces the images to 6 components. It should be noted that even with just these 6 components, the SVM gives very good classification. The standard PCA reduced raw images did not give good classification. However, with Gabor pre-processed faces followed by standard

PCA reduction gave much better results. Interestingly, Gabor pre-processing does not help the non-linear CCA method.

The classification results for the Neutral and Angry face images shown in Table 2, indicates the overall classification accuracy is not as good as with the happy versus neutral dataset. Classifying angry faces is a difficult task for computation models and can be seen from these results. Nevertheless, the SVM performs well with 79.54% accuracy with raw faces. There is not much difference in the classification accuracy with raw faces reduced in dimensionality with PCA and CCA.

The results with both sets of data suggest that the raw face images give the best classification results. From some of the examples of misclassifications shown in Figure 7, it is not clear which feature has caused misclassification. Hence, we are currently undertaking further experiments with human subjects. We are attempting to see if there are any associations between the computational model and human performance.

4 Conclusions

Identifying facial expressions is a challenging and interesting task. Our experiment shows that identification from raw images can be performed very well with happy faces and angry faces. However, with a larger data set, it may be computationally intractable to use the raw images. It is therefore important to reduce the dimensionality of the data. The dimensionality reduction methods do fairly well. A linear method such as PCA does not appear to be sufficiently tunable to identify features that are relevant for facial expression characterization. However, performing Gabor pre-processing on the images increases the classification accuracy of the data after performing PCA in the case of happy and neutral face images. This, however, does not apply to images that are subjected to dimensionality reduction with CCA. Gabor pre-processed PCA data with just 23 components is capable of performing well in comparison to the raw images reduced with PCA. The Gabor pre-processed CCA images, however, with just 5 components does not yield such comparable results. With the second model, classifying angry with neutral faces, the raw faces manage to give just 35 out of 44 faces correct (79.54%) and indicates the difficulty of classifying angry faces. Though the results of the classification for PCA and CCA processed raw images are comparable, it can be noted that Gabor pre-processing has managed to provide good classification with PCA reduced data and with CCA with just 23 and 6 components respectively. Future work will include extending the experiment to other four expressions and comparing the performance of the computational model with performance by human subjects.

References

1. Ekman, P., Friesen, W.V.: Constants across cultures in the face of the emotion. Journal of Personality and Social Psychology 17 (1971)
2. Batty, B., Taylor, M.J.: Early processing of the six basic facial emotional expressions. In: Cognitive Brain Research, p. 17 (2003)

3. Demartines, P., de Hérault, J.: Curvilinear component analysis: A self-organizing neural network for nonlinear mapping of data sets. IEEE Transactions on Neural Networks 8(1), 148–154 (1997)
4. Grassberger, P., Proccacia, I.: Measuring the strangeness of strange attractors. Physica D, 9 (1983)
5. Jain, A.K., Farrokhnia, F.: Unsupervised texture segmentation using Gabor filters. Pattern Recognition 24(12) (1991)
6. Movellan, J.R.: Tutorial on Gabor Filters (2002)
7. Chang, C.-C., Lin, C.-J.: LIBSVM: a library for support vector machines (2001)
8. Zheng, D., Zhao, Y., Wang, J.: Features Extraction using A Gabor Filter Family. In: Proceedings of the sixth Lasted International conference, Hawaii (2004)
9. Daugman, J.G.: Uncertainty relation for resolution in space, spatial frequency and orientation optimized by two dimensional visual cortical filters. Journal of Optical.Society of America A 2(7) (1985)
10. Kulikowski: Theory of spatial position and spatial frequency relations in the receptive fields of simple cells in the visual cortex. Biological Cybernetics 43(3), 187–198 (1982)
11. Cortes, C., Vapnik, V.: Support Vector Networks. Machine Learning 20, 273–297 (1995)
12. Blanz, V., et al.: Comparison of view-based object recognition algorithms using realistic 3D models. In: Proc. Int. Conf. on Artificial Neural Networks 1996, pp. 251–256 (1996)
13. Yin, L., Wei, X., Sun, Y., Wang, J., Rosato, M.J.: A 3D Facial Expression Database For Facial Behavior Research. In: 7th International Conference on Automatic Face and Gesture Recognition, FGR 2006 (2006)

Modeling and Forecasting CAT and HDD Indices for Weather Derivative Pricing

Achilleas Zapranis and Antonis Alexandridis*

Department of Accounting and Finance, University of Macedonia of Economics and Social
Studies, 156 Egnatia St., P.O. 54006, Thessaloniki, Greece
zapranis@uom.gr, alex@uom.gr, aalex@uom.gr

Abstract. In this paper, we use wavelet neural networks in order to model a
mean-reverting Ornstein-Uhlenbeck temperature process, with seasonality in
the level and volatility. We forecast up to two months ahead out of sample daily
temperatures and we simulate the corresponding Cumulative Average Temperature and Heating Degree Day indices. The proposed model is validated in 8
European and 5 USA cities all traded in Chicago Mercantile Exchange. Our results suggest that the proposed method outperforms alternative pricing methods
proposed in prior studies in most cases. Our findings suggest that wavelet networks can model the temperature process very well and consequently they constitute a very accurate and efficient tool for weather derivatives pricing. Finally,
we provide the pricing equations for temperature futures on Heating Degree
Day index.

Keywords: Weather Derivatives, Pricing, Forecasting, Wavelet Networks.

1 Introduction

Recently a new class of financial instruments -weather derivatives- has been introduced. The purpose of weather derivatives is to allow business to insure themselves
against fluctuations in the weather. According to [1, 2] nearly $1 trillion of the US
economy is directly exposed to weather risk. Just as traditional contingent claims,
whose payoffs depend upon the price of some fundamental, a weather derivative has
its underlying measure such as: rainfall, temperature, humidity or snowfall. Weather
derivatives are used to hedge volume risk, rather than price risk,

The Chicago Mercantile Exchange (CME) reports that the estimated value of its
weather products reached $22 billion through September 2005, with more than
600,000 contracts traded. This represents sharp rise comparing with 2004 in which
notional value was $2.2 billion [3]. Moreover, it is anticipated that the weather market
will continue to develop, broadening its scope in terms of geography, client base and
inter-relationship with other financial and insurance markets. In order to fully exploit
all the advantages that this market offers, adequate pricing approach is required.

Weather risk is unique in that it is highly localized, and despite great advances in
meteorological science, still cannot be predicted precisely and consistently. Weather

---

* Corresponding author.

D. Palmer-Brown et al. (Eds.): EANN 2009, CCIS 43, pp. 210–222, 2009.

derivatives are also different than other financial derivatives in that the underlying weather index, like Heating Degree Days (HDD), Cooling Degree Days (CDD), Cumulative Average Temperature (CAT), etc. cannot be traded. Mathematical explanations of these indices are given in section 4. Furthermore, the corresponding market is relatively illiquid. Consequently, since weather derivatives cannot be cost-efficiently replicated with other weather derivatives, arbitrage pricing cannot directly apply to them. The weather derivatives market is a classic incomplete market, because the underlying weather variables are not tradable

The first and simplest method that has been used in weather derivative pricing is burn analysis. Burn analysis is just a simple calculation of how a weather derivative would perform in the past years. By taking the average of these values an estimate of the price of the derivative is obtained. Burn analysis is very easy in calculation since there is no need to fit the distribution of the temperature or to solve any stochastic differential equation. Moreover, burn analysis is based in very few assumptions. First, we have to assume that the temperature time-series is stationary. Next, we have to assume that the data for different years are independent and identically distributed. For a detailed explanation of Burn Analysis we refer to [4].

A closer inspection of a temperature time series shows that none of these assumptions are correct. It is clear that the temperature time-series is not stationary since it contains seasonalities, jumps and trends [5, 6]. Also the independence of the years is under question. [4] show that these assumptions can be used if the data can be cleaned and detrended. Although their results show that the pricing still remains inaccurate. Other methods as index and daily modeling are more accurate but still burn analysis is usually a good first approximation of the derivative's price.

In contrast to the previous methods, a dynamic model can be used which directly simulates the future behavior of temperature. Using models for daily temperatures can, in principle, lead to more accurate pricing than modeling temperature indices. On the other hand, deriving an accurate model for the daily temperature is not a straightforward process. Observed temperatures show seasonality in all of the mean, variance, distribution and autocorrelations and long memory in the autocorrelations. The risk with daily modeling is that small misspecifications in the models can lead to large mispricing in the contracts.

The continuous processes used for modeling daily temperatures usually take a mean-reverting form, which has to be descretized in order to estimate its various parameters. Previous works suggest that the parameter of the speed of mean reversion, α, is constant. The work of [5] indicates exactly the opposite. In addition, α is important for the correct and accurate pricing of temperature derivatives, [7]. In [5] the parameter a was modeled by a Neural Network (NN). More precisely wavelet analysis was used in order to identify the trend and the seasonal part of the temperature signal and then a NN was used in the detrended and deseasonalized series. However WA is limited to applications of small input dimension since its constructing wavelet basis of a large input dimension is computationally expensive, [8]. In NN framework the initial values of the weights are randomly chosen which usually leads to large training times and to convergence to a local minimum of a specified loss function. Finally, the use of sigmoid NN does not provide any information about the network construction. In this study we expand the work of [5] by combining these two steps. To overcome these problems we use networks with wavelets as activation functions, namely Wavelet Networks, (WN). More precisely we use truncated Fourier series to

remove the seasonality and the seasonal volatility of the temperature in various locations as in [9]. Next a wavelet network is constructed in order to fit the daily average
temperature in 13 cities and to forecast daily temperature up to two months hoping
that the waveform of the activation functions of the feedforward network will fit
much better the temperature process than the classical sigmoid functions. For a con-
cise treatment of wavelet analysis we refer to [10-12] while for wavelet networks we
refer to the works of [8, 13, 14]. The forecasting accuracy from the proposed methodology is validated in a two month ahead out of sample window. More precisely the
proposed methodology is compared against historical burn analysis and the Benth &
Benth's model in forecasting CAT and HDD indices. Finally, we extend the work of
[5] by presenting the pricing equations for future HDD contracts when the speed of
mean reversion is not constant.

The rest of the paper is organized as follows. In section 2, we describe the process
used to model the average daily temperature. In section 3 a brief introduction to wavelet
networks is given. In section 4 we describe our data and apply our model to real data. In
section 5 we discuss CAT and HDD derivatives pricing and finally, in section 6 we
conclude.

2 Modeling Temperature Process

Many different models have been proposed in order to describe the dynamics of a
temperature process. Early models were using AR(1) processes or continuous equivalents [7, 15, 16]. Others like [17] and [18] have suggested versions of a more general
ARMA(p,q) model. [19] have shown, however, that all these models fail to capture
the slow time decay of the autocorrelations of temperature and hence lead to significant underpricing of weather options. Thus more complex models were proposed, like
an Ornstein-Uhlenbeck process [20]. Also in the noise part of the process, the
Brownian noise was at first replaced by a fractional Brownian noise and then by a
Levy process [21]. A temperature Ornstein-Uhlenbeck process is:

$$dT(t) = dS(t) - \kappa(T(t) - S(t))dt + \sigma(t)dB(t) \qquad (1)$$

where, $T(t)$ is the daily average temperature, $B(t)$ is a standard Brownian motion, $S(t)$
is a deterministic function modelling the trend and seasonality of the average temperature, while $\sigma(t)$ is the daily volatility of temperature variations. In [9] both $S(t)$
and $\sigma^2(t)$ were modeled as truncated Fourier series:

$$S(t) = a + bt + a_0 + \sum_{i=1}^{I_1} a_i \sin(2i\pi(t - f_i)/365) + \sum_{j=1}^{J_1} b_j \cos(2 j\pi(t - g_j)/365) \qquad (2)$$

$$\sigma^2(t) = c + \sum_{i=1}^{I_2} c_i \sin(2i\pi t/365) + \sum_{j=1}^{J_2} d_j \cos(2 j\pi t/365) \qquad (3)$$

From the Ito formula an explicit solution for (1) can be derived:

$$T(t) = s(t) + (T(t-1) - s(t-1))e^{-\kappa\tau} + \int_{t-1}^{t} \sigma(u)e^{-\kappa(t-u)}dB(u) \qquad (4)$$

According to this representation $T(t)$ is normally distributed at t and it is reverting to a mean defined by $S(t)$. A discrete approximation to the Ito formula, (4), which is the solution to the mean reverting Ornstein-Uhlenbeck process (1), is:

$$T(t+1)-T(t) = S(t+1)-S(t)-(1-e^{-k})\{T(t)-S(t)\}+\sigma(t)\{B(t+1)-B(t)\} \quad (5)$$

which can be written as:

$$\tilde{T}(t+1) = a\tilde{T}(t)+\tilde{\sigma}(t)\varepsilon(t) \quad (6)$$

where

$$\tilde{T}(t) = T(t)-S(t) \quad (7)$$

$$a = e^{-k} \quad (8)$$

In order to estimate model (6) we need first to remove the trend and seasonality components from the average temperature series. The trend and the seasonality of daily average temperatures is modeled and removed as in [9]. Next a wavelet neural network is used to model and forecast daily detrended and deseasonalized temperatures. Hence, equation (6) reduces to:

$$T(t) = \varphi(T(t-1))+e_t \quad (9)$$

where $\varphi(\bullet)$ is estimated non-parametrically by a wavelet network. Hence the parameter α is not constant. Once we have the estimator of the underlying function φ, then we can compute the daily values of α as follows:

$$a = d\tilde{T}(t+1)/d\tilde{T}(t) = d\varphi/d\tilde{T} \quad (10)$$

The analytic expression for the wavelet network derivative $d\varphi/d\tilde{T}$ can be found in [14]. Due to space limitations we will refer to the works of [5, 9] for the estimation of parameters in equations (2), (3), (6) and (8).

3 Wavelet Neural Networks for Multivariate Process Modeling

Here the emphasis is in presenting the theory and mathematics of wavelet neural networks. So far in literature various structures of a WN have been proposed [8, 13, 22]. In this study we use a multidimensional wavelet neural network with a linear connection of the wavelons to the output. Moreover in order for the model to perform well in linear cases we use direct connections from the input layer to the output layer. A network with zero hidden units (HU) is the linear model.

The network output is given by the following expression:

$$\hat{y}(\mathbf{x}) = w_{\lambda+1}^{[2]} + \sum_{j=1}^{\lambda} w_j^{[2]} \cdot \Psi_j(\mathbf{x}) + \sum_{i=1}^{m} w_i^{[0]} \cdot x_i \quad (11)$$

In that expression, $\Psi_j(\mathbf{x})$ is a multidimensional wavelet which is constructed by the product of m scalar wavelets, $\mathbf{x}$ is the input vector, m is the number of network inputs,

λ is the number of hidden units and w stands for a network weight. Following [23] we use as a mother wavelet the Mexican Hat function. The multidimensional wavelets are computed as follows:

$$\Psi_j(\mathbf{x}) = \prod_{i=1}^{m} \psi(z_{ij}) \qquad (12)$$

where ψ is the mother wavelet and

$$z_{ij} \quad \frac{x_i - w^{[1]}_{(\xi)ij}}{w^{[1]}_{(\zeta)ij}} \qquad (13)$$

In the above expression, $i = 1, \ldots, m$, $j = 1, \ldots, \lambda+1$ and the weights w correspond to the translation ($w^{[1]}_{(\xi)ij}$) and the dilation ($w^{[1]}_{(\zeta)ij}$) factors. The complete vector of the network parameters comprises:

$$w = \left(w_i^{[0]}, w_j^{[2]}, w_{\lambda+1}^{[2]}, w^{[1]}_{(\xi)ij}, w^{[1]}_{(\zeta)ij} \right) \qquad (14)$$

There are several approaches to train a WN. In our implementation we have used ordinary back-propagation which is less fast but also less prone to sensitivity to initial conditions than higher order alternatives. The weights $w_i^{[0]}$, $w_j^{[2]}$ and parameters $w^{[1]}_{(\xi)ij}$ and $w^{[1]}_{(\zeta)ij}$ are trained for approximating the target function.

In WN, in contrast to NN that use sigmoid functions, selecting initial values of the dilation and translation parameters randomly may not be suitable, [22]. A wavelet is a waveform of effectively limited duration that has an average value of zero and localized properties hence a random initialization may lead to wavelons with a value of zero. Also random initialization affects the speed of training and may lead to a local minimum of the loss function, [24]. In literature more complex initialization methods have been proposed, [13, 23, 25]. All methods can be summed in the following three steps.

1. Construct a library W of wavelets
2. Remove the wavelets that their support does not contain any sample points of the training data.
3. Rank the remaining wavelets and select the best regressors.

The wavelet library can be constructed either by an orthogonal wavelet or a wavelet frame. However orthogonal wavelets cannot be expressed in closed form. It is shown that a family of compactly supported non-orthogonal wavelets is more appropriate for function approximation, [26]. The wavelet library may contain a large number of wavelets. In practice it is impossible to count infinite frame or basis terms. However arbitrary truncations may lead to large errors, [27].

In [23] three alternative methods were proposed in order to reduce and rank the wavelet in the wavelet library namely the Residual Based Selection (RBS) a Stepwise Selection by Orthogonalization (SSO) and a Backward Elimination (BE) algorithm. In this study we use the BE initialization method that proved in previous studies to outperform the other two methods, [14, 23].

All the above methods are used just for the initialization of the dilation and translation parameters. Then the network is further trained in order to obtain the vector of the parameters $w = w_0$ which minimizes the cost function.

4 Modeling and Forecasting CAT and HDD Indices

In this section real weather data will be used in order to validate our model. The data set consists of 4015 values, corresponding to the average daily temperatures of 11 years (1995-2005) in Paris, Stockholm, Rome, Madrid, Barcelona, Amsterdam, London and Oslo in Europe and New York, Atlanta, Chicago, Portland and Philadelphia in USA. Derivatives on the above cities are traded in CME. In order for each year to have equal observations the 29th of February was removed from the data. Finally, the model was validated in data consisting of 2 months, January – February, of daily average temperatures (2005-2006) corresponding to 59 values. Note that meteorological forecasts over 10 days are not considered accurate.

The list of traded contracts in weather derivatives market is extensive and constantly evolving. However over 90% of the contracts are written on temperature CAT and HDD indices. In Europe, CME weather contracts for the summer months are based on an index of CAT. The CAT index is the sum of the daily average temperatures over the contract period. The average temperature is measured as the simple average of the minimum and maximum temperature over one day. The value of a CAT index for the time interval $[\tau_1, \tau_2]$ is given by the following expression:

$$\int_{\tau_1}^{\tau_2} T(s)ds \tag{15}$$

where the temperature is measured in degrees of Celsius. In USA, CME weather derivatives are based on HDD or CDD index. A HDD is the number of degrees by which daily temperature is below a base temperature, while a CDD is the number of degrees by which the daily temperature is above the base temperature,

i.e., Daily HDD = max (0, base temperature – daily average temperature),

Daily CDD = max (0, daily average temperature – base temperature).

The base temperature is usually 65 degrees Fahrenheit in the US and 18 degrees Celsius in Europe. HDDs and CDDs are usually accumulated over a month or over a season. At the end of 2008, at CME were traded weather derivatives for 24 US cities[1], 10 European cities[2], 2 Japanese cities[3] and 6 Canadian cities[4].

---

[1] Atlanta, Detroit, New York, Baltimore, Houston, Philadelphia, Boston, Jacksonville, Portland, Chicago, Kansas City, Raleigh, Cincinnati, Las Vegas, Sacramento, Colorado Spring, Little Rock, Salt Lake City, Dallas, Los Angeles, Tucson, Des Moines, Minneapolis-St. Paul, Washington, D.C.

[2] Amsterdam, Barcelona, Berlin, Essen, London, Madrid, Paris, Rome, Stockholm, Oslo.

[3] Tokyo, Osaka.

[4] Calgary, Montreal, Vancouver, Edmonton, Toronto, Winnipeg.

Table 1 shows the descriptive statistics of the temperature in each city for the past 11 years. The mean and standard deviation HDD represent the mean and the standard deviation of the HDD index for the past 11 years for a period of two months, January and February. For consistency all values are presented in degrees Fahrenheit. It is clear that the HDD index exhibits large variability. Similar the difference between the maximum and minimum is close to 70 degrees Fahrenheit in average for all cities while the standard deviation of temperature is close to 15 degrees Fahrenheit. Also for all cities there is kurotsis significant smaller than 3 and with exceptions of Barcelona Madrid and London there is negative skewness.

Table 1. Descriptive Statistics of temperature in each city

	Mean	St.Dev	Max	Min	Skewness	Kurtosis	Mean HDD	std. HDD
Paris	54.38	12.10	89.90	13.80	-0.04	2.50	1368.40	134.95
Rome	60.20	11.37	85.80	31.10	-0.04	1.96	1075.00	121.52
Stockholm	45.51	14.96	79.20	-5.00	-0.09	2.33	2114.36	197.93
Amsterdam	51.00	11.00	79.90	12.20	-0.18	2.54	1512.07	189.05
Barcelona	61.56	10.59	85.70	32.60	0.09	2.03	899.23	102.75
Madrid	58.61	13.84	89.80	24.90	0.17	1.94	1262.56	128.53
New York	55.61	16.93	93.70	8.50	-0.15	2.08	1783.44	207.13
London	52.87	10.03	83.00	26.70	0.02	2.36	1307.11	98.80
Oslo	41.47	15.65	74.60	-8.70	-0.31	2.50	2404.53	236.19
Atlanta	62.18	14.52	89.60	13.70	-0.45	2.25	1130.75	121.19
Chicago	50.61	19.40	91.40	-12.90	-0.25	2.17	2221.33	211.89
Portland	46.80	17.36	83.20	-3.70	-0.22	2.22	2382.04	192.47
Philadelphia	56.02	17.13	90.50	9.50	-0.19	2.03	1766.98	211.65

Next we forecast the two months, 59 days, ahead out-of-sample forecasts for the CAT and cumulative HDD indices. Our method is validated and compared against two forecasting methods proposed in prior studies, the historical burn analysis (HBA) and the Benth's & Saltyte-Benth's (B-B) model which is the starting point for our methodology.

Table 2 shows the relative (percentage) errors for the CAT index of each method. It is clear that the proposed method using WN outperforms both HBA and B-B. More precisely the WN give smaller out-of-sample errors in 9 out of 13 times while it outperforms B-B in 11 out of 13 times. It is clear that the WN can be used with great success in European cities where the WN produces significant smaller errors than the alternative methods. Only in Oslo and Amsterdam WN performs worse than the HBA method but still the forecasts are better than the B-B. In USA cities WN produces the smallest out of sample error in three cases while HBA and B-B produce the smaller out of sample error in one and two cases respectively. Observing Table 1 again, we notice that when the temperature shows large negative skewness, with exception of New York, Portland and Philadelphia, the proposed method is outperformed either by HBA or by B-B. On the other hand in the cases of Barcelona and Madrid where the skewness is positive the errors using the wavelet network method are only 0.03% and 0.74% and significant smaller than the errors produced by the other two methods. Table 3 shows the relative (percentage) errors for the HDD index of each method. The results are similar.

Finally we examine the fitted residuals in model (6). Note that the B-B model, in contrast to the wavelet network model, is based on the hypothesis that the remaining residuals follow the normal distribution. It is clear from Table 4 that only in Paris the normality hypothesis marginally accepted. The Jarque-Bera statistic is slightly higher than 0.05. In every other case the normality test is rejected. More precisely the Jarque-Bera statistics are very large and the p-values are close to zero. Hence, alternative methods like wavelet analysis, must be used to capture the seasonal part of the data, [5].

Table 2. Relative errors for the three forecasting models. CAT index.

Errors	HBA	B-B	WNN
Paris	10.63%	8.34%	7.12%
Rome	4.49%	4.39%	3.93%
Stockholm	10.34%	9.47%	9.29%
Amsterdam	7.40%	8.60%	8.55%
Barcelona	1.46%	0.19%	0.03%
Madrid	7.18%	2.10%	0.74%
New York	10.63%	9.02%	8.76%
London	7.99%	6.07%	5.75%
Oslo	2.87%	5.62%	4.53%
Atlanta	2.21%	1.83%	2.58%
Chicago	15.55%	10.22%	10.90%
Portland	14.02%	8.87%	8.32%
Philadelphia	8.45%	5.95%	5.92%

Table 3. Relative errors for the three forecasting models. HDD index.

	HBA	B-B	WNN
Paris	14.77%	11.58%	9.88%
Rome	9.94%	9.71%	8.70%
Stockholm	7.09%	6.49%	6.37%
Amsterdam	9.57%	11.13%	11.06%
Barcelona	4.50%	0.57%	0.10%
Madrid	12.00%	3.51%	1.24%
New York	15.86%	13.45%	13.07%
London	12.52%	9.51%	9.01%
Oslo	1.64%	3.20%	2.58%
Atlanta	5.71%	4.74%	6.67%
Chicago	15.45%	10.15%	10.82%
Portland	11.05%	6.99%	6.56%
Philadelphia	12.11%	8.53%	8.49%

Table 4. Normality test for the B-B residuals

	Jarque-Bera	P-Value
Paris	5.7762	0.054958
Rome	170.12339	0.001
Stockholm	60.355	0.001
Amsterdam	44.6404	0.001
Barcelona	685.835	0.001
Madrid	69.52	0.001
New York	53.91428	0.001
London	11.66163	0.003947
Oslo	37.272738	0.001
Atlanta	403.0617	0.001
Chicago	44.329798	0.001
Portland	21.91905	0.001
Philadelphia	89.54923	0.001

5 Temperature Derivative Pricing

So far, we modeled the temperature using an Ornstein-Uhlenbeck process [9]. We have shown in [5] that the mean reversion parameter α in model (6) is characterized by significant daily variation. Recall that parameter α is connected to our initial model with $a = e^{-\kappa}$ where κ is the speed of mean reversion. It follows that, the assumption of a constant mean reversion parameter introduces significant error in the pricing of weather derivatives. In this section, we give the pricing formula for a future contract written on the HDD index that incorporate the time dependency of the speed of the mean reversion parameter. The corresponding equations for the CAT index already presented in [5].

The CDD, HDD indices over a period $[\tau_1, \tau_2]$ are given by

$$HDD = \int_{\tau_1}^{\tau_2} \max\left(c - T(s), 0\right) ds \tag{16}$$

$$CDD = \int_{\tau_1}^{\tau_2} \max\left(T(s) - c, 0\right) ds \tag{17}$$

Hence, the pricing equations are similar for both indices.

First, we re-write (1) where parameter κ, now is a function of time t, $\kappa(t)$.

$$dT(t) = dS(t) + \kappa(t)\left(T(t) - S(t)\right) + \sigma(t) dB(t) \tag{18}$$

From the Ito formula an explicit solution can be derived:

$$T(t) = S(t) + e^{\int_0^t \kappa(u)du}\left(T(0) - S(0)\right) + e^{\int_0^t \kappa(u)du} \int_0^t \sigma(s) e^{-\int_0^s \kappa(u)du} dB(s) \tag{19}$$

Note that $\kappa(t)$ is bounded away from zero.

Our aim is to give a mathematical expression for the HDD future price. It is clear that the weather derivative market is an incomplete market. Cumulative average temperature contracts are written on a temperature index which is not a tradable or storable asset. In order to derive the pricing formula, first we must find a risk-neutral probability measure $Q \sim P$, where all assets are martingales after discounting. In the case of weather derivatives any equivalent measure Q is a risk neutral probability. If Q is the risk neutral probability and r is the constant compounding interest rate then the arbitrage free future price of a HDD contract at time $t \leq \tau_1 \leq \tau_2$ is given by:

$$e^{-r(\tau_2 - t)} E_Q \left[\int_{\tau_1}^{\tau_2} \max\left(0, c - T(\tau)\right) d\tau - F_{HDD}(t, \tau_1, \tau_2) \mid \mathcal{F}_t \right] = 0 \qquad (20)$$

and since F_{HDD} is $\mathcal{F}_t$ adapted we derive the price of a HDD futures to be

$$F_{HDD}(t, \tau_1, \tau_2) = E_Q \left[\int_{\tau_1}^{\tau_2} \max\left(0, c - T(\tau)\right) d\tau \mid \mathcal{F}_t \right] \qquad (21)$$

Using the Girsanov's Theorem, under the equivalent measure Q, we have that
$$dW(t) = dB(t) - \theta(t)dt \qquad (22)$$
and note that $\sigma(t)$ is bounded away from zero. Hence, by combining equations (18) and (22) the stochastic process of the temperature in the risk neutral probability Q is:

$$dT(t) = dS(t) + \left(\kappa(t)\left(T(t) - s(t)\right) + \sigma(t)\theta(t) \right) dt + \sigma(t)dW(t) \qquad (23)$$

where $\theta(t)$ is a real-valued measurable and bounded function denoting the market price of risk. The market price of risk can be calculated by historical data. More specifically $\theta(t)$ can be calculated by looking the market price of contracts. The value that makes the price of the model fits the market price is the market price of risk. Using Ito formula, the solution of equation (23) is:

$$T(t) = S(t) + e^{\int_0^t \kappa(u)du} \left(T(0) - S(0)\right) + e^{\int_0^t \kappa(u)du} \int_0^t \sigma(s)\theta(s)e^{-\int_0^s \kappa(u)du} ds$$
$$+ e^{\int_0^t \kappa(u)du} \int_0^t \sigma(s)e^{-\int_0^s \kappa(u)du} dB(s) \qquad (24)$$

By replacing this expression to (21) we find the price of future contract on HDD index at time t where $0 \leq t \leq \tau_1 \leq \tau_2$. Following the notation of [28] we have the following proposition.

Proposition 1. The HDD future price for $0 \leq t \leq \tau_1 \leq \tau_2$ is given by

$$F_{HDD}(t, \tau_1, \tau_2) = E_Q \left[\int_{\tau_1}^{\tau_2} \max\left(c - T(s)\right) ds \mid \mathcal{F}_t \right] = \int_{\tau_1}^{\tau_2} v(t, s) \Psi\left(\frac{m(t, s)}{v(t, s)}\right) ds \qquad (25)$$

where,

$$m(t, s) = c - S(s) - e^{\int_t^s \kappa(z)dz} \tilde{T}(t) - e^{\int_t^s \kappa(z)dz} \int_t^s \sigma(u)\theta(u)e^{-\int_t^u \kappa(z)dz} du \qquad (26)$$

$$v^2(t, s, x) = e^{2\int_t^s \kappa(z)dz} \int_t^s \sigma^2(u)\theta(u)e^{-2\int_t^u \kappa(z)dz} du \qquad (27)$$

and $\Psi(x) = x\Phi(x) + \Phi'(x)$ where Φ is the cumulative standard normal distribution function.

Proof. From equation (21) and (24) we have that:

$$F_{HDD}(t, \tau_1, \tau_2) = E_Q\left[\int_{\tau_1}^{\tau_2} \max\left(c - T(s)\right) ds \,|\mathcal{F}_t\right]$$

and using Ito's Isometry we can interchange the expectation and the integral

$$E_Q\left[\int_{\tau_1}^{\tau_2} \max\left(c - T(s)\right) |\mathcal{F}_t\right] = \int_{\tau_1}^{\tau_2} E_Q\left[\max\left(c - T(s)\right) | \mathcal{F}_t\right] ds$$

$T(s)$ is normally distributed under the probability measure Q with mean and variance given by:

$$E_Q[T(s)|\mathcal{F}_t] = S(s) + e^{\int_t^s \kappa(z)dz}\tilde{T}(t) + e^{\int_t^s \kappa(z)dz}\int_t^s \sigma(u)\theta(u)e^{-\int_t^u \kappa(z)dz}\,du$$

$$Var_Q[T(s)|\mathcal{F}_t] = e^{2\int_t^s \kappa(z)dz}\int_t^s \sigma^2(u)\theta(u)e^{-2\int_t^u \kappa(z)dz}\,du$$

Hence, $c - T(s)$ is normally distributed with mean given by $m(t, s)$ and variance given by $v^2(t, s)$ and the proposition follows by standard calculations using the properties of the normal distribution.   □

6 Conclusions

This paper proposes and implements a modeling and forecasting framework for temperature based weather derivatives. The proposed method is an extension of the works proposed by [5] and [9]. Here the speed of mean reversion parameter is considered to be a time varying parameter and it is modeled by a wavelet neural network. It is proved that the waveform of the activation function of the proposed network provides a better fit of the data.

Our method is validated in a two month ahead out of sample forecast period. Moreover the relative errors produced by the wavelet network are compared against the original B-B model and historical burn analysis. Results show that the wavelet network outperforms the other methods. More precisely the wavelet network forecasting ability is better than the B-B and HBA in 11 times out of 13. Finally testing the fitted residuals of B-B we observe that the normality hypothesis is rejected in almost every case. Hence, B-B cannot be used for forecasts. Finally, we provided the pricing equations for temperature futures of a HDD index derivative when a is time depended.

The results in this study are preliminary and can be improved. More precisely the number of sinusoids in equations (2) and (3) in B-B framework, representing the seasonal part of the temperature and the variance of residuals, are chosen according to [9]. Alternative methods can improve the fitting in the original data. Hence a better training set is expected for the wavelet network and more accurate forecasts.

Another important aspect is to test the largest forecasting window of each method. Meteorological forecasts of a window larger than 10 days considered inaccurate. Hence, it is important to develop a model than can accurately predict daily average temperatures in larger windows. Also, this analysis will let us use the best model according to the desired forecasting interval.

References

1. Challis, S.: Bright Forecast for Profits, Reactions. June edn. (1999)
2. Hanley, M.: Hedging the Force of Nature. Risk Professional 1, 21–25 (1999)
3. Ceniceros, R.: Weather derivatives running hot. Business Insurance 40 (2006)
4. Jewson, S., Brix, A., Ziehmann, C.: Weather Derivative Valuation: The Meteorological, Statistical, Financial and Mathematical Foundations. Cambridge University Press, Cambridge (2005)
5. Zapranis, A., Alexandridis, A.: Modelling Temperature Time Dependent Speed of Mean Reversion in the Context of Weather Derivetive Pricing. Applied Mathematical Finance 15, 355–386 (2008)
6. Zapranis, A., Alexandridis, A.: Weather Derivatives Pricing: Modelling the Seasonal Residuals Variance of an Ornstein-Uhlenbeck Temperature Process With Neural Networks. Neurocomputing (accepted, to appear)
7. Alaton, P., Djehince, B., Stillberg, D.: On Modelling and Pricing Weather Derivatives. Applied Mathematical Finance 9, 1–20 (2000)
8. Zhang, Q., Benveniste, A.: Wavelet Networks. IEEE Trans. Neural Networks 3, 889–898 (1992)
9. Benth, F.E., Saltyte-Benth, J.: The volatility of temperature and pricing of weather derivatives. Quantitative Finance 7, 553–561 (2007)
10. Daubechies, I.: Ten Lectures on Wavelets. SIAM, Philadelphia (1992)
11. Mallat, S.G.: A Wavelet Tour of Signal Processing. Academic Press, San Diego (1999)
12. Zapranis, A., Alexandridis, A.: Wavelet analysis and weather derivatives pricing. HFFA, Thessaloniki (2006)
13. Oussar, Y., Dreyfus, G.: Initialization by Selection for Wavelet Network Training. Neurocomputing 34, 131–143 (2000)
14. Zapranis, A., Alexandridis, A.: Model Identification in Wavelet Neural Networks Framework. In: Iliadis, L., Vlahavas, I., Bramer, M. (eds.) Artificial Intelligence Applications and Innovations III. IFIP, vol. 296, pp. 267–277. Springer, New York (2009)
15. Cao, M., Wei, J.: Pricing the weather. In: Risk Weather Risk Special Report, Energy And Power Risk Management, pp. 67–70 (2000)
16. Davis, M.: Pricing weather derivatives by marginal value. Quantitative Finance 1, 1–4 (2001)
17. Dornier, F., Queruel, M.: Caution to the wind. Weather risk special report. In: Energy Power Risk Management, pp. 30–32 (2000)
18. Moreno, M.: Riding the temp. Weather Derivatives. FOW Special Support (2000)
19. Caballero, R., Jewson, S., Brix, A.: Long Memory in Surface Air Temperature: Detection Modelling and Application to Weather Derivative Valuation. Climate Research 21, 127–140 (2002)

20. Brody, C.D., Syroka, J., Zervos, M.: Dynamical Pricing of Weather Derivatives. Quantitave Finance 2, 189–198 (2002)
21. Benth, F.E., Saltyte-Benth, J.: Stochastic Modelling of Temperature Variations With a View Towards Weather Derivatives. Applied Mathematical Finance 12, 53–85 (2005)
22. Oussar, Y., Rivals, I., Presonnaz, L., Dreyfus, G.: Trainning Wavelet Networks for Nonlinear Dynamic Input Output Modelling. Neurocomputing 20, 173–188 (1998)
23. Zhang, Q.: Using Wavelet Network in Nonparametric Estimation. IEEE Trans. Neural Networks 8, 227–236 (1997)
24. Postalcioglu, S., Becerikli, Y.: Wavelet Networks for Nonlinear System Modelling. Neural Computing & Applications 16, 434–441 (2007)
25. Xu, J., Ho, D.W.C.: A Basis Selection Algorithm for Wavelet Neural Networks. Neurocomputing 48, 681–689 (2002)
26. Gao, R., Tsoukalas, H.I.: Neural-wavelet Methodology for Load Forecasting. Journal of Intelligent & Robotic Systems 31, 149–157 (2001)
27. Xu, J., Ho, D.W.C.: A constructive algorithm for wavelet neural networks. In: Wang, L., Chen, K., S. Ong, Y. (eds.) ICNC 2005. LNCS, vol. 3610, pp. 730–739. Springer, Heidelberg (2005)
28. Benth, F.E., Saltyte-Benth, J., Koekebakker, S.: Putting a price on temperature. Scandinavian Journal of Statistics 34, 746–767 (2007)

Using the Support Vector Machine as a Classification Method for Software Defect Prediction with Static Code Metrics

David Gray, David Bowes, Neil Davey, Yi Sun, and Bruce Christianson

Science and Technology Research Institute,
University of Hertfordshire, UK
{d.gray,d.h.bowes,n.davey,y.2.sun,b.christianson}@herts.ac.uk

Abstract. The automated detection of defective modules within software systems could lead to reduced development costs and more reliable software. In this work the static code metrics for a collection of modules contained within eleven NASA data sets are used with a Support Vector Machine classifier. A rigorous sequence of pre-processing steps were applied to the data prior to classification, including the balancing of both classes (defective or otherwise) and the removal of a large number of repeating instances. The Support Vector Machine in this experiment yields an average accuracy of 70% on previously unseen data.

1 Introduction

Software defect prediction is the process of locating defective modules in software and is currently a very active area of research within the software engineering community. This is understandable as "Faulty software costs businesses $78 billion per year" ([1], published in 2001), therefore any attempt to reduce the number of latent defects that remain inside a deployed system is a worthwhile endeavour.

Thus the aim of this study is to observe the classification performance of the Support Vector Machine (SVM) for defect prediction in the context of eleven data sets from the NASA Metrics Data Program (MDP) repository; a collection of data sets generated from NASA software systems and intended for defect prediction research. Although defect prediction studies have been carried out with these data sets and various classifiers (including an SVM) in the past, this study is novel in that thorough data cleansing methods are used explicitly.

The main purpose of static code metrics (examples of which include the number of: lines of code, operators (as proposed in [2]) and linearly independent paths (as proposed in [3]) in a module) is to give software project managers an indication toward the quality of a software system. Although the individual worth of such metrics has been questioned by many authors within the software engineering community (see [4], [5], [6]), they still continue to be used.

Data mining techniques from the field of artificial intelligence now make it possible to predict software defects; undesired outputs or effects produced by

D. Palmer-Brown et al. (Eds.): EANN 2009, CCIS 43, pp. 223–234, 2009.

software, from static code metrics. Views toward the worth of using such metrics for defect prediction are as varied within the software engineering community as those toward the worth of static code metrics. However, the findings within this study suggest that such predictors are useful, as on the data used here they correctly classify modules with an average accuracy of 70%.

2   Background

2.1   Static Code Metrics

Static code metrics are measurements of software features that may potentially relate to quality. Examples of such features and how they are often measured include: size, via lines of code (LOC) counts; readability, via operand and operator counts (as proposed by [2]) and complexity, via linearly independent path counts (as proposed by [3]).

Consider the C program shown in Figure 1. Here there is a single function called main. The number of lines of code this function contains (from opening to closing bracket) is 11, the number of arguments it takes is 2, the number of linearly independent paths through the function (also known as the cyclomatic complexity [3]) is 3. These are just a few examples of the many metrics that can be statically computed from source code.

```
#include <stdio.h>

int main(int argc, char *argv[])
{
int return_code = 0;
if (argc < 2) {
  printf("No Arguments Given\n");
  return_code = -1;
}
int x;
for(x = 1; x < argc; x++)
  printf("'%s'\n", argv[x]);
return return_code;
}
```

Fig. 1. An example C program

Because static code metrics are calculated through the parsing of source code their collection can be automated. Thus it is computationally feasible to calculate the metrics of entire software systems, irrespective of their size. [7] points out that such collections of metrics can be used in the following contexts:

- **To make general predictions about a system as a whole.** For example, has a system reached a required quality threshold?

- **To identify anomalous components.** Of all the modules within a software system, which ones exhibit characteristics that deviate from the overall average? Modules highlighted as such can then be used as pointers to where developers should be focusing their efforts. [8] points out that this is common practice amongst several large US government contractors.

2.2 The Support Vector Machine

A Support Vector Machine (SVM) is a supervised machine learning algorithm that can be used for both classification and regression [9]. SVMs are known as maximum margin classifiers as they find the best separating hyperplane between two classes. This process can also be applied recursively to allow the separation of any number of classes. Only those data points that are located nearest to this dividing hyperplane, known as the *support vectors,* are used by the classifier. This enables SVMs to be used successfully with both large and small data sets. Moreover, the process of finding the decision boundary is a convex optimisation problem, so there are no problems with local minima.

Although maximum margin classifiers are strictly intended for linear classification, they can also be used successfully for non-linear classification (such as the case here) via the use of a kernel function. A kernel function is used to implicitly map the data points into a higher-dimensional feature space, and to take the inner-product in that feature space [10]. The benefit of using a kernel function is that the data is more likely to be linearly separable in the higher feature space. Additionally, the actual mapping to the higher-dimensional space is never needed.

There are a number of different kinds of kernel functions (any continuous symmetric positive semi-definite function will suffice) including: linear, polynomial, Gaussian and sigmoidal. Each have varying characteristics and are suitable for different problem domains. The one used here is the *Gaussian radial basis function* (RBF), as it can handle non-linear problems, requires fewer parameters than other non-linear kernels and is computationally less demanding than the polynomial kernel [11]. In fact, this kernel implicitly maps the data into an infinite dimensional feature space.

When an SVM is used with a Gaussian RBF kernel, there are two user-specified parameters, C and γ. C is the error cost parameter; a variable that determines the trade-off between minimising the training error and maximizing the margin (see Fig. 2). γ controls the width / radius of the Gaussian RBF. The performance of an SVM is largely dependant on these parameters, and the optimal values need to be determined *for each training set* via a systematic search.

2.3 Data

The data used within this study was obtained from the NASA Metrics Data Program (MDP) repository[1]. This repository currently contains thirteen data

---

[1] http://mdp.ivv.nasa.gov/

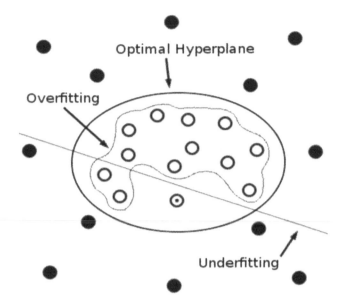

Fig. 2. The importance of optimal parameter selection. The solid and hollow dots represent the training data for two classes. The hollow dot with a dot inside is the test data. Observe that the test dot will be misclassified if too simple (underfitting, the straight line) or too complex (overfitting, the jagged line) a hyperplane is chosen. The optimum hyperplane is shown by the oval line.

Table 1. The eleven NASA MDP data sets that were used in this study. Note that KLOC refers to thousand lines of code.

Name	Language	Total KLOC	No. of Modules	% Defective Modules
CM1	C	20	505	10
KC3	Java	18	458	9
KC4	Perl	25	125	49
MC1	C & C++	63	9466	0.7
MC2	C	6	161	32
MW1	C	8	403	8
PC1		40	1107	7
PC2	C	26	5589	0.4
PC3		40	1563	10
PC4		36	1458	12
PC5	C++	164	17168	3

sets, each of which represent a NASA software system / subsystem and contain the static code metrics and corresponding fault data for each comprising module. Note that a module in this domain can refer to a function, procedure or method. Eleven of these thirteen data sets were used in this study: brief details of each are shown in Table 1. A total of 42 metrics and a unique module identifier comprise each data set (see Table 5, located in the appendix), with the exception of MC1 and PC5 which do not contain the decision density metric.

All the metrics shown within Table 5 with the exception of *error count* and *error density*, were generated using McCabeIQ 7.1; a commercial tool for the automated collection of static code metrics. The *error count* metric was calculated by the number of error reports that were issued for each module via a bug tracking system. Each error report increments the value by one. *Error density* is derived from *error count* and *LOC total*, and describes the number of errors per thousand lines of code (KLOC).

3 Method

3.1 Data Pre-processing

The process for cleansing each of the data sets used in this study is as follows:

Initial Data Set Modifications. Each of the data sets initially had their module identifier and *error density* attribute removed, as these are not required for classification. The *error count* attribute was then converted into a binary target attribute for each instance by assigning all values greater than zero to defective, non-defective otherwise.

Removing Repeated and Inconsistent Instances. Repeated feature vectors, whether with the same (repeated instances) or different (inconsistent instances) class labels, are a known problem within the data mining community [12].

Ensuring that training and testing sets do not share instances guarantees that all classifiers are being tested against *previously unseen data*. This is very important as testing a predictor upon the data used to train it can greatly overestimate performance [12]. The removal of inconsistent items from training data is also important, as it is clearly illogical in the context of binary classification for a classifier to associate the same data point with both classes.

Carrying out this pre-processing stage showed that some data sets (namely MC1, PC2 and PC5) had an overwhelmingly high number of repeating instances (79%, 75% and 90% respectively, see Table 2). Although no explanation has yet been found for these high number of repeated instances, it appears highly unlikely that this is a true representation of the data, i.e. that 90% of modules within a system / subsystem could possibly have the same number of: lines, comments, operands, operators, unique operands, unique operators, conditional statements, etc.

Table 2. The result of removing all repeated and inconsistent instances from the data

Name	Original Instances	Instances Removed	% Removed
CM1	505	51	10
KC3	458	134	29
KC4	125	12	10
MC1	9466	7470	79
MC2	161	5	3
MW1	403	27	7
PC1	1107	158	14
PC2	5589	4187	75
PC3	1563	130	8
PC4	1458	116	8
PC5	17186	15382	90
Total	**38021**	**27672**	**73**

Removing Constant Attributes. If an attribute has a fixed value throughout all instances then it is obviously of no use to a classifier and should be removed.

Each data set had between 1 and 4 attributes removed during this phase with the exception of KC4 that had a total of 26. Details are not shown here due to space limitations.

Missing Values. Missing values are those that are unintentionally or otherwise absent for a particular attribute in a particular instance of a data set. The only missing values within the data sets used in this study were within the *decision density* attribute of data sets CM1, KC3, MC2, MW1, PC1, PC2, and PC3.

Manual inspection of these missing values indicated that they were almost certainly supposed to be representing zero, and were replaced accordingly.

Balancing the Data. All the data sets used within this study, with the exception of KC4, contain a much larger amount of one class (namely, non-defective) than they do the other. When such *imbalanced* data is used with a supervised classification algorithm such as an SVM, the classifier will be expected to over predict the majority class [10], as this will produce lower error rates in the test set.

There are various techniques that can be used to balance data (see [13]). The approach taken here is the simplest however, and involves randomly undersampling the majority class until it becomes equal in size to that of the minority class. The number of instances that were removed during this undersampling process, along with the final number of instances contained within each data set, are shown in Table 3.

Table 3. The result of balancing each data set

Name	Instances Removed	% Removed	Final no. of Instances
CM1	362	80	**92**
KC3	240	74	**84**
KC4	1	1	**112**
MC1	1924	96	**72**
MC2	54	35	**102**
MW1	320	85	**56**
PC1	823	87	**126**
PC2	1360	97	**42**
PC3	1133	79	**300**
PC4	990	74	**352**
PC5	862	48	**942**

Normalisation. All values within the data sets used in this study are numeric, so to prevent attributes with a large range dominating the classification model all values were normalised between -1 and +1. Note that this pre-processing stage was performed just prior to training for each training and testing set, and that each training / testing set pair were scaled in the same manner [11].

Randomising Instance Order. The order of the instances within each data set were randomised to defend against *order effects*, where the performance of a predictor fluctuates due to certain orderings within the data [14].

3.2 Experimental Design

When splitting each of the data sets into training and testing sets it is important to ameliorate possible anomalous results. To this end we use five-fold cross-validation. Note that to reduce the effects of sampling bias introduced when randomly splitting each data set into five bins, the cross-validation process was repeated 10 times for each data set in each iteration of the experiment (described below).

As mentioned in Section 2.2, an SVM with an RBF kernel requires the selection of optimal values for parameters C and γ for maximal performance. Both values were chosen for each training set using a five-fold *grid search* (see [11]), a process that uses cross-validation and a wide range of possible parameter values in a systematic fashion. The pair of values that yield the highest average accuracy are then taken as the optimal parameters and used when generating the final model for classification.

Due to the high percentage of information lost when balancing each data set (with the exception of KC4), the experiment is repeated fifty times. This is in order to further minimise the effects of sampling bias introduced by the random undersampling that takes place during balancing.

Pseudocode for the full experiment carried out in this study is shown in Fig 3. Our chosen SVM environment is LIBSVM [15], an open source library for SVM experimentation.

```
M = 50      # No. of times to repeat full experiment
N = 10      # No. of cross-validation repetitions
V = 5       # No. of cross-validation folds

DATASETS = ( CM1, KC3, KC4, MC1, MC2, MW1, PC1, PC2, PC3, PC4, PC5 )

results = ( )       # An empty list

repeat M times:
  for dataSet in DATASETS:
    dataSet = pre_process(dataSet)     # As described in Section 3.1
    repeat N times:
      for i in 1 to V:
        testerSet = dataSet[i]
        trainingSet = dataSet - testerSet
        params = gridSearch(trainingSet)
        model = svm_train(params, trainingSet)
        results += svm_predict(model, testerSet)

FinalResults = avg(results)
```

Fig. 3. Pseudocode for the experiment carried out in this study

4   Assessing Performance

The measure used to assess predictor performance in this study is *accuracy*. Accuracy is defined as the ratio of instances correctly classified out of the total number of instances. Although simple, accuracy is a suitable performance measure for this study as each test set is balanced. For imbalanced test sets more complicated measures are required.

5   Results

The average results for each data set are shown in Table 4. The results show an average accuracy of 70% across all 11 data sets, with a range of 64% to 82%. Notice that there is a fairly high deviation shown within the results. This is to be expected due to the large amount of data lost during balancing and supports the decision for the experiment being repeated fifty times (see Fig. 3). It is notable that the accuracy for some data sets is extremely high, for example with data set PC4, four out of every five modules were being correctly classified.

The results show that all data sets with the exception of PC2 have a mean accuracy greater than two standard deviations away from 50%. This shows the statistical significance of the classification results when compared to a dumb classifier that predicts all one class (and therefore scores an accuracy of 50%).

Table 4. The results obtained from this study

Name	% Mean Accuracy	Std.
CM1	68	5.57
KC3	66	6.56
KC4	71	4.93
MC1	65	6.74
MC2	64	5.68
MW1	71	7.3
PC1	71	5.15
PC2	64	9.17
PC3	76	2.15
PC4	82	2.11
PC5	69	1.41
Total	**70**	**5.16**

6 Analysis

Previous studies ([16], [17], [18]) have also used data from the NASA MDP repository and an SVM classifier. Some of these studies briefly mention data pre-processing, however we believe that it is important to explicitly carry out all of the data cleansing stages described here. This is especially true with regard to the removal of repeating instances, ensuring that all classifiers are being tested against previously unseen data.

The high number of repeating instances found within the MDP data sets was surprising. Brief analysis of other defect prediction data sets showed a repeating average of just 1.4%. We are therefore suspicious of the suitability of the data held within the MDP repository for defect prediction and believe that previous studies which have used this data and not carried out appropriate data cleansing methods may be reporting inflated performance values.

An example of such a study is [18], where the authors use an SVM and four of the NASA data sets, three of which were used in this study (namely CM1, PC1 and KC3). The authors make no mention of data pre-processing other than the use of an attribute selection algorithm. They then go on to report a minimum average *precision*, the ratio of correctly predicted defective modules to the total number of modules predicted as defective, of 84.95% and a minimum average *recall*, the ratio of defective modules detected as such, of 99.4%. We believe that such high classification rates are highly unlikely in this problem domain due to the limitations of static code metrics and that not carrying out appropriate data cleansing methods may have been a factor in these high results.

7   Conclusion

This study has shown that on the data studied here the Support Vector Machine can be used successfully as a classification method for defect prediction. We hope to improve upon these results in the near future however via the use of a one-class SVM; an extension to the original SVM algorithm that trains upon only defective examples, or a more sophisticated balancing technique such as SMOTE (Synthetic Minority Over-sampling Technique).

Our results also show that previous studies which have used the NASA data may have exaggerated the predictive power of static code metrics. If this is not the case then we would recommend the explicit documentation of what data pre-processing methods have been applied. Static code metrics can only be used as probabilistic statements toward the quality of a module and further research may need to be undertaken to define a new set of metrics specifically designed for defect prediction.

The importance of data analysis and data quality has been highlighted in this study, especially with regard to the high quantity of repeated instances found within a number of the data sets. The issue of data quality is very important within any data mining experiment as poor quality data can threaten the validity of both the results and the conclusions drawn from them [19].

References

1. Levinson, M.: Lets stop wasting $78 billion per year. CIO Magazine (2001)
2. Halstead, M.H.: Elements of Software Science (Operating and programming systems series). Elsevier Science Inc., New York (1977)
3. McCabe, T.J.: A complexity measure. In: ICSE 1976: Proceedings of the 2nd international conference on Software engineering, p. 407. IEEE Computer Society Press, Los Alamitos (1976)
4. Hamer, P.G., Frewin, G.D.: M.H. Halstead's Software Science - a critical examination. In: ICSE 1982: Proceedings of the 6th international conference on Software engineering, pp. 197–206. IEEE Computer Society Press, Los Alamitos (1982)
5. Shen, V.Y., Conte, S.D., Dunsmore, H.E.: Software Science Revisited: A critical analysis of the theory and its empirical support. IEEE Trans. Softw. Eng. 9(2), 155–165 (1983)
6. Shepperd, M.: A critique of cyclomatic complexity as a software metric. Softw. Eng. J. 3(2), 30–36 (1988)
7. Sommerville, I.: Software Engineering, 8th edn. International Computer Science Series. Addison Wesley, Reading (2006)
8. Menzies, T., Greenwald, J., Frank, A.: Data mining static code attributes to learn defect predictors. IEEE Transactions on Software Engineering 33(1), 2–13 (2007)
9. Schölkopf, B., Smola, A.J.: Learning with Kernels: Support Vector Machines, Regularization, Optimization, and Beyond. In: Adaptive Computation and Machine Learning. The MIT Press, Cambridge (2001)

10. Sun, Y., Robinson, M., Adams, R., Boekhorst, R.T., Rust, A.G., Davey, N.: Using sampling methods to improve binding site predictions. In: Proceedings of ESANN (2006)
11. Hsu, C.W., Chang, C.C., Lin, C.J.: A practical guide to support vector classification. Technical report, Taipei (2003)
12. Witten, I.H., Frank, E.: Data Mining: Practical Machine Learning Tools and Techniques, 2nd edn. Morgan Kaufmann Series in Data Management Systems. Morgan Kaufmann, San Francisco (2005)
13. Wu, G., Chang, E.Y.: Class-boundary alignment for imbalanced dataset learning. In: ICML 2003 Workshop on Learning from Imbalanced Data Sets, pp. 49–56 (2003)
14. Fisher, D.: Ordering effects in incremental learning. In: Proc. of the 1993 AAAI Spring Symposium on Training Issues in Incremental Learning, Stanford, California, pp. 34–41 (1993)
15. Chang, C.C., Lin, C.J.: LIBSVM: a library for support vector machines (2001), http://www.csie.ntu.edu.tw/~cjlin/libsvm
16. Li, Z., Reformat, M.: A practical method for the software fault-prediction. In: IEEE International Conference on Information Reuse and Integration, 2007. IRI 2007, pp. 659–666 (2007)
17. Lessmann, S., Baesens, B., Mues, C., Pietsch, S.: Benchmarking classification models for software defect prediction: A proposed framework and novel findings. IEEE Transactions on Software Engineering 34(4), 485–496 (2008)
18. Elish, K.O., Elish, M.O.: Predicting defect-prone software modules using support vector machines. J. Syst. Softw. 81(5), 649–660 (2008)
19. Liebchen, G.A., Shepperd, M.: Data sets and data quality in software engineering. In: PROMISE 2008: Proceedings of the 4th international workshop on Predictor models in software engineering, pp. 39–44. ACM, New York (2008)

Appendix

Table 5. The 42 metrics originally found within each data set

Metric Type	Metric Name
	Metric Name
McCabe	01. Cyclomatic Complexity
	02. Cyclomatic Density
	03. Decision Density
	04. Design Density
	05. Essential Complexity
	06. Essential Density
	07. Global Data Density
	08. Global Data Complexity
	09. Maintenance Severity
	10. Module Design Complexity
	11. Pathological Complexity
	12. Normalised Cyclomatic Complexity
Raw Halstead	13. Number of Operators
	14. Number of Operands
	15. Number of Unique Operators
	16. Number of Unique Operands
Derived Halstead	17. Length (N)
	18. Volume (V)
	19. Level (L)
	20. Difficulty (D)
	21. Intellegent Content (I)
	22. Programming Effort (E)
	23. Error Estimate (B)
	24. Programming Time (T)
LOC Counts	25. LOC Total
	26. LOC Executable
	27. LOC Comments
	28. LOC Code and Comments
	29. LOC Blank
	30. Number of Lines (opening to closing bracket)
Misc.	31. Node Count
	32. Edge Count
	33. Branch Count
	34. Condition Count
	35. Decision Count
	36. Formal Parameter Count
	37. Modified Condition Count
	38. Multiple Condition Count
	39. Call Pairs
	40. Percent Comments
Error	41. Error Count
	42. Error Density

Adaptive Electrical Signal Post-processing with Varying Representations in Optical Communication Systems

Stephen Hunt[1], Yi Sun[1], Alex Shafarenko[1], Rod Adams[1], Neil Davey[1],
Brendan Slater[2], Ranjeet Bhamber[2], Sonia Boscolo[2], and Sergei K. Turitsyn[2]

[1] Biological and Neural Computation Research Group, School of Computer Science
University of Hertfordshire, Hatfield, Herts. AL10 9AB UK
{s.p.hunt,y.2.sun,a.shafarenko,r.g.adams,n.davey}@herts.ac.uk
http://homepages.feis.herts.ac.uk/~nngroup/
[2] Photonics Research Group, School of Engineering and Applied Science
Aston University, Birmingham B4 7ET, UK
{slaterbm,bhambers,s.a.boscolo,s.k.turitsyn}@aston.ac.uk
http://www.ee.aston.ac.uk/research/prg/

Abstract. Improving bit error rates in optical communication systems is a difficult and important problem. Error detection and correction must take place at high speed, and be extremely accurate. Also, different communication channels have different characteristics, and those characteristics may change over time. We show the feasibility of using simple artificial neural networks to address these problems, and examine the effect of using different representations of signal waveforms on the accuracy of error correction. The results we have obtained lead us to the conclusion that a machine learning system based on these principles can improve on the performance of existing error correction hardware at the speed required, whilst being able to adapt to suit the characteristics of different communication channels.

Keywords: Error correction, classification, optical communication, adaptive signal processing.

1 Introduction

High-speed and long-distance data communications make extensive use of fibre-optic links. Performance of a fibre-optic communication link is typically affected by a complex combination of random processes (amplified spontaneous emission noise, polarization mode dispersion, and so on) and of deterministic or quasi-deterministic effects (such as nonlinear inter- and intra-channel signal interactions, dispersive signal broadening, and various forms of cross-talk), resulting from the design of the communication system and the regime under which it operates. Any installed fibre link has its own individual specific transmission impairments: its own signature of how the transmitted signal is corrupted and

D. Palmer-Brown et al. (Eds.): EANN 2009, CCIS 43, pp. 235–245, 2009.

distorted, and its own characteristic pattern of errors introduced into the digital data stream.

There is great value in a signal post-processing system that can undo some of these signal distortions, or that can separate line-specific distortions from non-recoverable errors. Signal post-processing in optical data communication can offer new margins in system performance in addition to other enabling techniques. A variety of post-processing techniques have been already used to improve overall system performance, such as tunable dispersion compensation and electronic equalization (see e.g. [3], [4], [9], [11], and references therein). Note that post-processing can be applied both in the optical and electrical domain (after conversion of the optical field into electrical current). Application of electronic signal processing for compensation of transmission impairments is an attractive technique that has become quite popular thanks to recent advances in high-speed electronics. An adaptive system, as proposed here, is of even greater value, because it may be tuned to the specific charateristics of each data transmission link, and re-tuned as the characteristics of the link change, which they inevitably will over time.

In this work we apply machine learning techniques to adaptive signal post-processing in optical communication systems. We adopt several different representations of signal waveforms, including the discrete wavelet transform, and independent components from independent components analysis. To the best of our knowledge this is the first time that such techniques have been applied in this area. One key feature of this problem domain is that the trainable classifier must perform at an extremely high speed, because optical communication systems typically operate at bit rates of around 40GHz. We demonstrate the feasibility of bit-error-rate improvement by adaptive post-processing of received electrical signals.

2   Background to the Problem

In optical data communications a digital data stream is transmitted over a fibre-optic link as a continuously varying optical signal. At the receiver (typically after filtering) the optical signal is converted by a photodiode into an electric current. Detection of the digital signal requires discrimination of the logical 1s and 0s using some decision threshold. This can be done in different ways (e.g. by considering currents at a certain optimized sample point within the bit time slots or by analyzing current integrated over some time interval) and is determined by a specific design of the receiver. Here without loss of generality we assume that discrimination is made using current integrated over the whole time slot. Note that the approach proposed in this paper and described in detail below is very generic and can easily be adapted to any particular receiver design. To improve system performance and minimize the bit-error-rate, we propose here to use a method to adjust the receiver to cope with transmission impairments specific for a given line. This is achieved by applying learning algorithms based on the analysis of sampled currents within bit time slots and adaptive correction of the decisions taking into account accumulated information gained from analysis of the signal waveforms.

3   Description of the Data

The data represents the received signal taken in the electrical domain after conversion of the optical signal into an electrical current. The data consists of a large number of received bits with the waveforms represented by 32 floating point numbers corresponding to values of electrical current at each of 32 equally spaced sample points within a bit time slot. A sequence of 5 consecutive bits is shown in Figure 1. As already explained the pulse can be classified according to the current integrated over the width of a single bit. For each of the time slots in our data we have the original bit that was transmitted. Therefore the data consists of 32-ary vectors each with a corresponding binary label.

In all we have a stream of 65536 bits to classify. Categorising the vast majority of these bits is straightforward. In fact with an optimally set electrical current integrated over the whole time slot (*energy threshold*) we can correctly classify all but 1842 bits correctly. We can therefore correctly classify 97.19% of the data, an error rate of 2.81%. This error rate is, however, significantly too high. The target error rate is less than one bit in a thousand, or 0.1%. Figure 2 (a) gives an example of a misclassification. The middle bit of the sequence is a 0 but is identified from its energy as a 1. This is due to the presence of two 1's on either side and to distortion of the transmitted signal. It would be difficult for any classifier to rectify this error.

However other cases can be readily identified by the human eye and therefore could be amenable to automatic identification. Figure 2 (b) shows an example

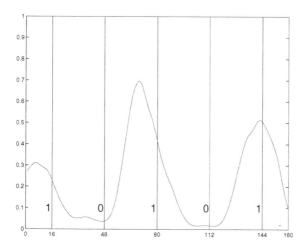

Fig. 1. An example of the electrical signal for a stream of 5 bits - 1 0 1 0 1

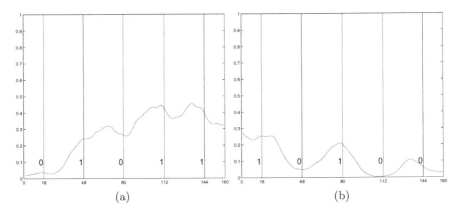

Fig. 2. (a) An example of a difficult error to identify. The middle bit is meant to be a 0, but jitter has rendered it very hard to see; (b) The central bit has been dragged down by the two 0s surrounding it and is classified as a 0 from its energy. However to the human eye the presence of a 1 is obvious.

where the bit pattern is obvious to the eye but where a misclassification actually occurs. The central bit is a 1 but is misclassified as a 0 from its energy alone.

3.1   Representation of the Data

Different datasets can be produced depending on how the values representing the sampled electrical current are processed. As well as representing a single bit as a 32-ary vector (called the *Waveform-1* dataset), it can also be represented as a single energy value (the sum of the 32 values, *Energy-1*).

As described above, of the 65536 bits, all but 1842 are correctly identified by an energy threshold. The ease with which a bit can be classified may be influenced by its context, so we examine each bit in the data stream as the central item in a 5 bit string, in order to determine the effect of context on classification difficulty. There are 32 possible strings of 5 bits, and 9 of these are represented in the set of bits that are difficult to classify with a frequency from 5.86% to 21.17% among the 1842 misclassified cases. These nine sequences are shown in Table 1.

As can be seen the majority of these involve a 1 0 1 or 0 1 0 sequence around the middle bit, and these are the patterns for which difficulties are most likely to occur. Therefore we may also want to take advantage of any information that may be present in adjacent bits. To this end we can form windowed inputs, in which the 3 vectors representing 3 contiguous bits are concatenated together using the label of the central bit as the target output (*Waveform-3*). It is also possible that using adjacent bit information by simply taking 3 energy values instead of the full waveform (*Energy-3*), or using information from a window of 3 bits, with 1 energy value either side of the target bit expressed as a vector (*Energy-Waveform-Energy*), see Table 2.

Table 1. Nine sequences for which difficulties are most likely to occur

$$
\begin{array}{ccccc}
0 & 0 & 1 & 0 & 0 \\
0 & 0 & 1 & 0 & 1 \\
0 & 1 & 0 & 1 & 0 \\
0 & 1 & 0 & 1 & 1 \\
1 & 0 & 0 & 1 & 1 \\
1 & 0 & 1 & 0 & 0 \\
1 & 0 & 1 & 0 & 1 \\
1 & 1 & 0 & 1 & 0 \\
1 & 1 & 0 & 1 & 1 \\
\end{array}
$$

Table 2. The different datasets used in experiments

Name	Arity	Description
Energy-1	1	The energy of the target bit
Energy-3	3	The energy of the target bit and one bit either side
Waveform-1	32	The waveform of the target bit
Waveform-3	96	The waveform of the target bit and the waveforms of the bits on either side
Energy-Waveform-Energy (E-W-E)	34	The waveform of the target bit and the energy of one bit either side of the target bit
CA_3	18	Approximation coefficients with wavelet 'db4' at level 3
CD_1	51	Detail coefficients with wavelet 'db4' at level 1
CD_2	29	Detail coefficients with wavelet 'db4' at level 2
CD_3	18	Detail coefficients with wavelet 'db4' at level 3
ICA	19	ICA representations

3.2   The Discrete Wavelet Transform

Wavelet transforms are similar to Fourier transforms, and are particularly useful for deconstructing non-periodic and/or non-stationary signals. Wavelet analysis has been applied successfully to areas of signal processing. The goal of the wavelet transform is to turn the information contained in a signal into a set of coefficients, which can be analysed.

An efficient way to implement the discrete wavelet transform using filters was developed by Mallat [6]. In wavelet analysis, a signal is decomposed to *approximation* and *detail* components. The approximations are the high-scale, low frequency components of the signal, which are also called *smoothed* signals. The details are the low-scale, high-frequency components.

Denote g as a high pass filter, and h a low pass filter. The original signal c_0 of length N is passed through the two complementary filters. The transform consists of a convolution of c_0 with each of the filters, where every other element is discarded (a process known as dyadic decimation), and produces the lower resolution approximation coefficients CA_1 and detail coefficients CD_1 at level 1.

One important property of the wavelet coefficients is that the decomposition process is recursive with successive approximations being decomposed iteratively, as shown in the following equations:

$$CA_{j-1,k} = \sum_n h_{n-2k} CA_{j,n} \, , \tag{1}$$

and

$$CD_{j-1,k} = \sum_n g_{n-2k} CA_{j,n} \, , \tag{2}$$

where $1 \leq n \leq N$, k is an integer that controls the dilation of the generated wavelets, and j denotes the level of multiresolution decomposition. More details about wavelets can be found in [10].

Here the signal (Waveform-3) is decomposed by the *db4* wavelet at level 3 [10].

3.3 Independent Component Analysis

Independent component analysis (ICA) is a statistical method, which can be used to find a suitable representation of multivariate data. Denote $\mathbf{x}$ as a $\mathcal{D}$-dimensional vector of observed data. Our aim is to seek the representation $\mathbf{s}$ (also called source signals) as a linear transformation of the observed data $\mathbf{x}$, as follows:

$$\mathbf{s} = \mathbf{W}\mathbf{x} \, , \tag{3}$$

where $\mathbf{W}$ is a constant (weight) matrix.

There are many different approaches to estimate $\mathbf{W}$. One popular way of formulating the ICA problem is to apply an information maximisation approach [1]. In [7], the generative model for ICA is given in detail. In this work, we employed fast and robust *fixed-point* algorithms for ICA (*FastICA*) proposed by Hyvärinen [5], where ICA was formulated as the search for a linear transformation that maximises the *differential entropy*. Readers who are interested in this technique can follow the reference [5] to learn more.

Table 2 also gives a summary of all the different datasets, formed by using approximation and detail coefficients from discrete wavelet analysis together with the dataset using the ICA representation.

4 Method

4.1 Easy and Hard Cases

One difficulty for the trainable classifier is that in this dataset the vast majority of examples are straightforward to classify. The hard cases are very sparsely represented, so that, in an unusual sense, the data is imbalanced. Figure 3 is a diagram of error rates of 0 and 1 as a function of the energy threshold. It shows that if the energy threshold is set to roughly 2.5, then all bits

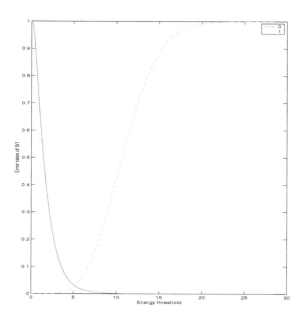

Fig. 3. A diagram of error rates of 0 and 1 as a function of an energy threshold

with energy less than this threshold are correctly classified into the 0 class; on the other hand, if the energy threshold is set to about 11, then all bits with energy greater than this are correctly classified into the 1 class. The optimal energy threshold to separate the two classes is 5.01, in which case, only 1842 of the bits 65532 are incorrectly classified - a bit error rate of 2.81%. Using this threshold we divide the data into easy and hard cases, that is, those classified correctly by the method are easy ones, otherwise they are hard cases.

4.2   Visualisation Using PCA

Before classifying bits into two classes, we first looked at the underlying data distribution by means of classical principal component analysis (PCA) [5], which linearly projects data into a two-dimensional space, where it can be visualised.

We visualise the easy cases using PCA, then project the hard cases into the same PCA projection space. The result is shown in Figure 4 (a). It shows that unsurprisingly the easy 0 and 1 classes are linearly separable. Interestingly, the hard 0 and 1 classes are for the most part also linearly separable. However, the hard 1s have almost complete overlap with the easy 0s, and the hard 0s have almost complete overlap with the easy 1s.

Figure 4 (b) is the eigenwave of the first component in the PCA analysis, which accounts for 86.5% of the total variance.

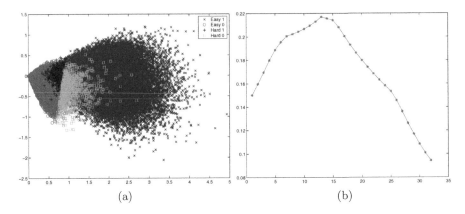

Fig. 4. (a) Projection of the easy set using PCA, where the hard patterns are also projected into the easy ones' first two principal components space; (b) Eigenwave of the first principal component

4.3   Single Layer Neural Network

As already described the classifiers need to be operationally very fast. Therefore the main classifier we use is a simple single layer neural network (SLN) [2]. Once trained (this is done off-line in advance) an SLN can be built in hardware and function with great speed. For comparison purposes a classifier that uses just an optimal energy threshold is implemented, where the threshold is the one giving the maximum accuracy rate (97.19%).

5   Performance Measures

We apply two performance measurements to analyse results. One is *classification accuracy*, that is the percentage of correctly classified cases; the other one is *information gain (IG)*. We follow [8] to give the definition of IG. Suppose given a set of cases S and a related attribute A, the $IG(S, A)$ is defined as:

$$IG(S, A) = Entropy(S) - \sum_{v \in Values(A)} \frac{|S_v|}{|S|} Entropy(S_v), \qquad (4)$$

where $Values(A)$ is the set of all possible values for attribute A, and S_v is the subset of S for which attribute A has value v. The value of IG is the number of bits saved when encoding the target value of an arbitrary member of S, by knowing the value of attribute A [8]. In this work, one can consider a set of labels of waveform signals as S, and predictions as A to obtain a value of IG.

6   Experiments

We segment the data into 10-fold cross-training/validation sets. Each distinct segment has 6369 easy cases and 184 hard ones. The results reported here are therefore evaluations on averages over the 10 different validation sets (test sets).

6.1   The First Experiment

The first experiment was carried out with Energy and Waveform representations. The main results are given in Table 3.

Table 3. The results of classifying the different test sets for the different data representations. We also give the standard deviation for the test sets.

Dataset	mean errors		mean	mean
	easy set	hard set	accuracy (%)	IG
Energy-1	0	184	97.19	0.8163
Energy-3	27.5 ± 5.76	79.0 ± 6.65	98.37 ± 0.15	0.8802 ± 0.0086
Waveform-1	26.8 ± 5.90	63.1 ± 8.61	98.63 ± 0.16	0.8956 ± 0.0097
Waveform-3	26.0 ± 4.59	61.4 ± 7.23	98.67 ± 0.12	0.8979 ± 0.0075
E-W-E	25.5 ± 6.17	61.9 ± 7.95	98.67 ± 0.14	0.8980 ± 0.0085

The SLN classifier does give an improvement over the optimal energy threshold method (*Energy*-1), with the Waveform-3 and E-W-E datasets giving the best mean accuracy. There is 10% more information gained with E-W-E when compared to the optimal threshold method. Interestingly, the very simple SLN classifier using Energy-3 decreased the error rate on the test sets by nearly 42% when compared to the optimal threshold method. This classifier is simply a single unit with 3 weighted inputs.

6.2   The Second Experiment

The main objective in this experiment is to extract the maximum information in the wavelet domain (*wavelet: db4*) at different levels (3 *levels in total*) of resolution. The wavelet coefficients are extracted from Waveform-3, that is the 96ary vector consisting of 1 bit either side of the target. The experiment was implemented using the wavelet toolbox 4.3 in MATLAB. We perform classification at each level. Table 4 also contains the result using majority voting of the previous 4 results.

Table 4 shows that the best result is obtained with the approximation of signals (at level 3), although all the results are similar. Using majority voting does not further improve the classification performance. This suggests that in all cases the errors are the same. Again, the classifiers do give an improvement over the optimal energy threshold method (*Energy*-1). Looking at the result of SLN/CA_3 in Table 4 and the one from SLN/Waveform-3 in Table 3, one can see

Table 4. The results of classifying the different test sets for the wavelet representations

Dataset	mean errors		mean	mean
	easy set	hard set	accuracy (%)	IG
CA_3	25.0 ± 5.33	61.2 ± 6.61	98.68 ± 0.12	0.8991 ± 0.0073
CD_1	25.9 ± 5.65	61.2 ± 6.55	98.67 ± 0.11	0.8982 ± 0.0068
CD_2	25.4 ± 5.08	61.6 ± 7.04	98.67 ± 0.12	0.8983 ± 0.0077
CD_3	25.5 ± 5.04	61.7 ± 7.93	98.67 ± 0.12	0.8981 ± 0.0073
$MajV$	24.9 ± 5.32	61.3 ± 6.57	98.68 ± 0.12	0.8991 ± 0.0073

there is no big difference between them. Actually, the information gained with SLN/CA_3 is only 0.0012 more than the one obtained with SLN/Waveform-3. However, we must point out that the number of inputs to SLN with CA_3 (18) is much smaller than the one with Waveform-3 (96).

6.3   The Third Experiment

This experiment was implemented using the *FastICA package* for MATLAB, which can be obtained from http://www.cis.hut.fi/projects/ica/fastica/. Default parameters were used in this work. We assume the observed signals are in Waveform-3. The results are shown in Table 5.

Table 5. The results of classifying the different test sets for the ICA representations

Dataset	mean errors		mean	mean
	easy set	hard set	accuracy (%)	IG
ICA	25.2 ± 5.27	61.4 ± 7.78	98.68 ± 0.13	0.8987 ± 0.0079

The result of SLN/ICA in Table 5 is much the same as the one from SLN/Waveform-3 in Table 3. However, again, the number of inputs to SLN with ICA (19) is much smaller than the one with Waveform-3 (96).

7   Discussion

The fast decoding of a stream of data represented as pulses of light is a commercially important and challenging problem. Computationally, the challenge is in the need for a classifier that is highly accurate, yet is sufficiently simple that it can be made to operate extremely quickly. We have therefore restricted our investigation, for the most part, to SLNs, and used data is either a sampled version of the light waveform or just the energy of the pulse. Experiment 1 showed that an SLN trained with the 96-ary representation of the waveform (using 1 bit either side of the target) gave the best performance, reducing the bit error rate from 2.81% to 1.33%. This figure is still quite high and we hypothesised that this could be explained by the fact that despite the data set being very large (65532 items), the number of difficult examples (those misclassified by the threshold method) was very small and dominated by the number of straightforward examples. We undertook experiments 2 and 3 to see if we could construct a classifier that can correctly identify a significant number of these infrequent but difficult examples with a much small number of feature representations. Although the improvement obtained by using wavelet coefficients and independent components is minor, the number of inputs to the SLN is much smaller than the one from Waveform-3. One of the most interesting features of these results is that they suggest all the wavelet coefficients capture the same information with respect to this classification task. Further work is needed on much larger datasets.

This is early work and much of interest is still to be investigated, such as threshold band sizes, and other methods to identify difficult cases.

References

1. Bell, A.J., Sejnowski, T.J.: AN information-maxmization approach to blind separation and blind deconvolution. Neural Computation 7, 1129–1159 (1995)
2. Bishop, C.M.: Neural Networks for Pattern Recognition. Oxford University Press, New York (1995)
3. Bulow, H.: Electronic equalization of transmission impairments. In: OFC, Anaheim, CA, Paper TuE4 (2002)
4. Haunstein, H.F., Urbansky, R.: Application of Electronic Equalization and Error Correction in Lightwave Systems. In: Proceedings of the 30th European Conference on Optical Communications (ECOC), Stockholm, Sweden (2004)
5. Hyvärinen, A.: Fast and robust fixed-point algorithms for independent component analysis. IEEE Trans. on Neural Networks 10(3), 626–634 (1999)
6. Mallat, S.: A theory for multiresolution signal decomposition: the wavelet representation. In: IEEE Transactions of Pattern Analysis and Machine Intelligence, pp. 674–693 (1989)
7. Mackay, D.: Information Theory, Inference, and Learnig Algorithms. Cambridge University Press, Cambridge (2003)
8. Mitchell, T.M.: Machine Leanring. The McGraw-Hill Companies, Inc., New York (1997)
9. Rosenkranz, W., Xia, C.: Electrical equalization for advanced optical communication systems. AEU - International Journal of Electronics and Communications 61(3), 153–157 (2007)
10. Vidakovic, B.: Statistical Modelling by Wavelets. John Wiley & Sons, Inc., Chichester (1999)
11. Watts, P.M., Mikhailov, V., Savory, S., Bayvel, P., Glick, M., Lobel, M., Christensen, B., Kirkpatrick, P., Shang, S., Killey, R.I.: Performance of single-mode fiber links using electronic feed-forward and decision feedback equalizers. IEEE Photon. Technol. Lett. 17(10), 2206–2208 (2005)

Using of Artificial Neural Networks (ANN) for Aircraft Motion Parameters Identification

Anatolij Bondarets and Olga Kreerenko

Beriev Aircraft Company, Aviatorov Square 1,
347900 Taganrog, Russia
anatolij_bondarets@beriev.com, olgadmk@yandex.ru

Abstract. The application of neural networks to solve an engineering problem is introduced in the paper. Artificial neural networks (ANN) are used for model parameters identification of aircraft motion. Unlike conventional identification methods, neural networks have memory, so results are verified and accumulated during repeated "training" cycles (when new samples of initial data are used). The DCSL (Dynamic Cell Structure) neural network from "Adaptive Neural Network Library" is selected as the identification tool. The problem is solved using Matlab Simulink tool. The program includes math model of aircraft motion along runway. The data accumulated from flight tests in real conditions were used to form samples for training of neural networks.. The math modeling results have been tested for convergence with experimental data.

Keywords: Artificial neural networks, math model, identification.

1 Introduction

Calculations and mathematical modeling are essential for the aircraft development and determination of its operating limitations, including estimation of aircraft behavior safety limits. The recent development of aircraft modeling applied for real time analysis of flight data enables to prevent accidents [1,2] and demand the reliable mathematical presentation of an aircraft and its systems behavior. However, application of computational methods requires compliance between the computation (math modeling) results and the experimental data, i.e. it is necessary to identify the math model parameters from the experimental data of real object behavior.

The experience in development and application of the procedure for the flight dynamics math model parameters estimation according to flight tests data [3, 4] has shown that the most complicated element of practical identification tasks is the adjustment of identification results obtained from different samples of initial data. The effort to solve this problem was made in the procedure [3] by identification of corrections for aerodynamic coefficients with "parallel" optimization of disagreement criteria in two (some) flight test data fragments obtained in similar conditions at similar speed and altitudes (for example: maneuvers with stick "to the left" and "to the right", "forwards" and "backwards"). However, when solving the problem [4], main efforts were made to adjust the corrections determined by identification procedure on different data samples.

D. Palmer-Brown et al. (Eds.): EANN 2009, CCIS 43, pp. 246–256, 2009.

Artificial neural networks (ANN) allow determination of the required relations between input and output parameters of the object. Moreover, unlike the traditional identification methods, neural networks have a memory: it means that the results could be verified and accumulated during repeated "training" cycles (during the processing of new samples of initial data). Thus, neural networks allow getting the required relations for wide range of conditions at once and, as a result, to align random factors, which are unavoidable for experimental data.

This paper presents the use of neural network to solve the problem of identification (during the aircraft takeoff or landing run on a concrete runway) of rolling resistance and wheels braking coefficients. In recent years international aviation community has paid much attention to the analysis of aircraft behavior during its motion on runway, in particular during take-off and landing on precipitation-covered runway. For instance, since 1996 NASA, FAA and Transport Canada have been performed JW/RFMP program (The Joint Winter Runway Friction Measurement Program). Present-day knowledge about aircraft behavior on the contaminated runway are generalized in the amendment to European certification requirements (NPA No. 14/2004), developed by JAA, but the conclusion on the necessity of further investigation of this problem was made. Reports on flight accidents, related to aircraft overrun the runway indicate the necessity of such research.

At this stage we considered to estimate the efficiency of neural networks as the means of identification of the aircraft motion math model parameters. For this purpose the simple enough but practically important task was selected: evaluation of wheels compression and their rotational speed influence on the friction coefficient value (for dry concrete runway). This task is presented here in details.

A more complicated procedure based on ANN application, which is presented here in main results, is intended for identification of dependencies describing wheels resistance and braking performances on precipitation-covered runway (see [5]).

2   Longitudinal Forces during Takeoff Run

The given task is related to rolling friction for the entire aircraft wheel system, i.e. difference in conditions of separate wheels rolling is not taken into account. It is supposed that under these conditions the load on main landing gear (MLG) wheels is distributed uniformly, and the load on nose landing gear (NLG) wheels is low, i.e. the error due to such simplification can not be large because of the inessential share of NLG wheels in the total resistance to rolling.

In the given task, projections of the forces to longitudinal axis (along runway) are presented as follows:

$$F_x = P - G*sin(i) - X_{aero} - F_R, \text{ where} \qquad (1)$$

G – aircraft weight; i – runway slope angle (uphill >0); P – engines thrust; $X_{aero} = C_D*q*S$; q – dynamic pressure; S – wing area; $F_R = \mu_R * F_y$; $F_y = G*cos(i) - Y_{aero}$; $Y_{aero} = C_L*q*S$.

Nature of rolling friction coefficient μ_R is shown in Fig. 1 taken from [6]. At this figure, parameter "a" determines the resultant normal force displacement from the wheel axis. In compliance with this figure the value of force resisting to rolling is equal to $F_y*(a/r)$, i. e. $\mu_R = a/r$ (see a and r in Fig. 1). It is clear that on non-contaminated flat surface, when other conditions are the same, the value "a/r" depends on wheel compression and wheel rotational speed. For dry, rigid horizontal surfaces, accordingly [6] $\mu_R=0.005\text{-:-}0.03$.

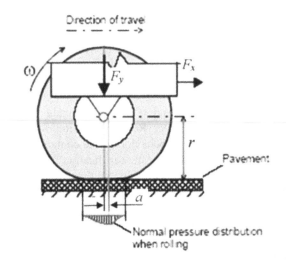

Fig. 1.

3 Problem Definition

Neural network was applied as a tool for identification of the aircraft motion math model parameters. Dependency of friction coefficient from the ground speed (V_{travel}) and the load on wheels (F_y) must be obtained during "training" process in the neural network (minimization of mismatch between empirical and calculated accelerations).

To solve this problem the simplified math model of aircraft motion along runway was developed. For identification purposes, math model is implemented in the form of longitudinal accelerations computation in the process of real aircraft run conditions reproduction

$$w_x^{calc}=F_x/m, \quad \text{where}$$

F_x – refer to formula (1); m – aircraft mass

In the process of identification, mismatch between empirical (w_x^{empir}) and calculated (w_x^{calc}) accelerations has to be minimized. Minimization of accelerations mismatch is performed by modification of friction forces (rolling friction coefficient values). Other components of calculated acceleration are supposed to be valid. w_x^{empir} values are determined as derivative of V_{travel} experimental data.

4 Identification Procedure

DCSL (Dynamic Cell Structure) neural network from "Adaptive Neural Network Library" [7] was selected as a tool of identification. The units from this library have a call interface format which seems to be the most suitable for our problem of identification. The DCSL unit scheme of data exchange in Matlab Simulink environment is shown in Fig. 2.

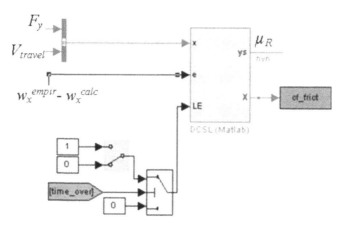

Fig. 2.

Input 'x' – arguments (in our task, these are F_y and V_{travel}) of the relation being determined by the neural network.

Output 'ys' – a function value, which is determined by neural network.

Input 'e' – a quality criterion (misalignment between the current calculated acceleration and the empirical acceleration).

Input 'LE' – an on-off switch (training process activation).

Output 'X' – internal parameters of neural network (data array). Training procedure changes this data array content.

The library includes several versions of neural networks. During this exercise other variants were tested as well. The trial runs show that DCSL neural network is the most suitable for our task, as it requires a relatively small number of repeated "training" cycles to get high convergence in the different samples of data. It is also very important that this neural network is converging well when returning to the initial sample after training with the other samples. Main elements of identification algorithm are shown in Fig. 3. The DCSL unit parameters, used during the given task performing, are shown in Fig. 4.

5 Identification Results

The identification has been performed using four samples of aircraft take-off run in actual conditions. The samples, describing takeoff run process for different take-off weights, were selected. About 300 training cycles (50-:-100 for every sample of take-off run) were done.

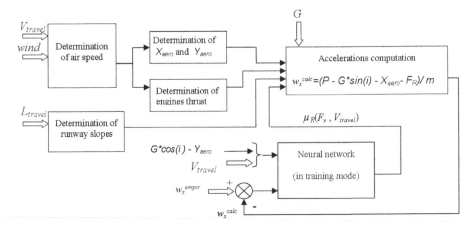

Fig. 3. The identification algorithm scheme (the shaped arrows indicate the data that are taken from the flight experiment records)

Problem of the describing of aircraft behavior is one of the most challenging due to complexity of the object. Recent scheme of aircraft description as a cortege of three mutually dependent matrix $<D_d, D_s, D_e>$ where D_d describe data dependency, D_s – state dependency and D_e element dependency proved to be efficient to analyze as various segments of the aircraft construction well as aircraft as a whole [1]. Using this approach an aircrafts air pressure system was simulated and analyzed in [2]. Applying such approach it is possible to assume that matrix of dependency is simulated using neural network.But at this stage of research we require a visual presentation of the $\mu_R\,(F_y\,,\,V_{travel})$ dependence received. To present the identification results visually, the neural network "probing" was performed. The results obtained by "probing" are given in a graphic form and corrected manually (smoothed). The obtained dependence is shown in Fig.5.

Block Parameters: DCSL (Matlab)   [x]

LDCS NN (mask) (link)

Self Adaptive Discrete Time Piecewise Linear DCS Neural Network

Parameters
[Ni No]

[2 1]

[Nmax RsThr Lambda Alpha Theta]

[40   0.001 100   0.5 0.001]

[Epsb Epsn etaW]

[0.001 0.0005 0.01]

Initial Condition , size=(Nmax*(Nmax+Ni+4)+2)*No

cf_frict(end.)%zeros(1*(40*(40+2+4)+2),1)%

Sample Time

0

OK   Cancel   Help   Apply

Fig. 4.

It is worth to mention that not all of the "load-speed" combinations, which are given in Fig. 5, were available in the flight tests data. At low speeds (about zero), the aerodynamic unloading is not available, and the loads were not lower than 140 tons. At high speeds, the loads grew smaller (high loads were absent).

6 Check Modeling

To analyze the validity of obtained rolling friction coefficient estimations, the check modeling task has been developed. In this task, the determination of friction coefficient is provided using three methods:

- by direct usage of neural network;
- by the use of $\mu_R (F_y , V_{travel})$ dependence obtained by neural network "probing";
- by usage of constant value of $\mu_R = 0.02$.

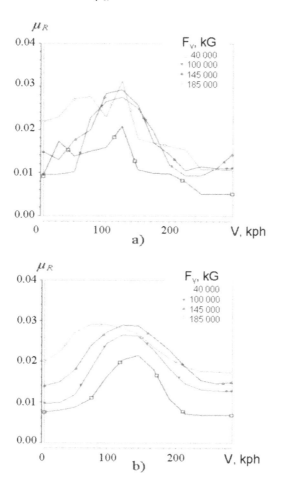

Fig. 5. The $\mu_R (F_y , V_{travel})$ dependency obtained by a) neural network "probing" and b) smoothed relationship

The results of this paper demonstrate that influence on μ_R of wheels compression and wheels rotational speed really exists. Though, one should accept that for the computation of takeoff run characteristics in general conditions, this influence appeared to be inessential. Satisfactory convergence with experimental data is also observed using a constant (averaged over the obtained relation) value of coefficient.

The test consists of L_{travel} (V_{air}) comparison of dependences (obtained in the experiments and by computations). The test was performed for four data samples that were used for identification (neural network training) and also for three additional samples of data, which were not taken into account during neural network training.

Test results have shown the acceptable convergence of experimental and computational distances traveled during take-off run versus air speed reached. The satisfactory convergence is observed also for samples of initial data that were not used during neural network training.

The example of computations results in which the determination of μ_R value was done by different methods are compared and shown in Fig. 6. Fig. 6 shows that form of $\mu_R(F_y, V_{travel})$ dependence practically has no effect on the calculated takeoff run length. In other words, to calculate the takeoff run length, the identification of rolling friction coefficient could be reduced to simple selection of constant value of μ_R. Assuming that friction coefficient μ_R dependence on the load on wheels and their rotational speed must become more obvious when the significant difference in loads on landing gear lags (if run is performed within strong cross wind conditions, for example).

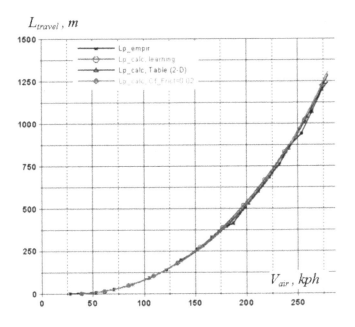

Fig. 6.

7 Estimation of Actual Aircraft Braking Characteristics under Different Runway Conditions

Similar approach of ANN application has been used for identification of dependencies describing Be-200 amphibian aircraft wheels resistance and braking performances on precipitation-covered runway (see Fig.7). This task demands to develop algorithm including more sophisticated longitudinal forces formula (2) than (1) and two ANN blocks were trained simultaneously (see [5]). In first ANN rolling resistance coefficient was estimated for sum: rolling itself plus contamination drag (respectively to wheels load) in dependency μ_{R+D} (δ/r, V) from wheel inflation and aircraft ground speed. Braking coefficient (second ANN) was estimated as a dependency $\mu_{t/g}^{MAX}(V)$ from aircraft ground speed.

Fig. 7.

$$F_x = P_x - G * sin(i) \; - \; X_{aero} - (\; F_R + \; F_D + \; F_B \;), \qquad \text{where} \qquad (2)$$

$F_R + F_D = F_{R+D}$ – rolling resistance plus contamination drag force; $F_{R+D} = \mu_{R+D} \, (\delta^{MLG}/r^{MLG}, \; V, d_c) * F_y^{MLG}$ $+ \mu_{R+D} \, (\delta^{NLG}/r^{NLG}, \; V, d_c) * F_y^{NLG}; \; r^{MLG}$ and r^{NLG} – MLG wheel and NLG wheel radius; δ^{MLG} and δ^{NLG} MLG wheel and NLG wheel inflation; d_c – contaminant thickness ; $F_B = \mu_B \, (V) * F_y^{MLG}$ – braking force; F_y^{MLG} and F_y^{NLG} – load on MLG wheels and on NLG wheels;

$$\mu_B = \begin{cases} M_B \, / [(r^{main} - \delta^{main}) \bullet Fy^{main}], & if \quad \mu_B \leq \mu_{t/g}^{MAX} \\ \mu_{t/g}^{MAX}(V) \end{cases}$$

Fig.8 shows the final stage of identification process: how neural networks changes the braking coefficient (left) and how it influences the ground speed time histories. Fig.9 shows identification results: a) for dry and the compacted snow covered runway (no contamination drag exists: d_c=0), b) for runway covered with slush ($d_c \approx 0.02$m). Dependencies are defined by "probing" of neural networks (ANN) and further smoothing (sm).

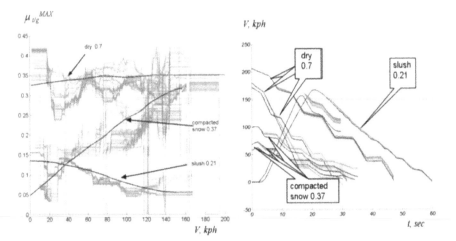

Fig. 8.

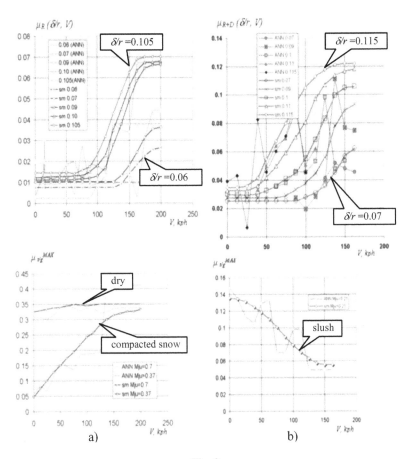

a)

b)

Fig. 9.

About 200 training cycles for every sample of aircraft run were done. Data samples were alternated during training process – about 20 cycles for one sample, then 20 cycles for another, and so on.

In Fig.10 the received braking coefficient dependency for slush covered runway is compared with dependency recommended in NPA-14 [8]. Unfortunately so far we have a single example of empiric data for such conditions. And that data are not quite suitable for identification purposes – pilot turned brakes on and off very frequently and never used full braking pressure. So, additional empiric data sets processing is required to make any more detailed conclusions.

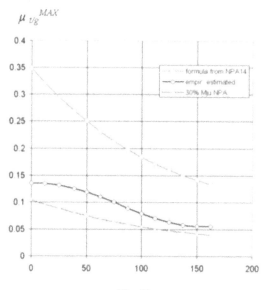

Fig. 10.

8   Further Solutions for ANN-Based Identification Tasks

The accomplished research shows the possibility of neural networks application for the aircraft math model parameters identification. It is shown that ability to deal with dependencies instead of separate values and to align random factors in experimental data sets are provided using ANN-based approach.

Advanced algorithms and estimation procedures will be required for aerodynamic characteristics identification. The algorithms, assumed in this paper, are based on the direct monitoring (minimization) of the calculated and empirical accelerations mismatch. Hypothetically, such approach can be used only for estimation of aerodynamic parameters, which characteristics become apparent in static or almost static flight conditions (during trimming, smooth approaching to the runway surface in order to estimate the ground effect, etc).

Taking into account the measurements uncertainty and errors it is reasonable to use neural networks using approach presented in [3, 4], and estimate the actual aerodynamic parameters in transient processes (stick displacements, etc). The features of the above-mentioned approach include:

– an assumption that transient process is reproduced in the flight dynamics math model, and the control signals, recorded during experiment, are supplied;

– a priori known aerodynamic data (from wind-tunnel tests) are included initially into math model. The math model includes too the corrections to aerodynamic coefficients. The identification task is to optimize these corrections values;

– total mismatch (of transient process in all) of kinematical and dynamic parameters are the criteria for the selection of corrections.

Applying this approach, another concept of neural networks application will be required. As far as monitoring of integral (in all transient process) criterion of mismatching is supposed, one cycle of training for all transient process is possible, and during modeling of transient process, selection of corrections from the "frozen" neural network is to be used. Practice of aerodynamic data identification [3, 4] has shown that the successful solving of this problem could not be fully settled by data processing itself. The participation of the human-expert is necessary for the on-line data analysis and the processing procedure options tuning.

Matlab Simulink provides the ability of algorithm variants testing and it is efficient when "on-the-fly" improvement of approaches and computations is necessary. Practice of flight dynamics math model parameters identification assumes processing and analysis of a large number of experimental data samples. For such task it is important to have the reliably checked initial data controls as well as the means of results recording and on-line graphical analysis. Data controls, which are presented by Matlab Simulink, are not satisfactory for such tasks. Solution of more complicated identification problem requires a development of special software (for example, using C++).

References

1. Bukov, V., Kirk, B., Schagaev, I.: Applying the Principle of Active Safety to Aviation. In: 2nd European Conference for Aerospace Sciences EUCASS, Brussels (July 2007)
2. Bukov, V., Schagaev, I., Kirk, B.: Analytical synthesis of aircraft control law. In: 2nd European Conference for Aerospace Sciences EUCASS, Brussels (July 2007)
3. Bondarets, A.Ya.: The flight dynamics math model refinement procedure based on flight tests data. In: The reports of 3-d Scientific Conference on Hydroaviation "Gidroaviasalon 2000" (2000) (in Russian)
4. Bondarets, A.Y., Ogolev, Y.A.: The results of Be-200 amphibian aircraft flight dynamics math model refinement based on flight tests data. In: The reports of 4-th Scientific Conference on Hydroaviation "Gidroaviasalon-2002" (2002) (in Russian)
5. Bondarets, A.Y., Kreerenko, O.D.: The neural networks application for estimation of wheels braking actual parameters for an airplane on the runway covered with precipitations. In: The reports of 5-th scientific conference Control and information technologies, Saint Petersburg (2008) (in Russian)
6. Andresen, A., Wambold, J.C.: Friction Fundamentals, Concepts and Methodology, TP 13837E Prepared for Transportation Development Centre Transport Canada (October 1999)
7. Campa, G., Fravolini, M.L.: Adaptive Neural Network Library, Version 3.1 (Matlab R11.1 through R13), West Virginia University (July 2003)
8. JAA NPA. No 14/2004 on certification specifications for large airplanes (CS-25) Operation on Contaminated Runways

Ellipse Support Vector Data Description[*]

Mohammad GhasemiGol, Reza Monsefi, and Hadi Sadoghi Yazdi

Department of Computer Engineering, Ferdowsi University of Mashhad (FUM),
Mashhad, Iran
ghasemigol@wali.um.ac.ir, monsefi@um.ac.ir, sadoghi@um.ac.ir

Abstract. This paper presents a novel Boundary-based approach in one-class classification that is inspired by support vector data description (SVDD). The SVDD is a popular kernel method which tries to fit a hypersphere around the target objects and of course more precise boundary is relied on selecting proper parameters for the kernel functions. Even with a flexible Gaussian kernel function, the SVDD could sometimes generate a loose decision boundary. Here we modify the structure of the SVDD by using a hyperellipse to specify the boundary of the target objects with more precision, in the input space. Due to the usage of a hyperellipse instead of a hypersphere as the decision boundary, we named it "Ellipse Support Vector Data Description" (ESVDD). We show that the ESVDD can generate a tighter data description in the kernel space as well as the input space. Furthermore the proposed algorithm boundaries on the contrary of SVDD boundaries are less influenced by change of the user defined parameters.

Keywords: Mapping functions, One-class classification, Outlier detection, Support vector data description.

1 Introduction

The one-class classification problem is an interesting field in pattern recognition and machine learning researches. In this kind of classification, we assume the one class of data as the target class and the rest of data are classified as the outlier. One-class classification is particularly significant in applications where only a single class of data objects is applicable and easy to obtain. Objects from the other classes could be too difficult or expensive to be made available. So we would only describe the target class to separate it from the outlier class. Three general approaches have been proposed to resolve the one-class classification problems [1]:

1) The most straightforward method to obtain a one-class classifier is to estimate the density of the training data and to set a threshold on its density. Several distributions can be assumed, such as a Gaussian or a Poisson distribution. The most pop-

---
[*] This work has been partially supported by Iran Telecommunication Research Center (ITRC), Tehran, Iran. Contract No: T/500/1640.

D. Palmer-Brown et al. (Eds.): EANN 2009, CCIS 43, pp. 257–268, 2009.

ular three density models are the Gaussian model, the mixture of Gaussians and the Parzen density [2, 3].

2) In the second method a closed boundary around the target set is optimized. K-centers, nearest neighborhood method and support vector data description (SVDD) are example of the boundary methods [4, 5].

3) Reconstruction methods are another one-class classification method which have not been primarily constructed for one-class classification, but rather to model the data. By using prior knowledge about the data and making assumptions about the generating process, a model is chosen and fitted to the data. Some types of reconstruction methods are: the k-means clustering, learning vector quantization, self-organizing maps, PCA, a mixture of PCAs, diabolo networks, and auto-encoder networks.

The SVDD is a kind of one-class classification method based on Support Vector Machine [6]. It tries to construct a boundary around the target data by enclosing the target data within a minimum hypersphere. Inspired by the support vector machines (SVMs), the SVDD decision boundary is described by a few target objects, known as support vectors (SVs). A more flexible boundary can be obtained with the introduction of kernel functions, by which data are mapped into a high-dimensional feature space. The most commonly used kernel function is Gaussian kernel [7].

This method has attracted many researchers from the various fields. Some of whom consider SVDD applications and the others try to improve the method for generating better results. For example Yan Liu et al. apply the SVDD techniques for novelty detection as part of the validation on an Intelligent Flight Control System (IFCS) [11]. Ruirui Ji et al. discussed the SVDD application in gene expression data clustering [12]. Xiaodong Yu et al. used SVDD for image categorization from internet images [13]. Recently, some efforts have been expended to improve the SVDD method. S.M. Guo et al. proposed a simple post-processing method which tries to modify the SVDD boundary in order to achieve a tight data description [14]. As another example Hyun-Woo Cho apply the orthogonal filtering as a preprocessing step is executed before SVDD modeling to remove the unwanted variation of data [15].

In this paper a new boundary method has been proposed, which tries to fit a better boundary around the target data. The SVDD method sometimes could not obtain an appropriate decision boundary in the input space and the good result is depending on the proper kernel function. For example when we have an elliptical dataset as a target class, the sphere cannot be a proper decision boundary for this kind of data set and it also covers a large amount of the outlier space. When we describe the target class with more precision in the input space, we can get better results. Therefore, we define a hyperellipse around the target class to describe a tighter boundary. Indisputably the hyperellipse constructs a better decision boundary from the hypersphere in the input space. The experiments confirm the power of ESVDD against the standard SVDD in the feature space.

The paper is organized as follows. In the next section we review the support vector data description (SVDD). The proposed method (ESVDD) is explained in Section 3. Finally, in the last section the experimental results are presented.

2 Support Vector Data Description

The support vector data description was presented by Tax and Duin [6] and again in Tax and Duin [5] with extensions and a more thorough treatment. The SVDD is a one-class classification method that estimates the distributional support of a data set. A flexible closed boundary function is used to separate trustworthy data on the inside from outliers on the outside.

The basic idea of SVDD is to find a minimum hypersphere containing all the objective samples and none of the nonobjective samples. The hypersphere is specified by its center a and its radius R. The data description is achieved by minimizing the error function:

$$F(R, a) = R^2 , \tag{1}$$

$$\text{s.t.} \quad \|x_i - a\|^2 \leq R^2, \ \forall i . \tag{2}$$

In order to allow for outliers in the training data set, the distance of each training sample x_i to the center of the sphere should not be strictly smaller than R^2. However, large distances should be penalized. Therefore, after introducing slack variables $\xi_i \geq 0$ the minimization problem becomes:

$$F(R, a) = R^2 + C \sum_i \xi_i , \tag{3}$$

$$\text{s.t.} \quad \|x_i - a\|^2 \leq R^2 + \xi_i, \quad \forall i . \tag{4}$$

The parameter C gives the tradeoff between the volume of the description and the errors. The constraints can be incorporated into the error function by introducing Lagrange multipliers and constructing the Lagrangian.

$$L(R, a, \alpha_i, \gamma_i, \xi_i) = R^2 + C \sum_i \xi_i$$
$$- \sum_i \alpha_i \{R^2 + \xi_i - (\|x_i\|^2 - 2a.x_i + \|a\|^2)\} - \sum_i \gamma_i \xi_i . \tag{5}$$

With the Lagrange multipliers $\alpha_i \geq 0$ and $\gamma_i \geq 0$. Setting partial derivatives to 0 gives these constraints:

$$\frac{\partial L}{\partial R} = 0 : \quad \sum_i \alpha_i = 1 , \tag{6}$$

$$\frac{\partial L}{\partial a} = 0 : \quad a = \frac{\sum_i \alpha_i x_i}{\sum_i \alpha_i} = \sum_i \alpha_i x_i , \tag{7}$$

$$\frac{\partial L}{\partial \xi_i} = 0 : \quad C - \alpha_i - \gamma_i = 0 . \tag{8}$$

From the above equations and the fact that the Lagrange multipliers are not all negative, when we add the condition $0 < \alpha_i < C$, Lagrange multipliers γ_i can be safely removed. So the problem can be transformed into maximizing the following function L with respect to the Lagrange multipliers α_i:

$$L = \sum_i \alpha_i(x_i. x_i) - \sum_{i,j} \alpha_i \alpha_j (x_i. x_j),$$ (9)

$$\text{s.t} \quad 0 < \alpha_i < C.$$ (10)

Note that from Eq. (7), the center of the sphere is a linear combination of the training samples. Only those training samples x_i which satisfy Eq. (4) by equality are needed to generate the description since their coefficients are not zero. Therefore these samples are called Support Vectors. The radius can be computed using any of the support vectors:

$$R^2 = (x_k. x_k) - 2 \sum_i \alpha_i(x_i. x_k) - \sum_{i,j} \alpha_i \alpha_j (x_i. x_j).$$ (11)

To judge a test sample z whether is in the target class, its distance to the center of sphere is computed and compared with R, if satisfies Eq. (12), it will be accepted, and otherwise, rejected.

$$\|z - a\|^2 = (z.z) - 2 \sum_i \alpha_i(z. x_i) - \sum_{i,j} \alpha_i \alpha_j (x_i. x_j) \le R^2.$$ (12)

SVDD is stated in terms of inner products. For more flexible boundaries, therefore, inner products of samples $(x_i.x_j)$ can be replaced by a kernel function $K(x_i,x_j)$, where $K(x_i,x_j)$ satisfies Mercer's theorem [8]. This implicitly, maps samples into a nonlinear feature space to obtain a more tight and nonlinear boundary. In this context, the SVDD problem of Eq. (9) can be expressed as:

$$L = \sum_i \alpha_i K(x_i. x_i) - \sum_{i,j} \alpha_i \alpha_j K(x_i. x_j).$$ (13)

Several kernel functions have been proposed for the SV classifier. Not all kernel functions are equally useful for the SVDD. It has been demonstrated that using the Gaussian kernel:

$$K(x, y) = \exp(-\frac{\|x-y\|}{s^2})^2,$$ (14)

results in tighter description. By changing the value of S in the Gaussian kernel, the description transforms from a solid hypersphere to a Parzen density estimator.

3   The Proposed Method

The SVDD method does not seem to get a good decision boundary in the input space. For example Fig. 1 shows the different datasets in the 2-dimensional space. The SVDD method fits a sphere around the target class as a separator in the input space. In this situation, the generated boundary covers the large amount of outlier space incorrectly. So it just generates good results when we use spherical datasets (Fig 1 (b)).

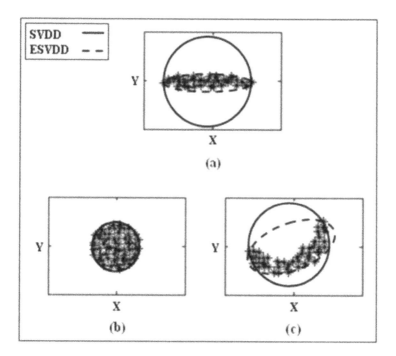

Fig. 1. SVDD and ESVDD boundary in the input space

Therefore we want to define a new method which generates a better decision boundary around the target class in the input space. In this space a more precision boundary is reached if we use an ellipse instead of a sphere which is presented in the SVDD method. In the high dimensional input space, we can also use a hyperellipse as a substitute for a hypersphere.

Although this technique is more useful for the elliptical datasets (Fig 1 (a)), it also generates noticeable results on the other datasets such as banana datasets (Fig 1 (c)). On the other hand an ellipse is a general form of a sphere; so in the worst state it transforms to a sphere and we obtain the same decision boundary as the SVDD method (Fig1 (b)). Here we try to find a hyperellipse with a minimum volume which encloses all or most of these target objects. We demonstrate this technique result in better decision boundary in the feature space as well as the input space. The problem of finding the minimum hyperellipse around n samples with d dimensions represented by a center a and the radii R_j which can be formulated into:

$$\min F = \sum_j R_j^2 , \tag{15}$$

$$\text{s.t. } \sum_j \left(\frac{x_{i,j} - a_j}{R_j}\right)^2 \le 1, \quad \forall (i = 1, \dots, n), (j = 1, \dots, d). \tag{16}$$

Corresponding to the presented SVDD to allow for outliers in the training data set, each training sample x_i should not be strictly into the hyperellipse. However, large distances should be penalized. Therefore, after introducing slack variables $\xi_i \geq 0$ the minimization problem becomes:

$$\min F = \sum_j R_j^2 + C \sum_i \xi_i, \tag{17}$$

$$\text{s.t. } \sum_j \left(\frac{x_{i,j} - a_j}{R_j}\right)^2 \leq 1 + \xi_i, \quad \xi_i \geq 0, \quad \forall (\, i = 1, \dots, n\,), (\, j = 1, \dots, d\,). \tag{18}$$

where C controls the trade-off between the hyperellipse volume and the description error. In order to solve the minimization problem in Eq. (17), the constraints of Eq. (18) are introduced to the error function using Lagrange multipliers:

$$L(R_j, \alpha_i, \gamma_i, \xi_i) = \sum_j R_j^2 + C \sum_i \xi_i - \sum_i \alpha_i \left\{ \sum_j \left(\frac{x_{i,j} - a_j}{R_j}\right)^2 - 1 - \xi_i \right\} - \sum_i \gamma_i \xi_i. \tag{19}$$

where $\alpha_i \geq 0$ and $\gamma_i \geq 0$ are Lagrange multipliers. Note that for each object x_i in dimension j a corresponding α_i and γ_i are defined. L has to be minimized with respect to R_j, ξ_i and maximized with respect to α_i and γ_i. A test object z is accepted when it satisfies the following inequality.

$$\sum_j \left(\frac{z_{i,j} - a_j}{R_j}\right)^2 \leq 1. \tag{20}$$

Analogous to SVDD, we might obtain a better fit between the actual data boundary and the hyperellipse model. Assume we are given a mapping Φ of the data which improves this fit. We can apply this mapping to Eq. (17) and we obtain:

$$L(R_j, \alpha_i, \gamma_i, \xi_i) = \sum_j R_j^2 + C \sum_i \xi_i - \sum_i \alpha_i \left\{ \sum_j \left(\frac{\Phi(x_i)_j - a_j}{R_j}\right)^2 - 1 - \xi_i \right\} - \sum_i \gamma_i \xi_i. \tag{21}$$

According to [9] we can get these Φ functions from the standard kernels which is been proposed for the support vector classifier. For example to find the Φ function for polynomial kernel with d=2, in the 2-dimensional space, we should do the following procedure:

$$x = [\, x_1, x_2\,]$$
$$y = [\, y_1, y_2\,]$$
$$K(x, y) = (1 + x^T y)^2$$
$$K(x, y) = 1 + x_1^2 y_1^2 + 2x_1 x_2 y_1 y_2 + x_2^2 y_2^2 + 2x_1 y_1 + 2x_2 y_2$$
$$K(x, y) = \phi(x)^T \phi(y)$$
$$\phi(x) = [\, 1, x_1^2, \sqrt{2} x_1 x_2, x_2^2, \sqrt{2} x_1, \sqrt{2} x_2\,].$$

Here we run SVDD and ESVDD on a simple ellipsoid dataset by using the above polynomial kernel. Fig 2 shows that the better boundary is obtained when we use the ESVDD algorithm by the same parameters. In other words, the proposed algorithm

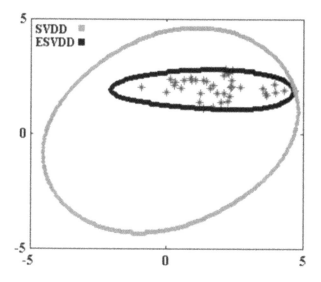

Fig. 2. SVDD and ESVDD boundary by using the polynomial kernel

boundaries on the contrary of SVDD boundaries are less influenced by changing the user defined parameters.

Using the Gaussian kernel instead of the polynomial kernel results in tighter descriptions in the SVDD method. So we use this kernel for a random ellipsoid dataset. First we extract the Φ function from the Gaussian kernel. For this reason we solve the following problem:

$$K(x,y) = exp(-\frac{\|x-y\|}{S^2})^2 = \phi(x)^T \phi(y).$$

If we suppose S=1 then:

$$K(x,y) = e_{i=1,\dots,n}^{-\|x_i-y_i\|} = e^{-[(x_1-y_1)^2+(x_2-y_2)^2+\cdots+(x_n-y_n)^2]}$$

$$= e^{-(x_1-y_1)^2} * e^{-(x_2-y_2)^2} * \dots * e^{-(x_n-y_n)^2}$$

$$= e^{-x_1^2} * e^{-y_1^2} * e^{2x_1y_1} * \dots * e^{-x_n^2} * e^{-y_n^2} * e^{2x_ny_n}$$

$$= e^{-(x_1^2+\cdots+x_n^2)} * e^{-(y_1^2+\cdots+y_n^2)} * e^{2x_1y_1} * \dots * e^{2x_ny_n}.$$

The Taylor formula for $e^{2x_iy_i}$ is

$$e^{2x_iy_i} = 1 + 2x_1y_1 + \frac{4x_1^2y_1^2}{2!} + \cdots + \frac{2^n x_1^2 y_1^2}{n!}$$

By using proper substitutions we can get desired Φ function. For example in the 1-dimensional input space and using four terms to compute the Taylor formula, the following Φ function is obtained.

$$K(x_1, y_1) = e^{-x_1^2} * e^{-y_1^2} * e^{2x_1 y_1} = \phi(x)^T \phi(y)$$

$$= e^{-x_1^2} * e^{-y_1^2} * \left(1 + 2x_1 y_1 + 2x_1^2 y_1^2 + \frac{4x_1^3 y_1^3}{3}\right).$$

$$\phi(x) = \left[e^{-x_1^2}, \sqrt{2}x_1 e^{-x_1^2}, \sqrt{2}x_1^2 e^{-x_1^2}, \frac{2}{\sqrt{3}}x_1^3 e^{-x_1^2}\right]. \tag{22}$$

For the Gaussian kernel no finite mapping $\Phi(x)$ of object x can be given. But we can get an approximation of Φ functions by using the Taylor formula. So we can use this function for mapping input space into feature space.

Fig. 3 compares the SVDD boundary to the ESVDD results. The proposed method generates the better boundary when the same parameters are used. In this experiment the Taylor formula is expand until the first four terms.

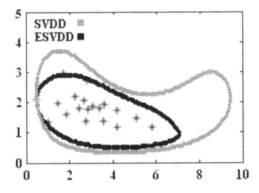

Fig. 3. SVDD and ESVDD boundary with the Gaussian kernel (By using ten terms of Taylor formula)

3.1 Φ Functions Characteristics

Now we confronted a new difficulty about the Φ functions. As mentioned in the previous section, the Gaussian kernel has no finite mapping $\Phi(x)$ of object x. So we use an approximation of Φ functions. If we consider these functions with more attention, interesting results are obtained.

For example in the 4-dimensional space we get a Φ function which is mapping this space into 10000-dimensional space. Although many of these dimensions contained very small coefficient which can be connived them. So eliminating of these dimensions does not impose a critical error in the final results. Fig. 4 shows the logarithm of coefficients for each dimension. Just few dimensions have considerable coefficients which can be efficient in space transformation.

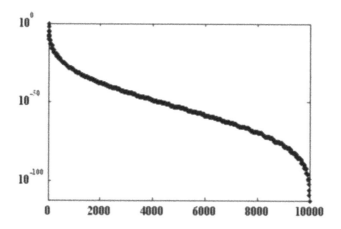

Fig. 4. The logarithm of coefficients for each dimension

Therefore we can map 4-dimensional space into a smaller feature space with fewer dimensions. These dimensions are selected from 10000 dimensions which have considerable coefficients (bigger than 10^{-6}). Hence many of these dimensions are useless and can be eliminated. In the next section we do simple experiments for proving this claim.

4 Experiments

We compare the performances of the SVDD and ESVDD methods with a synthetic dataset and some of datasets taken from UCI Machine Learning Dataset Repository [10]. Table 1 provides details about the datasets used here.

Table 1. UCI datasets used for the evaluation of the outlier detection methods

Dataset	No. of Objects	No. of Classes	No. of Features
Iris	150	3	4
haberman	306	2	3
balance-scale	625	3	4

In Iris and balance-scale datasets three classes with four features are existed. Since, to use them for outlier detection, two of the classes are used as the target class while the remainder class is considered as outlier. Haberman's Survival dataset has two classes and uses three features. In this dataset we have 225 samples in one class and 81 samples in the other one. So we can use it easily for a one class classification problem. In this situation, the class with more samples supposed as the target class.

In the first step we should create some Φ functions for mapping the various input space into a high dimension feature space.

Table 2. Mapping functions characteristics. (a) Φ functions characteristics. (b) Selected dimensions from Φ functions.

<table>
<tr><td colspan="3">(a)</td><td colspan="3">(b)</td></tr>
<tr><th>Φ functions</th><th>Input Space Dimensions</th><th>Feature Space Dimensions</th><th>Φ' functions</th><th>Input Space Dimensions</th><th>Feature Space Dimensions</th></tr>
<tr><td>Φ_1</td><td>2</td><td>100</td><td>Φ_1'</td><td>2</td><td>6</td></tr>
<tr><td>Φ_2</td><td>3</td><td>1000</td><td>Φ_2'</td><td>3</td><td>10</td></tr>
<tr><td>Φ_3</td><td>4</td><td>10000</td><td>Φ_3'</td><td>4</td><td>15</td></tr>
</table>

Here we connection with datasets by 2, 3 or 4 dimensions. The related mapping function characteristics are presented in Table 2. We repeat all of the experiments with some of the selected dimensions of Φ functions (Φ'). According to the previous section we claim that reducing the dimensions in the feature space has not any critical effects in the classification. Even in some cases fewer dimensions lead to better results. Table 3 compares the performance of SVDD and ESVDD methods with Iris, balance-scale and haberman datasets in 10 Iterations.

Table 3. Recognition Rate in SVDD and ESVDD methods for the Iris, balance-scale and heberman datasets

Experiment Conditions	SVDD	Mapping function	ESVDD
Iris Dataset (Target class= Class 1 & Class 2) (Outliers= Class 3) Random Learning Samples=50 Random Testing Samples=100 (50 target + 50 outlier)	85.3 %	Φ_3	89.0 %
		Φ_3'	87.7 %
balance-scale Dataset (Target class=Class L & Class R) (Outliers= Class B) Random Learning Samples=40 Random Testing Samples=98 (49 target + 49 outlier)	65.3 %	Φ_3	67.4 %
		Φ_3'	66.2 %
heberman Dataset (Target class= Class 1) (Outliers= Class 2) Random Learning Samples=40 Random Testing Samples=100 (50 target + 50 outlier)	69.7 %	Φ_2	78.6 %
		Φ_2'	78.9 %

By running the proposed method on an elliptical synthetic dataset, the following results are obtained (see Table 4). In these experiments, we use the same user defined parameters. Thus ESVDD generates better decision boundary than SVDD in the non-spherical datasets specially.

Table 4. Recognition Rate in SVDD and ESVDD methods for an elliptical Synthetic dataset

Experiment Conditions	SVDD	Mapping function	ESVDD
Synthetic Dataset Iteration=100 (Target class=Class 1) (Outliers= Class 2)		Φ_1	96.917 %
	92.560 %		
Learning Samples=150 Testing Samples=300 (150 target + 150 outlier)		$\Phi_1{}'$	97.667 %

5 Conclusion

In this paper, we propose a new approach to make the SVDD boundary closely fit the contour of the target data. The SVDD method uses a hypersphere which cannot be a good decision boundary for the target data, in the input space. So we define a hyperellipse instead of a hypersphere and resolve the equations by applying this alteration.

On the other hand the SVDD tries to improve the results by using the kernel functions. In this state, data are mapped to a high dimensional space. Then we use a hypersphere around the data in the new space. Experiments show that using a hyperellipse lead to better results in the feature space beside the input space. Furthermore as an important benefit, it is less influenced by changing the user defined parameters and we even obtained acceptable results with the inappropriate parameters.

Acknowledgment. This work has been partially supported by Iran Telecommunication Research Center (ITRC), Tehran, Iran (Contract No: T/500/1640). This support is gratefully acknowledged.

References

1. Tax, D.M.J.: One-class classification concept learning in the absence of counter-examples. Technische Universiteit Delft, Netherlands 65 (2001)
2. Parzen, E.: On estimation of a probability density function and mode. Annals of Mathenatical Statistics 33, 1065–1076 (1962)
3. Bishop, C.: Neural Networks for Pattern Recognition. Oxford University Press, Oxford (1995)
4. Ypma, A., Duin, R.: Support objects for domain approximation. In: Proceedings of Int. Conf. on Artificial Neural Networks (ICANN 1998), Skovde, Sweden (1998)
5. Tax, D.M.J., Duin, R.P.W.: Support Vector Data Description. Machine Learning 54, 45–66 (2004)
6. Tax, D.M.J., Duin, R.P.W.: Support vector domain description. Pattern Recognition Letters 20, 1191–1199 (1999)
7. Guo, S.M., Chen, L.C., Tsai, J.S.H.: A boundary method for outlier detection based on support vector domain description. Pattern Recognition 42, 77–83 (2009)
8. Scholkopf, B., Smola, A.J., Muller, K.: Nonlinear component analysis as a kernel eigenvalue problem. Neural Computation 10, 1299–1319 (1999)

9. Haykin, S.: Nerual Networks a Comprehensive Foundation. Prentic Hall International, Inc., Englewood Cliffs (1999)
10. Blake, C.L., Merz, C.J.: UCI repository of machine learning databases. Department of Information and Computer Sciences, University of California, Irvine, http://www.ics.uci.edu/~mlearn/MLRepository.html
11. Liu, Y., Gururajan, S., Cukic, B., Menzies, T., Napolitano, M.: Validating an Online Adaptive System Using SVDD. In: Proceedings of the 15th IEEE International Conference on Tools with Artificial Intelligence, ICTAI 2003 (2003)
12. Ji, R., Liu, D., Wu, M., Liu, J.: The Application of SVDD in Gene Expression Data Clustering. In: The 2nd International Conference on Bioinformatics and Biomedical Engineering (ICBBE 2008), pp. 371–374 (2008)
13. Yu, X., DeMenthon, D., Doermann, D.: Support Vector Data Description for Image Categorization From Internet Images. In: The 19th International Conference on Pattern Recognition, ICPR 2008 (2008)
14. Guo, S.M., Chen, L.C., Tsai, J.S.H.: A boundary method for outlier detection based on support vector domain description. Pattern Recognition 42, 77–83 (2009)
15. Cho, H.W.: Data description and noise filtering based detection with its application and performance comparison. Expert Systems with Applications 36, 434–441 (2009)

Enhanced Radial Basis Function Neural Network Design Using Parallel Evolutionary Algorithms

Elisabet Parras-Gutierrez[1], Maribel Isabel Garcia-Arenas[2],
and Victor M. Rivas-Santos[1]

[1] Department of Computer Sciences,
Campus Las Lagunillas s/n, 23071, Jaen, Spain
[2] Department of Computer Architecture and Technology
C/Periodista Daniel Saucedo s/n, 18071, Granada, Spain
{eparrasg,vrivas}@vrivas.es,
maribel@atc.ugr.es

Abstract. In this work SymbPar, a parallel co-evolutionary algorithm for automatically design the Radial Basis Function Networks, is proposed. It tries to solve the problem of huge execution time of Symbiotic_CHC_RBF, in which method are based. Thus, the main goal of SymbPar is to automatically design RBF neural networks reducing the computation cost and keeping good results with respect to the percentage of classification and net size. This new algorithm parallelizes the evaluation of the individuals using independent agents for every individual who should be evaluated, allowing to approach in a future bigger size problems reducing significantly the necessary time to obtain the results. SymbPar yields good results regarding the percentage of correct classified patterns and the size of nets, reducing drastically the execution time.

Keywords: Neural networks, evolutionary algorithms, parallelization, co-evolution.

1 Introduction

Radial Basis Function Networks (RBFNs) have shown their capability to solve a wide range of problems, such as classification, function approximation, and time series prediction, among others.

RBFNs represent a special kind of nets since, once their structure has been fixed, the optimal sets of weights linking hidden to outputs neurons can be analytically computed. For this reason, many data mining algorithms have been developed to fully configure RBFNs, and researchers have applied data mining techniques to the task of finding the optimal RBFN that solves a given problem. Very often, these algorithms need to be given a set of parameters for every problem they face, thus methods to automatically find these parameters are required.

One the methods present in literature is Symbiotic_CHC_RBF [11],[12], a co-evolutionary algorithm which tries to automatically design RBF neural network, establishing an optimal set of parameters for the method EvRBF [14], [15], (an evolutionary algorithm developed to automatically design asymmetric RBFNs).

D. Palmer-Brown et al. (Eds.): EANN 2009, CCIS 43, pp. 269–280, 2009.

In this paper the method SymbPar, a co-evolutionary and parallel algorithm, is proposed. It is a parallel approach based on the method Symbiotic_CHC_RBF, resolving its huge computation time problem. Thus, the main goal of SymbPar is to automatically design RBF neural networks reducing the time cost. To do this, the method tries to find an optimal configuration of parameters for the method EvRBF, adapted automatically to every problem. So, SymbPar is a good example of the capacity of co-evolution to increase the usability of traditional techniques, one of the future trends in data mining [7].

The term of co-evolution can be defined as "the evolution of two or more species that do not cross among them but that possess a narrow ecological relation, across reciprocal selection pressures, the evolution of one of the species depends on the evolution of other one" [19]. More specifically, symbiont organisms are extreme cases in which two species are so intimately integrated that they act as an alone organism.

The symbiotic co-evolution approach used in this work makes every population to contribute to the evolution of the other and vice versa. The phenomenon of co-evolution has allowed at the same time to consider the parallel approach of it. This way, each population can be evolved in an independent form, though sharing the necessary information to influence in the development of other one.

The rest of the paper is organized as follows: section 2 briefly describes the state of the art, section 3 introduces the method SymbPar, section 4 shows the experimentation carried out and the results obtained, and finally, section 5 describes some conclusions and future works.

2   State of the Art

Harpham et al. reviewed in [5] some of the best known methods that apply evolutionary algorithms to RBFNs design. As for others kind of nets, they concluded that methods tend to concentrate in only one aspect when designing RBFNs. Nevertheless, there also exist methods intended to optimize the whole net, such as [9] for Learning Vector Quantization (LVQ) nets or [3] for multilayer perceptrons.

Recent applications try to overcome the disadvantages of the preceding methods. Rivera [16] made many neurons compete, modifying them by means of *fuzzy evolution*. Ros et al.'s method [17] automatically initializes RBFNs, finding a good set of initial neurons, but relying in some other method to improve the net and to set its widths.

On the other hand, EvRBF [14], [15] is an evolutionary algorithm designed to fully configure an RBFNs, since it searches for the optimal number of hidden neurons, as well as their internal parameters (centers and widths). EvRBF is a steady state evolutionary algorithm that includes elitism; it follows a Pittsburgh scheme, in which each individual is a full RBFNs whose size can change, while population size remains equal.

Different kinds of co-evolution have also been used to design ANN, as can be found in literature. Paredis [10] proposed a general framework for the use of co-evolution to boost the performance of genetic search, combining co-evolution with life-time fitness evaluation. Potter and De Jong [13] proposed Cooperative Co-evolutionary Algorithms (CCGAs) for the evolution of ANN of cascade network topology.

Schwaiger and Mayer [18] employed a genetic algorithm for the parallel selection of appropriate input patterns for the training datasets (TDs). After this, Mayer [8], in order to generate co-adapted ANN and TDs without human intervention, investigated the use of Meta (cooperative) co-evolution. Independent populations of ANN and TDs were evolved by a genetic algorithm, where the fitness of an ANN was equally credited to the TDs it had been trained with.

3   Method Overview

This section firstly describes EvRBF, and after this describes Symbiotic_CHC_RBF and SymbPar. These last two evolutionary algorithms have been developed to find a suitable configuration of parameters necessary for EvRBF.

3.1   Description of EvRBF

EvRBF [14,15] is a steady state evolutionary algorithm that includes elitism. It follows the Pittsburgh scheme, in which each individual is a full RBFN whose size can vary, while population size remains equal.

EvRBF codifies in a straightforward way the parameters that define each RBFN, using an object-oriented paradigm; for this reason, it includes operators for recombination and mutation that directly deal with the neurons, centers and widths themselves. Recombination operator, X_FIX, interchanges information between individuals, trying to find the building blocks of the solution. More precisely, X_FIX replaces a sequence of hidden neurons in RBFN R_1 by a sequence of the same size taken from RBFN R_2.

Mutation operators (centers and width modification: C_RANDOM and R_RANDOM) use randomness to increase diversity generating new individuals so that local minima can be avoided. Furthermore, EvRBF tries to determine the correct size of the hidden layer using the operators ADDER and DELETER to create and remove neurons, respectively. The exact number of neurons affected by these operators (except ADDER) is determined by their internal application probabilities.

EvRBF incorporates tournament selection for reproduction. The skeleton of EvRBF, showed in figure 1, is commonly used in evolutionary algorithms. The *fitness function* measures the generalization capability of each individual as the percentage of samples it correctly classifies. When comparing two individuals, if and only if both two individuals have exactly the same error rate, the one with less neurons is said to be better. In order to train individuals, the LMS algorithm is used.

```
1. Create, train, evaluate and set fitness of first generation.
2. Until stop condition is reached
   (a) Select and copy individuals from current population.
   (b) Modify, train, and set fitness of the copies.
   (c) Replace worst individuals by the copies.
3. Train last generation with training and validation data
4. Use test data set to obtain the generalization ability of each individual.
```

Fig. 1. General skeleton of EvRBF

3.2   Description of Symbiotic_CHC_RBF

EvRBF is able to build neural networks in order to solve different sort of problems. Nevertheless, everytime it is going to be executed a big set of parameters has to be fixed by hand. The initial set[14] includes up to 25 different parameters, that could be reduced to 8 while maintaining a tradeoff between generalization ability and network sizes [15].

In order to fully automatize the design of the RBFNN, Symbiotic_CHC_RBF was developed. Symbiotic_CHC_RBF is an evolutionary algorithm which tries to find an optimal configuration of parameters for the method EvRBF. The parameter setting can be then automatically adapted to every problem.

This method uses the CHC algorithm [4], an evolutionary approach which introduces an appropriate balance between diversity and convergence. The CHC algorithm was developed in order to solve the problems of premature convergence that genetic algorithms frequently suffer, and it uses a conservative strategy of selection.

Therefore, it uses a high selective pressure based on an elitist scheme in combination with a highly disruptive crossover and a restart when the population is stagnated. Then, it is based on four components [4]: elitist selection, HUX crossover operator, incest prevention and re-inicialization of the population.

Every individual of the Symbiotic_CHC_RBF algorithm is a binary string representing a set of 8 parameters for the method EvRBF, as the size of the population, the size of the tournament for the selection of the individuals or the maximum number of generations.

In order to set the *fitness* of an individual, the chromosome is decoded into the set of parameters it represents. Then, these parameters are used to perform a complete execution of EvRBF. Once EvRBF has finished, the percentage of training patterns correctly classified by the best net found, is used as *fitness* for the individual. The representation of Symbiotic_CHC_RBF can be seen in figures 2 and 4.

3.3   Description of SymbPar

Symbiotic_CHC_RBF could be enhanced if some of the tasks it has to carry out could be executed in parallel. For this reason, SymbPar has been developed. Thus, SymbPar is a parallel evolutionary algorithm, based on Symbiotic_CHC_RBF. As

1. Create, train, evaluate and set fitness of first generation
2. Set threshold for cross over
3. Until stop condition is reached
   (a) Select and copy individuals from current population
   (b) Modify, train, and set fitness of the copies
   (c) Replace worst individuals by the copies
   (d) Reduce the threshold if the population is the same
   (e) If threshold < 0 restart the population
4. Train last generation with training and validation data
5. Use test data set to obtain generalization ability of each individual

Fig. 2. Execution scheme of Symbiotic_CHC_RBF

this algorithm, SymbPar's main goal is to automatically design the RBFNs finding a suitable configuration of parameters for the method EvRBF.

The task of parallelizing an algorithm involves to find the critical points of the sequential evolutionary algorithm in order to carry out a successful parallelization. There are two important parameters of sequential algorithm, which set the speed of the algorithm. The first one is the population size, many times, this parameter is high to get a suitable diversity, so the speed of the algorithm is minor. And the other one is *fitness* function of the individuals of population, thus with complex evaluation functions the execution time can rocket.

Therefore, sequential evolutionary algorithms sometimes need a huge amount of physical memory, to keep big populations, and an elevated processing time. Both the requirement of memory and computation time are two good reasons to attempt to speed up the execution of an evolutionary algorithm by means of parallelization. In Symbiotic_CHC_RBF, good solutions are achieved with relatively small populations (around 50 individuals), so the reason of its parallelization is the total execution time of the algorithm.

Candidates operations to be parallelized are those that can be independently applied to every individual, or a small set of individuals, for example mutation or evaluation. However, operations which need the whole population, like selection, can't be parallelized using this technique. The operation more commonly used is the individual evaluation, when it is independent from the rest of individuals [20], since it is the most difficult operation of the whole sequent evolutionary algorithm.

SymbPar tries to solve the problem that Symbiotic_CHC_RBF has when it works with large databases and the computation time is huge, owing that each individual evaluation be made in a sequential way. The evolution schemes of Symbiotic_CHC_RBF and SymbPar are the same,nevertheless, there is an important difference between them, the execution scheme of evaluations in Symbiotic_CHC_RBF is sequential. It means that until the evaluation of the algorithm EvRBF with the first individual does not finish, does not begin with the second individual, and so on. On the other hand, SymbPar carries out every individual evaluation in a parallel form, following a *Master-Worker* process scheme [2].

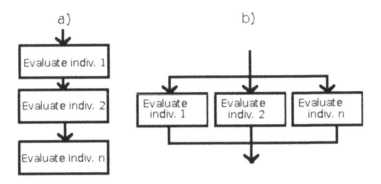

Fig. 3. Differences in the evaluation of individuals when (a) Symbiotic_CHC_RBF is used and (b) SymbPar is used

In order to distribute the computation load SymbPar uses the communication abilities provided by the framework DRM. DRM follows the *Master-Worker* scheme and has been deeply described in [6]. Briefly, SymbPar defines two important classes: *Master* class, whose functions are to organize the work to do the evaluation of all individuals, to create the independent tasks that make this evaluation and to collect the results once the tasks have ended; and *Worker* class, whose functions are to carry out the task ordered by *Master* and to send the obtained result. Then, the algorithm SymbPar creates a type *Master* instance for every generation, just before executing the evaluation of each individual of the population. The *Master* object is an independent agent who provokes the stop of the algorithm and creates as many objects of the type *Worker* as necessary to evaluate every individual. Figure 3 shows that individuals in Symbiotic_CHC_RBF are sequentially evaluated, while SymbPar is able to create a different Worker for each individual to be evaluated.

4   Experiments and Results

In order to test the performance of Symbiotic_CHC_RBF and SymbPar, the algorithms have been evaluated with the following data sets: Flag, German, Glass, Haberman, Ionosphere, New-thyroid, Pima, Postoperative, Sonar, Vehicle, and WDBC from UCI data set repository[1].

In addition, for Ionosphere, Sonar, and WDBC databases, 7 different data sets generated using various feature selection methods, have been used in order to test Symbiotic_CHC_RBF and SymbPar with a higher number of datasets. So that the methods have been tested over 29 different classification data sets. Table 1 shows features, instances and model parameters of data sets.

Then, a 10-crossfold validation method has been used for every data set, so that every one has been divided into 10 different sets of training-test patterns.

---

[1] http://www.ics.uci.edu/~mlearn/MLRepository.html

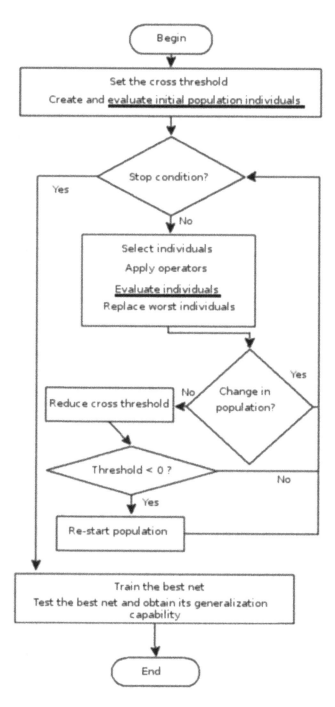

Fig. 4. Execution scheme of Symbiotic_CHC_RBF for every generation. SymbPar execute the evaluations steps (underlined in this figure) in parallel, while Symbiotic_CHC_RBF executes them sequentially.

Table 1. Characteristics of the databases: number of features, number of instances, and model size (features by instances)

Database	Features	Instances	Size	Database	Features	Instances	Size
S.FS-Forward	5	61	305	I.FS-SSGA	10	351	3510
S.FS-Focus	7	61	427	W.FS-GGA	8	570	4560
Postoperative	9	90	810	W.FS-LFV	9	570	5130
S.FS-GGA	15	61	915	W.FS-SSGA	9	570	5130
S.FS-LFV	18	61	1098	Flag	29	194	5626
S.FS-SSGA	18	61	1098	W.FS-Focus	11	570	6270
Haberman	4	306	1224	Pima	9	768	6912
New-thyroid	6	215	1290	I.FS-Backward	30	351	10530
I.FS-Focus	4	351	1404	Ionosphere	35	351	12285
I.FS-Forward	5	351	1755	Sonar	61	208	12688
Glass	10	214	2140	W.FS-Backward	28	570	15960
S.FS-Backward	40	61	2440	Vehicle	19	846	16074
I.FS-GGA	8	351	2808	WDBC	31	570	17670
W.FS-Forward	6	570	3420	German	21	1000	21000
I.FS-LFV	10	351	3510				

For every training-set couple, the algorithms have been executed three times, and the results show the average over these 30 executions. So, since we have 29 data sets, a total of 870 executions have been performed.

The set of experiments have been executed using the Scientific Computer Center of Andalucia's cluster[2]. In case of SymbPar, a single machine with four cores has been used. The work was randomly distributed among the cores, by the operating system itself (Ubuntu 7.10 with a kernel Linux 2.6.22-14-generic SMP). Moreover, the function library JEO [1], which uses support for distribution of tasks in a distributed virtual machine named DRM [6], has been used.

In table 2 can be seen the obtained results for the two methods. This table is organized by columns, showing for both algorithms and from left to right, the number of neurons (Nodes), the percentage of correct classified patterns (Test), and the computation time represented in minutes (T). For every measure, the average and standard deviation are shown. In bold are marked the best results with respect to the execution time with every database, that is, the fastest comparing the two algorithms.

As can be observed in the table, the quality of the solutions yielded by Symbiotic_CHC_RBF is neither improved nor worsened by SymbPar. The sizes of the nets founds by both methods are quite similar and also the percentages of classification. On the other hand, there are obvious differences between the sequential and parallel algorithms regarding the execution time, since SymbPar drastically reduces the computation in almost every data set. Figure 5 shows the ratio between the results yielded by Symbiotic_CHC_RBF and SymbPar for both classification percentages and executiontime. Every ratio is calculated dividing Symbiotic_CHC_RBF

---

[2] CICA: http://www.cica.es/condor.html

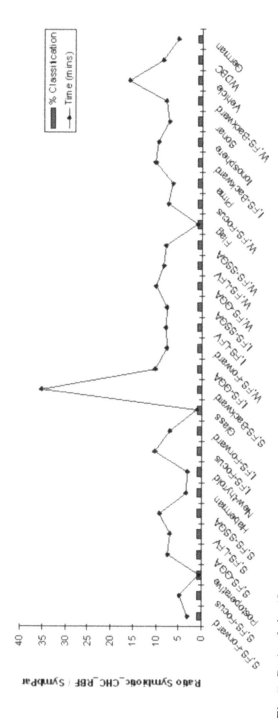

Fig. 5. Ratio of classification percentages versus ratio of time executions. Ratio is calculated dividing the results yielded by Symbiotuc_CHC_RBF by the ones yielded by SymbPar.

Table 2. Results of Symbiotic_CHC_RBF and SymbPar

Database	Symbiotic_CHC_RBF			SymbPar		
	Nodes	Test(%)	T(min)	Nodes	Test(%)	T(min)
S.FS-Forward	06 ± 06	81.16 ± 7.97	37 ± 09	05 ± 04	82.76 ± 8.70	**12 ± 0**
S.FS-Focus	08 ± 05	85.52 ± 7.68	68 ± 11	06 ± 06	84.72 ± 7.90	**14 ± 0**
Postoperative	02 ± 01	82.96 ± 6.99	**05 ± 01**	02 ± 02	82.22 ± 8.56	10 ± 0
S.FS-GGA	15 ± 13	85.25 ± 7.68	172 ± 35	15 ± 09	86.03 ± 5.84	**23 ± 2**
S.FS-LFV	11 ± 07	82.81 ± 8.34	145 ± 33	10 ± 07	82.13 ± 8.19	**21 ± 2**
S.FS-SSGA	16 ± 13	84.70 ± 9.03	268 ± 82	12 ± 09	83.93 ± 9.85	**29 ± 3**
Haberman	07 ± 03	82.55 ± 5.28	57 ± 12	06 ± 03	81.66 ± 4.67	**17 ± 1**
New-thyroid	07 ± 03	99.22 ± 1.78	60 ± 08	06 ± 03	98.44 ± 2.57	**20 ± 2**
I.FS-Focus	17 ± 10	94.90 ± 3.85	418 ± 81	20 ± 13	95.27 ± 3.55	**41 ± 6**
I.FS-Forward	10 ± 07	94.87 ± 2.42	161 ± 36	09 ± 06	95.07 ± 2.36	**23 ± 1**
Glass	01 ± 01	92.22 ± 2.19	**11 ± 02**	01 ± 00	92.08 ± 2.11	11 ± 0
S.FS-Backward	15 ± 12	84.77 ± 7.82	1200 ± 148	14 ± 08	85.10 ± 7.09	**34 ± 4**
I.FS-GGA	18 ± 10	95.89 ± 3.92	357 ± 56	18 ± 10	95.61 ± 3.42	**35 ± 4**
W.FS-Forward	07 ± 04	95.14 ± 3.21	249 ± 50	07 ± 04	95.21 ± 3.41	**32 ± 2**
I.FS-LFV	11 ± 04	94.03 ± 4.26	228 ± 42	12 ± 08	93.94 ± 3.69	**29 ± 2**
I.FS-SSGA	11 ± 06	95.83 ± 2.44	217 ± 82	10 ± 03	95.82 ± 2.34	**28 ± 5**
W.FS-GGA	10 ± 05	96.72 ± 2.43	503 ± 134	11 ± 07	96.44 ± 2.42	**50 ± 10**
W.FS-LFV	07 ± 04	92.21 ± 1.94	259 ± 219	08 ± 09	92.63 ± 2.22	**31 ± 13**
W.FS-SSGA	07 ± 04	94.83 ± 3.35	220 ± 97	07 ± 06	94.91 ± 3.16	**28 ± 7**
Flag	01 ± 00	87.39 ± 15.56	**10 ± 02**	02 ± 01	87.39 ± 16.15	11 ± 0
W.FS-Focus	09 ± 04	93.84 ± 1.77	237 ± 48	11 ± 06	93.68 ± 1.98	**32 ± 2**
Pima	09 ± 06	80.06 ± 3.24	309 ± 53	16 ± 11	80.99 ± 2.90	**49 ± 5**
I.FS-Backward	19 ± 14	97.64 ± 2.47	459 ± 92	17 ± 09	96.91 ± 2.88	**45 ± 5**
Ionosphere	13 ± 07	97.64 ± 2.47	316 ± 82	12 ± 06	97.46 ± 2.82	**33 ± 4**
Sonar	06 ± 04	71.79 ± 6.66	130 ± 32	06 ± 04	70.99 ± 8.86	**18 ± 2**
W.FS-Backward	06 ± 04	94.79 ± 3.55	298 ± 53	07 ± 05	95.21 ± 3.14	**38 ± 5**
Vehicle	49 ± 20	94.99 ± 1.80	1903 ± 442	44 ± 21	93.77 ± 2.24	**119 ± 16**
WDBC	08 ± 06	95.02 ± 3.29	310 ± 55	09 ± 07	95.27 ± 3.36	**36 ± 3**
German	06 ± 05	73.20 ± 2.43	259 ± 36	08 ± 07	73.47 ± 2.39	**50 ± 7**

by SymbPar. Bars in the figure show that the ratio for classifications keeps close to 1 in all the considered problems. Lines in the figure show the ratio for time executions, proving that Symbiotic_CHC_RBF can be up to 37 times slower than SymbPar.

The above observations have been corroborated by the Wilcoxon test. For both nodes and test measures, the statistical p-value shows that there are no differences between the algorithms. In the case of time, Wilcoxon test shows that SymbPar is definitively better than Symbiotic_CHC_RBF, since the p-value gets a value of $2.625e - 06$.

5   Conclusions and Future Research

In this work SymbPar, a parallel evolutionary algorithm for automatically design the RBFNs, is proposed.

The algorithm SymbPar presents an alternative with threads of execution in parallel to the problem of the high execution time of the algorithm Symbiotic_CHC_RBF reducing drastically the computation times in the great majority of the cases.

This opens the possibility of continuing with a research line developing the original idea of the algorithm Symbiotic_CHC_RBF with bigger databases without having to wait for results during weeks even months, when the size of the used database grows.

In the execution times presented in this work, the parallel executions have been carried out in a computer with four processors, where the threads were executed under the administration of the operating system.

A future work could be driven to make a study of the scalability and profit of the parallel version of the algorithm SymbPar, executing the experiments in different number of machines to see its behave with regard to the execution time when the number of machines changes.

Acknowledgment

This work has been partially supported by the Spanish project MYCYT NoHNES (Ministry of Education and Science - TIN2007-6083), the excellent project of Junta de Andalucia PC06-TIC-02025, and the project UJA_08_16_30 of the University of Jaen.

References

1. Arenas, M.G., Dolin, B., Merelo, J.J., Castillo, P.A., Fernandez, I., Schoenauer, M.: JEO: Java evolving objects. In: GECCO2: Proceedings of the Genetic and Evolutionary Computation Conference (2002)
2. Bethke, A.D.: Comparison of genetic algorithms and gradient-based optimizers on parallel processors: Efficiency of use of processing capacity. Tech. Rep., University of Michigan, Ann Arbor, Logic of Computers Group (1976)
3. Castillo, P.A., et al.: G-Prop: Global optimization of multilayer perceptrons using GAs. Neurocomputing 35, 149–163 (2000)
4. Eshelman, L.J.: The CHC adptive search algorithm: How to have safe search when engaging in nontraditional genetic recombination. In: First Workshop on Foundations of Genetic Algorithms, pp. 265–283. Morgan Kaufmann, San Francisco (1991)
5. Harpham, C., et al.: A review of genetic algorithms applied to training radial basis function networks. Neural Computing & Applications 13, 193–201 (2004)
6. Jelasity, M., Preub, M., Paechter, B.: A scalable and robust framework for distributed application. In: Proc. on Evolutionary Computation, pp. 1540–1545 (2002)
7. Kriegel, H., Borgwardt, K., Kroger, P., Pryakhin, A., Schubert, M., Zimek, A.: Future trends in data mining. Data Mining and Knowledge Discovery: An International Journal 15(1), 87–97 (2007)
8. Mayer, A.H.: Symbiotic Coevolution of Artificial Neural Networks and Training Data Sets. LNCS, pp. 511–520. Springer, Heidelberg (1998)

9. Merelo, J., Prieto, A.: G-LVQ, a combination of genetic algorithms and LVQ. In: Artificial Neural Nets and Genetic Algorithms, pp. 92–95. Springer, Heidelberg (1995)
10. Paredis, J.: Coevolutionary Computation. Artificial Life, 355–375 (1995)
11. Parras-Gutierrez, E., Rivas, V.M., Merelo, J.J., del Jesus, M.J.: Parameters estimation for Radial Basis Function Neural Network design by means of two Symbiotic algorithms. In: ADVCOMP 2008, pp. 164–169. IEEE computer society, Los Alamitos (2008)
12. Parras-Gutierrez, E., Rivas, V.M., Merelo, J.J., del Jesus, M.J.: A Symbiotic CHC Co-evolutionary algorithm for automatic RBF neural networks design. In: DCAI 2008, Advances in Softcomputing, Salamanca, pp. 663–671 (2008) ISSN: 1615-3871
13. Mitchell Potter, A., De Jong, K.A.: Evolving Neural Networkds with Collaborative Species. In: Proc. of the Computer Simulation Conference (1995)
14. Rivas, V.M., Merelo, J.J., Castillo, P.A., Arenas, M.G., Castellanos, J.G.: Evolving RBF neural networks for time-series forecasting with EvRBF. Information Sciences 165(3-4), 207–220 (2004)
15. Rivas, V.M., Garcia-Arenas, I., Merelo, J.J., Prieto, A.: EvRBF: Evolving RBF Neural Networks for Classification Problems. In: Proceedings of the International Conference on Applied Informatics and Communications, pp. 100–106 (2007)
16. Rivera Rivas, A.J., Rojas Ruiz, I., Ortega Lopera, J., del Jesus, M.J.: Coevolutionary Algorithm for RBF by Self-Organizing Population of Neurons. In: Mira, J., Álvarez, J.R. (eds.) IWANN 2003. LNCS, vol. 2686, pp. 470–477. Springer, Heidelberg (2003)
17. Ros, F., Pintore, M., Deman, A., Chrtien, J.R.: Automatic initialization of RBF neural networks. In: Chemometrics and intelligent laboratory systems, vol. 87, pp. 26–32. Elsevier, Amsterdam (2007)
18. Schwaiger, R., Mayer, H.A.: Genetic algorithms to create training data sets for artificial neural networks. In: Proc. of the 3NWGA, Helsinki, Finland (1997)
19. Thompson, J.N.: The Geographic Mosaic of Coevolution. University of Chicago Press, Chicago (2005)
20. Tomassini, M.: Parallel and distributed evolutionary algorithms: A review. In: Miettinen, K., et al. (eds.) Evolutionary Algorithms in Engineering and Computer Science, pp. 113–133. J. Wiley and Sons, Chichester (1999)

New Aspects of the Elastic Net Algorithm for Cluster Analysis

Marcos Lévano and Hans Nowak

Escuela de Ingeniería Informática, Universidad Católica de Temuco
Av. Manuel Montt 56, Casilla 15-D, Temuco, Chile
mlevano@inf.uct.cl, hans.nowak@gmail.com

Abstract. The elastic net algorithm, formulated by Durbin-Willshaw as an heuristic method and initially applied to solve the travelling salesman problem, can be used as a tool for data clustering in n-dimensional space. With the help of statistical mechanics it can be formulated as an deterministic annealing method in which a chain of nodes interacts at different temperatures with the data cloud. From a given temperature on the nodes are found to be the optimal centroid's of fuzzy clusters, if the number of nodes is much smaller then number of data points.

We show in this contribution that for this temperature the centroid's of hard clusters, defined by the nearest neighbor clusters of every node, are in the same position as the optimal centroid's of the fuzzy clusters. This result can be used as a stopping criterion for the annealing process. The stopping temperature and the number and size of the hard clusters depend on the number of nodes in the chain.

Test were made with homogeneous and inhomogeneous artificial clusters in two dimensions.

Keywords: Statistical mechanics, deterministic annealing, elastic net, clusters.

1 Introduction

Problems of classification of groups of similar objects exist in different areas of science. Classification of patterns of genes expressions in genetics and botanic taxonomy are two out of many examples. The objects are described as data points in a multivariate space and similar objects form clusters in this space. In order to find clusters without the knowledge of a prior distribution function, one uses an arbitrary partition of the data points in clusters and defines a cost functional which will then be minimized. From the minimization results a distribution function which gives the optimal clusters and there positions.

One of the most successful method which uses this procedure is the Elastic Net Algorithm (ENA) of Durbin-Willshaw [1] which was initially formulated as a heuristic method and applied to the travelling salesman problem. Later it was given an mechanical statistic foundation [2], [3] and used for diverse problems in pattern research [4], [5], [6]. In this formulation a chain of nodes interact with a given set of data points by a square distance cost-or energy function. For the distribution of the data points one uses the principle of maximum entropy which gives a probability to find the distance between data points and nodes as a function of a parameter which, in a physical analogy,

D. Palmer-Brown et al. (Eds.): EANN 2009, CCIS 43, pp. 281–290, 2009.

is the inverse temperature. Additionally one introduces a square distance interaction between the nodes. The total cost function is then minimized with respect to the node positions. The resulting non-linear set of equations can be iterated with the steepest descent algorithm.

If the number of data points is much larger then the number of nodes and for a not to large node-node interaction, one can treat the interaction energy between the nodes as a perturbation. For a typical temperature one finds then a partition of the whole cluster into a number of fuzzy clusters. In this work we use additionally to the fuzzy clusters hard clusters, defined by the nearest neighbors of every node, and derive a stopping criterion for the annealing process which gives the most optimal fuzzy-and associated hard clusters for a given number of nodes. We test the results by applying the annealing process and the criterion to a homogeneous and a nonhomogeneous two-dimensional cluster.

2   Some New Aspects on the Elastic Net Algorithm

For a given number of data points x_i in n-dimensional space we define a chain of nodes y_j in the same space and connect the nodes with the data points by a square distance energy function.

$$E_{ij} = \frac{1}{2}|x_i - y_j|^2 \tag{1}$$

Data and nodes are normalized and without dimensions. The total energy is then given by [7]

$$E = \sum_i \sum_j P_{ij} E_{ij} \tag{2}$$

with the unknown probability distribution P_{ij} to find the distance $|x_i - y_j|$. Putting the system in contact with a heat reservoir and using the maximum entropy principle, one finds with the constraint (2) a Gaussian probability distribution [7],[2]:

$$P_{ij} = \frac{e^{-\beta E_{ij}}}{Z_i} \tag{3}$$

with the partition function for the data point i

$$Z_i = \sum_j e^{-\beta E_{ij}}. \tag{4}$$

The parameter β is inversely proportional to the temperature and shows a probability which for high temperature associates a data point equally with all nodes and localizes it to one node for low temperature. Using independent probabilities for the interaction of every data point with the nodes, one finds for the total partition function:

$$Z = \prod_i Z_i \tag{5}$$

and the corresponding Helmhotz free energy [7], [8],

$$F = -\frac{1}{\beta}\ln Z + E_{nod} = -\frac{1}{\beta}\sum_i \ln \sum_j e^{-\beta E_{ij}} + E_{nod}, \tag{6}$$

where we have added an interaction energy between the nodes

$$E_{nod} = \frac{1}{2}\lambda \sum_j |y_j - y_{j-1}|^2 \qquad (7)$$

in form of a chain of springs, with the spring constant λ. Minimization of the free energy with respect to the node positions gives for every value of β a non-linear coupled equation for the optimal node positions:

$$\sum_i P_{ij}(x_i - y_j) + \lambda(y_{j+1} - 2y_j + y_{j-1}) = 0, \; \forall j. \qquad (8)$$

We solve the equation for different β values iteratively with the steepest descent algorithm [4]:

$$\triangle y_j = -\triangle \tau \frac{\partial F}{\partial y_j} = \triangle \tau \sum_i P_{ij}(x_i - y_j) + \triangle \tau \lambda(y_{j+1} - 2y_j + y_{j-1}) \; \forall j, \qquad (9)$$

with the time step $\triangle \tau$ chosen adequately.

In general we will have a large number of data points which interacts with a small number of nodes and for a not too large spring constant the node interaction energy will be small with respect to the interaction energy between nodes and data points and may be treated as a perturbation. In order to preserve the topology of the chain, we use a small node-node interaction. If we write equation (8) in the form

$$\frac{\sum_i P_{ij} x_i}{\sum_i P_{ij}} - y_j + \frac{\lambda}{\sum_i P_{ij}}(y_{j+1} - 2y_j + y_{j-1}) = 0, \; \forall j, \qquad (10)$$

we note that for $\frac{\lambda}{\sum_i P_{ij}} \ll 1$ the second term in equation (10) is negligible. This is the case if around the node y_j are sufficient data points which contribute with a large probability. If there are only few data points which contribute, then some next neighbor nodes must be present at a small distance and again the second term is negligible because of the difference of the nodes. That means that in a good approximation equation (10) reduces to

$$y_j = \frac{\sum_i P_{ij} x_i}{\sum_i P_{ij}} \; \forall j, \qquad (11)$$

which represent the optimal centroid's of fuzzy clusters around every node [4].

In the *deterministic annealing* method one solves the equation (8) firstly for a low value of β (high temperature). In this case one should find only one cluster, the whole cluster, equation (11) gives its centroid and the free energy is in its global minimum. Then one solves equation (8) successively for increasing β, tracing the global minimum.

Rose et.al.[2] have shown that for the case where no node-node interaction ($E_{nod} = 0$) is taken in account, the whole clusters splits into smaller clusters and one finds some of the nodes y_j as optimal centroid's of the clusters. They interpret the forming of different numbers of clusters as a phase transition. In every phase one has a well defined number of fuzzy clusters and one can estimate their mean size, using the Gaussian form of the probability distribution, given by equations (3),(4) and (1). One expects that only those data points for which on the average $\frac{1}{2}\beta\sigma_c^2 \approx 1$ contribute to the clusters, with σ_c the average value of the standard deviation. This gives an average value for the size of the fuzzy clusters:

$$\sigma_c \approx \sqrt{2/\beta} \tag{12}$$

The same situation is found when one takes in account the node-node interaction as a perturbation. If there are more nodes then clusters, the rest of the nodes will be found near the nodes which correspond to the clusters. For more clusters then nodes these will not be detected anymore. There exist a typical value of β, where the number of nodes are the same as the forming clusters. These clusters are optimal, e.g. there nodes are optimal centroid's of the corresponding clusters. One would expect that these clusters are *well defined* in the sense that there probability distribution has large values for the data points of the clusters and goes fast to zero for other data points.

In order to establish a criterion for finding the β value for this situation and stop the annealing process, we define a *hard cluster* for every node, determined by the nearest neighbor data points of the node. The centroid of the hard cluster, the *hard centroid*, is given by

$$\hat{y}_j = \frac{1}{n_j} \sum_{i \in C_j} x_i, \tag{13}$$

where C_j forms the cluster of nearest neighbors of node y_j with n_j members. The hard clusters can be compared with the fuzzy clusters of the optimal situation and one expects that fuzzy clusters and hard clusters have their best approach. For this stopping criterion we use the centroid's and the standard deviation:

$$y_j \approx \hat{y}_j, \quad \text{and} \quad \sigma_j \approx \hat{\sigma}_j \;\; \forall j, \tag{14}$$

where σ_j and $\hat{\sigma}_j$ are the standard deviations of the fuzzy clusters and the hard clusters.

During the annealing process one passes a series of phase transitions which are characterized by a certain number of clusters and a deformation of the chain of nodes. As the energy of the chain E_{nod} is generally much smaller then the interaction energy of the nodes with the data points, the change of the deformation should be of the order of the energy of the chain and E_{nod} shows clearly this phase changes.

3   Application to Artificial Created Two-Dimensional Clusters

As an illustration of the algorithm and the stopping criterion, we apply the ENA to two artificial created clusters. In the first example, cluster $C1$, we create a two-dimensional rectangular homogeneous cluster and in the second example, cluster $C2$, we create a two-dimensional rectangular nonhomogeneous cluster with different density regions. Fig. 1 shows the two clusters.

We begin with a chain of 20 nodes which crosses the cluster diagonally. The β-dependent node positions and energy E_{nod} of the chain are then calculated with the steepest descent algorithm (9) with a time step $\triangle\tau = 0.001$. Table 1 shows the relevant data for the two clusters.

Fig. 2 shows the node interaction energy E_{nod} for the cluster $C1$ as a function of β.

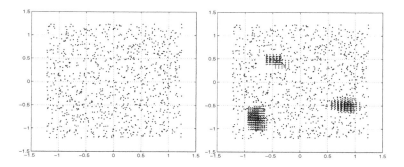

Fig. 1. Homogeneous clusters $C1$ and nonhomogeneous cluster $C2$

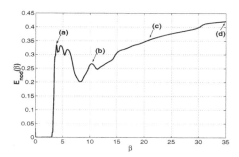

Fig. 2. Node Interaction energy E_{nod} as a function of β for cluster C1. The arrows indicate the phases a,b,c,d for data registrations.

Table 1. Number of data points and nodes, initial value β_i, final value β_f, iteration step $\triangle \beta$ and spring constant λ for the cluster calculations

cluster	data points	nodes	β_i	β_f	$\triangle\beta$	λ
C1	1000	20	0.001	35.0	0.001	0.1
C2	1673	20	0.001	70.0	0.001	0.1

The arrows indicate the phases for which we registered the node positions, the hard centroid's and the standard deviations. Fig. 3 shows data points, chain of nodes and hard centroid's for the phases a: $\beta = 3.8$, b: $\beta = 10.3$, c: $\beta = 25$ and d: $\beta = 35$. Fig. 4 shows the standard deviations of the fuzzy clusters, hard clusters and the mean size σ_c of estimation (12) for the different phases of $C1$.

For values of $\beta < 3$ in fig. 2, $E_{nod} \approx 0$, because the initial chain reduces to nearly a point. We see a first strong transition at point a, where the nodes extend to a first configuration of the chain. This corresponds to a connection between few nodes which form at this value of β. Fig. 3a shows nodes and hard centroid's for this phase and fig. 4a their standard deviations. The nodes accumulate in 4 points which should correspond to the optimal centroid's of 4 fuzzy clusters. The chain is large and has a high E_{nod}, because its connects the different nodes in the 4 clusters multiple times. We have not connected the hard centroid's, because they show large dispersions. The standard deviations for the fuzzy clusters are much larger then the standard deviations for the hard clusters. The estimation of mean value σ_c of formula (12) for the fuzzy clusters fits remarkably well the average fuzzy clusters size.

For increasing β the nodes separate and the chain contracts slightly. In phase b, fig.3b, the nodes accumulate in 9 clusters, lowering in this way the interaction energy between nodes and data points. The hard centroid's are now nearer to the nodes and fig. 4b shows that the difference between the standard deviations gets smaller. In phase c the separation between the 20 nodes are nearly equidistant and the chain covers the whole cluster. The hard centroid's are approximately in the same position as the nodes and one expects that 20 fuzzy clusters exist. They are not yet well defined. This is seen in fig. 4c, where the difference between the standard deviations is around 0.1. The stopping criterion (14) is nearly fulfilled.

There is another marked change in E_{nod} between phase c and phase d. In fig. 3d one notes that the nodes and hard centroid's coincide. The standard deviations in fig. 4d differ around 0.05. One node and its hard centroid have changed position such that the nodes cover better the area of the data points. The criterion (14) is much better fulfilled for phase d and one could stop the annealing process.

Finally we note that the estimation for the mean size σ_c of the clusters in phases a,b,c and d are good approximations of the mean standard deviation of the fuzzy clusters.

For the second example, cluster $C2$, we see in fig. 5 that the node interaction energy E_{nod} shows a much clearer distinction for the different phases. This is a consequence

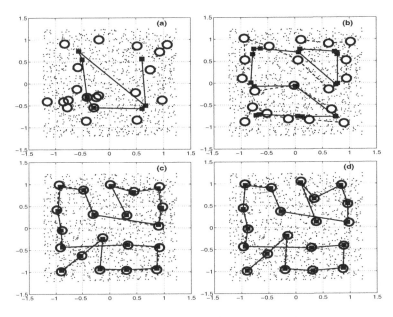

Fig. 3. Data points (·), nodes (■) and hard centroid's (○) for phases a: $\beta = 3.8$, b: $\beta = 10.3$, c: $\beta = 25$, d: $\beta = 35$ of cluster $C1$

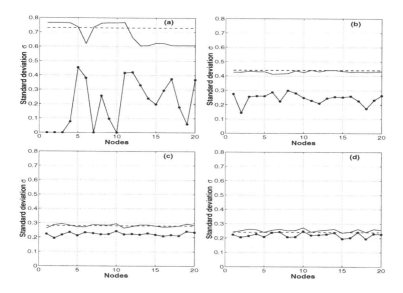

Fig. 4. Standard deviations for fuzzy clusters $(-)$, hard clusters $(-\bullet)$ and the estimation σ_c $(--)$ of the mean size for the phases a, b, c, d of cluster $C1$

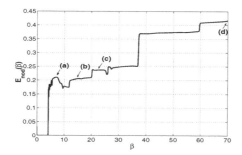

Fig. 5. Node Interaction energy E_{nod} as a function of β for cluster $C2$. The arrows indicate the phases a, b, c, d for data registrations.

of the three region of higher data point density in the inhomogeneous cluster which attracts more nodes. Fig. 6 shows nodes and hard centroid's for the 4 phases a: $\beta = 7.0$, b: $\beta = 15$, c: $\beta = 23$ and d: $\beta = 70$ and fig. 7 the standard deviations and the estimation (12) of mean fuzzy cluster size σ_c for these phases.

As in the homogeneous cluster $C1$, one notes in fig. 6a that in cluster $C2$, after the first transition, all nodes are nearly in the same position of 4 sub clusters, which should correspond to 4 fuzzy clusters formed in this phase, with the difference that three of the positions are now near the *high density regions*, HDR. Hard centroid's and nodes are different and in fig. 7a one notes a large difference between the standard deviations. The standard deviation of the fuzzy clusters is constant for 3 group of nodes, which

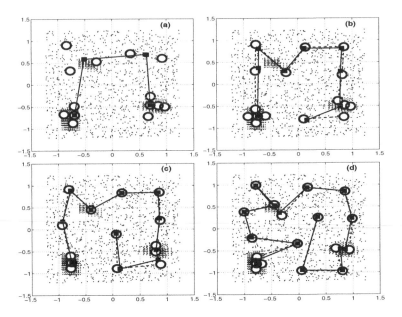

Fig. 6. Data points (·), nodes (■) and hard centroid's (○) for phases a: $\beta = 7$, b: $\beta = 15$, c: $\beta = 23$, d: $\beta = 70$ of cluster $C2$

correspond to the nodes in the HDR. The estimation for the mean cluster size gives a larger value then the standard deviation for the fuzzy clusters.

Fig. 6b shows the next phase which corresponds approximately to phase b of cluster $C1$. One notes, too, 9 positions for the 20 nodes which should correspond to 9 fuzzy clusters, but in this case some of them group around the HDR. The hard centroid's for the nodes which are not near the HDR are slightly different from their nodes, but the hard centroid's for the nodes near or inside the HDR have larger deviations from the corresponding nodes. This seems to be a consequence of the close proximity of the nodes in this region. Their fuzzy clusters are not well defined because their members are from low and high density regions. The problem may be solved by using a larger number of nodes for the chain. This difference is seen, too, in the standard deviations in figure 7b for the nodes near or in the HDR. The mean cluster size approaches the fuzzy cluster standard deviation.

Phase c shows an intermediate situation with 3 group of nodes in the HDR and a smaller difference between the nodes and hard centroid's and between the standard deviations (fig.6c, fig.7c). The mean cluster size is well approximated by the mean value of the fuzzy cluster standard deviation.

In phase d (fig. 6d) one finds a larger number of separate nodes and a node concentration in the HDR. Nodes and hard centroid's are now in the same positions. The standard deviations of fig. 7d are the same for the nodes outside the HDR and gets smaller for nodes in the HDR, in comparison with phases a, b and c. The mean cluster size is a little bit smaller then the fuzzy cluster standard deviation. For this case one can determine

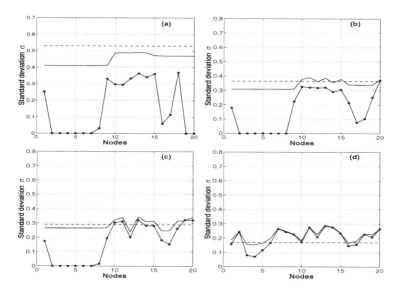

Fig. 7. Standard deviations for fuzzy clusters ($-$), hard clusters ($-\bullet$) and the estimation σ_c ($- -$) of the mean size for the phases a, b, c, d of cluster $C2$

the data points of the hard clusters and analyze its properties which should be in good accordance with the optimal fuzzy clusters for the given number of nodes.

4 Conclusions

We use a statistical mechanics formulation of the elastic net algorithm of Durbin-Willshaw [1], [2] with the interaction between the nodes as a perturbation in order to get in a n-dimensional space a variable number of fuzzy clusters which depend on a temperature parameter, or the inverse parameter β, of the annealing process. For a given number of nodes of the elastic chain we find a value for β, for which the number of optimal fuzzy clusters is the same as the number of nodes and any further annealing with more smaller optimal fuzzy clusters cannot detect these smaller clusters.

In order to determine the value of this β for which one can stop the annealing process, we define a *hard centroid* as the centroid of the cluster of nearest neighbors of every node, the *hard cluster*, and find for the stopping condition that the hard centroid associated to every node must be the same as the node position and the same goes for their standard deviations. This means that the fuzzy clusters for this number of nodes are best approximated by these hard clusters with its known members which can be analyzed for their properties. An estimation of the mean cluster size as a function of β is given and shows a good approximation to the fuzzy cluster standard deviation. This means that for a given value of β a mean cluster size is determined, but we do not know the minimal number of nodes which are necessary to give cluster of this size at the stopping condition.

We illustrate the found relations by 2 two-dimensional clusters, one homogeneous cluster, where for the stopping condition the nodes are mostly equally spaced and cover the whole area, and a nonhomogeneous cluster with regions of different point density, where for the stopping condition some nodes accumulate at the high density regions. This second example shows that a similar situation in multivariate space can give clear information about smaller regions of different density. Calculations of examples from genetics in multivariate space are in progress and will be published soon.

Acknowledgements

This work was supported by the project DGIPUCT No. 2008-2-01 of the "Dirección General de Investigación y Postgrado de la Universidad Católica de Temuco, Chile".

References

1. Durbin, R., Willshaw, D.: An Analogue Approach To The Traveling Salesman Problem Using An Elastic Net Method. Nature 326, 689–691 (1987)
2. Rose, K., Gurewitz, E., Fox, G.: Statistical Mechanics and Phase Transitions in Clustering. Physical Review Letters 65, 945–948 (1990)
3. Alan, Y.: Generalized Deformable Models, Statistical Physics, and Matching Problems. Journal Neural Computation 2, 1–24 (1990)
4. Duda, R., Hart, P., Stork, D.: Pattern Classification. Jonh Wiley and Sons Inc., Chichester (2001)
5. Gorbunov, S., Kisel, I.: Elastic net for standalone RICH ring finding. Proceedings - published in NIM A559, 139–142 (2006)
6. Salvini, R.L., Van de Carvalho, L.A.: neural net algorithm for cluster analysis. In: Neural Networks, Proceedings, pp. 191–195 (2000)
7. Reichl, L.E.: A Modern Course In Statistical Physics. Jonh Wiley and Sons Inc., Chichester (1998)
8. Ball, K., Erman, B., Dill, K.: The Elastic Net Algorithm and Protein Structure Prediction. J. Computational Chemistry 23, 77–83 (2002)

Neural Networks for Forecasting in
a Multi-skill Call Centre

Jorge Pacheco, David Millán-Ruiz, and José Luis Vélez

Telefónica Research & Development, Emilio Vargas, 6,
28043 Madrid, Spain
jorge.pacheco@altran.es, dmr@tid.es, jlvv@tid.es

Abstract. Call centre technology requires the assignment of a large volume of
incoming calls to agents with the required skills to process them. In order to
perform the right assignment of call types to agents in a production environment, an efficient prediction of call arrivals is needed. In this paper, we introduce a prediction approach to incoming phone calls forecasting in a multi-skill
call centre by modelling and learning the problem with an Improved Back-
propagation Neural Network which have been compared with other methods.
This model has been trained and analysed by using a real-time data flow in a
production system from our call centre, and the results obtained outperform
other forecasting methods. The reader can learn which forecasting method to
use in a real-world application and some guidelines to better adapt an improved
backpropagation neural network to his needs. A comparison among techniques
and some statistics are shown to corroborate our results.

Keywords: Neural networks, backpropagation, call centre, forecasting.

1 Introduction

Almost all large companies use call centres (CC) to assist with everything from customer service to the selling of products and services. Even though CCs have been
widely studied, there are some lacks on forecasting actions which may imply huge
losses of money and client dissatisfaction due to never-ending delays.

In a CC [1], the flow of calls is often divided into outbound and inbound traffic.
Our main concern in this paper is the prediction of inbound traffic. Inbound calls are
those that go from the client to the CC to contract a service, ask for information or
report a problem. This kind of calls is significantly different from outbound calls in
which the agents handle calls to potential clients mainly with commercial pretensions.
Inbound calls are modelled and classified into several call groups, CGs, in relation to
the nature. Once these CGs have been modelled, each call is assigned to a CG.

We assume that there are k types of calls, n customer calls and m agents that may
have up to i skills ($i \leq k$). This implies that an agent can handle different types of calls
and, given a type of call, it can be answered by several agents that have that skill.

As it can be expected, the mean arrival rate for each call type is not the same and
these calls have different modelling times. The load of call types, k, is the total amount
of time required for agents service. Note that the inbound flow in CCs is usually not a

D. Palmer-Brown et al. (Eds.): EANN 2009, CCIS 43, pp. 291–300, 2009.
© Springer-Verlag Berlin Heidelberg 2009

stationary Poisson process [2] and, the service times do not increase exponentially. Since calls arrive randomly according to a stochastic process, it would be desirable to have a very balanced distribution of the agents, who can be available or not, in order to handle the calls as soon as possible. Figure 1 illustrates the relationship among clients' calls, queues and agents.

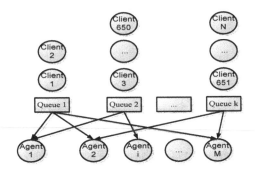

Fig. 1. Inbound scheme

This paper starts by explaining the classical problem of forecasting in CCs in Section 2. In Section 3, we give some guidelines to better adapt an improved backpropagation neural network (IBNN) to a multi-skill CC. Section 4 briefly describes different forecasting techniques which are compared and analysed. The comparative study is precisely the main contribution of this paper. Section 5 points out some overall ideas and prospects for future work on CC forecasting.

2   Forecasting in Call Centres

Forecasting refers to the estimation of values at certain specific future times. A precise prediction enables us to properly balance the workload among agents, giving better service levels and optimizing our resources. Traditionally, forecasting has been approached by a Poisson distribution. Assuming pure-chance arrivals and pure-chance terminations leads to the following probability distribution (1):

$$P(a) = \left(\frac{\mu^a}{a!}\right) e^{-\mu} \tag{1}$$

The number of call arrivals in a given time follows a Poisson distribution, where a is the number of call arrivals in an interval T, and μ is the mean of call arrivals in time T. For this reason, pure-chance traffic is also known as Poisson traffic. However, as mentioned previously, the prediction of call arrivals in a CC does not often follow a Poisson distribution with a deterministic rate [2]. In all studies, the arrival process agrees with a Poisson process only if the arrival rate of the Poisson process is itself a stochastic process. Characteristically, the variance of the incoming calls in a given interval is much larger than the mean. However, it should be equal to the mean for Poisson distributions. The mean arrival rate also depends strongly on the day time and often on the week day. Finally, there is positive stochastic dependence between arrival rates in successive periods within a day and arrival volumes of successive days. Taking into account all these premises, we can realise of the need of finding a more effective method to forecast which does not rely on the hypothesis of a simple Poisson arrival distribution. Section 3 explains the procedure to model an NN step-by-step to forecast unknown variables taking into account the nature of a real multi-skill CC environment rather than following a Poisson distribution when predicting.

3   Forecasting in CCs Using NNs

An artificial neural network (ANN) is a mathematical model based on the operation of biological NNs [5]. This model can be used as an effective method to forecast. Most statesmen are used to apply regression methods in which data are best–fitted to a specified relationship which is usually linear. However, these methods have several handicaps. For instance, relationships must be chosen in advance and these must be distinguished as linear or non–linear when defining the equation. NNs enable us to avoid all these problems. In regression, the objective is to forecast the value of a continuous variable which is the incoming flow rate in our case. The output required is a single numeric variable which has been normalized between 0 and 1. NNs can actually perform a number of regression tasks at once, although commonly each network performs only one.

3.1   Background

In this work, an Improved Backpropagation Neural Network (IBNN) [6] with a single hidden layer has been development in order to reduce complexity and computing times, making configurable the standard parameters. This type of ANN learns by reinforcement, in other words, it is necessary to provide a learning set with some input examples in which every output is well-known. The classical Backpropagation algorithm allows the network to adapt itself to the environment thanks to these example cases using a gradient-based learning. Our IBNN implements also the momentum algorithm [3]. The weights in the interval $t+1$ are determined by the previous weights, a percentage (η = learning rate) of change in the gradient in this interval and a percentage (μ = momentum) of change in the previous interval as follows (2):

$$w(t+1) = w(t) + \eta \frac{\partial E_{t+1}}{\partial w} + \mu \cdot \Delta w(t) \qquad (2)$$

In this implementation, all the parameters (number of neurons in the hidden layer, initial weights, learning rate and momentum) have been empirically determined to its optimum value and used in the rest of section of this study. The implementation also makes configurable the batch learning (weights are changed at the end of each epoch) or online learning (weights are changed at the end of each pattern of the dataset). Although batch learning has its advantages (see [3]), the online or stochastic learning has improved the learning speed, the results and the computing times in our CC forecasting [3].

The main problem with gradient-descent methods is the convergence to local optimums instead of global optimums. Some local optimums can provide acceptable solutions although these often offer poor performance. This problem can be overcome by using global optimization techniques but these techniques require high computing times. Other improved gradient-based algorithms with more global information such as Rprop, iRprop or ARProp [10] were not appropriate because the training set was too large to be effectively applied (see Figure 2). Even if these algorithms could be applied correctly, the purpose of this paper is to demonstrate, by means of a comparative study, that even a "classical" NN can outperform other forecasting techniques by including minor changes in the learning rate.

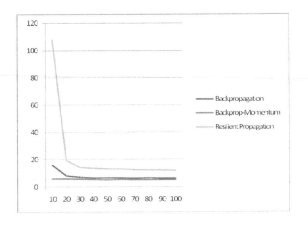

Fig. 2. NN Comparison (SSE x epochs)

3.2 Data Set

Once the basic aspects have been covered, how do we start the development of an NN? Firstly, it is essential to find a suitable dataset, and the best way to obtain this is to do a fair balance between the amount of data and a representative period of time measured in terms of days. The number of days chosen must be multiple of seven, because the week day has an important influence in the training and validation of the NN. Moreover, the number of days must be large enough to represent every possible situation but not too big because this makes the training slower.

Our implementation is completely configurable and this enables us to determine that the number of days to take into account should be, at least, 91 days in order to cover all possible patterns with the considerations previously explained.

The dataset has been split into subsets, which correspond to the CGs' data, to make a different model for each one. Each subset has been randomly divided into three sets, following the cross-validation structure [4]: Training (55%), generalization (20%) and validation (25%). In this strategy, the training dataset is used to make the NN learn, the generalization dataset is used to prevent overtraining [7] and the validation dataset, which has not been shown before to the NN, is used to obtain a measure of quality for the NN. The validation set covers the last two weeks as can be seen in Figure 3 and Figure 4.

3.3 Variables

Choosing the right inputs from all information we have is not trivial and is very important for obtaining a higher performance. Since the variables cannot be defined ad-hoc, the Mann–Whitney–Wilcoxon (MWW) test has been used to obtain a metric of the relevance of variables (see Table 1). MWW test is a non-parametric test for calculating whether two independent samples of observations have the same distribution. This test has been chosen because it is one of the best-known non-parametric significance ones.

Table 1. Chosen-variable relevance according to MWW

Variables	Relevance %
# Calls in Previous 0-5 Minutes	41,171 %
# Calls in Previous 5-10 Minutes	17,857 %
Night Shift Timetable	11,741 %
Week day	8,41 %
# Calls in Previous 10-15 Minutes	6,307 %
# Calls in Previous 15-20 Minutes	4,873 %
# Calls in Previous 20-25 Minutes	4,790 %
Minutes of the Day	3,433 %
Peak Time	1,415 %
Second Peak Time	1,398 %

Among all variables, the volume of incoming calls in previous intervals, night shift timetable (NST), week of the month, time, intervals of hours (2, 4 or 8 hours) and intervals of peak hours must be highlighted. For almost all CGs, the optimum number of previous intervals required is usually around 5-6 intervals. The NST offers an upgrading of the results for every CG. When splitting days up into intervals of hours, predictions are also improved. This division into intervals might guide us to a wrong decision because these variables are correlated but the causation comes from the night shift timetable and peak time variables (see Table 1). This happens because the correlation does not imply causality. The improvement is just obtained because these variables are correlated but only peak time intervals and night shift are useful when forecasting. Intervals of peak hours are interesting because these divisions clearly outperform the results for almost all CGs.

3.4 Metrics

To measure the quality of each result, several metrics have been selected. In order to make the process more understandable, the error is defined between the real result obtained in an interval and the result predicted by our NN. Some metrics are defined by the percentage of predictions whether error is inside a given interval that can be determined by an absolute value or by a percentage of incomings calls. These metrics are called right rate (RR), specifically absolute right rate (ARR) when X is defined as the absolute value used to define the interval (X-ARR) and percentage right rate (PRR) when the percentage $Y\%$ is used (Y-PRR). To valuate the model for each pattern, the mean absolute error (MAE), mean squared error (MSE), sum squared error

(SSE), mean absolute percentage error (MAPE), ARR and PRR have been considered. In the same way, the variance error, the maximum error and the minimum error have been measured with the aim of having a metric of the dispersion of the error. Although MAPE, as well as PRR, is measured, it is not a high-quality metric for those groups with a reduced volume of calls due to the real value is very small.

3.5 Adaptations

The number of groups to forecast (320 CGs) and its own behaviour make necessary to determine the initial parameters of the models, taking in consideration the behaviour of each one. To mitigate this problem, the CGs have been divided into sets according to the mean number of calls per day. This criterion has been taken because the behaviour is similar in others in which the volume of incoming calls is alike. A different learning rate has been assigned to each set because this improvement makes the system more adaptive to the behaviour of each CG.

CGs with less than 40 calls per day and non-stationary behaviour are very difficult to predict. When this happens, the error is minimized in order to avoid moving agents to these CGs without calls. Of course, this action improves the service rates but, obviously, a client cannot wait indefinitely. To solve this handicap, the NN returns the number of agents required in the last interval if the NN has predicted zero calls in the instant t_{-5}. With this modification, a client is always waiting less than the time interval (5 minutes) and the results remain committed to high-quality.

3.6 Adaptive Learning Rate

Due to the problem mentioned in the first paragraph of Section 3.5, a simple but efficient algorithm has been implemented in order to make the learning rate truly adaptable to the environmental circumstances and preventing from overtraining.

Before explaining the algorithm, it is necessary to justify the error that should be accepted in order to understand the key idea. The accepted error (ε) can be defined as a normalized MAE for the generalization data set between two consecutive epochs. The algorithm is based on four normalized parameters: initial accepted error ($\Lambda\varepsilon_0$), percentage of learning rate changed ($\Delta\eta$), percentage of accepted error ($\Delta\varepsilon$), and percentage of momentum changed ($\Delta\mu$) and also it uses the accepted error ($\Lambda\varepsilon$). For each epoch, the system is trained with the data set and, afterwards, the ε for the generalization data set is calculated and the algorithm updates the weights as follows (3):

$$w(t+1) = w(t) + \eta_t \frac{\partial E_{t+1}}{\partial w} + \mu_t \cdot \Delta w(t)$$

$$if \quad (\Lambda\varepsilon_t > \varepsilon_{t+1}) or (\varepsilon_{t+1} > 2 * \Lambda\varepsilon_t)$$

$$\eta_{t+1} = \eta_t * \Delta\eta /(1 + (\max(\varepsilon_{t+1} - 2 * \Lambda\varepsilon_t, 0)/(\varepsilon_{t+1} - 2 * \Lambda\varepsilon_t)))$$

$$\mu_{t+1} = \mu_t * \Delta\mu /(1 + (\max(\varepsilon_{t+1} - 2 * \Lambda\varepsilon_t, 0)/(\varepsilon_{t+1} - 2 * \Lambda\varepsilon_t))),_t \tag{3}$$

$$\Lambda\varepsilon_{t+1} = \Lambda\varepsilon_t * \Delta\varepsilon /(1 + (\max(\varepsilon_{t+1} - 2 * \Lambda\varepsilon_t, 0)/\varepsilon_{t+1} - 2 * \Lambda\varepsilon_t)),$$

$$else \quad \eta_{t+1} = \eta_t; \quad \mu_{t+1} = \mu_t; \quad \Lambda\varepsilon_{t+1} = \Lambda\varepsilon_t$$

The initial accepted error controls the initial interval so that this decides when to increase or decrease other parameters. This provokes a significant effect on the learning speed. If the initial accepted error had a big value, the risk of overtraining would be very low but, nevertheless, the learning speed would be too slow, making the results worse after several epochs. If computing time was infinite then the results would be better with a large value rather than a low value.

The percentage of learning rate that has been changed guides the increment or decrement of the learning rate. If it has a low value, the changes of the learning rate will be more gradual. It also prevents the NN from overtraining and makes the learning speed more acceptable when other parameters are well-configured. In other words, a big value makes abrupt and unpredictable changes on the behaviour of the NN. It is important to correctly initialise the learning rates for each set and the initial error accepted in order to be able to configure this parameter.

The percentage error of accepted change controls the variance of the accepted error. The accepted error should increase when the learning rate increases and it should decrease when learning rate decreases. The reason of this behaviour is that it changes the results that are elevated when the learning rate is bigger, and decreases the learning rate when it is lower. A small value makes the accepted interval constant. The system learns faster at the beginning but reaches overtraining at the end. To prevent the NN from overtraining, the value should be positive and bigger than the percentage of learning rate changed. The percentage of momentum changed controls the variance in the momentum. When the momentum is bigger, the learning speed increases although it might guide the NN to worse predictions because the change is propagated faster. This parameter is connected with all parameters of the algorithm but not linearly, making the configuration pretty difficult.

As mentioned at the beginning of Section 3, the most important problem with gradient descent methods is the premature convergence to local optimums which might be far from the global optimum. This problem can be solved by using global optimization techniques. However, these techniques require high computing times. As other improved gradient-based algorithms with more global information such as Rprop, iRprop or ARProp [10] were not appropriate because the training set was too large enough to be effectively applied (see Figure 2); an adaptive learning rate algorithm has been proposed. Our modification allows us to quickly analyse our CGs and outperforms other forecasting techniques.

4   Results

In this section, our results are analyzed and, then, compared with Weka [8] and R forecasting package [9]'s algorithms to quantify the efficiency of our IBNN and the convenience of its adaptive learning rate mechanism. Weka and R forecasting package have been chosen because their codes are well-implemented open-sources. Other data mining tools such as SPSS or SAS have potent algorithms too, but we would not have any insight about the algorithm behind them.

4.1   Analysis

The results of our network present stable behaviour for all CGs as Figure 3 demonstrates. CGs with less volume of calls, i.e. CG5, present better performance when the threshold is equal to five calls, reaching a level of accuracy up to 100%. Logically, those groups with huge volumes of incoming calls have greater absolute errors; while the error, in terms of percentage, is alike (see Figure 3). For this reason, other authors use a percentage dependant on the size to measure the quality of the prediction. As Figure 3 illustrates, the system learns pretty fast at the beginning due to its adaptive learning rate mechanism. Passed some epochs, it learns more slowly because it is harder to minimize the error since there are many surrounding local optimums. The risk of overfitting because of the fast learning is covered with the generalization data-set, like other traditional methodologies do. This strategy is the correct one to optimize the computing times, which is a critical factor in this problem, allowing us to reach solutions close to the global optimum in a few epochs.

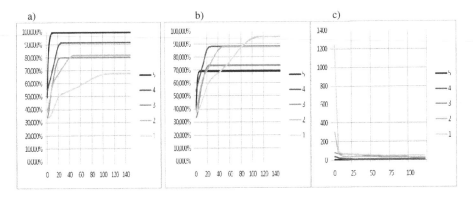

Fig. 3. (a) 5-ARR x Income calls; (b) 10-PRR x Income calls (c) MSE x epochs

4.2 Comparative

Weka and R [8] and R forecasting package [9]'s algorithms have been selected and adapted to be analysed and compared to our implementation. The same parameters, number of epochs, and values for these parameters have been considered in order to carry out a fair analysis. In Figure 4, we compare regression models (RM), exponential time series (ETS), ARIMA models (AM) and NNs with our implementation, highlighting the performance of our model. RMs present excellent results for some groups but these cannot tackle well-enough others, i.e. CG1 and CG4. In the same way, AM and ETS present accurate results for CGs with stable behaviour and/or low incoming flow but, compared to ANN, the results are poor for CGs with complex behaviour, i.e. CG2 and CG3. The ANN models usually present better results than ETS for CGs with a large volume of calls, i.e. CG1 and CG2. Our adaptive learning rate algorithm makes the IBNN more adaptable than a simple BNN for those CGs with low incoming flow, i.e. CG4 and CG5, making the prediction error closer to the ETS error.

Another consideration about the adaptive learning rate mechanism has to be done. The BNN parameters used by WEKA algorithms have been fixed to obtain an optimal model for each CG, but in practice, the model has to adapt its learning rate to obtain an optimal value because CGs can change dramatically in few months and new ones may appear. In real-time, the optimal parameter for each CG cannot be chosen every time the model is recalculated which makes necessary an algorithm that does it.

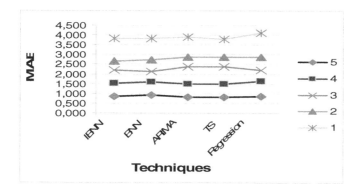

Fig. 4. Techniques comparative study (MAE given by different approaches)

Table 2. IBNN comparative

Call group	WEKA BNN MAE	Our IBNN MAE	Upgrading	Upgrading %
Group 1	3,82	3,82	0,003	0,078%
Group 2	2,72	2,65	0,065	2,385%
Group 3	2,11	2,21	-0,100	-4,735%
Group 4	1,59	1,54	0,047	2,975%
Group 5	0,92	0,85	0,071	7,779%

The implementation of the IBNN explained, when it is well configured, outperforms the results obtained by other techniques for most of the CGs (Table 2).

It is important to remark that the learning rate used in BNN is optimal for a given "picture of the situation" while IBNN auto-adapts its learning rate all the time. BNN has been configured to only obtain the best result possible at a given moment and IBNN has been fixed to a real-time environment whether CGs evolve. As it can be expected, it is impossible to determine, continuously, the optimal learning rate value because the situation changes very fast. IBNN sometimes gives worse results for some CGs than BNN because the optimal value is not obtained, but it makes the system feasible in real-time applications, obtaining learning rates values near the optimum without human participation even in hard-behaving CGs.

5 Conclusions

We have seen the problematic of forecasting in a CC and how NNs can help us. This paper also describes an efficient IBNN model and some upgrading adaptations. Our implementation is an efficient approach for real-time production environments such as a multi-skill CC. This solution is not universal and might offer worse results than others for environments in which time is not critical or conditions are more stable. Sometimes, a more sophisticated learning rate algorithm provokes a loss of computing time that cannot be allowed in a production environment. In our production environment, there are 320 CGs that are recalculated in less than two days for a 91-day data set running under a single processor, making the system truly adaptable to changes. Finally, we have analyzed our results and compared them to Weka and R's Algorithms which are outperformed. The future work should focus on the improvement of the adaptive algorithm to better control the learning process. In addition, other variables, which have not been considered, should be collected and meticulously analysed.

Acknowledgment

The authors would like to thank Severino F. Galán for his very useful comments.

References

1. Bhulaii, S., Koole, G., Pot, A.: Simple Methods for Shift Scheduling in Multiskill Call Centers. M&SOM 10(3), 411–420 (Summer 2008)
2. Ahrens, J.H., Ulrich, D.: Computer Methods for Sampling from Gamma, Beta, Poisson and Binomial Distributions. Computing 12(3), 223–246
3. Lecun, Y., Bottou, L., Orr, G., Muller, K.: Efficient BackProp. In: Orr-Muller (ed.) Neural Networks: Tricks of the trade, p. 44 (1998)
4. Black, J., Benke, G., Smith, K., Fritschi, L.: Artificial Neural Networks and Job-specific Modules to Assess Occupational Exposure. British Occupational Hygiene Society 48(7), 595–600 (2004)
5. Mandic, D., Chambers, J.: Recurrent Neural Networks for Prediction: Architectures. In: Learning algorithms and Stability. Wiley, Chichester (2001)
6. Bishop, C.: Neural Networks for Pattern Recognition, pp. 116–149. Oxford University Press, Oxford (1995)
7. Müller, K.R., Mika, S., Rätsch, G., Tsuda, K., Schölkopf, B.: An Introduction to Kernel-Based Learning Algorithms. IEEE Transactions on Neural Networks 12(2) (March 2001)
8. Weka project website: http://www.cs.waikato.ac.nz/ml/weka/
9. R Forecasting package website:
   http://cran.r-project.org/web/packages/forecasting/index.html
10. Igel, C., Husken, M.: Empirical Evaluation of the Improved Rprop Learning Algorithm. Neurocomputing 50(C), 105–123 (2003)

Relational Reinforcement Learning Applied to Appearance-Based Object Recognition

Klaus Häming and Gabriele Peters

University of Applied Sciences and Arts,
Computer Science, Visual Computing,
Emil-Figge-Str. 42, D-44221 Dortmund, Germany
gabriele.peters@fh-dortmund.de
http://www.inf.fh-dortmund.de/personen/professoren/peters/

Abstract. In this paper we propose an adaptive, self-learning system, which utilizes relational reinforcement learning (RRL), and apply it to a computer vision problem. A common problem in computer vision consists in the discrimination between similar objects which differ in salient features visible from distinct views only. Usually existing object recognition systems have to scan an object from a large number of views for a reliable discrimination. Optimization is achieved at most with heuristics to reduce the amount of computing time or to save storage space. We apply RRL in an appearance-based approach to the problem of discriminating similar objects, which are presented from arbitray views. We are able to rapidly learn scan paths for the objects and to reliably distinguish them from only a few recorded views. The appearance-based approach and the possibility to define states and actions of the RRL system with logical descriptions allow for a large reduction of the dimensionality of the state space and thus save storage and computing time.

Keywords: Relational reinforcement learning, computer vision, appearance-based object recognition, object discrimination.

1 Introduction

Relational reinforcement learning (RRL) is an attempt to extend the applicability of reinforcement learning (RL) by combining it with the relational description of states and actions. Disadvantages of RL are firstly its inability of handling large state spaces unless a regression technique is applied to approximate the Q-function. Secondly, the learned Q-function is neither easily interpretable nor easily extendable by a human, making it difficult to introduce a priori knowledge to support the learning process. And last, a generalization from problems the system was trained on to different but similar problems can hardly be done using classical RL. In [1], the relational representation of the Q-function was proposed and studied to overcome these issues. In this paper, we examine the applicability of the relational approach to a computer vision problem. The problem we consider is common in many computer vision applications (e.g., in manufacturing) and consists in the discrimination between similar objects which differ

D. Palmer-Brown et al. (Eds.): EANN 2009, CCIS 43, pp. 301–312, 2009.

slightly, e.g., in salient features visible from distinct views only. In our system a camera is rotated around an object (i.e., it is moved on an imaginary sphere around the object) until the system is able to reliably distinguish two similar objects. We want an agent to learn autonomously how to scan an object in the most efficient way, i.e., to find the shortest scan path which is sufficient to decide which object is presented. We want our approach to be general enough to extend from our simulated problem to a real world problem. Therefore we avoid using information that is uncertain or completely lacking in a real world application. This includes the camera parameters. Thus, in the state representation neither the camera's position nor its viewing direction are encoded, rather the design is purely appearance-based. The agent receives no information on the current 3D position of the camera. The only utilizable information is the appearance of the current view of the object in form of features visible in the view at hand. This will lead to learned rules such as: "seeing those features, an advantageous next action is to move the camera in that direction". This direction is encoded relatively to the current view of the agent.

2   Reinforcement Learning

RL [2] is a computational technique that allows an autonomous agent to learn a behavior via trial and error. A RL-problem is modeled by the following components: a set of states S, a set of actions A, a transition function $\delta : S \times A \to S$, and a reward function $r : S \times A \to \mathbb{R}$. The reward function is unknown to the agent, whose goal is to maximze the cumulated reward. In Q-learning, the agent attaches a Q(uality)-value to encountered state-action-pairs. After each transition, this Q-value of the state-action-pair is updated according to the update equation:

$$Q^{t+1}(s,a) = r(s,a) + \gamma \max_{a'} Q^t(s',a').$$

In this equation, $s' = \delta(s,a)$, while γ is a discount factor ensuring that the Q-values always stay finite even if the number of states is unbounded. As explained later, in our application it will also be helpful in search of a short scan path. During exploration, in each state the next action has to be chosen. This is done by a policy π. In Q-learning, its decision is based on the Q-values. We use the ϵ-greedy policy, which chooses with equal probability one of those actions that share the highest Q-value, except for a fraction of choices controlled by the parameter ϵ, in which the next action is drawn randomly from all actions.

3   Relational Reinforcement Learning

RLL uses propositional logic to encode states. Since this leads to a sparse representation of the Q-function, the number of state attributes for a given problem is smaller in the RRL approach than in the RL approach. Another advantage of RRL consists in the fact that slight changes in the state or action spaces do not necessarily lead to the mandatory relearning-from-scratch give in RL. In our application of learning short, discriminative scan paths we build a sparse representation of the Q-function by using a selection of encoutered sample views of the objects and apply a regression technique. There are different approaches to this, such as decision trees [1], kernel-based approaches [3], or k-nearest-neighbour approaches [4]. In this paper we adopt the k-nearest-neighbour approach of [4], which is now briefly reviewed. The basic idea is to approximate the Q-value q_i of state-action-pair i by comparing the logical description of that pair to the logical descriptions of the examples of state-action-pairs found so far. The n closest of these examples are used to derive the desired approximation $\hat{q}_i$. The formula is given by

$$\hat{q}_i = (\sum_{j=1}^{n} \frac{1}{dist_{i,j}} q_j)/(\sum_{j=1}^{n} \frac{1}{dist_{i,j}})$$

which is simply the weighted arithmetical mean of the nearest Q-values. The weight is given by a reciprocal distance measure $dist_{i,j}$, which depends on the logical modeling of the state-action-pairs and is defined in Sec. 4. We use a value of $n = 30$. The strategy to decide whether an example gets included into the Q-function is basically to test if this example contributes a significant amount of new information, the Q-function needs denser sampling, or none of these. The measure of information novelty is based on the local standard deviation σ_l of the Q-values in the vicinity of the current state-action-pair. If the difference of the Q-value q_i of the current example to its predicted value $\hat{q}_i$ exceeds σ_l, it is added to the database, which means that a function quite_new takes the value true:

$$\text{quite\_new}(q_i, c_1) = \begin{cases} \text{true}, & |\hat{q}_i - q_i| > c_1 \sigma_l \\ \text{false}, & else \end{cases}$$

for a constant c_1. To decide whether a denser sampling is needed, we relate σ_l to the global standard deviation σ_g of the Q-values of all examples constituting the Q-function, which means that we consider a function quite_sparse:

$$\text{quite\_sparse}(c_2) = \begin{cases} \text{true}, & \sigma_l > c_2 \sigma_g \\ \text{false}, & else \end{cases}$$

for a constant c_2. Both criteria are taken from [4].

4   Appearance-Based Modeling

In this section we first describe how we represent the states and actions in our application. Afterwards we explain the measure of the distance $dist_{i,j}$ between two state-action-pairs i and j.

Definition of States. In a RL-environment the design of the state representation can be difficult, if the states have to represent continuous information. As stated above, it is important to keep the state space small, because the number of possible paths from one state to the final state increases exponentially. The usual solution in an application with moving cameras is to limit the camera to certain positions on a sphere [5]. In contrast, our approach will allow arbitrary camera positions, because the positions will be encoded only implicitly by the perceived visual appearance and not by camera parameters. This appearance is captured in a set of features (e.g. interest points, but not necessarily). Each feature f_i is attached to the view in which it is detected. For each view we describe the *visibility* of a feature by the following expression: visibility(feature f_i) A *state* can now be defined by a list of visible features. The notation f_i stands for a feature and the index identifies this feature globally, i.e., across views. Since it is only necessary to identify whether or not two image-features represent the same global feature when computing the similarity between states, a pair-wise feature matching can be used to implement this global assignment. An encoding of a state can exemplarily look like this: visibility(f_1) $\wedge$ visibility(f_7) $\wedge$ visibility(f_{14}) This means that in this state the globally numbered features f_1, f_7, and f_{14} are visible. Thus, the current state of the RL system is defined by the features which are visible in the current view.

Definition of Actions. Actions of the agent are possible camera movements around an object. They are defined by their direction. As we want to proceed purely appearance-based we cannot utilize 3D information, thus these directions can be defined only in the 2D image plane of the current view. This is illustrated in the left diagram of Fig. 1. The expression which describes an action has the form to(α) where the parameter is simply an angle $\alpha \in [0; 2\pi]$ taken around the image center of the current view. Since the camera moves around the sphere of the object in 3D space, we have to derive from α a direction in 3D space. For this purpose we project the in-plane direction onto the object's sphere and choose a fixed step length as shown in the left diagram of Fig. 1. The choice of the next direction of movement is motivated by the idea that advantageous directions are those which promise a change in the density of the features. The left and right images of Fig. 2 show directions the agent is allowed to choose from for two example states. These directions are calculated by forming a linearly interpolated histogram of the feature point density. Each of 36 bins represents a range of 10 degrees, in which the number of features is counted. If bin_i is the number of features in bin i, the value of each bin is then replaced by $bin_i := 2 \cdot bin_i - bin_{i-1} - bin_{i+1}, i = 0, \ldots, 35$. Finally, the maxima of the resulting histogram define the valid directions for the next camera movement. The left part of Fig. 2 explains the derivation of actual camera movements in 3-space from their directions in 2-space of image planes. The centers C_1 and C_2 of two cameras are shown. After applying the movements determined by the actions to(α_1) and to(α_2), respectively, the cameras take their new positions C_1' and C_2'. These new positions are unambiguous, because the distance of the image plane to the center of the object remains constant. The distance d between the new camera positions determines the similarity of the actions to(α_1) and to(α_2). The right part of Fig. 2 illustrates the behavior of the state similarity for two examples. On the abscissa the angles in the range of $[-\pi; \pi]$ of a camera are logged, which is rotated around an object with a fixed axis through the center of the object and a step size of 0.1 radians. The ordinate represents the values

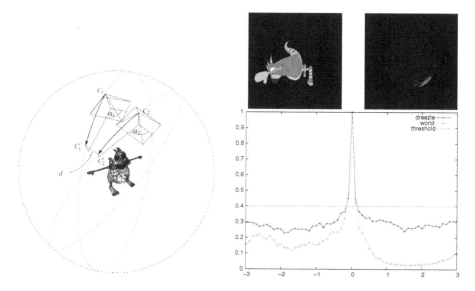

Fig. 1. Left: Derivation of actual camera movements in 3-space from their directions in 2-space of image planes. Right: Behavior of the state similarity for two examples.

of the state similarity function $s_{i,j,\texttt{visibility}}$. For each of the recorded views the state similarity to a reference view is calculated. The red curve represents the similarity values for the object "dreezle", the reference view of which is depicted in the upper left, the green curve does the same for the reference view of the object "world", shown in the upper right. Both reference views have been recorded at the position of zero degrees. The similarity function provides reliable values, even for the object "world", where the reference view contains very few features only. In addition, the inflection point p we use for the threshold function is drawn as dotted line at the similarity value of 0.4.

Distance Measure for State-Action-Pairs. We now examine the distance function $dist_{i,j}$ mentioned in Sec. 3. It is based on a similarity measure $sim_{i,j} : (i,j) \rightarrow [0; 1]$:

$$dist_{i,j} = 1 - sim_{i,j}.$$

This similarity measure is defined as a product of the state similarity and the action similarity. $sim_{i,j} = sim_{i,j}^S \cdot sim_{i,j}^A$. Both, sim^S and sim^A, are functions with range $[0; 1]$. They are defined as $sim_{i,j}^S = t(s_{i,j,\texttt{visibility}})$ and $sim_{i,j}^A = t(s_{i,j,\texttt{to}})$, where t is a sigmoidal function that maps its argument to $[0; 1]$. It imposes a soft threshold to punish poor values quickly while at the same time allowing small deviations from the optimum: $t(x) = \frac{1}{1+e^{-\mu(x-p)}}$ where μ is the steepness of the slope (set to 20 in our tests) and p is the inflection point (set to 0.4 for $sim_{i,j}^S$ and to 0.8 for $sim_{i,j}^A$). The unthresholded state similarity $s_{i,j,\texttt{visibility}}$ is based on the amount of overlapping between two views. A larger overlap leads to a larger similarity. The overlapping is measured by counting the features that reference the same global feature. This is measured

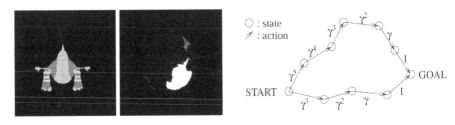

Fig. 2. Left: Appearance-based choice of actions for two objects. Red lines indicate possible directions for the next camera movement. Features are indicated by red dots on the object (close to the direction lines). The directions are computed as local maxima of the variation of the feature density. Right: Short scan paths are enforced by using a discount factor $\gamma \in]0;1[$ and giving zero rewards in all states except for the goal state. A shorter path leads to a less discounted reward. The exponents of γ are derived from the iterated application of the update equation.

by actually carrying out a feature matching between both views. The relation of the amount of same features to the amount of all features is used as the similarity measure. Let visible(x) be the set of all features visible in state x. Then

$$s_{i,j,\text{visibility}} = \frac{\text{visible}(i) \cap \text{visible}(j)}{\text{visible}(i) \cup \text{visible}(j)}. \tag{1}$$

The right diagram of Fig. 1 illustrates the property of this similarity measure by means of two examples. The term $s_{i,j,\text{to}}$ expresses the unthresholded similarity between two actions. It depends on the to-expressions introduced in Sec. 4. Two of these expressions point to the same state if the cameras capture the same features after moving in those directions. We use the distance of the camera centers after moving them along the directions the actions point to. The approach is also depicted in the left diagram of Fig. 1.

5   Application

The previously described framework has been used in an application that learns scan paths which are short but nevertheless allow the discrimination of two very similar objects. These objects look essentially the same except for a minor difference, for example in texture as seen in the left column of Fig. 3. This figure illustrates the following. Left: example objects and their differences. Upper row: The two versions of object "dreezle" can be distinguished by the starlike figure on its belly. Lower row: The two versions of object "world" can be distinguished by the yellow scribbling. Right: Phase 1 of the identification of the goal state (determining the most similar view for each object). Each row shows one example of the image retrieval process. The left column shows the current view of the agent. For *each* object in the database we determine the most similar view to the one in the left column. The right column shows the determined most similar view of only one special object, namely of that object which together with the object

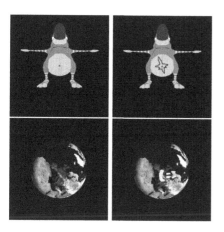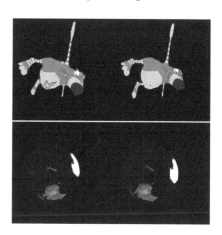

Fig. 3. Left: Example objects and their differences. Right: Phase 1 of the identification of the goal state (determining the most similar view for each object).

of the left column makes up the pair of similar objects. To set up the learning task for each pair of objects, both objects are scanned and their features are stored in a database. This database is used to generate the reward for the agent. One of these objects is then presented to the agent. The agent moves around the object, records images, and stops when she receives a reward of 1. Then the next episode starts. As a result, the agent learns to scan an object in such a way that she is able to distinguish it from a very similar object stored in the database.

Feature Detection and Calculation of Descriptors. To recognize and match feature points between images of an object, which have been taken from different camera positions, they are required to be reasonably stable. To comply with this requirement, we use a scale space Harris detector, combining the real time scale space of Lindeberg [6] with the Harris detector [7], while modifying the latter to make it work in a scale space pyramid [8]. To attach descriptors to these feature points we use Lowe's SIFT descriptors [9]. While he only uses those feature coordinates that pose maxima in scale space, we take all feature points into account as long as they are a local maximum in their own scale. This overcomes some stability deficiencies with the original approach that we experienced and that have been reported in [10] as well. As a result we get up to 500 features in each image. To reduce this amount, each camera takes a second image from a slightly different view point. Then, we apply a feature matching using a kd-tree [11]. The resulting correspondences are filtered by applying the epipolar constraint [12]. Only those feature points that survive this procedure are stored in our database. We aim at about 150 features for each view.

Rewards. We aim at learning the shortest scan path around the sphere of an object to a view that allows for the reliable discrimination from another similar object. As the agent aims to maximize the sum of all rewards, positive rewards in each step will keep the agent away from the goal state as long as possible. For this reason we do not reward a simple movement at all. The only action that receives a reward is a movement to the goal state. A shortest path can be made attractive to the agent by (ab)using the discount factor γ. If we set $\gamma := 1$, all Q-values will approach 1, since r is always 0. But with $\gamma \in]0;1[$ we will find a greater Q-value attached to states closer to the goal than to those farther away. This is illustrated in the right of Fig. 2 and leads to the desired preference of short paths towards the goal state.

Identification of the Goal State. As noted in the last subsection the reward is zero for all actions except those reaching the goal state. In fact, in our setup the goal state is *defined* as the state where the reward is one. In each step an image I is captured. After computing features and their descriptors of this image, they are used to identify the goal state. This identification proceeds in two phases. Phase 1 iterates through *all* objects inside the database (which consists of several pairs of pairwise similar objects) and identifies for each object the most similar view. This is basically an image retrieval task. Given one object O, our approach begins with building a kd-tree K_O with all descriptors belonging to O. For each descriptor $D_{I,i}$ of I, the most similar descriptor $D_{O,j}$ in K_O is identified and its Euclidean distance $d(D_{I,i}, D_{O,j})$ is computed. This distance d is used to vote for all views of object O, $D_{O,j}$ belongs to: $\texttt{score}(d) = \frac{1}{d}$ (taking the prevention of zero division into account). These scores are accumulated over all descriptors of I for each view separately. The one which receives the largest sum is taken as the most similar view of object O to I. This is done for all objects O in the database. The right column of Fig. 3 shows resulting pairs of images. Phase 2 aims at the decision whether the most similar images, e.g., I_{O1} and I_{O2}, of two candidate objects $O1$ and $O2$ show a significant difference. If so, we have reached our goal of finding the most discriminative views of two similar objects. Then we can mark the current state as a goal state by giving a reward of one. Finally we can find out which of the similar objects we have currently at hand. To do this, we reuse the similarity measure $s_{i,j,\texttt{visibility}}$ of Sec. 4. We compute the similarity of image I (corresponding to state i in (1)) to both candidates and take a normalized difference:

$$g = \frac{|s_{I,I_{O1},\texttt{visibility}} - s_{I,I_{O2},\texttt{visibility}}|}{\max(s_{I,I_{O1},\texttt{visibility}}, s_{I,I_{O2},\texttt{visibility}})} \qquad (2)$$

If this value g exceeds a certain threshold, the most discriminative view between the two similar objects is considered identified and the current episode ends. (Once this discriminative view is identified it is simple to determine which of both objects has been scanned.) For our learning scheme a threshold of 0.15 suffices. Fig. 4 shows results of this approach. Phase 2 of the identification of the goal state (finding the discriminative view) is illustrated here. On the abscissa the angles in the range of $[0; 2\pi]$ of a camera are logged, which is rotated around an object with a fixed axis through the center of the object. The ordinate represents the values of the normalized difference g (cf. (2)). Each value of the red curve is computed from three images: the test image I and the two best candidate images I_{O1} and I_{O2}. For the red curve image I shows the "dreezle" figure at a zero rotation angle, depicted on the left in the upper row of Fig. 3. One example

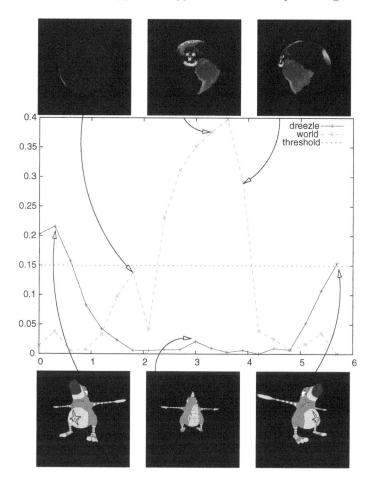

Fig. 4. Phase 2 of the identification of the goal state (finding the discriminative view)

for I_{O1} is the left image in the bottom row, which shows one view of the "dreezle" with the star on it belly, I_{O2} (which is not shown) is the same view of the object without the star. The resulting difference value g for this example is larger than the threshold of 0.15 marked by the dotted line. Thus this view is discriminative enough to tell both objects apart. In contrast, the view in the middle of the bottom row is not discriminative enough, as it does not reveal if the object has a star on its belly. This fact is represented well by the low value of the normalized difference function g. The upper row and green curve show more examples for the object "world". Here the reference view I has also been recorded at the position of zero degrees. This diagram illustrates that the objects can be distinguished more reliably with a larger visibility of their discriminative part.

Results. The left column of Fig. 5 illustrates the development of the set of samples of state-action-pairs constituting the Q-function. After the first episode only two samples are found inside the sample set. We will briefly examine these two

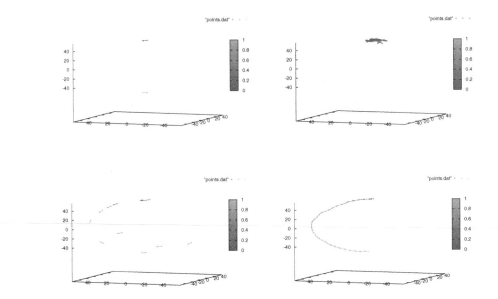

Fig. 5. Left: Learned rules in the form of state-action-pairs used to approximate the Q-function. Right: Application of the learned rules in a discrimination task.

samples to comprehend the constitution of the Q-function. The first reward the agent gets after the first episode is zero. Because nothing is known about the Q-function, this state-action-pair is immediately added to the set of samples. This sample predicts the reward of all following state-action-pairs also as zero. This holds true until the first step encounters a reward of 1. Then one additional sample is added to the Q-function: the state-action-pair that led to the goal state. The state-action-pair with the goal state is not inserted because it does not meet the conditions presented in Sec. 3. This is the end of the first episode. Further episodes insert samples according to the rules presented in the end of Sec. 3. Basically, it is tested if the Q-function needs denser sampling or not. The right column of Fig. 5 shows the paths an agent takes when she uses the Q-functions depicted in the left column of Fig. 5. It is obvious that the paths get shorter the more episodes have been completed. The examples clearly indicate the applicability of the RRL approach to computer vision learning tasks, especially because of its capability of handling continuous domains. Additionally, the Q-function consists of comprehensive state-action-pairs, where each pair encodes a rule that indicates the agent the merit of moving the camera in a certain direction with a given view at hand. This way, a human trainer can easily add information to the Q-function, e.g., by simply presenting a view of the object and a corresponding preferable direction. In addition, the relational encoding removes dependencies on coordinate systems, which may arise when using traditional RL approaches that use camera positions as their basis of a state's encoding. Fig. 5 in detail:

Left column: Learned rules in the form of state-action-pairs used to approximate the Q-function. The axes encode the 3D space with the object in the center of the coordinate system (not shown in these diagrams). A state is given as the origin of a vector and encodes a current view of the object, an action is given by the direction of a vector and encodes the next direction of the movement of the camera. (Top diagram: after 1 episode, bottom diagram: after 3 episodes.) The color encodes the assigned Q-value. Red stands for an expected accumulated reward of zero, while green indicates a reward of one. Right column: Application of the learned rules in a discrimination task. Here the paths are depicted an agent chooses based on the learned Q-function until a successful discrimination took place or a maximum number of steps has been exceeded. (Top diagram: using the Q-function learned after 1 episode (the agent has not found a path within 40 steps), bottom diagram: using the Q-function learned after 10 episodes.) Again, the color encodes the predicted Q-value of the path taken.

6   Conclusion

We proposed an adaptive, self-learning system which utilizes RRL and applied it to the problem of the discrimination between similar objects which differ, e.g., in salient features visible from distinct views only. Usually existing object recognition systems have to scan an object from a large number of views for a reliable recognition. In our RRL approach we introduced a representation of states and actions that are entirely based on the perceived appearance of an object. This enables the system to rapidly learn scan paths for objects and to reliably distinguish similar objects from only a few recorded views. The appearance-based approach and the possibility to define states and actions of the RRL-system with logical descriptions allow for a large reduction of the state space and thus save storage and computing time.

7   Future Research

This work leads to a number of possible future extensions. The image retrieval using a simple kd-tree is quite slow and can be accelerated. Using the best-bin-first-technique of [11] accelerates the process slightly, but the matching reliability deteriorates quickly and values below 20.000 comparisons are strongly discouraged. An integration of the generation of the object database into the learning algorithm would result in a system enabling the agent to explore her environment and constantly add features into the object database. Similar objects can share some of the feature sets. However, to generate representations of distinct objects, a criterion will have to be developed that groups feature sets according to their belonging to the same unique object type.

Acknowledgments. This research was funded by the German Research Association (DFG) under Grant PE 887/3-3.

References

1. Dzeroski, S., De Raedt, L., Driessens, K.: Relational reinforcement learning. In: Machine Learning, vol. 43, pp. 7–52 (2001)
2. Sutton, R.S., Barto, A.G.: Reinforcement Learning: An Introduction. MIT Press, Cambridge (1998)
3. Gartner, T., Driessens, K., Ramon, J.: Graph kernels and gaussian processes for relational reinforcement learning. In: Inductive Logic Programming, 13th International Conference, ILP (2003)
4. Driessens, K., Ramon, J.: Relational instance based regression for relational reinforcement learning. In: Proceedings of the Twentieth International Conference on Machine Learning, pp. 123–130 (2003)
5. Peters, G.: A Vision System for Interactive Object Learning. In: IEEE International Conference on Computer Vision Systems (ICVS 2006), New York, USA, January 5-7 (2006)
6. Lindeberg, T., Bretzner, L.: Real-time scale selection in hybrid multi-scale representations. In: Griffin, L.D., Lillholm, M. (eds.) Scale-Space 2003. LNCS, vol. 2695, pp. 148–163. Springer, Heidelberg (2003)
7. Harris, C., Stephens, M.: A Combined Corner and Edge Detector. In: 4th ALVEY Vision Conference, pp. 147–151 (1988)
8. Mikolajczyk, K., Schmid, C.: Scale and affine invariant interest point detectors. International Journal of Computer Vision 60(1), 63–86 (2004)
9. Lowe, D.G.: Distinctive image features from scale-invariant keypoints. Int. J. Comput. Vision 60(2), 91–110 (2004)
10. Baumberg, A.: Reliable feature matching across widely separated views. In: CVPR 2001, p. 1774 (2000)
11. Beis, J., Lowe, D.: Shape indexing using approximate nearest-neighbor search in highdimensional spaces (1997)
12. Hartley, R.I., Zisserman, A.: Multiple View Geometry in Computer Vision, 2nd edn. Cambridge University Press, Cambridge (2004)

Sensitivity Analysis of Forest Fire Risk Factors and Development of a Corresponding Fuzzy Inference System: The Case of Greece

Theocharis Tsataltzinos[1], Lazaros Iliadis[2], and Spartalis Stefanos[3]

[1] PhD candidate Democritus University of Thrace, Greece
tsataltzinos@yahoo.gr
[2] Associate Professor Democritus University of Thrace, Greece
liliadis@fmenr.duth.gr
[3] Professor Democritus University of Thrace, Greece
sspart@pme.duth.gr

Abstract. This research effort has two main orientations. The first is the sensitivity analysis performance of the parameters that are considered to influence the problem of forest fires. This is conducted by the Pearson's correlation analysis for each factor separately. The second target is the development of an intelligent fuzzy (Rule Based) Inference System that performs ranking of the Greek forest departments in accordance to their degree of forest fire risk. The system uses fuzzy algebra in order to categorize each forest department as "risky" or "non-risky". The Rule Based system was built under the MATLAB Fuzzy integrated environment and the sensitivity analysis was conducted by using SPSS.

1 Introduction

Forest fires have been a major issue for several countries of the world over the last century. Many countries have developed their own models of forecasting the location of the most severe forest fire incidents for the following fire season and the extent of the damages. Our research team has designed and implemented the FFR-FIS (Forest Fire Risk Inference System), which provides a ranking of the Greek forest departments according to their degree of forest fire risk. The system uses fuzzy logic and it can provide several results obtained by different scenarios, according to the optimism and to the perspective of its user. It is adjustable to any terrain and can be used under different circumstances. The first attempt to characterize the Greek forest departments as risky or not risky gave promising results that have proven to be a close approximation of reality. Initially, the first version of the system considered four factors that influence the degree of forest fire risk whereas this research effort proposes a restructured and adjusted system to use seven major factors. The determination of the optimal vector of factors that influence the overall degree of risk for every forest department is a difficult task and the gathering of data is a tedious procedure, expensive

D. Palmer-Brown et al. (Eds.): EANN 2009, CCIS 43, pp. 313–324, 2009.
© Springer-Verlag Berlin Heidelberg 2009

and time consuming. This study is a first attempt towards the development of an effective and simple Fuzzy Inference System (FIS) capable of considering the factors that are of great importance towards forest fire risk modeling in Greece.

2 Theoretical Framework

The FFR-FIS was designed and developed under the Matlab platform. It employees the concepts and the principles of fuzzy logic and fuzzy sets. The system assigns a degree of long term forest fire risk (LTDFFR) to each area by using Matlab's fuzzy toolbox and its integrated triangular membership function. See the following formula 1.

$$\mu_s(X) = \begin{cases} 0 \text{ if } X < a \\ (X-a)/(c-a) \text{ if } X \in [a,c] \\ (b-X)/(b-c) \text{ if } X \in [c,b] \\ 0 \text{ if } X > b \end{cases} \tag{1}$$

The first basic operation of the system is the determination of the main n risk factors (RF) affecting the specific risk problem. Three vectors of fuzzy sets (FS) can be formed for each RF:

1. $\tilde{S}_{1i}^{T} = \{(\mu j (A_j), X_i) \text{ (forest departments Aj of low risk) } / j = 1...N, i = 1... M\}$

2. $\tilde{S}_{2i}^{T} = \{(\kappa j (A_j), X_i) \text{ (forest departments A j of medium risk) } / j = 1...N, i = 1...M\}$

3. $\tilde{S}_{3i}^{T} = \{(\lambda j (A j), Xi) \text{ (forest departments A j of high risk) } / j = 1...N, i = 1...M \}$

where N is the number of forest departments under evaluation and M is the number of involved parameters. It should also be specified that X_i is a crisp value corresponding to a specific risk factor.

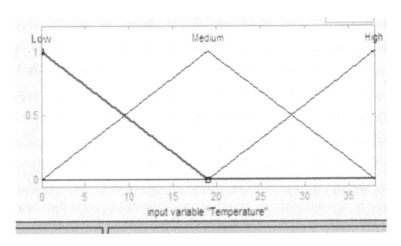

Fig. 1. Membership functions for the degree of risk related to "Temperature

The long term forest fire risk factors are distinguished in two basic categories; Human factors and Natural ones (Kailidis, 1990). Each one of these general risk types consists of several sub-factors that influence in their own way the overall risk degree (ORD). The factors under consideration are the following:

Table 1. Factors under consideration

a.	Average annual temperature
b.	Average annual humidity
c.	Average percentage of forest cover
d.	Average height
e.	Average annual wind speed
f.	Population Density
g.	Index of tourism-urban development (scale from 1 to 10)

Some of the above parameters are structural (they do not change overtime) and some are dynamic. The singleton values (minimum and maximum) that are needed in order to apply the fuzzyfication functions are calculated from the data inventories obtained by the Greek secretariat of forests and other public services. There might also be other factors with a potential influence in the annual number of fire incidents, but in this pilot attempt, the system has to be simple enough to provide clear results so that it becomes easier in the future to analyze all the data available. The FFR-FIS focuses in providing a LTDFFR index for the Greek forest departments related to their annual number of forest fires.

A critical step is the design and construction of the main rule set that will provide the reasoning of the system. Factors a, b, c, d and e in the above table 1 are the "*natural factors*" whereas the rest two are the "*human-related ones*". More specifically, the average annual temperature, humidity and wind speed belong to the "*meteorological parameters*" data group whereas the average height and the percentage of forest cover belong to the "*landscape parameters*" data group. This categorization and grouping of risk factors, enables the construction of a much smaller rule set for the estimation of the final ORD.

For example, in order to combine the Population Density (Pop) and the Tourism (Tour) factors in one subgroup named "*measurable human factors*" (MHF) the following 9 rules need to be applied:

1. If **Pop** is low and   **Tour** is low then   **MHF** is low
2. If **Pop** is average and **Tour** is average then **MHF** is average
3. If **Pop** is high and   **Tour** is high then   **MHF** is high
4. If **Pop** is low and   **Tour** is average then **MHF** is average
5. If **Pop** is low and   **Tour** is high then   **MHF** is high
6. If **Pop** is average and **Tour** is low then   **MHF** is average
7. If **Pop** is average and **Tour** is high then   **MHF** is high
8. If **Pop** is high and   **Tour** is average then **MHF** is high
9. If **Pop** is high and   **Tour** is low then   **MHF** is high

This makes it easier to combine all the factors with the use of proper fuzzy linguistics in order to produce one final degree of risk. The factors are combined in the way described in the following figure 2. The number of rules becomes even smaller when proper decision tables are used (Table 2).

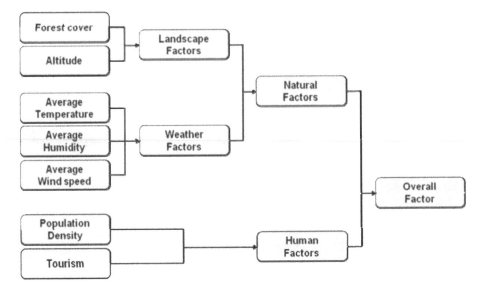

Fig. 2. Categorization of risk factors

Table 2. Example of a decision table used for the "Weather factors"

Temperature	L	L	L	L	L	L	L	L	L	M	M	M	M	M	M	M	M	M	H	H	H	H	H	H	H	H	H
Wind	L	L	L	M	M	M	H	H	H	L	L	L	M	M	M	H	H	H	L	L	L	M	M	M	H	H	H
Humidity	L	M	H	L	M	H	L	M	H	L	M	H	L	M	H	L	M	H	L	M	H	L	M	H	L	M	H
Low Danger	X	X	X		X	X		X	X																		
Medium Danger		X				X		X	X	X		X	X			X			X	X							
High Danger													X		X	X		X			X	X	X	X	X	X	X

Final Table		Group 1							Group 2							Group 3				
Temperature	L	L	L	L	L	L	L	M	M	M	M	M	M	M	H	H	H	H	H
Wind	L	M	M	M	H	H	H	L	M	M	M	H	H	H	L	L	L	M	H
Humidity	-	L	M	H	L	M	H	-	L	M	H	L	M	H	L	M	H	-	-
Low Danger	X		X	X		X	X												
Medium Danger		X			X				X	X		X	X		X	X			
High Danger							X				X	X		X			X	X	X

Legend
L = Low
M = Medium
H = High

The ORD is produced by applying various types of fuzzy relations, in order to perform fuzzy conjunction operations between the fuzzy sets (and consequently between partial risk indices due to separate risk parameters). The functions for the conjunction are called T-norms and for the disjunction T-conorms or S-norms (Kandel A., 1992). This project employees the Matlab's integrated Mamdani Inference method, which operates in a forward chaining mode (Cox, 2005). The Mamdani inference system comprises of five parts:

- Fuzzyfication of the input crisp values by using the Triangular membership function (1)
- Application of fuzzy conjunction/disjunction operators. The fuzzy OR operation is performed by employing the Maximum function whereas the fuzzy AND operation uses the Algebraic Product norm as seen in the following formulae 2 and 3 respectively.
- $\mu(x) = \max\{(\mu_{x_1}),(\mu_{x_2}),\ldots\ldots(\mu_{x_n})\}^T$ (2) $\mu(x) = \mu(x_1).\mu(x_2)\ldots\mu(x_n)$ (3)
- Application of the implication method Min using the following formula 4.

$$\mu(x) = \min\{(\mu_{x_1}),(\mu_{x_2}),\ldots\ldots(\mu_{x_n})\}^T \tag{4}$$

- Aggregation of output values with the use of max formula 2.
- Defuzzification on the output with the centroid method by applying formula 5 (below).

This method determines the centre of the area of the combined membership functions. It calculates the centroid or centre of gravity (COG) of the area under the membership function seen in formula 5. (G. W. Nurcahyo S. M. Shamsuddin, 2003 Vol3 no2)

$$f^{-1}\left(COG(\tilde{A})\right) = \frac{\int_\chi \mu\left(\tilde{A}(\chi)\right).\chi dx}{\int_\chi \mu\left(\tilde{A}(\chi)\right) dx} \tag{5}$$

3   Pearson Correlation Coefficients

Correlation analysis is being computed by researchers in order to find how two variables are related to each other. There are two fundamental characteristics of correlation coefficients that researchers care about. (Urdan, 2005)

1. Correlation coefficients can be either positive or negative
2. Correlation coefficients range in strength from -1.00 to +1.00

A perfect positive correlation of +1.00 reveals that for every member of the sample or population, a higher score on one variable is related to a higher score on the other variable, while a perfect negative to a lower score on the other variable. Correlation coefficients between -0.20 and +0.20 indicate a weak relation between two variables, those between 0.20 and 0.50 (either positive or negative) represent a moderate relationship, and those larger than 0.50 a strong relationship. (Urdan, 2005)

There are several different formulae that can be used to calculate Pearson correlation coefficients. The definitional function for calculating them is presented in formula 6 below.

$$r = \frac{\sum z_x z_y}{N} \tag{6}$$

Where   r = Pearson correlation coefficient
Z_x = z score for variable X
Z_y = z score for variable Y
N = the number of pairs of X and Y scores

To see whether this correlation coefficient is statistically significant, the table of *t* values is necessary. This helps in finding the probability value *p* and by using the traditional 0.05 *alpha* level it is possible to conclude whether the coefficient is statistically significant. In this study, correlation coefficients were calculated with the use of SPSS (Pearson, 1 tailed test).

4   Sensitivity Analysis

4.1   Analysis Related to the Case of Forest Fire Incidents

A more detailed study of the results show which factors are necessary and which are not. In this first approach, six of the factors used as inputs, have not proven to be significantly correlated to the number of forest fires according to Pearson's method as seen in table 4 below. These factors have very weak correlations to the annual number of fires that vary from -0.074 to just 0.171, and their significance scores are all high above the 0.05 level of confidence (Urdan 2005). The only significantly correlated factor is the population, with a moderate relationship to the annual number of fires. Population has an actual significant correlation due to its p value of 0.004 (Urdan 2005).

Despite the fact that most of the factors separately are not correlated to the final number of fires, when all of them are combined with the proper rule set, they provide an ORD that has statistically a moderate correlation degree of *0.347* which is statistically significant with a *p* value of *0.007*. Table 3 below presents the results of Pearson's correlation coefficients of the seven factors considered by the FFR-FIS

Table 3. Results with Pearson correlation coefficients

Correlations		Mean temperature	Mean humidity	Mean height	Mean density	Mean wind speed
Number of fires	Pearson Correlation	0,171	0,060	-0,072	-0,074	0,129
	Sig. (1-tailed)	0,118	0,340	0,310	0,304	0,186
Correlations		Population	Tourism	Human danger	Natural danger	Total danger
Number of fires	Pearson Correlation	,373(**)	0,108	,311(*)	0,010	,347(**)
	Sig. (1-tailed)	0,004	0,227	0,014	0,472	0,007

The terms *"human-caused risk"* and *"natural risk"* refer to unified risk indices that derive from human and natural related factors respectively. It is clear that the human degree of risk with a correlation degree of *0.311* and a *p* value of *0.014* is more significantly correlated to the number of fires than the natural degree of risk which has a 0.01 value for correlation degree and its *p* value is 0.472. This result is another indication of the major importance of the human factor in the determination of the number of forest fire instances.

The next step of this analysis is to find out which of the above factors are absolutely necessary in order to produce a good overall degree of risk and which of them can be ignored. The basic hypothesis made in this step is to assume that data related to a factor is not known and cannot be inferred. Starting from the first factor, its data is deleted and the average value of this factor for Greece is assigned to each forest department instead. The overall degree of risk is then recalculated and the correlations matrix is reproduced, in order to compare the new overall degree with the initial one. The same procedure is repeated iteratively for every factor that the system utilized. The results are shown in the table 4 below.

Table 4. Results with Pearson correlation coefficients, between the Overall degree of risk and the number of fire incidents, when each one of the factors is not known

	Correlations	Overall degree of risk
1	All factors known	,347(**) 0,007
2	Temperature uknown	0,092559 0,261309
3	Humidity uknown	,299(*) 0,017544
4	Altitude uknown	,249(*) 0,040906
5	Forest cover density uknown	,304(*) 0,015901
6	Wind speed uknown	0,230499 0,053653
7	Population uknown	-0,1072 0,229347
8	Tourism development index uknown	,349(**) 0,006449

These results show that the system provides an overall degree of risk which is in fact correlated to the number of fire instances, when all the seven factors are known, as explained above. Further analysis shows that even if there is no data about humidity, altitude, forest cover or tourism development index, the correlation degrees still are all above 0.249 and all have *p* values lower than 0.0409. This shows that the results still remain significantly associated to the number of forest fire instances. On the other hand it is not possible to provide an overall degree of risk when the temperature, the wind and the population of a forest department are not known As shown in Table 5 above, when the wind factor is missing, the correlation degree of the ORD with the annual number of forest fires is 0.23 and *p* is above 0.053. This is "almost" acceptable as value, whereas if the temperature or the population density indices are not known, the correlation degrees drop below 0.107 in absolute value and *p* rises above 0.22. These values are not acceptable due to the low degree of correlation, which means that there is no relation between the results and reality, and the fact that p is way above the 0.05 threshold, which makes the results statistically insignificant.

4.2 Analysis Related to the Case of the Burned Area

Of course one can argue that there are several other factors that might have a potential influence in the forest fire degree of risk. According to (Kailidis 1990), there are two different kinds of forest fire degrees of risk. More detailed analysis of forest fire risk needs the distinction between these two different types of forest fire degrees of risk as shown below:

- Forest fire degree of risk due to the total number of forest fire instances
- Forest fire degree of risk due to the total burned area

The factors that affect the first degree of risk have a different effect to the second degree. For example, Tourism is a great risk factor when examining the number of fire instances, but it does not influence the "acceleration risk degree" in a great way. Meteorological risk factors are those that mostly affect the second degree of danger, while human caused risk factors affect mostly the first risk degree. (Kailidis 1990) Thus, in order to produce a risk index due to the extent of the damages of each forest fire incident, different factors and a different rule set are required. Despite this fact, the overall risk index provided by the FFR-FIS is significantly associated to the second kind of risk degree. The following table 5 presents the correlation coefficients between the ORD and the degree of risk due to the extent of the burned area.

As seen in table 5 when all the factors are known, the correlation is significant at the 0.05 **alpha** level due to its low value of *p*=0.0467, whereas it has a moderate correlation degree of 0.24. By applying the previous method, assuming that every time one factor is not known, the results for temperature, humidity, altitude, forest cover density, wind and population produce correlation degrees between 0.052 and 0.221. Even though the value 0.221 (when the forest cover density is unknown) shows a moderate correlation, it cannot be taken under consideration due to its *p* value which is 0.061. The correlation degrees in every other case drop below 0.2 and *p* values are all above 0.17. On the contrary, when tourism development index is treated as not known, the correlation becomes greater (0.316) and the *p* value drops at 0.013. This proves that in order to produce significant results, all the factors, but "tourism", are necessary. Thus, in order to produce a risk index that will try to evaluate the burned area, a different approach is needed.

Table 5. Results with Pearson correlation coefficients, between the Overall degree of risk and the extent of each fire incident, when each one of the factors is not known

	Correlations	Overall degree of risk
1	All factors known	0,240(*)
		0,046682
2	Temperature uknown	0,071871
		0,30995
3	Humidity uknown	0,090818
		0,265252
4	Altitude uknown	-0,05222
		0,359361
5	Forest cover density uknown	0,221249
		0,061282
6	Wind uknown	0,132014
		0,18039
7	Population uknown	0,137629
		0,170266
8	Tourism uknown	,316(*)
		0,012702

The factors have to be studied carefully, in order to use the most appropriate of all and the rule set needs to be rebuilt so that the correlations become stronger and have a greater significance. More detailed analysis reveals that in order to produce better results, the method should be focused on "*natural factors*". As seen in the following table 6, the extent of forest fires is significantly correlated to the mean forest cover density (correlation degree=-0.307 and p=0.015) and to the population density (correlation degree=0.401 and p=0.02) of each forest department.

Table 6. Results with Pearson correlation coefficients

Correlations		Mean Temperature	Mean Humidity	Mean Height	Mean Density	Mean Wind Speed
Extent of Burned area	Pearson Correlation	0,167	-0,230	-0,111	-,307(*)	0,149
	Sig. (1-tailed)	0,123	0,054	0,221	0,015	0,151

Correlations		Population	Tourism	Total Danger	Human Danger	Natural Danger
Extent of Burned area	Pearson Correlation	,401(**)	-0,161	,240(*)	0,007	,430(**)
	Sig. (1-tailed)	0,002	0,133	0,047	0,482	0,001

As it is shown in Table 6 above, the overall *"natural risk"* is even higher correlated to the extent of each forest fire (correlation degree=0.430 and *p*=0.001). On the other hand, the *human caused* unified degree of risk has almost no relation to the extent of fire incidents (correlation degree=0.007 and p=0.482). This means either that a totally different data index needs to be used in order to produce a reliable ORD or a different rule set has to be applied. Different human data is necessary and more natural factors have to be added, in order to use this system effectively towards ranking the forest departments according to their forest fire degree of risk due to the burned area.

5 Compatibility of the System's Output to the Actual Case

This study concerns Greece, which is divided into smaller parts called forest department. Data from 1983 to 2000 has been gathered. The first ten years of data have been used to produce the decade's mean values. These mean values have been cross-checked with the eleventh year's actual ranking of forest departments due to their forest fire frequencies. The next step was to estimate the mean values for a period of eleven years and to compare the results to the actual situation of the twelfth year and so on. The compatibility of the ORD output of the FFR-FIS to the actual annual forest fire situation varied from 52% to 70% as shown in table 7 below (based on the frequencies of forest fire incidents).

Table 7. Compatibility analysis according to the following year's total number of forest fire incidents

	93-94	94-95	95-96	96-97	97-98	98-99	99-00
Compatibility with the following year's actual ranking for the *risky area fuzzy set*	58%	52%	62%	62%	70%	64%	62%

Table 8. Factors that are absolutely necessary

Forest fire degree of risk due to total number of fire instances	Forest fire degree of risk due to the burned area
Temperature	Temperature
Wind	Humidity
Population	Altitude
	Forest cover density
	Wind
	Population

6 Conclusions and Discussion

In this first approach in order to produce a ranking of the Greek forest departments due to their annual number of fires, the results vary annually (Table 7). The factors, that were used, are capable of producing a ranking of the Greek forest departments that can be taken under consideration. The system was built based on the knowledge that meteorological factors do influence the fire degree of danger and that most of the forest fires in Greece are Human caused (Kailidis, 1990). This study indicates which of those factors have a significant correlation to the number of fires or to the extent of each forest fire and helps in deciding which factors can be overlooked and which cannot. Table 8 below shows which factors cannot be overlooked in each case so that the results remain statistically significant.

The correlation coefficients that were calculated reveal the fact that the FFR-FIS is not a tool for providing a universal degree of forest fire risk that can be translated into real number of fires. On the other hand it is clear that the relation between the overall degree of risk and the number of forest fires is significant, so it is possible to produce a long term based ranking.

The compatibility of the system's output (in the case of forest fire frequencies) and the actual number of forest fire incidents is quite high always more than 50% and it reaches as high as 70%. This is a very promising result considering the degree of difficulty of such a task. No matter what the statistics say, it is very important to be able to estimate roughly the number of the forest fire incidents in a major area one year earlier than the fire season.

Having in mind that the FFR-FIS provides a ranking of the Greek forest departments, it is safe to say that "*forest departments that appear as risky in the final ORD are more likely to be exposed into a greater number of fire instances than those that appear as non-risky and that the attention needs to be focused on them*". Additionally, the system is capable of producing a ranking of the Greek forest departments according to the "*acceleration risk degree*", despite the fact that this was not the main research field. A future research will involve the consideration of more risk factors and thus the construction of new main rule sets.

References

1. Kandel, A.: Fuzzy Expert Systems. CRC Press, USA (1992)
2. Leondes, C.T.: Fuzzy Logic and Expert Systems Applications. Academic Press, California (1998)
3. Christopher, F.H., Patil Sumeet, R.: Identification and review of sensitivity analysis methods. Blackwell, Malden (2002)
4. Cox, E.: Fuzzy modeling and genetic algorithms for data mining and exploration. Academic Press, London (2005)
5. Johnson, E.A., Miyanishi, K.: Forest Fire: Behavior and Ecological Effects (2001)
6. Nurcahyo, G.W., Shamsuddin, S.M., Alias, R.A., Sap, M.N.M.: Selection of Defuzzification Method to Obtain Crisp Value for Representing Uncertain Data in a Modified Sweep Algorithm. JCS&T 3(2) (2003)

7. Iliadis, L., Spartalis, S., Maris, F., Marinos, D.: A Decision Support System Unifying Trapezoidal Function Membership Values using T-Norms. In: ICNAAM (International Conference in Numerical Analysis and Applied Mathematics). J. Wiley-VCH Verlag GmbH Publishing co., Weinheim (2004)
8. Kahlert, J., Frank, H.: Fuzzy-Logik und Fuzzy-Control (1994)
9. Kailidis, D.: Forest Fires (1990)
10. Mamdani, E.H., Assilian, S.: An experiment in linguistic synthesis with a fuzzy logic controller. Man-Machine Studies, 1–13 (1975)
11. Nguyen, H., Walker, E.: A First Course in Fuzzy Logic. Chapman and Hall, Boca Raton (2000)
12. Tsataltzinos, T.: A fuzzy decision support system evaluating qualitative attributes towards forest fire risk estimation. In: 10th International Conference on Engineering Applications of Neural Networks, Thessaloniki, August 29-31 (2007)
13. Urdan, T.C.: Statistics in plain English. Routledge, New York (2005)
14. Kecman, V.: Learning and Soft Computing. MIT Press, London (2001)
15. Zhang, J.X., Huang, C.F.: Cartographic Representation of the Uncertainty related to natural disaster risk: overview and state of the art. Artificial Intelligence, 213–220 (2005)

Nonmonotone Learning of Recurrent Neural Networks in Symbolic Sequence Processing Applications

Chun-Cheng Peng and George D. Magoulas

School of Computer Science and Information Systems,
Birkbeck College, University of London,
Malet Street, Bloomsbury, London WC1E 7HX, UK
{ccpeng,gmagoulas}@dcs.bbk.ac.uk

Abstract. In this paper, we present a formulation of the learning problem that allows deterministic nonmonotone learning behaviour to be generated, i.e. the values of the error function are allowed to increase temporarily although learning behaviour is progressively improved. This is achieved by introducing a nonmonotone strategy on the error function values. We present four training algorithms which are equipped with nonmonotone strategy and investigate their performance in symbolic sequence processing problems. Experimental results show that introducing nonmonotone mechanism can improve traditional learning strategies and make them more effective in the sequence problems tested.

Keywords: BFGS, conjugate gradient, Levenberg-Marquardt, nonmonotone learning, recurrent neural networks, resilient propagation, training algorithms, symbolic sequences.

1 Introduction

Sequence processing involves several tasks such as clustering, classification, prediction, and transduction of sequential data which can be symbolic, non-symbolic or mixed. Examples of *symbolic* data patterns occur in modelling natural (human) language, while the prediction of water level of River Thames is an example of processing *non-symbolic* data. On the other hand, if the content of a sequence will be varying through different time steps, the sequence is called *temporal* or *time-series*.

In general, a *temporal* sequence consists of nominal symbols from a particular alphabet, while a *time-series* sequence deals with continuous, real-valued elements [1]. Processing both these sequences mainly consists of applying the current known patterns to produce or predict the future ones, while a major difficulty is that the range of data dependencies is usually unknown. Therefore, an intelligent system or approach with memorising and learning capabilities for previous information is crucial for effective and efficient sequence processing and modelling. In this work, the main focus is the problem of temporal sequence processing using Recurrent Neural Networks (RNNs).

Training an RNN can be considered in the framework of unconstrained optimisation as the following minimisation problem

D. Palmer-Brown et al. (Eds.): EANN 2009, CCIS 43, pp. 325–335, 2009.
© Springer-Verlag Berlin Heidelberg 2009

$$\min E(w), \ w \in R^n, \tag{1}$$

where $E : R^n \to R$ is the learning error function and its gradient, $g = g(w) = \nabla E(w)$, is available through the method of backpropagation through time (BPTT). This minimisation typically involves locating suitable values for more than several hundred free parameters (weights and biases).

This problem is typically solved by adopting an iterative method. Let the current approximation to the solution of the above problem be w_k, and if $g_k = \nabla E(w_k) \neq 0$, then, in some way, an iterative method finds a stepsize α_k along a search direction d_k, and computes the next approximation w_{k+1} as follows:

$$w_{k+1} = w_k + \alpha_k d_k. \tag{2}$$

Traditional optimisation strategies for RNNs are monotone ones, i.e. these strategies compute a step length that reduces the error function value at each iteration, as:

$$E_{k+1} \leq E_k, \tag{3}$$

which is the most straight-forward way to minimise the objective function. Unfortunately, even when an algorithm is proved to be globally convergent, there is no guarantee that the method will efficiently explore the search space in the sense that it may be trapped in a local minimum point early on and never jump out to a global one under ill conditions [2], such as poorly initialised weights.

Inspired by the nonmonotone way learning occurs in cognitive development [3], we propose here a nonmonotone formulation of the learning problem for RNNs developed in a deterministic framework of analysis. In the way of nonmonotone learning, the error value of the k-th epoch is allowed to be larger than or equal to the previous epoch. More precisely speaking, the monotone constraint shown in Eq. (3) would be relaxed and replaced by a nonmonotone constraint, such as in Eq. (4).

From deterministic optimisation perspective, algorithms with nonmonotone behaviour have been proposed in an attempt to better explore a search space and in certain cases accelerate the convergence rate [4]-[8]. They have been proved to have several advantages, such as global and superlinear convergence, requiring fewer numbers of line searches and function evaluations, and have demonstrated effectiveness for large-scale unconstrained problems.

We adopt here a nonmonotone strategy, such as the one introduced in [4], taking the M previous f values into consideration

$$E\left(w_k + \alpha_k d_k\right) \leq \max_{0 \leq j \leq M}\left\{E\left(w_{k-j}\right)\right\} + \delta\alpha_k g_k^T d_k, \tag{4}$$

where M is a nonnegative integer, named nonmonotone learning horizon [9], and constant $\delta \in (0,1)$. Also, we exploit the morphology of the error function as represented by a local estimation of the Lipschitz constant to determine the size of M dynamically and thus avoid using a poorly user-defined value for the nonmonotone learning horizon.

It is worth mentioning that learning algorithms with momentum terms, such as the well known momentum backpropagation, do not belong by default to the class of nonmonotone algorithms discussed here. Although momentum backpropagation may occasionally exhibit nonmonotone behaviour, it does not formally apply a non-monotone strategy, such as the one derived in Eq. (4).

The rests of this paper are organised as following. Section 2 provides a brief over-view of four recently proposed nonmonotone training algorithms for RNNs, while Section 3 presents simulation results using symbolic sequences. Section 4 concludes the paper.

2 Nonmonotone Training Algorithms

In this section, we present first-order and second-order methods that employ non-monotone strategy. They are nonmonotone modifications of the Resilient Propagation (Rprop), Conjugate Gradient (CG), BFGS quasi-Newton, and Levenberg-Marquardt (LM) algorithms. Examples of learning behaviours are also provided (see Figures 1 and 2), showing the changes of the Mean-squared-error (MSE), stepsize and learning horizon M and scale factor (if applicable) for NARX recurrent networks, [10][11], in the learning of sequences from parity-5 (P5) and parity-10 (P10).

Aiming to overcome the disadvantages of the pure gradient descent based back-propagation procedure, the first version of *Resilient Propagation* (Rprop, so-called *Resilient Backpropagation*) algorithm was proposed in [12]. The "*Improved Rprop*" (*iRprop*) method proposed by [13] is considered an alternative of the original method. More recently, [14] revised the backtracking step of the iRprop proposing a more powerful alternative, the so-called JRprop, and showed that this method is superior to both the original Rprop and iRprop. We have extended this work here by developing a nonmonotone version of the JRprop, whose key loop in shown in Table 1. This method performs one-step backtracking depending on whether the nonmonotone condition is satisfied or not.

Table 1. Key loop of the Adaptive Non-Monotone JRprop algorithm

Loop of ANM-JRprop
If $E_k \leq E_{k-1}$ then{update w_{k+1} by Rprop; set $q = 1;$ }
Else {
if $E_k > \max_{0 \leq j \leq M_k} \left(E_{k-j} \right) + \delta \alpha_k g_k d_k$ { $w_{k+1} = w_k + \frac{1}{2^q} \Delta w_{k-1};$ $q = q + 1;$ }
}

Another approach is based on the Conjugate Gradient methods, which are in principle approaches suitable for large-scale problems [2]. In [15], we proposed a nonmonotone version of CG methods, which is summarised in Table 2; example of behaviour is shown in Figure 1. In Step 2, the local estimation of the Lipschitz is calculated in order to control the size of M at each iteration.

Table 2. Algorithm: Advanced ANM-CG Algorithm (A2NM-CG)

Algorithm: A2NM-CG

STEP 0.   Initialise w_0, $k=0$, $M^0=0$, $M^{\max}$ is an upper
           boundary for the learning horizon M_k, $l_0 = 0$,
           $a_0, \sigma, \delta \in (0,1)$ and $d_0 = -g_0$;

STEP 1.   If $g_k = 0$, then stop;

STEP 2.   If $k \geq 1$, calculate a local Lipschitz as

$$\Lambda_k = \frac{\|g_k - g_{k-1}\|}{\|w_k - w_{k-1}\|}, \text{ and adapt } M_k :$$

$$M_k = \begin{cases} M_{k-1}+1, & \text{if } \Lambda_k < \Lambda_{k-1} < \Lambda_{k-2} \\ M_{k-1}-1, & \text{if } \Lambda_k > \Lambda_{k-1} > \Lambda_{k-2} , \\ M_{k-1}, & \text{otherwise,} \end{cases}$$

           where $M_k = \min\{M_k, M^{\max}\}$;

STEP 3.   For all $k \geq 1$, find a stepsize $\alpha_k = (2\Lambda_k)^{-1} \cdot \sigma^{l_k}$
           satisfying the following condition:

$$E(w_k + \alpha_k d_k) \leq \max_{0 \leq j \leq M_k} \left[E(w_{k-j}) \right] + \delta \cdot \alpha_k \cdot g_k^T \cdot d_k,$$

           where $l_k = l_k + 1$;

STEP 4.   Generate a new point by $w_{k+1} = w_k + \alpha_k d_k$;

STEP 5.   Update search direction $d_k = -g_k + \beta_{k-1} d_{k-1}$,
           where $\beta_k = \beta_k^{PR}$ or β_k^{FR} is greater than zero and can
           be calculated using the Polak-Ribière or
           Fletcher-Reeves rule respectively;

STEP 6.   Let $k = k+1$, go to STEP 1.

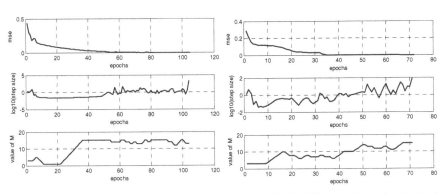

Fig. 1. Convergence behaviours of P5 (left) and P10 (right): NARX networks, trained with the A2NM-CG algorithm

In the context of second-order methods, Quasi-Newton (QN) methods are well-known algorithms for finding local minima of functions in the form of Eq. (1). QN methods exploit the idea of building up curvature information as the iterations of the training method are progressing. This is achieved by using the objective function values and its gradient to estimate the Hessian matrix B_k. The main step of our nonmonotone BFGS approach, which incorporates a self-scaling factor, is in Table 3; further details can be found in [16]. Figures 2a-2b provides examples of behaviour from training NARX networks using the parity-5 and 10 sequences respectively.

Table 3. Main step of the Adaptive Self-scaling Non-monotone BFGS Algorithm

---

Algorithm: ASCNM-BFGS

If $k \geq 1$,

    Calculate a local approximation of the Lipschitz constant and adapt M_k as in A2NM-CG

    Check that stepsize α_k satisfies the nonmonotone condition;
    otherwise, find stepsize that satisfies it;
    Generate a new weight vector $w_{k+1} = w_k + \alpha_k d_k$;

    Update the search direction $d_k = -B_k^{-1} g_k$, using

$$B_{k+1} = \rho_k \left[B_k - \frac{B_k s_k s_k^T B_k}{s_k^T B_k s_k} \right] + \frac{y_k y_k^T}{y_k^T s_k},$$

where $s_k = w_{k+1} - w_k$, $y_k = g_{k+1} - g_k$ and $\rho_k = \frac{y_k^T s_k}{s_k^T B_k s_k}$;

Another second-order method is the Levenberg-Marquardt (LM) method [17][18]. Since the first attempt [19] to train static neural networks, the LM method had been revised to include adaptive-momentum terms [20], namely the LMAM method, and exhibited improved performance in training static neural networks. Our improved nonmonotone LMAM algorithm is shown in Table 4, where the same notation as in [20] is used.

3 Experiments and Results

The nonmonotone training algorithms have been tested on a set of real-world symbolic sequence applications: a sequence classification (SC) problem [21], the sequence learning (SL) problem [22], and the reading aloud (RA) [23]) problem using three different RNN architectures, i.e. Feed-Forward Time-Delayed Network (FFTD, [24][25]), Layered Recurrent Network (LRN, [26][27]), and Nonlinear Auto-Regressive network with eXogenous Inputs (NARX, [10][11]). Details of numbers of adjustable weights and biases are summarised in Table 5. All simulations were coded in Matlab 7.0 on Windows platform, while results reported here are average of 100 randomly initialized runs. Due to page limitation we provide here a summary of our results reporting on the performance of NARX networks in the SC, SL and RA sequences. Results for the monotone and nonmonotone JRprop, CG, BFGS and LM methods are shown in Tables 6-8, 9-11, 12-14, and 15-16, respectively. In the tables, *MSE* means the mean-squared-error in training and *CE* is the classification error in testing with unknown data. In general, the proposed monotone versions of the four training algorithms are superior to the original monotone methods. They also appear to be more effective when compared with mainstream methods used in the literature for the tested applications [22]-[23]. For example, the best result reported in the literature for the SL application is an MSE of 25% in training and 22% in testing using LRN with 10 hidden nodes. In the RA application, RNNs with 100 hidden nodes which should be trained for 1900 epochs are needed to achieve results similar to ours.

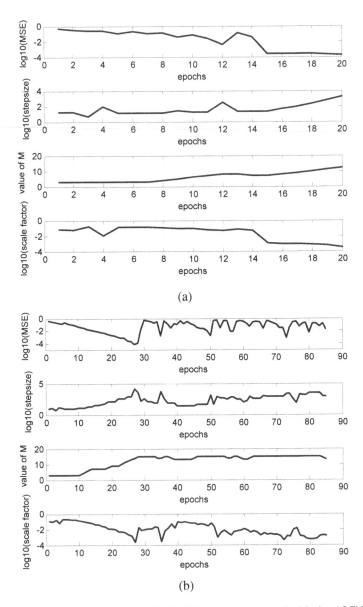

Fig. 2. (a) Convergence behaviours of P5: NARX networks, trained with the ASCNM-BFGS. (b) Convergence behaviours of P10: NARX networks, trained with the ASCNM-BFGS.

Table 4. Adaptive Non-monotone Levenberg-Marquardt Algorithm with adaptive momentum

Algorithm: ANM-LMAM

STEP 0.   Initialize $\tau > 1$, , w_0, D_0, M^{max}, Φ, $\delta \in (0,1)$, $\Theta \in (0,1)$, c, and $k = 0$;

STEP 1.   If $g_k = J_k e_k \neq 0$, calculate $H_k = J_k^T J_k$; otherwise, stop;

STEP 2.   Compute β_k by $\beta_k = -\dfrac{\psi_1}{2\psi_2}\left[(H_k + \alpha_k D_k)\right]^{-1} g_k + \dfrac{1}{2\psi_2}\beta_{k-1}$, where

$$\psi_1 = \frac{-2\psi_2\varepsilon + g_k^T\beta_k}{g_k^T H_k^{-1} g_k}, \quad \psi_2 = \frac{1}{2}\left[\frac{\beta_{k-1}^T H_k \beta_k \cdot g_k^T H_k^{-1} g_k - \left(g_k^T\beta_{k-1}\right)^2}{g_k^T H_k^{-1} g_k \Theta^2 - \varepsilon^2}\right]^{1/2}, \text{and}$$

$$\varepsilon = -c\Theta\left(g_k^T H_k^{-1} g_k\right)^{1/2};$$

STEP 3.   If $k \geq 1$, calculate the local Lipschitz estimation Λ_k and update M_k;

STEP 4.   If the nonmonotone condition is not satisfied, $\alpha_k = \tau\alpha_k$ and go to Step 2; Else, $\alpha_k = \alpha_k/\tau$ and $w_{k+1} = w_k + \beta_k$;

STEP 5. $k = k+1$, go to Step 1.

Table 5. Total number of free parameter for the architectures used in the simulations

Problem (Inputs/Outputs)	# hidden nodes	RNN Architecture		
		FFTD	LRN	NARX
SC (36/3)	5	383	228	304
	10	763	503	433
	15	1143	828	648
SL (4/4)	1	17	14	25
	2	30	26	46
	5	69	74	109
	7	95	116	151
	10	134	194	214
RA (105/61)	5	1421	921	2031
	10	2781	1831	4001

Table 6. Average performance for JRprop-trained NARX nets in the SC problem

Algorithm	#hid	MSE (%)		CE (%)	
		Train	Test	Train	Test
JRprop	5	10.701	19.593	36.330	30.712
	10	7.661	17.314	26.813	27.315
	15	7.397	17.576	26.803	27.740
ANM-JRprop	5	2.8427	17.091	13.394	26.192
	10	2.6338	17.326	13.034	25.781
	15	2.6050	17.410	13.069	25.767

Table 7. Average performance for JRprop-trained NARX nets in the SL problem

Algorithm	#hid	MSE (%)	
		Train	Test
JRprop	1	35.077	36.660
	2	26.814	28.495
	5	18.274	21.897
	10	14.030	17.839
ANM-JRprop	1	33.932	35.493
	2	25.431	27.345
	5	16.809	20.733
	10	12.027	16.160

Table 8. Average performance for JRprop-trained NARX nets in the RA problem

Algorithm	#hid	MSE (%)	
		Train	Test
JRprop	5	12.608	24.215
	10	13.843	22.800
ANM-JRprop	5	3.750	17.663
	10	3.009	18.055

Table 9. Results for conjugate gradient-trained NARX networks in the SC problem

Algorithms	Conv (%)	Epo	MSE (%) Training	CE (%) Testing
CG	58	470	2.816	38.986
A2NM-CG	91	452	0.940	36.986

Table 10. Results for conjugate gradient-trained NARX networks in the SL problem

Algorithms	#hid	MSE (%)	
		Training	Testing
CG	2	16.534	16.761
	5	16.921	17.025
	10	18.321	18.146
A2NM-CG	2	21.951	22.689
	5	13.449	14.206
	10	10.859	11.819

Table 11. Results for conjugate gradient-trained NARX networks in the RA problem

Algorithms	#hid	MSE (%)	
		Training	Testing
CG	5	8.294	7.849
	10	4.796	5.824
A2NM-CG	5	8.942	4.610
	10	7.879	2.182

Table 12. Average performance for BFGS-trained NARX networks in the SC problem

Algorithms	MSE (%) Training	CE (%) Testing
BFGS	7.496	27.247
ASCNM-BFGS	7.100	27.082

Table 13. Average performance for BFGS-trained NARX networks in the SL problem

Algorithms	#hid	MSE (%)	
		Training	Testing
BFGS	2	20.196	20.824
	5	11.029	12.119
	10	8.991	9.846
ASCNM-BFGS	2	19.972	20.792
	5	9.740	10.360
	10	7.584	8.313

Table 14. Average performance for BFGS-trained NARX networks in the RA problem

Algorithms	#hid	MSE (%)	
		Training	Testing
BFGS	5	8.5899	16.6255
	10	6.2481	15.4072
ASCNM-BFGS	5	7.0804	16.1067
	10	5.1843	14.5477

Table 15. Average improvement of ANM-LMAM trained RNNs compared with LMAM trained ones in the SC problem

RNN	MSE (%)		CE (%)	
	Train	Test	Train	Test
FFTD	16.648	18.222	23.355	27.224
LRN	17.125	20.096	23.813	35.502
NARX	27.783	22.953	20.783	24.338

Table 16. Average improvement of ANM-LMAM trained RNNs compared with LMAM trained ones in the SL problem

RNN	MSE (%)	
	Train	Test
FFTD	1.306	1.286
LRN	3.594	3.549
NARX	1.172	1.201

4 Conclusions

Sequence processing applications involve several tasks, such as clustering, classification, prediction and transduction. One of the major challenges is that the data dependencies are usually unknown, and in order to be verified a trial-and-error approach is usually implemented, i.e. changing of the number of delays and/or the number of hidden nodes in an RNN. Effective training algorithms can facilitate this task, especially when training high-dimensional networks.

In this paper, we provided an overview of approaches that employ nonmonotone learning for recurrent neural networks. These consist of nonmonotone first-order (JRprop and conjugate gradient) and second-order (BFGS and Levenberg-Marquardt) algorithms, which were tested in symbolic sequence processing applications. One of the features of our algorithms is that they incorporate an adaptive schedule for determining the length of the nonmonotone learning horizon and the stepsize. As a result, the influence of application-dependent settings can be reduced to some extend. Our simulation results, briefly reported here, show that introducing the nonmonotone strategy could generally improve the performances of training algorithms in terms of smaller training and classification errors, namely MSE and CE. Nonmonotone methods appear to outperform previously reported results in the sequence processing applications tested and are able to train effectively RNNs of various architectures using smaller number of hidden nodes than the original methods.

References

1. Antunes, C.M., Oliveira, A.L.: Temporal data mining: an overview. In: Proc. KDD Workshop on Temporal Data Mining, San Francisco, CA, August 26, 2001, pp. 1–13 (2001)
2. Gill, P.E., Murray, W., Wright, M.H.: Practical Optimization. Academic Press, London (1981)
3. Elman, J.L., Bates, E.A., Johnson, M.H., Karmiloff-Smith, A., Parisi, D., Plunkett, K.: The shape of change. In: Rethinking Innateness: A Connectionist Perspective on Development, ch. 6. MIT Press, Cambridge (1997)
4. Grippo, L., Lampariello, F., Lucidi, S.: A nonmonotone line search technique for Newton's method. SIAM J. Numerical Analysis 23, 707–716 (1986)
5. Grippo, L., Lampariello, F., Lucidi, S.: A quasi-discrete Newton algorithm with a nonmonotone stabilization technique. J. Optimization Theory and Applications 64(3), 495–510 (1990)
6. Grippo, L., Lampariello, F., Lucidi, S.: A class of nonmonotone stabilization methods in unconstrained optimization. Numerische Mathematik 59, 779–805 (1991)
7. Grippo, L., Sciandrone, M.: Nonmonotone globalization techniques for the Barzilai-Borwein gradient method. Computational Optimization and Applications 23, 143–169 (2002)
8. Fasano, G., Lampariello, F., Sciandrone, M.: A truncated nonmonotone Gauss-Newton method for large-scale nonlinear least-squares problems. Computational Optimization and Applications 34, 343–358 (2006)
9. Plagianakos, V.P., Magoulas, G.D., Vrahatis, M.N.: Deterministic nonmonotone strategies for effective training of multi-layer perceptrons. IEEE Trans. Neural Networks 13(6), 1268–1284 (2002)

10. Medsker, L.R., Jain, L.C.: Recurrent neural networks: design and applications. CRC Press, Boca Raton (2000)
11. Nelles, O.: Nonlinear System Identification. Springer, Berlin (2000)
12. Riedmiller, M., Braun, H.: Rprop – a fast adaptive learning algorithm. In: Proc. Int'l Symposium on Computer and Information Sciences, Antalya, Turkey, pp. 279–285 (1992)
13. Igel, C., Hüsken, M.: Empirical evaluation of the improved Rprop learning algorithms. Neurocomputing 50, 105–123 (2003)
14. Anastasiadis, A., Magoulas, G.D., Vrahatis, M.N.: Sign-based Learning Schemes for Pattern Classification. Pattern Recognition Letters 26, 1926–1936 (2005)
15. Peng, C.-C., Magoulas, G.D.: Advanced Adaptive Nonmonotone Conjugate Gradient Training Algorithm for Recurrent Neural Networks. Int'l J. Artificial Intelligence Tools (IJAIT) 17(5), 963–984 (2008)
16. Peng, C.-C., Magoulas, G.D.: Adaptive Self-scaling Non-monotone BFGS Training Algorithm for Recurrent Neural Networks. In: de Sá, J.M., Alexandre, L.A., Duch, W., Mandic, D.P. (eds.) ICANN 2007. LNCS, vol. 4668, pp. 259–268. Springer, Heidelberg (2007)
17. Levenberg, K.: A method for the solution of certain problems in least squares. Quart. Applied Mathematics 5, 164–168 (1944)
18. Marquardt, D.: An algorithm for least squares estimation of nonlinear parameters. J. Society for Industrial and Applied Mathematics 11(2), 431–441 (1963)
19. Hagan, M.T., Menhaj, M.B.: Training feedforward networks with the Marquardt algorithm. IEEE Trans. Neural Networks 5, 989–993 (1994)
20. Ampazis, N., Perantonis, S.J.: Two highly efficient second-order algorithms for training feedforward networks. IEEE Trans. Neural Networks 13, 1064–1074 (2002)
21. Magoulas, G.D., Chen, S.Y., Dimakopoulos, D.: A personalised interface for web directories based on cognitive styles. In: Stary, C., Stephanidis, C. (eds.) UI4ALL 2004. LNCS, vol. 3196, pp. 159–166. Springer, Heidelberg (2004)
22. McLeod, P., Plunkett, K., Rolls, E.T.: Introduction to connectionist modelling of cognitive processes, pp. 148–151. Oxford University Press, Oxford (1998)
23. Plaut, D., McClelland, J., Seidenberg, M., Patterson, K.: Understanding normal and impaired reading: computational principles in quasi-regular domains. Psychological Review 103, 56–115 (1996)
24. Waibel, A.: Modular construction of time-delay neural networks for speech recognition. Neural Computation 1(1), 39–46 (1989)
25. Waibel, A., Hanazawa, T., Hilton, G., Shikano, K., Lang, K.J.: Phoneme recognition using time-delay neural networks. IEEE Transactions on Acoustics, Speech, and Signal Processing 37, 328–339 (1989)
26. Elman, J.L.: Finding structure in time. Cognitive Science 14, 179–211 (1990)
27. Hagan, M.T., Demuth, H.B., Beale, M.H.: Neural Network Design. PWS Publishing, Boston (1996)

Indirect Adaptive Control Using Hopfield-Based Dynamic Neural Network for SISO Nonlinear Systems

Ping-Cheng Chen[1], Chi-Hsu Wang[2], and Tsu-Tian Lee[1]

[1] Department of Electrical Engineering, National Taipei University of Technology
pcchen@ntut.edu.tw, ttlee@ntut.edu.tw
[2] Department of Electrical and Control Engineering, National Chiao Tung University

Abstract. In this paper, we propose an indirect adaptive control scheme using Hopfield-based dynamic neural network for SISO nonlinear systems with external disturbances. Hopfield-based dynamic neural networks are used to obtain uncertain function estimations in an indirect adaptive controller, and a compensation controller is used to suppress the effect of approximation error and disturbance. The weights of Hopfield-based dynamic neural network are on-line tuned by the adaptive laws derived in the sense of Lyapunov, so that the stability of the closed-loop system can be guaranteed. In addition, the tracking error can be attenuated to a desired level by selecting some parameters adequately. Simulation results illustrate the applicability of the proposed control scheme. The designed parsimonious structure of the Hopfield-based dynamic neural network makes the practical implementation of the work in this paper much easier.

Keywords: Hopfield-based dynamic neural network, dynamic neural network, Lyapunov stability theory, indirect adaptive control.

1 Introduction

Recently, static neural networks (SNNs) and dynamic neural networks (DNNs) are wildly applied to solve the control problems of nonlinear systems. Some static neural networks, such as feedforward fuzzy neural network (FNN) or feedforward radius basis function network (RBFN), are frequently used as a powerful tool for modeling the ideal control input or nonlinear functions of systems [1]-[2]. However, the complex structures of FNNs and RBFNs make the practical implementation of the control schemes infeasible, and the numbers of the hidden neurons in the NNs' hidden layers (in general more than the dimension of the controlled system) are hard to be determined. Another well-known disadvantage is that SNNs are quite sensitive to the major change which has never been learned in the training phase. On the other hand, DNNs have a advantage that a smaller DNN is possible to provide the functionality of a much larger SNN [3]. In addition, SNNs are unable to represent dynamic system mapping without the aid of tapped delay, which results in long computation time, high sensitivity to external noise, and a large number of neurons when high dimensional systems are considered [4]. This drawback severely affects the applicability of SNNs to system identification, which is the central part in some control techniques for

D. Palmer-Brown et al. (Eds.): EANN 2009, CCIS 43, pp. 336–349, 2009.
© Springer-Verlag Berlin Heidelberg 2009

nonlinear systems. On the other hand, owing to their dynamic memory, DNNs have good performance on identification, state estimation, trajectory tracking, etc., even with the unmodeled dynamics. In [5]-[7], researchers first identify the nonlinear system according to the measured input and output, and then calculate the control low based on the NN model. The output of the nonlinear system is forced by the control law to track either a given trajectory or the output of a reference model. However, there are still some drawbacks. In [5], although both identification and tracking errors are bounded, the control performance shown in the simulations is not satisfactory. In [6], two DNNs are utilized in the iterative learning control system to approximate the nonlinear system and mimic the desired system output and thus increase the complexity of the control scheme and computation loading. The work in [7] requires a prior knowledge of the strong relative degree of the controlled nonlinear system. Besides, an additional filter is needed to obtain higher derivatives of the system output. These drawbacks restrict the applicability of the above works to practical implementation.

Hence, we try to fix the above drawbacks by an indirect adaptive control scheme using Hopfield-based DNNs. Hopfield model was first proposed by Hopfield J.J. in 1982 and 1984 [8]-[9]. Because a Hopfiled circuit is quite easy to be realized and has the property of decreasing in energy by finite number of node-updating steps, it has many applications in different fields. In this paper, a so-called indirect adaptive control scheme using Hopfield-based dynamic neural network (IACHDNN) for SISO nonlinear systems is proposed. The Hopfield-based DNN can be viewed as a special kind of DNNs. The control object is to force the system output to track a given reference signal. The uncertain parameters of the controlled plant are approximated by the internal states of Hopfield-based DNNs, and a compensation controller is used to dispel the effect of the approximation error and bounded external disturbance. The synaptic weights of the Hopfiled-based DNNs are on-line tuned by adaptive laws derived in the Lyapunov sense. The control law and adaptive laws provide stability for the closed-loop system with external disturbance. Furthermore, the tracking error can be attenuated to a desired level provided that the parameters of the control law are chosen adequately. The main contributions of this paper are summarized as follows. 1) The structure of the used Hopfield-based DNN contains only one neuron, which is much less than the number of neurons in SNNs or other DNNs for nonlinear system control. 2) The simple Hopfield circuit greatly improves the applicability of the control scheme to practical implements. 3) No strong assumptions or prior knowledge of the controlled plant are needed in the development of IACHDNN.

2 Hopfield-Based Dynamic Neural Model

2.1 Descriptions of the DNN Model

Consider a DNN described by the following nonlinear differential equation [5]

$$\dot{\chi} = A\chi + BW\sigma(V_1\chi) + B\Psi\varphi(V_2\chi)\gamma(\overline{u}) \qquad (1)$$

where $\chi = [\chi_1\ \chi_2\ \cdots\ \chi_n]^T \in R^n$ is the state vector, $\overline{u} = [\overline{u}_1\ \overline{u}_2\ \cdots\ \overline{u}_m]^T \in R^m$ is the input vector, $\sigma : R^r \to R^k$, $A \in R^{n \times n}$ is Hurwitz matrix, $B = \text{diag}\{b_1, b_2, \cdots, b_n\} \in R^{n \times n}$, $W \in R^{n \times k}$, $V_1 \in R^{r \times n}$, $\Psi \in R^{n \times l}$, $V_2 \in R^{s \times n}$, $\varphi : R^s \to R^{l \times n}$, and $\gamma : R^m \to R^n$. In (1),

χ is the state of the DNN, $\mathbf{W}$ and $\mathbf{\Psi}$ are the weight matrices describing output layer connections, $\mathbf{V}_1$ and $\mathbf{V}_2$ are the weight matrices describing the hidden layer connections, $\mathbf{\sigma}(\cdot)$ is a sigmoid vector function responsible for nonlinear state feedbacks, and $\mathbf{\gamma}(\cdot)$ is a differentiable input function. A DNN in (1) satisfying

$$r = s = l = n, \quad \mathbf{V}_1 = \mathbf{V}_2 = I_{n \times n}, \quad \varphi(\cdot) = I_{n \times n} \tag{2}$$

is the simplest DNN without any hidden layers and can be expressed as

$$\dot{\chi} = \mathbf{A}\chi + \mathbf{B}\mathbf{W}\mathbf{\sigma}(\chi) + \mathbf{B}\mathbf{\Psi}\mathbf{\gamma}(\bar{\mathbf{u}}) \tag{3}$$

To simply our further analysis, we follow the literatures [5] to choose $k = n$, $\mathbf{A} = diag\{-a_1 \ -a_2 \ \cdots -a_n\}$ with $a_i > 0$, $i=1, 2, ..., n$, and $\mathbf{\gamma}(\bar{\mathbf{u}}) = [\bar{\mathbf{u}} \ \mathbf{o}]^T \in R^n$ with $n \geq m$ and $\mathbf{o} \in R^{n-m}$ being a zero vector. Let $\mathbf{\Psi} = [\mathbf{\Theta} \ \mathbf{\Theta}_r]^T$, where $\mathbf{\Theta} \in R^{n \times m}$ and $\mathbf{\Theta}_r \in R^{n \times (n-m)}$. Then, the expression in (3) can be modified as

$$\dot{\chi} = \mathbf{A}\chi + \mathbf{B}\mathbf{W}\mathbf{\sigma}(\chi) + \mathbf{B}\mathbf{\Theta}\bar{\mathbf{u}} \tag{4}$$

From (4), we have

$$\dot{\chi}_i = -a_i\chi_i + b_iW_i^T\mathbf{\sigma}(\chi) + b_i\Theta_i^T\bar{\mathbf{u}}, \quad i = 1, 2, \cdots, n \tag{5}$$

where $W_i^T = [w_{i1} \ w_{i2} \cdots w_{in}]$ and $\Theta_i^T = [\theta_{i1} \ \theta_{i2} \cdots \theta_{im}]$ are the ith rows of $\mathbf{W}$ and $\mathbf{\Theta}$, respectively. Solving the differential equation (5), we obtain

$$\chi_i = b_i\left(W_i^T\xi_{W,i} + \Theta_i^T\xi_{\Theta,i}\right) + e^{-a_it}\chi_i(0) - e^{-a_it}b_i\left[W_i^T\xi_{W,i}(0) + \Theta_i^T\xi_{\Theta,i}(0)\right], \quad i = 1, 2, \cdots, n. \tag{6}$$

where $\chi_i(0)$ is the initial state of χ_i; $\xi_{W,i} \in R^n$ and $\xi_{\Theta,i} \in R^m$ are the solutions of

$$\dot{\xi}_{W,i} = -a_i\xi_{W,i} + \mathbf{\sigma}(\chi) \tag{7}$$

and

$$\dot{\xi}_{\Theta,i} = -a_i\xi_{\Theta,i} + \bar{\mathbf{u}} \tag{8}$$

respectively; $\xi_{W,i}(0)$ and $\xi_{\Theta,i}(0)$ are initial states of $\xi_{W,i}$ and $\xi_{\Theta,i}$, respectively. Note that the terms $e^{-a_it}\chi_i(0)$ and $e^{-a_it}b_i\left[W_i^T\xi_{W,i}(0) + \Theta_i^T\xi_{\Theta,i}(0)\right]$ in (6) will exponentially decay with time owing to $a_i > 0$.

2.2 Hopfied-Based DNN Approximator

A DNN approximator for continuous functions can be defined as

$$\chi_i = b_i\left(\hat{W}_i^T\xi_{W,i} + \hat{\Theta}_i^T\xi_{\Theta,i}\right) + e^{-a_it}\chi_i(0) - e^{-a_it}b_i\left[\hat{W}_i^T\xi_{W,i}(0) + \hat{\Theta}_i^T\xi_{\Theta,i}(0)\right], \quad i = 1, 2, \cdots, n \tag{9}$$

where $\hat{W}_i$ and $\hat{\Theta}_i$ are the estimations of W_i and Θ_i, respectively. For a continuous vector function $\mathbf{\Phi} = [\Phi_1 \ \Phi_2 \ \cdots \Phi_n]^T \in R^n$, we first define optimal vectors W_i^* and Θ_i^* as

$$
(W_i^*, \Theta_i^*) = \arg \min_{\hat{W}_i \in \Omega_{W_i}, \hat{\Theta}_i \in \Omega_{\Theta_i}} \left\{ \sup_{\chi \in D_\chi, \bar{u} \in D_{\bar{U}}} \left| b_i \left(\hat{W}_i^T \xi_{W,i} + \hat{\Theta}_i^T \xi_{\Theta,i} \right) + e^{-a_i t} \chi_i(0) \right. \right.
$$
$$
\left. \left. - e^{-a_i t} b_i \left[\hat{W}_i^T \xi_{W,i}(0) + \hat{\Theta} \xi_{\Theta,i}(0) \right] \right| \right\}
\tag{10}
$$

where $D_\chi \subset R^N$ and $D_{\bar{U}} \subset R^m$ are compact sets; $\Omega_{W_i} = \left\{ \hat{W}_i : \left\| \hat{W}_i \right\| \leq M_{W_i} \right\}$ and $\Omega_{\Theta_i} = \left\{ \hat{\Theta}_i : \left\| \hat{\Theta}_i \right\| \leq M_{\Theta_i} \right\}$ are constraint sets for $\hat{W}_i$ and $\hat{\Theta}_i$. Then, $\mathbf{\Phi}$ can be expressed as

$$
\Phi_i = b_i \left(W_i^{*T} \xi_{W,i} + \Theta_i^{*T} \xi_{\Theta,i} \right) + e^{-a_i t} \chi_i(0) - e^{-a_i t} b_i \left[W_i^{*T} \xi_{W,i}(0) + \Theta_i^{*T} \xi_{\Theta,i}(0) \right] + \Delta_i \quad i = 1, 2, \cdots, n \tag{11}
$$

where Δ_i is the approximation error. Note that the optimal vectors W_i^* and Θ_i^* are difficult to be determined and might not be unique. The modeling error vector $\tilde{\chi} = [\tilde{\chi}_1 \ \tilde{\chi}_2 \ \cdots \tilde{\chi}_n]^T$ can be defined from (9) and (11) as

$$
\tilde{\chi}_i = \Phi_i - \chi_i
$$
$$
= b_i \left(\tilde{W}_i^T \xi_{W,i} + \tilde{\Theta}_i^T \xi_{\Theta,i} \right) - e^{-a_i t} b_i \left[\tilde{W}_i^T \xi_{W,i}(0) + \tilde{\Theta}_i^T \xi_{\Theta,i}(0) \right] + \Delta_i \quad i = 1, 2, \cdots, n \tag{12}
$$

where $\tilde{W}_i = W_i^* - \hat{W}_i$, and $\tilde{\Theta}_i = \Theta_i^* - \hat{\Theta}_i$.

In this paper, a Hopfield-based dynamic neural network is adopted as the approximator. It is known as a special case of DNN with $a_i = 1/(R_i C_i)$ and $b_i = 1/C_i$, where $R_i > 0$ and $C_i > 0$ representing the resistance and capacitance at the ith neuron, respectively [7]. The sigmoid function vector $\boldsymbol{\sigma}(\chi) = [\sigma_1(\chi_1) \ \sigma_2(\chi_2) \cdots \ \sigma_n(\chi_n)]^T$ is defined by a hyperbolic tangent function as

$$
\sigma(\chi_i) = \tanh(\kappa_i \chi_i), \quad i = 1, 2, \cdots, n \tag{13}
$$

where κ_i is the slope of $\tanh(\cdot)$ at the origin. It is known that tangent function is bounded by $-1 < \tanh(\cdot) < 1$.

3 Problem Formulation

Let $S \subset R^n$ and $Q \subset R^n$ be an open sets, $D_S \subset S$ and $D_Q \subset Q$ be and compact sets. Consider the nth-order nonlinear dynamic system of the form

$$
x^{(n)} = f(\mathbf{x}) + g(\mathbf{x})u + d
$$
$$
y = x \tag{14}
$$

where $\mathbf{x} = [x_1 \ x_2 \ \cdots x_n]^T = [x \ \dot{x} \cdots x^{(n-1)}]^T$ is the state vector; $f : D_s \to R$ and $g : D_Q \to R$ are uncertain continuous functions; $u \in R$ and $y \in R$ are the continuous control input and output of the system, respectively; $d \in R$ is a bounded external disturbance. We consider only the nonlinear systems which can be represented as (14). It is required that $g \neq 0$ so that (14) is controllable. Without losing generality, we assume that $0 < g < \infty$. The control objective is to force the system output y to follow a given bounded reference signal $y_r \in C^h$, $h \geq n$. The error vector $\mathbf{e}$ is defined as

$$\mathbf{e} = [e, \dot{e}, \cdots, e^{(n-1)}]^T \in R^n \tag{15}$$

with $e = y_r - x = y_r - y$.

If $f(\mathbf{x})$ and g are known and the system is free of external disturbance, the ideal controller can be designed as

$$u_{ideal} = \frac{1}{g(\mathbf{x})}\left[-f(\mathbf{x}) + y_r^{(n)} + \mathbf{k}_c^T \mathbf{e}\right] \tag{16}$$

where $\mathbf{k}_c = [k_n \ k_{n-1} \cdots k_1]^T$. Applying (16) to (14), we have the following error dynamics

$$e^{(n)} + k_1 e^{(n-1)} + \cdots + k_n e = 0. \tag{17}$$

If k_i, $i=1$, 2, ..., n are chosen so that all roots of the polynomial $H(s) \underline{\underline{\Delta}} s^n + k_1 s^{n-1} + \cdots + k_n$ lie strictly in the open left half of the complex plane, we have $\lim\limits_{t \to \infty} \lim e(t) = 0$ for any initial conditions. However, since the system dynamics may be unknown, the ideal feedback controller u_{ideal} in (16) cannot be implemented.

4 Design of IACHDNN

To solve this problem, a new indirect adaptive control scheme using Hopfield-based dynamic neural network (IACHDNN) for SISO nonlinear systems is proposed. Two Hopfield-based DNNs are used to estimate the uncertain continuous functions f and g, respectively. The indirect adaptive controller u_{id} takes the following form

$$u_{id} = \frac{1}{\hat{g}}\left[-\hat{f} + y_r^{(n)} + \mathbf{k}_c^T \mathbf{e} - u_c\right] \tag{18}$$

where $\hat{f}$ and $\hat{g}$ are the estimations of f and g, respectively; u_c is the compensation controller employed to compensate the effects of external disturbance and the approximation error introduced by the Hopfield-based DNN approximations (described later). Substituting (18) into (14) and using (16) yield

$$\begin{aligned}\dot{\mathbf{e}} &= \mathbf{A}_c \mathbf{e} - \mathbf{B}_c\left[(f - \hat{f}) + (g - \hat{g})u_{id}\right] + \mathbf{B}_c u_c - \mathbf{B}_c d \\ &= \mathbf{A}_c \mathbf{e} - \mathbf{B}_c(\tilde{f} + \tilde{g}u_{id}) + \mathbf{B}_c u_c - \mathbf{B}_c d\end{aligned} \tag{19}$$

$$\text{where } \mathbf{A}_c = \begin{bmatrix} 0 & 1 & 0 & \cdots & 0 \\ \vdots & \ddots & \ddots & \ddots & 0 \\ 0 & \cdots & & 0 & 1 \\ -k_n & -k_{n-1} & \cdots & \cdots & -k_1 \end{bmatrix} \in R^{n \times n} ; \ \mathbf{B}_c = \begin{bmatrix} 0 & 0 & 0 & 1 \end{bmatrix}^T \in R^n ;$$

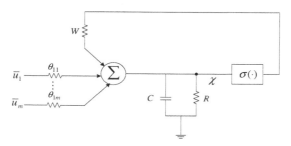

Fig. 1. Electric circuit of the Hopfield-based DNN containing only a single neuron

$\tilde{f} = f - \hat{f}$; $\tilde{g} = g - \hat{g}$. According to the discussion in Sec. 2.2, the Hopfield-based DNNs used to approximate f and g containing only a single neuron and can be expressed as

$$\hat{f} = \frac{1}{C_f}\left(\hat{W}_f \xi_{W_f} + \hat{\Theta}_f^T \xi_{\Theta_f} \right) + e^{-\frac{1}{R_f C_f}t} \hat{f}(0) - \frac{1}{C_f} e^{-\frac{1}{R_f C_f}t}\left[\hat{W}_f \xi_{W_f}(0) + \hat{\Theta}_f^T \xi_{\Theta_f}(0) \right] \quad (20)$$

and

$$\hat{g} = \frac{1}{C_g}\left(\hat{W}_g \xi_{W_g} + \hat{\Theta}_g^T \xi_{\Theta_g} \right) + e^{-\frac{1}{R_g C_g}t} \hat{g}(0) - \frac{1}{C_g} e^{-\frac{1}{R_g C_g}t}\left[\hat{W}_g \xi_{W_g}(0) + \hat{\Theta}_g^T \xi_{\Theta_g}(0) \right] \quad (21)$$

where $\hat{f}(0)$ and $\hat{g}(0)$ are the initial value of $\hat{f}$ and $\hat{g}$; the subscripts (and the secondary subscripts) f and g indicate the variables corresponding to the estimations $\hat{f}$ and $\hat{g}$ in this paper. Note that $\hat{W}_f$, $\hat{W}_g$, ξ_{W_f} , and ξ_{W_g} are scalars, and the input signals of the Hopfield-based DNNs are $\bar{u} = [x \ \dot{x} \cdots x^{(n-1)}]^T$. Fig. 1. shows the electric circuit of the Hopfield-based DNN containing only a single neuron. Substituting (20) and (21) into (19) yields

$$\dot{\mathbf{e}} = \mathbf{A}_c \mathbf{e} - \mathbf{B}_c (\tilde{f} + \tilde{g} u_{id}) + \mathbf{B}_c u_c - \mathbf{B}_c \varepsilon \quad (22)$$

where

$$\tilde{f} = \frac{1}{C_f}\tilde{W}_f\left[\xi_{W_f} - e^{-\frac{1}{R_f C_f}t}\xi_{W_f}(0) \right] + \frac{1}{C_f}\tilde{\Theta}_f^T\left[\xi_{\Theta_f} - e^{-\frac{1}{R_f C_f}t}\xi_{\Theta_f}(0) \right],$$

$$\bar{g} = \frac{1}{C_g}\tilde{W}_g\left[\xi_{W_g} - e^{-\frac{1}{R_g C_g}t}\xi_{W_g}(0)\right] + \frac{1}{C_g}\tilde{\Theta}_g^T\left[\xi_{\Theta_g} - e^{-\frac{1}{R_g C_g}t}\xi_{\Theta_g}(0)\right],$$

and

$\varepsilon = \Delta_f + \Delta_g u_{id} + d$, where Δ_f and Δ_g are the approximation errors defined as the same way in (11). In order to derive the main theorem in this paper, the following assumption and lemma is required.

Assumption: Assume that there exists a finite constant μ so that

$$\int_0^t \varepsilon^2 d\tau \le \mu, \ 0 \le t < \infty. \tag{23}$$

Lemma: Suppose $\mathbf{P} = \mathbf{P}^T > 0$ satisfies

$$\mathbf{A}_c^T\mathbf{P} + \mathbf{P}\mathbf{A}_c + \mathbf{Q} + \mathbf{P}\mathbf{B}_c(\frac{1}{\rho^2} - \frac{1}{\delta})\mathbf{B}_c^T\mathbf{P} = 0 \tag{24}$$

where $\mathbf{Q} = \mathbf{Q}^T > 0$; $\rho > 0$ and $\delta > 0$ satisfies $\frac{1}{\rho^2} - \frac{1}{\delta} \le 0$. Let $\hat{W}_f(0) \in \Omega_{W_f}$, $\hat{W}_g(0) \in \Omega_{W_g}$, $\hat{\Theta}_f(0) \in \Omega_{\Theta_f}$, and $\hat{\Theta}_g(0) \in \Omega_{\Theta_g}$, where $\hat{W}_f(0)$, $\hat{W}_g(0)$, $\hat{\Theta}_f(0)$ and $\hat{\Theta}_g(0)$ are the initial values of $\hat{W}_f$, $\hat{W}_g$, $\hat{\Theta}_f$, and $\hat{\Theta}_g$, respectively. For simplifying the mathematical expressions in the rest of the paper, here we define eight conditions as follows.

Condition A: $\left|\hat{W}_f\right| < M_{W_f}$ or $\left(\left|\hat{W}_f\right| = M_{W_f} \text{ and } \mathbf{e}^T\mathbf{PB}_c\hat{W}_f\left[\xi_{W_f} - e^{-\frac{1}{R_f C_f}t}\xi_{W_f}(0)\right] \ge 0\right)$.

Condition B: $\left|\hat{W}_f\right| = M_{W_f}$ and $\mathbf{e}^T\mathbf{PB}_c\hat{W}_f\left[\xi_{W_f} - e^{-\frac{1}{R_f C_f}t}\xi_{W_f}(0)\right] < 0$.

Condition C: $\left|\hat{W}_g\right| < M_{W_g}$ or $\left(\left|\hat{W}_g\right| = M_{W_g} \text{ and } \mathbf{e}^T\mathbf{PB}_c\hat{W}_g\left[\xi_{W_g} - e^{-\frac{1}{R_g C_g}t}\xi_{W_g}(0)\right] \ge 0\right)$.

Condition D: $\left|\hat{W}_g\right| = M_{W_g}$ and $\mathbf{e}^T\mathbf{PB}_c\hat{W}_g\left[\xi_{W_g} - e^{-\frac{1}{R_g C_g}t}\xi_{W_g}(0)\right] < 0$.

Condition E: $\left\|\hat{\Theta}_f\right\| < M_{\Theta_f}$ or $\left(\left\|\hat{\Theta}_f\right\| = M_{\Theta_f} \text{ and } \mathbf{e}^T\mathbf{PB}_c\hat{\Theta}_f\left[\xi_{\Theta_f} - e^{-\frac{1}{R_f C_f}t}\xi_{\Theta_f}(0)\right]u_{id} \ge 0\right)$.

Condition F: $\left\|\hat{\Theta}_f\right\| = M_{\Theta_f}$ and $\mathbf{e}^T\mathbf{PB}_c\hat{\Theta}_f\left[\xi_{\Theta_f} - e^{-\frac{1}{R_f C_f}t}\xi_{\Theta_f}(0)\right]u_{id} < 0$.

Condition G: $\left\|\hat{\Theta}_g\right\| < M_{\Theta_g}$ or $\left(\left\|\hat{\Theta}_g\right\| = M_{\Theta_g} \text{ and } \mathbf{e}^T\mathbf{PB}_c\hat{\Theta}_g\left[\xi_{\Theta_g} - e^{-\frac{1}{R_g C_g}t}\xi_{\Theta_g}(0)\right]u_{id} \ge 0\right)$.

Condition H: $\left\|\hat{\Theta}_g\right\| = M_{\Theta_g}$ and $\mathbf{e}^T \mathbf{PB}_c \hat{\Theta}_g \left[\xi_{\Theta_g} - e^{-\frac{1}{R_g C_g}t} \xi_{\Theta_g}(0) \right] u_{id} < 0$

If the adaptive laws are designed as

$$
\dot{\hat{W}}_f = -\dot{\tilde{W}}_f =
\begin{cases}
-\dfrac{\beta_{W_f}}{C_f} \mathbf{e}^T \mathbf{PB}_c \left[\xi_{W_f} - e^{-\frac{1}{R_f C_f}t} \xi_{W_f}(0) \right] & \text{if } \textit{Condition A} \\[3mm]
\mathbf{Pr}\left\{ \dfrac{\beta_{W_f}}{C_f} \mathbf{e}^T \mathbf{PB}_c \left[\xi_{W_f} - e^{-\frac{1}{R_f C_f}t} \xi_{W_f}(0) \right] \right\} & \text{if } \textit{Condition B}
\end{cases}
\tag{25}
$$

$$
\dot{\hat{W}}_g = -\dot{\tilde{W}}_g =
\begin{cases}
-\dfrac{\beta_{W_g}}{C_g} \mathbf{e}^T \mathbf{PB}_c \left[\xi_{W_g} - e^{-\frac{1}{R_g C_g}t} \xi_{W_g}(0) \right] & \text{if } \textit{Condition C} \\[3mm]
\mathbf{Pr}\left\{ \dfrac{\beta_{W_g}}{C_g} \mathbf{e}^T \mathbf{PB}_c \left[\xi_{W_g} - e^{-\frac{1}{R_g C_g}t} \xi_{W_g}(0) \right] \right\} & \text{if } \textit{Condition D}
\end{cases}
\tag{26}
$$

$$
\dot{\hat{\Theta}}_f = -\dot{\tilde{\Theta}}_f =
\begin{cases}
-\dfrac{\beta_{\Theta_f}}{C_f} \mathbf{e}^T \mathbf{PB}_c \left[\xi_{\Theta_f} - e^{-\frac{1}{R_f C_f}t} \xi_{\Theta_f}(0) \right] u_{id} & \text{if } \textit{Condition E} \\[3mm]
\mathbf{Pr}\left\{ \dfrac{\beta_{\Theta_f}}{C_f} \mathbf{e}^T \mathbf{PB}_c \left[\xi_{\Theta_f} - e^{-\frac{1}{R_f C_f}t} \xi_{\Theta_f}(0) \right] u_{id} \right\} & \text{if } \textit{Condition F}
\end{cases}
\tag{27}
$$

$$
\dot{\hat{\Theta}}_g = -\dot{\tilde{\Theta}}_g =
\begin{cases}
-\dfrac{\beta_{\Theta_g}}{C_g} \mathbf{e}^T \mathbf{PB}_c \left[\xi_{\Theta_g} - e^{-\frac{1}{R_g C_g}t} \xi_{\Theta_g}(0) \right] u_{id} & \text{if } \textit{Condition G} \\[3mm]
\mathbf{Pr}\left\{ \dfrac{\beta_{\Theta_g}}{C_g} \mathbf{e}^T \mathbf{PB}_c \left[\xi_{\Theta_g} - e^{-\frac{1}{R_g C_g}t} \xi_{\Theta_g}(0) \right] u_{id} \right\} & \text{if } \textit{Conditon H}
\end{cases}
\tag{28}
$$

where β_{W_f}, β_{W_g}, β_{Θ_f}, and β_{Θ_g} are positive learning rates; the projection operators $\mathbf{Pr}\{*\}$ are defined as

$$
\mathbf{Pr}\left\{ \frac{\beta_{W_f}}{C_f} \mathbf{e}^T \mathbf{PB}_c \left[\xi_{W_f} - e^{-\frac{1}{R_f C_f}t} \xi_{W_f}(0) \right] \right\} = \frac{\beta_{W_f}}{C_f} \mathbf{e}^T \mathbf{PB}_c \left\{ -\left[\xi_{W_f} - e^{-\frac{1}{R_f C_f}t} \xi_{W_f}(0) \right] + \frac{\hat{W}_f \left[\xi_{W_f} - e^{-\frac{1}{R_g C_g}t} \xi_{W_f}(0) \right]}{\left|\hat{W}_f\right|^2} \hat{W}_f \right\},
$$

$$
\mathbf{Pr}\left\{ \frac{\beta_{W_g}}{C_g} \mathbf{e}^T \mathbf{PB}_c \left[\xi_{W_g} - e^{-\frac{1}{R_g C_g}t} \xi_{W_g}(0) \right] \right\} = \frac{\beta_{W_g}}{C_g} \mathbf{e}^T \mathbf{PB}_c \left\{ -\left[\xi_{W_g} - e^{-\frac{1}{R_g C_g}t} \xi_{W_g}(0) \right] + \frac{\hat{W}_g \left[\xi_{W_g} - e^{-\frac{1}{R_g C_g}t} \xi_{W_g}(0) \right]}{\left|\hat{W}_g\right|^2} \hat{W}_g \right\},
$$

$$
\mathbf{Pr}\left\{ \frac{\beta_{\Theta_f}}{C_f} \mathbf{PB}_c \left[\xi_{\Theta_f} - e^{-\frac{1}{R_f C_f}t} \xi_{\Theta_f}(0) \right] u_{id} \right\} = \frac{\beta_{\Theta_f}}{C_f} \mathbf{e}^T \mathbf{PB}_c \left\{ -\left[\xi_{\Theta_f} - e^{-\frac{1}{R_f C_f}t} \xi_{\Theta_f}(0) \right] u_{id} + \frac{\hat{\Theta}_f^T \left[\xi_{\Theta_f} - e^{-\frac{1}{R_f C_f}t} \xi_{\Theta_f}(0) \right] u_{id}}{\left\|\hat{\Theta}_f\right\|^2} \hat{\Theta}_f \right\},
$$

$$\mathbf{Pr}\left\{\frac{\beta_{\Theta_g}}{C_g}\mathbf{PB}_c\left[\xi_{\Theta_g}-e^{-\frac{1}{R_gC_g}}\xi_{\Theta_g}(0)\right]u_{id}\right\}=\frac{\beta_{\Theta_g}}{C_g}\mathbf{e}^T\mathbf{PB}_c\left\{-\left[\xi_{\Theta_g}-e^{-\frac{1}{R_gC_g}}\xi_{\Theta_g}(0)\right]u_{id}+\frac{\hat{\Theta}_g^T\left[\xi_{\Theta_g}-e^{-\frac{1}{R_gC_g}}\xi_{\Theta_g}(0)\right]u_{id}}{\left\|\hat{\Theta}_g\right\|^2}\hat{\Theta}_g\right\},$$

then $\hat{W}_f$, $\hat{W}_g$, $\hat{\Theta}_f$ and $\hat{\Theta}_g$ are bounded by $\left|\hat{W}_f\right|\le M_{W_f}$, $\left|\hat{W}_g\right|\le M_{W_g}$, $\left\|\hat{\Theta}_f\right\|\le M_{\Theta_f}$, and $\left\|\hat{\Theta}_g\right\|\le M_{\Theta_g}$ for all $t\ge 0$ [10]. Now we are prepared to state the main theorem of this paper.

Theorem: Suppose the **Assumption** (23) holds. Consider the plant (14) with the control law (18). The function estimations $\hat{f}$ and $\hat{g}$ are given by (25) and (21) with the adaptive laws (25)-(28). The compensation controller u_s is given as

$$u_c=-\frac{1}{2\delta}\mathbf{B}_c^T\mathbf{Pe} \tag{29}$$

Then, the overall control scheme guarantees the following properties:

i) $\dfrac{1}{2}\displaystyle\int_0^t \mathbf{e}^T\mathbf{Qe}\,d\tau \le \dfrac{1}{2}\mathbf{e}(0)^T\mathbf{Pe}(0)+\dfrac{\tilde{W}_f(0)\dot{\tilde{W}}_f(0)}{2\beta_{W_f}}+\dfrac{\tilde{W}_g(0)\dot{\tilde{W}}_g(0)}{2\beta_{W_g}}+\dfrac{\tilde{\Theta}_f^T(0)\dot{\tilde{\Theta}}_f(0)}{2\beta_{\Theta_f}}$

$$+\dfrac{\tilde{\Theta}_g^T(0)\dot{\tilde{\Theta}}_g(0)}{2\beta_{\Theta_g}}+\dfrac{\rho^2}{2}\int_0^t \varepsilon^2\,d\tau \tag{30}$$

for $0\le t<\infty$, where $\mathbf{e}(0)$, $\tilde{W}_f(0)$, $\tilde{W}_g(0)$ and $\tilde{\Theta}_f(0)$, $\tilde{\Theta}_g(0)$ are the initial values of $\mathbf{e}$, $\tilde{W}_f$, $\tilde{W}_g$, $\tilde{\Theta}_f$, and $\tilde{\Theta}_g$, respectively.

ii) The tracking error $\|\mathbf{e}\|$ can be expressed in terms of the lumped uncertainty as

$$\|\mathbf{e}\|\le\sqrt{\frac{2V(0)+\rho^2\mu}{\lambda_{\min}(\mathbf{P})}} \tag{31}$$

where $V(0)$ is the initial value of a Lyapunov function candidate defined later and $\lambda_{\min}(\mathbf{P})$ is the minimum eigenvalue of $\mathbf{P}$.

Proof

i) Define a Lyapunov function candidate as

$$V=\frac{1}{2}\mathbf{e}^T\mathbf{Pe}+\frac{1}{2\beta_{W_f}}\tilde{W}_f^2+\frac{1}{2\beta_{W_f}}\tilde{W}_f^2+\frac{1}{2\beta_{\Theta_f}}\tilde{\Theta}_f^T\dot{\tilde{\Theta}}_f+\frac{1}{2\beta_{\Theta_g}}\tilde{\Theta}_g^T\dot{\tilde{\Theta}}_g \tag{32}$$

Differentiating (32) with respect to time and using (22) yield

$$\dot{V} = \frac{1}{2}\mathbf{e}^T\mathbf{P}\dot{\mathbf{e}} + \frac{1}{2}\dot{\mathbf{e}}^T\mathbf{P}\mathbf{e} + \frac{1}{\beta_{W_f}}\tilde{W}_f\dot{\tilde{W}}_f + \frac{1}{\beta_{W_g}}\tilde{W}_g\dot{\tilde{W}}_g + \frac{1}{\beta_{\Theta_f}}\tilde{\Theta}_f^T\dot{\tilde{\Theta}}_f + \frac{1}{\beta_{\Theta_g}}\tilde{\Theta}_g^T\dot{\tilde{\Theta}}_g$$

$$= \frac{1}{2}\mathbf{e}^T(\mathbf{A}_c^T\mathbf{P}+\mathbf{P}\mathbf{A}_c)\mathbf{e} - \mathbf{e}^T\mathbf{P}\mathbf{B}_c(\bar{f}+\bar{g}u_{id}) + \mathbf{e}^T\mathbf{P}\mathbf{B}_c u_c - \mathbf{e}^T\mathbf{P}\mathbf{B}_c\varepsilon + \frac{1}{\beta_{W_f}}\tilde{W}_f\dot{\tilde{W}}_f + \frac{1}{\beta_{W_g}}\tilde{W}_g\dot{\tilde{W}}_g$$

$$+ \frac{1}{\beta_{\Theta_f}}\tilde{\Theta}_f^T\dot{\tilde{\Theta}}_f + \frac{1}{\beta_{\Theta_g}}\tilde{\Theta}_g^T\dot{\tilde{\Theta}}_g$$

$$= \frac{1}{2}\mathbf{e}^T(\mathbf{A}_c^T\mathbf{P}+\mathbf{P}\mathbf{A}_c)\mathbf{e} + \mathbf{e}^T\mathbf{P}\mathbf{B}_c u_c - \mathbf{e}^T\mathbf{P}\mathbf{B}_c\varepsilon + V_{W_f} + V_{W_g} + V_{\Theta_f} + V_{\Theta_g} \qquad (33)$$

where

$$V_{W_f} = \tilde{W}_f\left\{-\frac{1}{C_f}\mathbf{e}^T\mathbf{P}\mathbf{B}_c\left[\xi_{W_f} - e^{-\frac{1}{R_f C_f}t}\xi_{W_f}(0)\right] + \frac{1}{\beta_{W_f}}\dot{\hat{W}}_f\right\},$$

$$V_{W_g} = \tilde{W}_g\left\{-\frac{1}{C_g}\mathbf{e}^T\mathbf{P}\mathbf{B}_c\left[\xi_{W_g} - e^{-\frac{1}{R_g C_g}t}\xi_{W_g}(0)\right] + \frac{1}{\beta_{W_g}}\dot{\hat{W}}_g\right\},$$

$$V_{\Theta_f} = \tilde{\Theta}_f^T\left\{-\frac{1}{C_f}\mathbf{e}^T\mathbf{P}\mathbf{B}_c\left[\xi_{\Theta_f} - e^{-\frac{1}{R_f C_f}t}\xi_{\Theta_f}(0)\right]u_{id} + \frac{1}{\beta_{\Theta_f}}\dot{\hat{\Theta}}_f\right\},$$

$$V_{\Theta_g} = \tilde{\Theta}_g^T\left\{-\frac{1}{C_g}\mathbf{e}^T\mathbf{P}\mathbf{B}_c\left[\xi_{\Theta_g} - e^{-\frac{1}{R_g C_g}t}\xi_{\Theta_g}(0)\right]u_{id} + \frac{1}{\beta_{\Theta_g}}\dot{\hat{\Theta}}_g\right\}.$$

Substituting (29) into (33) and using (24), we have

$$\dot{V} = \frac{1}{2}\mathbf{e}^T(\mathbf{A}_c^T\mathbf{P}+\mathbf{P}\mathbf{A}_c)\mathbf{e} - \frac{1}{2\delta}(\mathbf{e}^T\mathbf{P}\mathbf{B}_c)(\mathbf{B}_c\mathbf{P}\mathbf{e}) + \mathbf{e}^T\mathbf{P}\mathbf{B}_c\varepsilon + V_{W_f} + V_{W_g} + V_{\Theta_f} + V_{\Theta_g}$$

$$= \frac{1}{2}\mathbf{e}^T(\mathbf{A}_c^T\mathbf{P}+\mathbf{P}\mathbf{A}_c - \frac{1}{\delta}\mathbf{P}\mathbf{B}_c\mathbf{B}_c^T\mathbf{P})\mathbf{e} + \mathbf{e}^T\mathbf{P}\mathbf{B}_c\varepsilon + V_{W_f} + V_{W_g} + V_{\Theta_f} + V_{\Theta_g}$$

$$= \frac{1}{2}\mathbf{e}^T(-\mathbf{Q} - \frac{1}{\rho^2}\mathbf{P}\mathbf{B}_c\mathbf{B}_c^T\mathbf{P})\mathbf{e} + \mathbf{e}^T\mathbf{P}\mathbf{B}_c\varepsilon + V_{W_f} + V_{W_g} + V_{\Theta_f} + V_{\Theta_g}$$

$$= -\frac{1}{2}\mathbf{e}^T\mathbf{Q}\mathbf{e} - \frac{1}{2}\left[\frac{1}{\rho}\mathbf{B}_c^T\mathbf{P}\mathbf{e} - \rho\varepsilon\right]^2 + \frac{1}{2}\rho^2\varepsilon^2 + V_{W_f} + V_{W_g} + V_{\Theta_f} + V_{\Theta_g} \qquad (34)$$

Using (25), we have

$$V_w = \begin{cases} 0 & \text{if } Condition\ A \\ -\frac{1}{C_f}\mathbf{e}^T\mathbf{P}\mathbf{B}_c\dfrac{\hat{W}_f\left[\xi_{W_f} - e^{-\frac{1}{R_f C_f}t}\xi_{W_f}(0)\right]}{\left|\hat{W}_f\right|^2}\tilde{W}_f\hat{W}_f & \text{if } Condition\ B \end{cases} \qquad (35)$$

For the second condition in (35), we have $\left|\hat{W}_f\right| = M_{W_f} \ge \left|W_f^*\right|$ because W_f^* belongs to the constraint set Ω_{W_f}. Using this fact, we obtain $\tilde{W}_f \hat{W}_f = \frac{1}{2}(W_f^{*2} - \hat{W}_f^{\,2} - \tilde{W}_f^{\,2}) \le 0$. Thus, the second line of (35) can be rewritten as

$$V_w = -\frac{1}{2C_f}\mathbf{e}^T\mathbf{Pb}\frac{\hat{W}_f^T\left[\xi_{W_f} - e^{-\frac{1}{R_f C_f}t}\xi_{W_f}(0)\right]}{\left|\hat{W}_f\right|^2}(\left|W_f^*\right|^2 - \left|\hat{W}_f\right|^2 - \left|\tilde{W}_f\right|^2) \le 0. \quad (36)$$

Thus, in (35), we obtain $V_{W_f} \le 0$. Similarly, we can also obtain that $V_{W_g} \le 0$, $V_{\Theta_f} \le 0$, and $V_{\Theta_g} \le 0$. Using this knowledge, we can further rewrite (34) as

$$\dot{V} \le -\frac{1}{2}\mathbf{e}^T\mathbf{Q}\mathbf{e} + \frac{1}{2}\rho^2\varepsilon^2 \quad (37)$$

Integrating both sides of the inequality (37) yields

$$V(t) - V(0) \le -\frac{1}{2}\int_0^t \mathbf{e}^T\mathbf{Q}\mathbf{e}\,d\tau + \frac{\rho^2}{2}\int_0^t \varepsilon^2\,dt \quad (38)$$

for $0 \le t < \infty$. Since $V(t) \ge 0$, we obtain

$$\frac{1}{2}\int_0^t \mathbf{e}^T\mathbf{Q}\mathbf{e}\,d\tau \le V_0 + \frac{\rho^2}{2}\int_0^t \varepsilon^2\,dt. \quad (39)$$

Substituting (32) into (39), we obtain (30).

ii) From (37) and since $\int_0^t \mathbf{e}^T\mathbf{Q}\mathbf{e}\,dt \ge 0$, we have

$$2V(t) \le 2V(0) + \rho^2\mu, \quad 0 \le t < \infty \quad (40)$$

From (32), it is obvious that $\mathbf{e}^T\mathbf{P}\mathbf{e} \le 2V$, for $t \ge 0$. Because $\mathbf{P} = \mathbf{P}^T \ge 0$, we have

$$\lambda_{\min}(\mathbf{P})\|\mathbf{e}\|^2 = \lambda_{\min}(\mathbf{P})\mathbf{e}^T\mathbf{e} \le \mathbf{e}^T\mathbf{P}\mathbf{e} \quad (41)$$

Thus, we obtain

$$\lambda_{\min}(\mathbf{P})\|\mathbf{e}\|^2 \le \mathbf{e}^T\mathbf{P}\mathbf{e} \le 2V(t) \le 2V(0) + \rho^2\mu \quad (42)$$

from (39)-(40). Therefore, from (42), we can easily obtain (31), which explicitly describe the bound of tracking error $\|\mathbf{e}\|$. If initial state $V(0) = 0$, tracking error $\|\mathbf{e}\|$ can be made arbitrarily small by choosing adequate ρ. Equation (31) is very crucial to show that the proposed IACHDNN will provide the closed-loop stability rigorously in the Lyapunov sense under the **Assumption** (23).   **Q. E. D.**

The block diagram of IACHDNN is shown in Fig. 2.

Remark: Equation (31) describes the relations among $\|\mathbf{e}\|$, ρ, and $\lambda_{\min}(\mathbf{P})$. To get more insight of (31), we first choose $\rho^2 = \delta$ in (24) to simplify the analysis. Thus, from (24), we can see that $\lambda_{\min}(\mathbf{P})$ is fully affected by the choice of $\lambda_{\min}(\mathbf{Q})$ in the way that a larger $\lambda_{\min}(\mathbf{Q})$ leads to a larger $\lambda_{\min}(\mathbf{P})$, and vice versa. Now, one can easily observe form (31) that the norm of tracking error can be attenuated to any desired small level by choosing ρ and $\lambda_{\min}(\mathbf{Q})$ as small as possible. However, this may lead to a large control signal which is usually undesirable in practical systems.

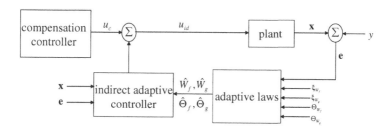

Fig. 2. The Block diagram of IACHDNN

5   Simulation Results

Example: Consider an inverted pendulum system. Let x_1 (rad) be the angle of the pendulum with respect to the vertical line. The dynamic equations of the inverted pendulum are [10]

$$\dot{x}_1 = x_2$$

$$\dot{x}_2 = \frac{g_v \sin x_1 - \dfrac{mlx_2^2 \cos x_1 \sin x_1}{m_c + m}}{l\left(\dfrac{4}{3} - \dfrac{m\cos^2 x_1}{m_c + m}\right)} + \frac{\dfrac{\cos x_1}{m_c + m}}{l\left(\dfrac{4}{3} - \dfrac{m\cos^2 x_1}{m_c + m}\right)} u + d \tag{43}$$

where $g_v = 9.8 \text{ m/s}^2$ is the acceleration due to gravity; m_c is the mass of the cart; m is the mass of the pole; l is the half-length of the pole; u is the applied force (control) and d is the external disturbance. The reference signal here is $y_r = (\pi/30)\sin t$, and d is a square wave with the amplitude ± 0.05 and period 2π. Also we choose $m_c = 1$, $m = 0.5$, and $l = 0.5$. The initial states are $\begin{bmatrix} x_1(0) & x_2(0) \end{bmatrix}^T = [0.2 \ 0.2]^T$. The learning rates of weights adaption are selected as $\beta_{w_f} = \beta_{\Theta_f} = 0.2$ and $\beta_{w_g} = \beta_{\Theta_g} = 0.005$; the slope of tanh($\cdot$) at the origin are selected as $\kappa_f = k_g = 1$ and $\delta = 0.1$ for the compensation controller. The resistance and capacitance are chosen as $R_f = R_g = 10\Omega$ and $C_f = C_g = 0.001F$. For a choice of $\mathbf{Q} = 15\mathbf{I}$, $\mathbf{k}_c = \begin{bmatrix} 2 & 1 \end{bmatrix}^T$, we

solve the Riccati-like equation (24) and obtain $\mathbf{P} = \begin{bmatrix} 22.5 & 7.5 \\ 7.5 & 7.5 \end{bmatrix}$. The simulation

results for are shown in Figs. 3, where the tracking responses of state x_1 and x_2 are shown in Figs. 3(a) and 3(b), respectively, the associated control inputs are shown Fig. 3(c). From the simulation results, we can see that the proposed IACHDNN can achieve favorable tracking performances with external disturbance. This fact shows the strong disturbance-tolerance ability of the proposed system.

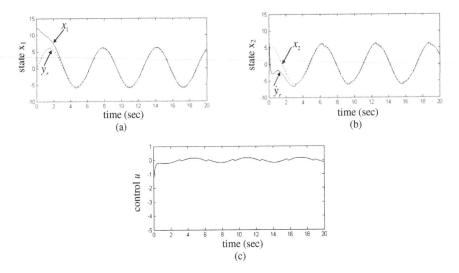

Fig. 3. Simulation results

6   Conclusions

An indirect adaptive control scheme using Hopfield-based dynamic neural networks for SISO nonlinear systems is proposed in this paper. The simple Hopfield-based DNNs are used to approximate the uncertain parameters of the controlled plant and the synaptic weights Hopfield-based DNNs are tuned on-line by the adaptive laws. A compensation controller is merged into control law to compensate the effect of modeling error and external disturbance. By the Lyapunov stability analysis, we prove that the closed-loop system is stable, and the tracking error can be attenuated to a desired level. Note that no strong assumptions and prior knowledge of the controlled plant are needed in the development of IACHDNN. Simulation results demonstrate the effectiveness and robustness of the proposed IACHDNN in the presence of external disturbance. The parsimonious structure of the Hopfield-based DNN (only one neuron is contained) and the simple Hopfield circuit make the IACHDNN much easier to implement and more reliable in practical purposes.

References

1. Wang, C.H., Lin, T.C., Lee, T.T., Liu, H.L.: Adaptive hybrid Intelligent control for uncertain nonlinear dynamical systems. IEEE Trans. Syst., Man, Cybern. B 5(32), 583–597
2. Li, Y., Qiang, S., Zhang, X., Kaynak, O.: Robust and adaptive backstepping control for nonlinear systems using RBF neural networks. IEEE Trans. Neural Networks 15, 693–701 (2004)
3. Lin, C.T., George Lee, C.S.: Neural fuzzy systems: a neuro-fuzzy synergism to intelligent systems. Prentice-Hall, Englewood Cliffs (1996)
4. Yu, D.L., Chang, T.K.: Adaptation of diagonal recurrent neural network model. Neural Comput. & Applic. 14, 189–197 (2005)
5. Poznyak, A.S., Yu, W., Sanchez, D.N., Perez, J.P.: Nonlinear adaptive trajectory tracking using dynamic neural networks. IEEE Trans. Neural Networks 10, 1402–1411 (1999)
6. Chow, T.W.S., Li, X.D., Fang, Y.: A real-time learning control approach for nonlinear continuous-time system using recurrent neural networks. IEEE Trans. Ind. Electronics. 47, 478–486 (2000)
7. Ren, X.M., Rad, A.B., Chan, P.T., Lo, W.L.: Identification and control of continuous-time nonlinear systems via dynamic neural networks. IEEE Trans. Ind. Electronics 50, 478–486 (2003)
8. Hopfield, J.J.: Neural Networks and Physical Systems with Emergent Collective Computational Abilities. Proceedings of National Academy of sciences, USA 79, 2554–2558 (1982)
9. Hopfield, J.J.: Neurons with graded response have collective computational properties like those of two-state neurons. Proceedings of National Academy of sciences, USA 81, 3088–3092 (1984)
10. Wang, L.X.: Adaptive Fuzzy Systems and Control - Design and Stability Analysis. Prentice-Hall, Englewood Cliffs (1994)

A Neural Network Computational Model of Visual Selective Attention

Kleanthis C. Neokleous[1], Marios N. Avraamides[2], Costas K. Neocleous[3], and Christos N. Schizas[1]

[1] Department of Computer Science
[2] Department of Psychology, University of Cyprus,
75 Kallipoleos, 1678, POBox 20537, Nicosia, Cyprus
[3] Department of Mechanical Engineering, Cyprus University of Technology,
Lemesos, Cyprus

Abstract. One challenging application for Artificial Neural Networks (ANN) would be to try and actually mimic the behaviour of the system that has inspired their creation as computational algorithms. That is to use ANN in order to simulate important brain functions. In this report we attempt to do so, by proposing a Neural Network computational model for simulating visual selective attention, which is a specific aspect of human attention. The internal operation of the model is based on recent neurophysiologic evidence emphasizing the importance of neural synchronization between different areas of the brain. Synchronization of neuronal activity has been shown to be involved in several fundamental functions in the brain especially in attention. We investigate this theory by applying in the model a correlation control module comprised by basic integrate and fire model neurons combined with coincidence detector neurons. Thus providing the ability to the model to capture the correlation between spike trains originating from endogenous or internal goals and spike trains generated by the saliency of a stimulus such as in tasks that involve top – down attention [1]. The theoretical structure of this model is based on the temporal correlation of neural activity as initially proposed by Niebur and Koch [9]. More specifically; visual stimuli are represented by the rate and temporal coding of spiking neurons. The rate is mainly based on the saliency of each stimuli (i.e. brightness intensity etc.) while the temporal correlation of neural activity plays a critical role in a later stage of processing were neural activity passes through the correlation control system and based on the correlation, the corresponding neural activity is either enhanced or suppressed. In this way, attended stimulus will cause an increase in the synchronization as well as additional reinforcement of the corresponding neural activity and therefore it will "win" a place in working memory. We have successfully tested the model by simulating behavioural data from the "attentional blink" paradigm [11].

Keywords: Neural Network, coincidence detector neurons, visual selective attention.

1 Introduction

Due to the great number of sensory stimuli that a person experiences at any given point of conscious life, it is practically impossible to integrate available information

D. Palmer-Brown et al. (Eds.): EANN 2009, CCIS 43, pp. 350–358, 2009.

into a single perceptual event. This implies that a selective mechanism must be present in the brain to effectively focus its resources on specific stimuli; otherwise we would have been in constant distraction by irrelevant information. Attention can be guided by top-down or via bottom-up processing as cognition can be regarded as a balance between internal motivations and external stimulations. Volitional shifts of attention or endogenous attention results from "top-down" signals originating in the prefrontal cortex while exogenous attention is guided by salient stimuli from "bottom-up" signals in the visual cortex [1]. In this paper we emphasize and try to simulate the behaviour of selective attention, especially in top-down tasks, mostly based on the theoretical background behind neural mechanisms of attention as it is explained in the field of neuroscience.

The underlying mechanisms of the neuronal basis of attention are supported by two main hypotheses. The first is known as "biased competition" [8] and it originated from studies with single-cell recordings. These studies have shown that attention enhances the firing rates of the neurons that represent the attended stimuli and suppresses the firing rates of the neurons encoding unattended stimuli. The second more recent hypothesis, places emphasis on the synchronization of neural activity during the process of attention. The second hypothesis stems from experiments showing that neurons selected by the attention mechanism have enhanced gamma-frequency synchronization [14, 3]. More specifically, Fries et al [3] measured activity in area V4 of the brain of macaque monkeys while they were attending behaviorally relevant stimuli and observed increased gamma frequency synchronization of attended stimuli compared to the activity elicited by distractors.

The proposed computational model for endogenous and exogenous visual attention is based on the second hypothesis for the neural mechanisms behind attention. The basic functionality of the model is based on the assumption that the incoming visual stimulus will be manipulated by the model based on its rate and temporal coding. The rate of the visual stimuli will have important role in the case of exogenous attention since this type of attention is mainly affected by the different features of the visual stimuli. More salient stimuli will have an advantage to pass in a further stage of processing and finally to access working memory. On the other hand, endogenous or top-down attention is mainly affected by the synchronization of the corresponding neural activity that represents the incoming stimuli with the neural activity initiated by internal goals that are setup when the individual is requested to carry out a specific task. These goals are possibly maintained in the prefrontal cortex of the brain. The direct connection of top-down attention with synchronization is supported by many recent studies [9, 5]. For example, Saalmann et al [12] recorded neural activity simultaneously from the posterior parietal cortex and an earlier area in the visual pathway of the brain of macaques while they were performing a visual matching task. Their findings revealed that there was synchronization of the timing activities of the two regions when the monkeys selectively attended to a location. Thus, it seems that parietal neurons which presumably represent neural activity of the endogenous goals may selectively increase activity in earlier sensory areas. Additionally, the adaptive resonance theory by Grossberg [4] implies that temporal patterning of activities could be ideally suited to achieve matching of top–down predictions with bottom–up inputs, while Engel et al [2] in their review (p.714) have noted that "If top–down effects induce a particular pattern of subthreshold fluctuations in dendrites of the target population, these could be 'compared' with temporal patterns arising from peripheral input".

2 Proposed Computational Model of Visual Selective Attention

Based on the above theories for visual selective attention, we suggest that a correlation control module responsible for comparing temporal patterns arising from top-down information and spike trains initiated by the characteristics of each incoming stimuli could be applied in the computational model of visual selective attention. If we extend this assumption based on relevant anatomical areas of the brain then the possible existence of such a correlation control module, would more ideally fit somewhere in the area V4 of the visual cortex where synchronization of neural activity has mostly been observed as can be seen in figure 1 below.

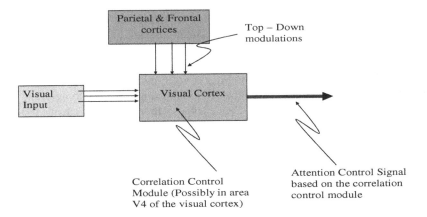

Fig. 1. Neural activity that corresponds to a specific visual input propagates along the visual cortex and initially to area V1. From there, the corresponding neural activity continues into the visual hierarch and specifically to area V4. Additionally, top-down signals originate from "higher" brain areas such as parietal and frontal lobe where possibly interact with the neural activity from the incoming stimuli. Correlation between these two streams of information could be examined in area V4 of the visual cortex.

The schematic representation of the proposed computational model is depicted in Figure 2. Specifically, each stimulus that enters the visual field is represented by a stream of binary events. Part of the stream represents no action potential occurrence ('zeros') and an isolated '1' that represents an action potential or spike. These binary waveforms are generated in order to represent the different spike trains initiated by each incoming stimulus. However, two important factors define the generation of these spike trains. The first is the firing rate or the frequency of spikes which is mainly based on the saliency of each stimulus, and the second factor is the exact timing that each spike appears. This means that in the "race" between the different visual stimuli to access working memory, both of these characteristics will contribute [10]. The model can be seen as a two stage model where in the first stage, spike-trains representing each incoming stimulus enters into a network comprised by integrate and fire neurons. As a result, the corresponding neural activity will propagate along the network with the task to access a working memory node. Based on the firing rate of each incoming stimulus, a different neural activation will reach the working memory node and if the corresponding neural activity is strong enough to cause the working memory node to fire, then what can be inferred is that the specific stimulus that caused this activation has accessed working memory and thus it has been attended. However, in a later stage of processing, top- down signals coming from parietal and frontal lobes enter the network and try to influence the selection based on internal goals. For example, suppose that a person is asked to identify and respond if the letter A appears in the visual field. Then, information represented by spike trains that encode how letter A is stored in long term memory will enter the network as top – down signals. As a result, if a visual stimulus enters the visual field and has strong correlations with the corresponding top-down information, it will be aided in its attempt to access working memory.

The interaction between top-down information and the neural activity generated by each incoming stimulus is performed in the correlation control module which is the major component of the model (Figure 2).

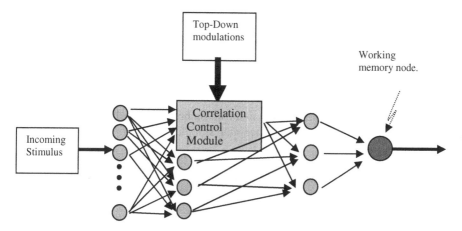

Fig. 2. A schematic representation of the proposed computational mode of visual selective attention

One possible explanation of the mechanism behind the correlation control theory proposed in this report can be made based on coincidence detector neurons.

3   Coincidence Detector Neurons and the Correlation Control Module

Coincidence detection is a very simplified model of neuron, however there is considerable experimental evidence signifying that under certain conditions, such as high background synaptic activity, neurons can function as coincidence detectors [10, 7].

Specifically as far as the neurophysiology of vision is concern, the main neurons found in several layers of the visual cortex are the **Pyramidal cells**. More importantly though is a recent theory about the function of pyramidal neurons which implies that the neurons responds best to coincident activation of multiple dendritic compartments. An interesting review about coincidence detection in pyramidal neurons is presented by Spruston [13].

A plausible way to model coincidence detection can be based on separate inputs converging on a common target. For example let consider a basic neural circuit of two input neurons with excitatory synaptic terminals, A and B converging on a single output neuron, C (Figure 3). The output target neuron C will only fire if the two input neurons fire synchronously. Thus it can be inferred that a coincidence detector is a neuron model that can detect synchronization of pulses from distinct connections.

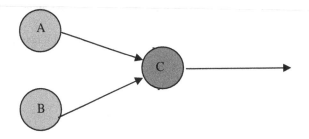

Fig. 3. A coincidence detector neuron C will fire only if the two input neurons A and B fire synchronously

In Figure 4 a simple representation is shown on how the correlation control module adjusts the neural activation of each incoming stimulus.

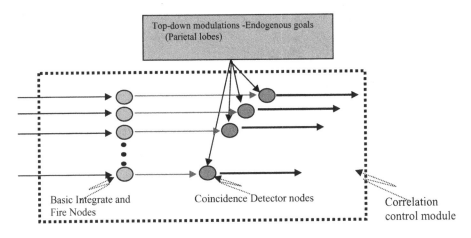

Fig. 4. Correlation control mechanism between the endogenous goals and the incoming stimuli

4 Simulations and Evaluation of the Model

In order to examine the models validity we have attempted to reproduce behavioral data from a famous attention related task named the attentional blink which it is explained with more detail in the next section.

4.1 Attentional Blink Explanation – Theory

Attentional blink (AB) is a phenomenon observed in the rapid serial visual presentation (RSVP) paradigm and refers to the finding that when 2 targets are presented among distractors in the same spatial location, correct identification of the 1st target, usually results in a deficit for identification of a 2nd target if it appears within a brief temporal window of 200-500 ms. When the 2nd target appears before or after this time window it is identified normally (Figure 5.b). More specifically, in the original experiment [11], participants were requested to identify two letter targets T1 and T2 among digit distractors while every visual stimulus appeared for about 100ms as shown in Figure 5.a.

Another important finding from the AB paradigm is that if T1 is not followed by a mask (distractor), the AB impairment is significantly reduced. That is if lag 2 (t=100ms) and/or lag 3 (t=300ms) are replaced by a blank then the AB curve takes the form shown in Figure 5 by series 2 and 3.

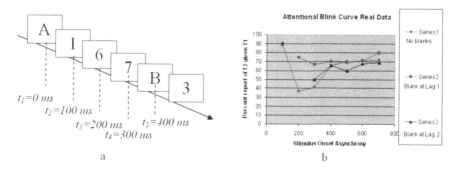

Fig. 5. Presentation of the RSVP for the "attentional blink" experiment (Figure5.a) and the attentional blink basic curve with no blanks (series 1), with blank at lag 1 (series 2) and blank at lag 2 (series 3) based on the behavioral data of Raymond and Sapiro (1992) (Figure5.b)

One possible explanation for the classic U-shaped curve of Figure 5.b (red series) is based on Electroencephalography (EEG) measurements and more importantly on two attention related Event Related Potentials (ERPs). The first ERPs appear at about 180-240 ms post-stimulus and are referred to as the P2/N2 signals. These signals have been proposed as control signals for the movement of attention [6, 15]. The second component is the P300 signal at about 350–600 ms post-stimulus which is associated with the working memory sensory buffer site and is taken to be the signal of the availability for report. Therefore, the explanation for the U- shaped curve lies in the assumption that the P300 signal generated by the first target falls into the time window that the P2/N2 component of the second target was about to be generated. However due to this interaction, the P2/N2 component of the second target is inhibited.

The explanation behind the curves of figure 5.b (with blank at lag 1 (green series) and blank at lag 2 (black series)) is based on the neural mechanisms behind selection at attentional tasks. Mostly, is based in the competition process between various stimuli in order to access working memory. This competition is reflected through relevant inhibition between the neural activities that corresponds to each stimulus.

The proposed computational model has been implemented in the Matlab-Simulink environment. Each of the visual stimulus has been represented by a 10 ms sequence of spikes and in each ms there is a one (spike) or a zero (no-spike) as seen in Figure 6. For coding both the distractors and the targets, the same firing rate has been used since both (targets and distractors) have the same effect from the salience filters (same brightness, intensity etc.). However, the difference between the spike trains generated by the targets and the spike trains generated by distractors is in the temporal patterns. Therefore, it is possible through the coincidence detector module to capture the correlation between the spike trains generated by the targets and spike trains initiated by internal goals if those two sources have similar temporal patterns in their spike trains. Based on the degree of correlation between the incoming stimulus and the internal goals, a relevant control signal is generated that could be associated with the N2/P2 component explained in the previous section. Additionally, once a specific working memory node that corresponds to a specific stimulus fires, then another signal is generated that inhibits at that timing any attempt for the coincidence control module to generate a new control signal.

As a consequence, three important features of the model that rely on neurophysiologic evidence have given the ability to reproduce the behavioural data from the attentional blink experiment as shown in Figure 7 below. These important features of

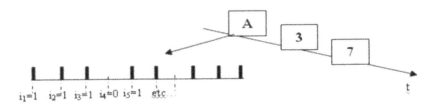

Fig. 6. Coding of the incoming visual stimuli

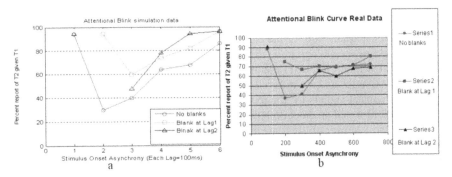

Fig. 7. Comparison between simulation data (7.a) and experimental data (7.b)

the model are: a) The correlation control module that generates a control signal relevant to the degree of correlation b) the interaction between the signals related to identification and response (P300) with the control signal and c) the competitive inhibition between each incoming stimuli.

5 Discussion

The main advantages of the implementation of a computational model of specific brain functions can be seen in a twofold manner. First, a biologically plausible model will give the ability to perform appropriate and detailed simulations in order to study the most important aspects of the specific brain function as well as to magnify or weaken related theories. On the other hand, the detailed study of the psychological and neurophsiological approach will aim into an improved understanding of the specific functionality in the brain. This, combined with knowledge from computer science, will provide the potentials to advance in neurally inspired computing and information processing. Robots and other engineered systems that mimic biological capabilities as well as brain-computer interfaces are some of the potential applications that can be benefit and improved by this knowledge.

Therefore, in terms of the first issue mentioned in the previous paragraph, based on the results that we got in our attempt to simulate behaviour data from the attentional blink experiment, we emphasize the importance of temporal coding and as a consequence, temporal coincidence, as an important mechanism for regulating information through the brain. Thus, we can propose that this way of controlling streams of information as well as transferring information between different modules could be applied in control system engineering.

References

1. Corbetta, M., Shulman, G.L.: Control of goal-directed and stimulus-driven attention in the brain. Nature R. Neuroscience 3, 201–215 (2002)
2. Engel, A.K., Fries, P., Singer, W.: Dynamic predictions: Oscillations and synchrony in top–down processing. Nature 2, 704–716 (2001)
3. Fries, P., Reynolds, J.H., Rorie, A.E., Desimone, R.: Modulation of oscillatory neuronal synchronization by selective visual attention. Science 291, 1560–1563 (2001)
4. Grossberg, S.: The link between brain learning, attention, and consciousness. Conscious. Cogn. 8, 1–44 (1999)
5. Gross, J., Schmitz, F., Schnitzler, I., et al.: Modulation of long-range neural synchrony - reflects temporal limitations of visual attention in humans. PNAS 101(35), 13050–13055 (2004)
6. Hopf, J.-M., Luck, S.J., Girelli, M., Hagner, T., Mangun, G.R., Scheich, H., Heinze, H.-J.: Neural sources of focused attention in visual Search. Cereb. Cortex 10, 1233–1241 (2000)
7. Kempter, R., Gerstner, W., van Hemmen, J.: How the threshold of a neuron determines its capacity for coincidence detection. Biosystems 48(1-3), 105–112 (1998)
8. Moran, J., Desimone, R.: Selective attention gates visual processing in the extrastriate cortex. Science 229, 782–784 (1985)
9. Niebur, E., Hsiao, S.S., Johnson, K.O.: Synchrony: a neuronal mechanism for attentional selection? Cur.Op. in Neurobio. 12, 190–194 (2002)

10. Niebur, E., Koch, C.: A Model for the Neuronal Implementation of Selective Visual Attention Based on Temporal Correlation Among Neurons. Journal of Computational Neuroseience 1, 141–158 (1994)
11. Raymond, J.E., Shapiro, K.L., Arnell, K.M.: Temporary suppression of visual processing in an RSVP task: an attentional blink? J. of exp. psyc. Human perc., and performance 18(3), 849–860 (1992)
12. Saalmann, Y.B., Pigarev, I.N., et al.: Neural Mechanisms of Visual Attention: How Top-Down Feedback Highlights Relevant Locations. Science 316, 1612 (2007)
13. Spruston, N.: Pyramidal neurons: dendritic structure and synaptic integration. Nature Reviews Neuroscience 9, 206–221 (2008)
14. Steinmetz, P.N., Roy, A., et al.: Attention modulates synchronized neuronal firing in primate somatosensory Cortex. Nature 404, 187–190 (2000)
15. Taylor, J.G., Rogers, M.: A control model of the movement of attention. Neural Networks 15, 309–326 (2002)

Simulation of Large Spiking Neural Networks on Distributed Architectures, The "DAMNED" Simulator

Anthony Mouraud and Didier Puzenat

GRIMAAG laboratory, Université Antilles-Guyane, Pointe-à-Pitre, Guadeloupe
{amouraud,dpuzenat}@univ-ag.fr

Abstract. This paper presents a spiking neural network simulator suitable for biologically plausible large neural networks, named DAMNED for "Distributed And Multi-threaded Neural Event-Driven". The simulator is designed to run efficiently on a variety of hardware. DAMNED makes use of multi-threaded programming and non-blocking communications in order to optimize communications and computations overlap. This paper details the even-driven architecture of the simulator. Some original contributions are presented, such as the handling of a distributed virtual clock and an efficient circular event queue taking into account spike propagation delays. DAMNED is evaluated on a cluster of computers for networks from 10^3 to 10^5 neurons. Simulation and network creation speedups are presented. Finally, scalability is discussed regarding number of processors, network size and activity of the simulated NN.

Introduction

Spiking neurons models open new possibilities for the Artificial Neural Networks (ANN) community, especially modelling biological architectures and dynamics with an accurate behaviour [6]. The huge computation cost of some individual neuron models [7, 9] and the interest of simulating large networks (regarding non-spiking ANN) make Spiking Neural Networks (SNN) good candidates for parallel computing [3]. However, if distributed programming has become easier, reaching good accelerations implies a huge work and a specific knowledge. Therefore we present a distributed simulator dedicated to SNN, with a minimal need of (sequential) programming to deal with a new project, eventually no programming at all. Our simulator is named **DAMNED** for **Distributed And Multi-threaded Neural Event-Driven** simulator. It is designed to run efficiently on various architectures as shown section 1, using an "event-driven" approach detailed in section 2. Sections 3 and 4 present two optimisations, optimized queues and a distributed approach of virtual time handling. Section 5 presents the creation of a SNN within DAMNED and section 6 analyses the scalability of DAMNED. Finally section 7 presents some future works.

D. Palmer-Brown et al. (Eds.): EANN 2009, CCIS 43, pp. 359–370, 2009.

1   MIMD-DM Architectures

Our goal is to build large SNN, so the parallel architecture must be scalable in computational power. Large SNN need large amount of memory so the architecture must be scalable in memory which is not possible with shared memory architectures. Thus, our simulator is optimized to run on MIMD-DM architectures. MIMD stands for *Multiple Instruction stream, Multiple Data stream* using Flynn's taxonomy [4], and means several processors working potentially independently; DM stands for *Distributed Memory* which means each processor accesses a private memory, thus communications between processors are necessary to access information located in non-local memory. Another advantage of using MIMD-DM architecture is the diversity of production hardware: from the most powerful parallel computers to the cheapest workstation clusters. The drawback of MIMD-DM architecture is the complexity of the program design; the need of a specific knowledge (message passing instead of well known shared variable programming); the complexity of the code (not to mention debugging and validation). However since DAMNED is end-user oriented, the distributed programming work is already done and validated. The user has no need to enter the parallel code or even to understand the design of the simulator.

Development has been done in C++ using the **MPI 2.0** library [5]. MPI stands for *Message Passing Interface* and provides methods to handle communications (blocking, non-blocking, and collectives) and synchronization. MPI can launch several tasks (the name of a process in MPI) on a single processor thanks to the scheduler of the host operating system. In such a case communications will not involve the physical network when occurring between two tasks hosted by the same processor. As a consequence, DAMNED runs out of the box on a SMP architecture or on a hyper-threaded architecture. Thereby DAMNED has been developed and tested on a single SMP dual-core ultra-portable laptop computer and validated on a MIMD-DM cluster of 35 dual-core workstations.

The DAMNED simulator is available free of charge; most implementations of MPI are free; the most suitable host operating system is Linux; and the target architecture can be a non-dedicated PC cluster with a standard ethernet network. As a consequence installing and running DAMNED potentially only cost electricity if it is used at time machines were used to be swhiched off.

2   Architecture of DAMNED

Simulating SNN implies dealing with time [1]. The simulator can divide the biological time in discrete steps and scan all the neurons and synapses of the network at each time step, this method is named "clock-driven". Our simulator uses another strategy named "event-driven" [17]. An event is a spike emission from pre-synaptic neurons towards post-synaptic neurons, and each event is stamped with the spike emission date. Thus, DAMNED is based on an infinite loop processing events. Each time an event is processed, the simulator (i) actualizes the state of the impacted neuron which eventually generates new events for post-synaptic neurons; and (ii) increases its virtual clock to the event time stamp. The event-driven strategy is suitable for spiking neuron simulation [12]. Indeed biological spike flows are generally irregular in time with a low average activity. Such irregular activities imply high and low activity periods. While a high activity period benefits to the clock-driven approach, all low activity periods advantage the event-driven approach. Furthermore, a clock-driven simulation looses the order of emission of spikes emitted at the same time step which can change the behaviour of the simulation [15]. In an event-driven simulation, temporal precision only depends on the precision of the variable used for time stamps and clocks, and even a simple 16 bits variable gives a suitable precision.

However, an event-driven simulation does have some drawbacks. The state of neurons is only actualized when events are processed, on reception of a spike on a target neuron, then this neuron decides to spike or not. However when the behaviour of the neuron is described by several differential equations and/or when synaptic impacts are not instantaneous. It is possible to make a prediction of the spike emission date [10]. Such a prediction implies heavy computational costs, and a mechanism must control *a posteriori* the correctness of the prediction.

To efficiently run variety of experiments, DAMNED relies on a single "**front-end task**" (typically running on a workstation) and on several "**simulation tasks**" (typically running on a cluster). The front-end sends inputs into the simulated spiking network and eventually receives outputs. The input can be a flow of stimuli, for example matching a biological experiment. In such a case the evolution of neurons can be monitored to compare the behaviour of the simulated network with biological recordings. The input can also be a physical device such as a camera, in such an example the output could be used to control a motor moving the camera, eventually producing new inputs. More generally, the front-end can host a virtual environment producing inputs for the SNN and eventually being modified by the output of the SNN.

The simulation tasks run the heart of the simulator, that is the processing of events; each task holds a part of the whole neural network. In a simulation of N neurons with P tasks, each task handles N/P neurons. A simulation task is composed of two threads named CMC and CPC, respectively for "**ComMunication Controller**" and "**ComPutation Controller**". The use of threads takes advantage of hyper-threaded and dual-core processors. More precisely, CPC and CMC share two priority queues to manage two types of events:

– incoming Events to be Processed (by local neurons) are in a so called "EtoP" queue, an EtoP is the result of an incoming spike and contains the target neuron ID, the source neuron ID, and the spike emission date;
– outgoing Events to be Emit are in a so called "EtoE" queue, an EtoE is the result of a local neuron spiking and contains the source neuron ID, the emission date, and a boolean flag used to eventually invalidate a predicted emission (see second paragraph of current section).

The simulation is an infinite loop where CPC processes the top event of the EtoP queue (step "CPC 1" on figure 1). The targeted neuron is activated and eventually spikes which generates an EtoE stored in the EtoE ordered queue according to the event emission date (step "CPC 2"). The "activation" of the target neuron implies changing its state according to its own dynamics. This processing can be achieved by CPC or by a dedicated thread as shown figure 1. Also within an infinite loop, CMC processes the top event of the EtoE queue (step "CMC 2"). CMC knows the tables of postsynaptic neurons ID for all its neurons so the processed EtoE is used to generate EtoPs in the EtoP queue (spikes emitted to local neurons) and to pack EtoPs in messages to be sent to other simulation tasks (spikes emitted to remote neurons). Some controls are performed by CMC and CPC, respectively "EC" and "PC" on figure 1, to keep the behaviour strictly conservative and avoid deadlocks (see section 4).

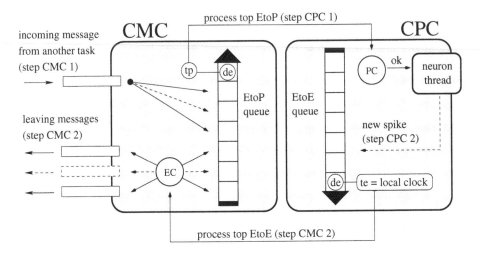

Fig. 1. The "ComMunication Controller" and "ComPutation Controller" threads

3   Delayed Queues of Events

In a first implementation, CMC and CPC were using priority queues from the C++ Standard Library (STL), thus accessing and deleting the top event had a null computational cost, but the insertion of an event had an $O(n \times \log(n))$ cost where n is the actual size of the queue. Such a cost becomes significant when queues contain tens of thousands of events, which can be an issue for DAMNED scalability.

To address this potential scalability issue, we have developed **"delayed queues"** (inspired from "calendar queues" [13]) taking into account the travel time of real (*i.e.* biological) spikes between source and target neurons (figure 2). Such a queue is based on two ordering criteria: (i) the time stamp of the event and (ii) the excitatory or inhibitory nature of the spike. Indeed, all inhibitory events for a given time increment has to be computed before excitatory ones to avoid the emission of erroneous events. According to biological knowledge [16], we consider that delays are finite and that the maximal delay δ_{max} is known at the beginning of the simulation. Furthermore we assume a discrete time whatever the scale could be. It is then possible to define a type of priority queue allowing no additional cost when inserting a new event nor accessing or deleting the next event. This delayed queue is a circular list containing as many sets of events as existing time stamps in the maximal delay δ_{max}. Each new event is inserted at the index corresponding to time $de+\delta$, where de is the actual time of emission and δ is the delay between source and target fields of the event. Knowing the maximal delay δ_{max} ensures that at time de no incoming event could be inserted later that $de + \delta_{max}$. When every event at time de are computed and the authorization to increase actual virtual time has been given (see next section) the delayed queue moves its first index to $de + \delta t$ where δt is the authorized time increment.

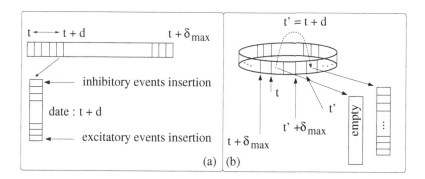

Fig. 2. Example of a delayed circular event queue

4   Conservative and Distributed Virtual Clock Handling

Communication protocols and MPI ensure that no message is lost. However, the distributed nature of DAMNED implies the risk of a message arriving late since each task runs the simulation at its own speed depending on the activity of local neurons (and eventually of the computer load in a non dedicated cluster). Taking into account such a late message would mean tracking changes made all over the simulation since the time stamp of the late events and restart the simulation from this date, which would be dramatically time consuming. An alternative is to introduce a synchronization protocol. A simple way to keep all tasks synchronous is to implement a centralized clock but this strategy may lead to a bottleneck and would be a limitation for the scalability of the simulator [11]. Therefore we introduce a **distributed clock handling method** [2] based on (i) the exchange of local virtual clocks between simulation tasks, and (ii) the implementation of controls to manage each task virtual clock incrementation making sure a task does not run too fast regarding others.

The local virtual clock on a given task T_i is CMC emission time te_i, that is the stamp of the top event of the EtoE queue. From there we introduce a global distributed clock as an array of te_j where te_j ($i \neq j$) are the last received values of all T_j local clock. Each time a task T_{src} sends an event message to a task T_{tgt}, the whole clock array of T_{src} is sent as a preamble. On T_{tgt}, the received te_i ($\forall i \neq tgt$) indicate T_i is granted to run the simulation till date te_i. The received clock array is used to actualize the local clock array by keeping the higher date between received and local elements, excepted for the local te. The CPC thread also handles its own clock standing for current event processing time and named tp, set from the time stamp of the top EtoP queue. Please note that, as described in following paragraphs, DAMNED will set te and tp to null or even negative value to carry special informations.

While no clock change occurs, CPC and CMC continue to respectively process EtoPs and send EtoEs. However when CMC is about to increment te (the task virtual clock), it is necessary to check if the task is not running the simulation too fast regarding other tasks with an "Emission Control" ("EC" on figure 1). This issue also exists for CPC regarding tp, thus a "Processed Control" is performed ("PC" on figure 1) when CPC is about to increment tp. An inaccurate element of the clock array on one of the task can lead to an erroneous check and blocks CPC or CMC resulting in a deadlock of the simulator. DAMNED uses a boolean array on each task T_i to ensure clock array accuracy: clock_sent[j] is set to *true* when the actual value of te_i is sent to task T_j (eventually without events); and each element of clock_sent is set to *false* when te_i increases.

When the EtoE queue is empty, CMC updates te_{src} as described by algorithm 1 where δ_{min} is the minimal "biological" delay within the SNN. The variable tp may has been set to null by CPC (as described in next paragraph), in such a case algorithm 1 enters line 2 if T_{src} has nothing to do, and sending a null clock value will inform other tasks not to wait T_{src}. Finally, if EtoE is empty but not EtoP (line 1), CMC sends $-|tp_{src} - \delta_{min}|$ which is the opposite value of the date untill which T_{src} is sure that other tasks can be granted to emit.

> **if** $tp_{src} \neq 0$ **then**
> 1 | $te_{src} \longleftarrow -|tp_{src} - \delta_{min}|$
>   **else**
> 2 | $te_{src} \longleftarrow 0$

Algorithm 1. Update of the actual virtual time

The control performed by CMC ("EC" on figure 1) on T_i is presented in algorithm 2 where de is the sending date of the processed EtoE. The first test (line 1 of algorithm 2) is a DAMNED mechanism that gives CPC the possibility to generate EtoEs that can be later invalidated, thus an invalidated EtoE treated by CMC is immediately destroyed (line 2) and emitting no more necessary. Such a mechanism is for example used if the emitting date is a prediction (see second paragraph of section 2). More tests are needed if an update of the virtual clock occurs (lines 3 to 6). First, if CPC has been authorized to process an event later than $|de + \delta_{min}|$ (line 3), checking all te_j (line 4) can be avoided which saves an O(P) loop. Otherwise, CMC checks two conditions (line 4): (i) the actual local clock must have been sent to all other tasks and (ii) the EtoE emission date (de) must be smaller than the possible processing date of a future received event (if any, *i.e.* if $te_j \neq 0$). If both conditions are fulfilled local clock can be incremented ($te_i \leftarrow de$) and emission granted (line 5), else emission is denied and te_i is set to $-de$ (line 6) meaning CMC is blocked till a future emission at de.

```
1 if EtoE not valid then
2 |   destroy EtoE
   |    → no emission necessary.
   else
      if de = te_i then
      |  → emission granted.
      else
3        if (tp_i > 0) and (tp_i ≥ de + δ_min) then
         |    te_i ⟵ de
         |    → emission granted.
         else
4           if clock_sent[j] and ((te_j = 0) or
            (de ≤ |te_j| + δ_min)), 0 ≤ j ≤ P − 1 (i ≠ j) then
5           |     te_i ⟵ de
            |     → emission granted
            else
6           |     te_i ⟵ −de
            |     → emission denied
```

Algorithm 2. CMC checking an EtoE emission on a task T_i

The control performed by CPC before activating the neuron targeted by the top EtoP ("PC" on figure 1) is presented in algorithm 3. CPC is free to process events while (i) tp_i does not have to be incremented (line 1) or (ii) the incrementation keeps tp_i in the range of the lookahead provided by δ_{min}: CPC must check that all tasks have an up-to-date te_i and that it is not any more possible to receive an anterior event that could void the neuron activation (line 2). If these conditions are fulfilled, tp_i can be incremented to de and EtoP can be processed (line 3). Otherwise CPC is blocked until the reception of a sooner event (*i.e.* until time de) and tp_i is set to the opposite of de (line 4). Furthermore, if the EtoE queue is empty at processing time (line 3) te_i is set to the opposite of tp_i.

```
1 if (tp_i = de) or
2 (clock_sent[j] and (de ≤ |te_j| + δ_min), 0 < j < P) or (te_j = 0)) then
   |   tp_i ⟵ de
3  |   if (te_i = 0) then
   |   └ te_i ⟵ −tp_i
   |   → processing granted
   else
4  |   tp_i ⟵ −de
   |   → processing denied
```

Algorithm 3. CPC checking an EtoP processing on a task T_i

5   Configuration of DAMNED and Definition of a SNN

The definition of a specific network to simulate is done though human friendly text files, with a mark-up language. The user can define "populations of neurons", each population can have a given size and be composed of a given proportion of excitatory or inhibitory neurons. A population can project synapses to other populations, and any type of projection are possible. Of course the density of connexions for a given projection can be chosen, as well as the range of weights and delays to be used. A point of significant interest is that each simulation task creates only its own part of the SNN. Network creation is then also distributed and speeded up (as presented new section).

The user can easily define an environment with input and output cells and stimulation protocols. The environment is a MPI task running on the front-end. It is indeed possible to perform measures of the activity of neurons and to plot classical biological experiments graphs such as "raster" (see figure 3), which facilitate comparisons with biological recordings.

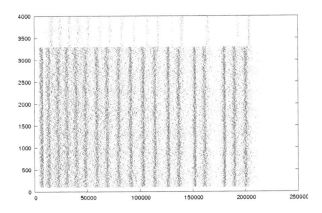

Fig. 3. Example of a "raster" output generated from a DAMNED simulation

However, for now the types of neurons used by a network is defined in the code and changing the type used imply to recompile DAMNED. Implemented models include LIF [8], GIF [14], and SRM0 [6]. The implementation of a new model implies writing a new C++ object but does not need parallel programming. The difficulty deeply depends of the type of neuron, and can be significant, for example if the spiking date must be predicted.

6   Results

The DAMNED simulator has been tested on a cluster, that is a classroom of the French West Indies University (*Université des Antilles et de la Guyane*) with 35 intel 2.2 GHz dual-core computers. Each machine has 1 gigabyte of RAM and is runing Mandriva GNU-Linux (a desktop oriented distribution, not a cluster optimized distribution). The network is based on a 100 Mb/s switch hosting 40 machines. The computers – and the network – were not dedicated for DAMNED so other processes may have influenced negatively the measured performances of DAMNED. However the simulator has been tested at night and all presented results are averaged values based on at least 5 executions making results quite trustful. Of course, DAMNED can run 24 hours a day at a low priority ("niced") without disturbing students, but our present concern is about accurate measures of DAMNED performances which rely on resources as dedicated as possible.

All experiments have been done for a 1 s duration of biological activity. Connectivity of the SNN is set to a biologically plausible 1000 connexions per neuron, meaning that what ever the size of the network each cell is connected to 1 000 other cells. Finally, activity has been tested at 1 Hz and 10 Hz. As an example, for 1000 neurons at 1 Hz, every neuron will spike (only one time) during the 1 s of the simulation, so DAMNED will handle 1000 spikes.

The best way to evaluate the performances of a distributed program is to measure its **speedup** for a given number of processors P as presented in equation 1 where T_i is the execution time on i processor(s). An accurate way to know the "best sequential time" is to develop a perfect simulator optimized for a single processor. However, in the present paper, the sequential time will be measured running DAMNED on a single machine (*i.e.* a single simulation task). Of course the machine is part of the cluster.

$$S_P = \frac{best\ T_1}{T_P} \tag{1}$$

However, measuring T_1 is not easy for large SNN since a single processor does not have enough memory. Therefore, we have been running a 10 000 and 100 000 neurons network but only small networks runs within the 1 GB of a single computer. As a consequence a theoretical sequential time has been extrapolated for large networks, according to a function $f(n) = \alpha\ n^2 + \beta\ n + \gamma$ where n is the size of the network. Such an estimation assumes an infinite memory for the sequential machine. The parameters α, β and γ have been fitted running DAMNED on a single machine for 1 000, 2 000, 3 000, 4 000, 7 000, and 10 000 neurons at 1 Hz and 10 Hz. The 4 000 neurons network is the largest before the OS starts to swap which leads to unusable simulation durations. For 7 000 and 10 000 neurons, the physical memory of the machine has been doubled with an identical memory module taken from an other machine of the cluster, reaching the maximum amount of memory possible for our machines regarding the available memory modules (only 2 memory slots, populated with 1 GB modules). The function f fits perfectly the data with $\alpha = 7.01\ 10^{-6}$, $\beta = 0.118$ and $\gamma = -116$ at 10 Hz. At 1 Hz we found $\alpha = 1.86\ 10^{-6}$, $\beta = 8.42\ 10^{-3}$ and $\gamma = -6.33$. The same methodology has been used to estimate the sequential time of the creation of the SNN, which does not depend of the frequency, giving $\alpha = 1.09\ 10^{-5}$, $\beta = -0.0346$ and $\gamma = 34$. Creation speedups are presented in the conclusion.

Figure 4 presents speedups as a function of the number of simulation tasks. No less than 15 tasks are launched for the 10 000 neurons network since enough machines are needed for the OS not to swap (see both dashed lines on figure 4). Making larger neural networks runnable is the main successful goal of DAMNED. Furthermore, results show that simulation times are significantly decreased, even with the low cost 100 Mb/s network used. The two 10 000 neurons simulations (dashed lines) are performed at 1 Hz and 10 Hz. Both frequencies are biologically plausible average values, however some cells can spike at a higher rate. Results show that a higher activity implies lower speedups. Indeed it is more difficult for computations to hide the increased number and size of messages. Thus dealing with high average frequencies would take advantage of a better network. Finally, the 100 000 neurons network (see plain line), involving at least 27 simulations tasks, validates the scalability of DAMNED for effective simulations.

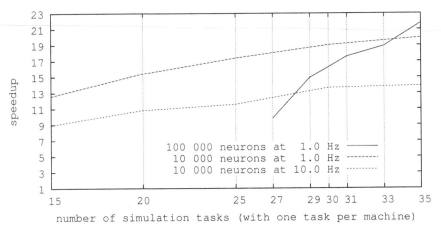

Fig. 4. Speedups as a function of the number of simulation tasks, with average spiking rates of 1 Hz and 10 Hz, with biologically plausible connectivity of respectively 0.1 % and 0.01 % of the network size for a 10 000 and a 100 000 neurons networks

7 Conclusion and Future Work

This paper has presented the DAMNED architecture and some original contributions to the field of distributed event-driven simulations, namely an optimised delayed event queue and efficient distributed virtual time handling algorithms. DAMNED is well adapted for biologically inspired protocols and can output valuable simulation data. Significant speedups have been achieved and prove that even using a non dedicated cluster of simple computers, the DAMNED simulator is able to handle large SNN simulation. Such results tend to validate the scalability of DAMNED. Lower speedups occur when the computational load is not high enough to overlap communications, thus complex neuron models would definitely lead to high speedups. Furthermore, results show that even when DAMNED is under-loaded, increasing of the number of processor does not slow the simulation which can in all cases reach a normal end. More evaluations will be done on dedicated parallel computers to simulate larger SNN.

Figure 5 presents the speedups of the creation of the SNN and shows that this step already takes advantage of the cluster. However the creation time remains high for large neural networks, about 2 000 s for a 100 000 neurons network using 35 machines while simulating 1 s of biological time takes about 150 s. We are currently working on this issue and significant improvements are anticipated.

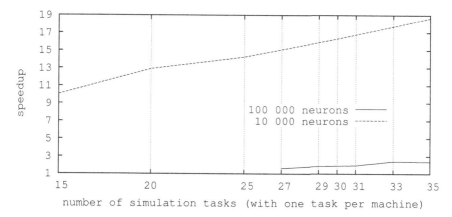

Fig. 5. Speedups of the creation of the SNN for a 10 000 and a 100 000 neurons network

Regarding usability, DAMNED is end user oriented and the creation of a SNN has been presented. A web based interface which enables to launch simulations with a simple navigator is already developed and will be improved. This interface shows available machines on the cluster, creates MPI configuration files, runs DAMNED and collects results. For now, the main remaining difficulty for an end-user is the modification of neuron models and implementation of new models, witch will be addressed by defining new neuron models from the web interface.

References

[1] Bohte, S.M.: The evidence for neural information processing with precise spike-times: A survey. Natural Computing 3(4), 195–206 (2004)
[2] Chandy, K.M., Misra, J.: Distributed simulation: A case study in design and verification of distributed programs. IEEE Transactions on Software Engineering, SE-5(5), 440–452 (1979)
[3] Ferscha, A.: Parallel and distributed simulation of discrete event systems. In: Parallel and Distributed Computing Handbook, pp. 1003–1041. McGraw-Hill, New York (1996)
[4] Flynn, M.J., Rudd, K.W.: Parallel architectures. ACM Computation Surveys 28(1), 67–70 (1996)
[5] Message Passing Interface Forum. MPI: A message-passing iterface standard. Technical Report UT-CS-94-230, University of Tennessee (1994)
[6] Gerstner, W., Kistler, W.M.: Spiking Neuron Models: An Introduction. Cambridge University Press, New York (2002)

[7] Izhikevich, E.M.: Simple model of spiking neurons. IEEE Transactions on Neural Networks 14(6), 1569–1572 (2003)

[8] Knight, B.W.: Dynamics of encoding in a population of neurons. Journal of General Physiology 59, 734–766 (1972)

[9] Lobb, C.J., Chao, Z.C., Fujimoto, R.M., Potter, S.M.: Parallel event-driven neural network simulations using the hodgkin-huxley model. In: Proceedings of the Workshop on Principles of Advanced and Distributed Simulations. PADS 2005, June 2005, pp. 16–25 (2005)

[10] Makino, T.: A discrete-event neural network simulator for general neuron models. Neural Computing and Applications 11(3-4), 210–223 (2003)

[11] Marin, M.: Comparative analysis of a parallel discrete-event simulator. In: SCCC, pp. 172–177 (2000)

[12] Mattia, M., Giudice, P.D.: Efficient event-driven simulation of large networks of spiking neurons and dynamical synapses. Neural Computation 12, 2305–2329 (2000)

[13] Morrison, A., Mehring, C., Geisel, T., Aertsen, A., Diesmann, M.: Advancing the boundaries of high-connectivity network simulation with distributed computing. Neural Computation 17, 1776–1801 (2005)

[14] Rudolph, M., Destexhe, A.: Analytical integrate-and-fire neuron models with conductance-based dynamics for event-driven simulation strategies. Neural Computation 18(9), 2146–2210 (2006)

[15] Shelley, M.J., Tao, L.: Efficient and accurate time-stepping schemes for integrate-and-fire neuronal network. Journal of Computational Neuroscience 11(2), 111–119 (2001)

[16] Swadlow, H.A.: Efferent neurons and suspected interneurons in binocular visual cortex of the awake rabbit: Receptive fileds and binocular properties. Journal of Neurophysiology 59(4), 1162–1187 (1988)

[17] Watts, L.: Event-driven simulation of networks of spiking neurons. In: Cowan, J.D., Tesauro, G., Alspector, J. (eds.) Advances in Neural Information Processing System, vol. 6, pp. 927–934. MIT Press, Cambridge (1994)

A Neural Network Model for the Critical Frequency of the F2 Ionospheric Layer over Cyprus

Haris Haralambous and Harris Papadopoulos

Computer Science and Engineering Department, Frederick University,
7 Y. Frederickou St., Palouriotisa, Nicosia 1036, Cyprus
{H.Haralambous,H.Papadopoulos}@frederick.ac.cy

Abstract. This paper presents the application of Neural Networks for the prediction of the critical frequency *foF2* of the ionospheric F2 layer over Cyprus. This ionospheric characteristic (*foF2*) constitutes the most important parameter in HF (High Frequency) communications since it is used to derive the optimum operating frequency in HF links. The model is based on ionosonde measurements obtained over a period of 10 years. The developed model successfully captures the variability of the *foF2* parameter.

Keywords: Ionosphere, HF communications, F2 layer critical frequency.

1 Introduction

Skywave HF communiations utilize the ability of the ionosphere to reflect waves up to 30 MHz to achieve medium to long-distance communication links with a minimum of infrastructure (figure 1). The ionosphere is defined as a region of the earth's upper atmosphere where sufficient ionisation can exist to affect the propagation of radio waves in the frequency range 1 to 30 MHz. It ranges in height above the surface of the earth from approximately 50 km to 600 km. The influence of this region on radio waves is accredited to the presence of free electrons.

The uppermost layer of the ionosphere is the F2 layer which is the principal reflecting region for long distance HF communications [1,2,3]. The maximum frequency that can be reflected at vertical incidence by this layer is termed the F2 layer critical frequency (*foF2*) and is directly related to the maximum electron density of the layer. The F2 layer critical frequency is the most important parameter in HF communication links since when multiplied by a factor which is a function of the link distance, it defines the optimum usable frequency of operation. The maximum electron density of free electrons within the F2 layer and therefore *foF2* depend upon the strength of the solar ionising radiation which is a function of time of day, season, geographical location and solar activity [1,2,3]. This paper describes the development of a neural network model to predict *foF2* above Cyprus. The model development is based on around 33000 hourly *foF2* measurements recorded above Cyprus from 1987 to 1997. The practical application of this model lies in the fact that in the absence of any real-time or near real-time information on *foF2* above Cyprus this model can provide an alternative method to predict its value under certain solar activity conditions.

D. Palmer-Brown et al. (Eds.): EANN 2009, CCIS 43, pp. 371–377, 2009.

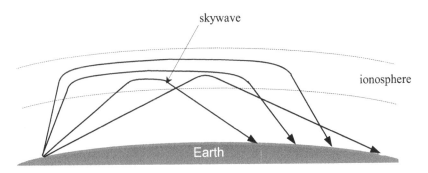

Fig. 1. Skywave communication modes via the ionosphere

2 Characteristics of the F2 Layer Critical Frequency

Measurements of *foF2* are conducted by ionosondes which are special types of radar used for monitoring the electron density at various heights in the ionosphere. Their operation is based on a transmitter sweeping through the HF frequency range transmitting short pulses. These pulses are reflected at various layers of the ionosphere, and their echoes are received by the receiver and analyzed to infer the ionospheric plasma frequency at each height. The maximum frequency at which an echo is received is called the critical frequency of the corresponding layer. Since the F2 layer is the most highly ionized ionosperic layer its critical frequency *foF2* is the highest frequency that can be reflected by the ionosphere. This implies that by operating close to that frequency when establishing oblique communication links will provide maximum range and at the same time favourable propagation conditions.

Solar activity has an impact on ionospheric dynamics which in turn influence the electron density of the ionosphere. The electron density of the F2 layer exhibits variability on daily, seasonal and long-term time scales in response to the effect of solar radiation. It is also subject to abrupt variations due to enhancements of geomagnetic activity following extreme manifestations of solar activity disturbing the ionosphere from minutes to days on a local or global scale.

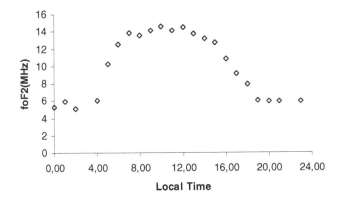

Fig. 2. Diurnal variation of *foF2*

The most profound solar effect on *foF2* is reflected on its daily variation as shown in figure 2. As it is clearly depicted, there is a strong dependency of *foF2* on local time which follows a sharp increase of *foF2* around sunrise and gradual decrease around sunset. This is attributed to the rapid increase in the production of electrons due to the photo-ionization process during the day and a more gradual decrease due to the recombination of ions and electrons during the night.

The long–term effect of solar activity on *foF2* follows an eleven-year cycle and is clearly shown in figure 3(a) where all the values of *foF2* are plotted against time as well as a modeled monthly mean sunspot number *R* which is a well established index of solar activity (figure 3(b)). We can observe a marked correlation of the mean level of *foF2* and modeled sunspot number.

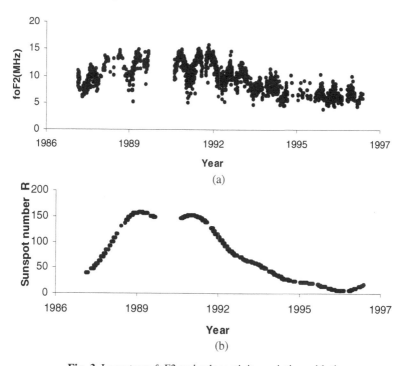

Fig. 3. Long-term *foF2* and solar activity variation with time

There is also a seasonal component in the variability of *foF2* which can be attributed to the seasonal change in extreme ultraviolet (EUV) radiation from the Sun. This can be clearly identified in figure 4 for noon values of *foF2* for year 1991 according to which in winter *foF2* tends to be higher than during the summer. In fact this variation reverses for night-time *foF2* values. This particular phenomenon is termed the winter anomaly. In addition to the effects of solar activity on *foF2* mentioned above we can also identify a strong effect on the diurnal variability as solar activity gradually increases through its 11-year cycle. This is demonstrated if figure 5 where the diurnal variation of *foF2* is plotted for three different days corresponding to low (1995), medium (1993) and high (1991) sunspot number periods. It is evident from this figure that the night to day variability in *foF2* increases as sunspot number increases.

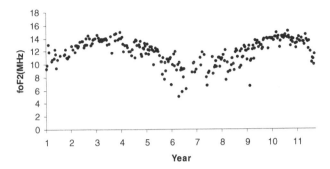

Fig. 4. Seasonal variation of *foF2* at 12:00

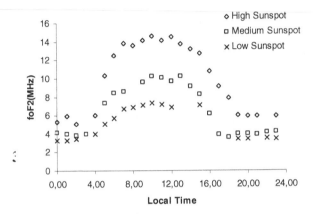

Fig. 5. Diurnal variability of *foF2* for low, medium and high solar activity

3   Model Parameters

The diurnal variation of *foF2* is clearly evident by observing figure 2 and figure 5. We therefore include hour number as an input to the model. The hour number, *hour*, is an integer in the range $0 \leq hour \leq 23$. In order to avoid unrealistic discontinuity at the midnight boundary, *hour* is converted into its quadrature components according to:

$$sinhour = \sin\left(2\pi \frac{hour}{24}\right) \tag{1}$$

and

$$coshour = \cos\left(2\pi \frac{hour}{24}\right) \tag{2}$$

A seasonal variation is also an underlying characteristic of *foF2* as shown in figure 4 and is described by day number *daynum* in the range $1 \leq daynum \leq 365$. Again to avoid unrealistic discontinuity between December 31st and January 1st *daynum* is converted into its quadrature components according to:

$$sinday = \sin\left(2\pi \frac{daynum}{365}\right) \qquad (3)$$

and

$$cosday = \cos\left(2\pi \frac{daynum}{365}\right) \qquad (4)$$

Long-term solar activity has a prominent effect on *foF2*. To include this effect in the model specification we need to incorporate an index, which represents a good indicator of solar activity. In ionospheric work the 12-month smoothed sunspot number is usually used, yet this has the disadvantage that the most recent value available corresponds to *foF2* measurements made six months ago. To enable *foF2* data to be modelled as soon as they are measured, and for future predictions of *foF2* to be made, the monthly mean sunspot number values were modeled using a smooth curve defined by a summation of sinusoids (figure 3(b)).

4   Experiments and Results

A Neural Network (NN) was trained to predict the *foF2* value based on *sinhour, coshour, sinday, cosday* and *R* (modeled sunspot number) model parameters. The 33149 values of the dataset recorded between 1987 and 1997 were used for training the NN, while the 3249 values of the more recent dataset recorded from 18.09.08 until 16.04.09 were used for testing the performance of the trained NN. The training set was sparse to a certain degree in the sense that many days had missing *foF2* hourly values and this did not allow the dataset to be approached as a time-series.

The network used was a fully connected two-layer neural network, with 5 input, 37 hidden and 1 output neuron. Both its hidden and output neurons had tan-sigmoid activation functions. The number of hidden neurons was determined by trial and error. The training algorithm used was the Levenberg-Marquardt backpropagation algorithm with early stopping based on a validation set created from the last 3000 training examples. In an effort to avoid local minima ten NNs were trained with different random initialisations and the one that performed best on the validation set was selected for application to the test examples. The inputs and target outputs of the network were normalized setting their minimum value to -1 and their maximum value to 1. The results reported here were obtained by mapping the outputs of the network for the test examples back to their original scale.

The RMSE of the trained NN on the test set was 0.688 MHz, which is considered acceptable for a prediction model [4,5,6]. To further evaluate the performance of the developed network, a linear NN was applied to the same data and the performance of the two was compared. The RMSE of the linear NN on the test set was 1.276 MHz, which is almost double that of the multilayer NN. Some examples of measured and predicted *foF2* values are given in figure 6. These demonstrate both the good performance of the developed NN and its superiority over the linear model.

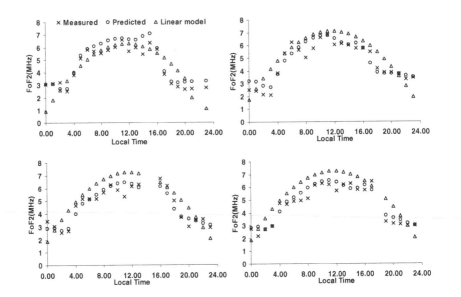

Fig. 6. Examples of measured and predicted *foF2* values

Despite the good agreement of measured and predicted values under benign geomagnetic conditions we have noticed several occasions where the discrepancy between them increases significantly. This is the case particularly during geomagnetic storms due to the impact of the disturbed magnetic field on the structure of the ionosphere causing rapid enhancements or depletions in its electron density. An example of such an occasion is given in figure 7. It is evident that around 15:00 the intense geomagnetic activity causes an increase in the error in the model due to excursions of *foF2* from its undisturbed behaviour.

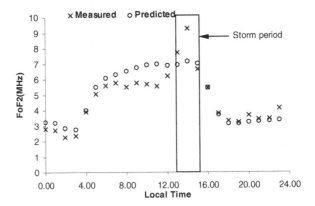

Fig. 7. Increase of model error under geomagnetically disturbed conditions

An alternative option would be to attempt a short-term forecasting model by using recent values of the modeled parameter as inputs to the model. However this would necessitate the operation of an ionosonde and the availability of near real-time data to the model user which is not very practical especially in the absence of an internet connection.

5 Conclusions and Future Work

In this paper we have presented the development of a neural network model for the long-term prediction of critical frequency of the F2 ionospheric layer (*foF2*) above Cyprus. The model has been developed based on a data set obtained during a period of ten years and tested on a dataset of seven months that was recently obtained. The model has produced a good approximation of the different time-scales in the variability of the modeled parameter. The next step will be to investigate the possibility to incorporate a geomagnetic index as an input parameter in order to represent better the geomagnetic variability of the electron density of the ionosphere. In addition the model will be further updated as we collect more *foF2* measurements in the next few years.

References

1. Goodman, J.: HF Communications, Science and Technology. Nostrand Reinhold (1992)
2. Maslin, N.: The HF Communications, a Systems Approach, San Francisco (1987)
3. McNamara, L.F.: Grid The Ionosphere: Communications, Surveillance, and Direction Finding. Krieger Publishing Company, Malabar (1991)
4. Altinay, O., Tulunay, E., Tulunay, Y.: Forecasting of ionospheric critical frequency using neural networks. Geophys. Res. Lett. 24, 1467–1470 (1997)
5. Cander, L.R., Lamming, X.: Forecasting Neural networks in ionospheric prediction and short-term forecasting. In: 10th International Conference Conferenceon Antennas and Propagation, Edinburgh, April 14-17, vol. 436, pp. 2.27-2.30. IEE Conference Publication (1997)
6. Wintoft, P., Cander, L.R.: Short term prediction of foF2 using time delay neural networks. Phys. Chem. Earth (C) 24, 343–347 (1999)

Dictionary-Based Classification Models. Applications for Multichannel Neural Activity Analysis

Vincent Vigneron[1,*], Hsin Chen[2], Yen-Tai Chen[3], Hsin-Yi Lai[3], and You-Yin Chen[3]

[1] IBISC CNRS FRE 3190, Université d'Evry, France
vincent.vigneron@ibisc.univ-evry.fr
[2] Dept. of Electrical Engineering, National Tsing Hwa University, Taiwan
hchen@ee.nthu.edu.tw
[3] Dept. of Electrical Engineering, National Chiao-Tung University, Taiwan
kenchen@cn.nctu.edu.tw

Abstract. We describe in this paper advanced protocols for the discrimination and classification of neuronal spike waveforms within multichannel electrophysiological recordings. The programs are capable of detecting and classifying the spikes from multiple, simultaneously active neurons, even in situations where there is a high degree of spike waveform superposition on the recording channels. Sparse Decomposition (SD) approach was used to define the linearly independent signals underlying sensory information in cortical spike firing patterns. We have investigated motor cortex responses recorded during movement in freely moving rats to provide evidence for the relationship between these patterns and special behavioral task. Ensembles of neurons were simultaneously recorded in this during long periods of spontaneous behaviour. Waveforms provided from the neural activity were then processed and classified. Typically, most information correlated across neurons in the ensemble were concentrated in a small number of signals. This showed that these encoding vectors functioned as a feature detector capable of selectively predicting significant sensory or behavioural events. Thus it encoded global magnitude of ensemble activity, caused either by combined sensory inputs or intrinsic network activity.

SD on an overcomplete dictionary has recently attracted a lot of attention in the literature, because of its potential application in many different areas including Compressive Sensing (CS). SD approach is compared to the generative approach derived from the likelihood-based framework, in which each class is modeled by a known or unknown density function. The classification of electroencephalographic (EEG) waveforms present 2 main statistical issues: high dimensional data and signal representation.

Keywords: Classification, Atomic Decomposition, Sparse Decomposition, Overcomplete Signal Representation.

1 Introduction

The analysis of continuous and multichannel neuronal signals is complex, due to the large amount of information received from every electrode. Neural spike recording have

---

* This project was supported in part by fundings from the Hubert Curien program of the Foreign French Minister and from the Taiwan NSC. The neural activity recordings were kindly provided by the Neuroengineering lab. of the National Chiao-Tung University.

D. Palmer-Brown et al. (Eds.): EANN 2009, CCIS 43, pp. 378–388, 2009.
© Springer-Verlag Berlin Heidelberg 2009

shown that the primary motor cortex (M1) encodes information about movement direction [25,24,23,1] limb velocity, force [4] and individual muscle activity [22,17]. To further investigate the relationship between cortical spike firing patterns and special behavioral task, we have investigated motor cortex responses recorded during movement in freely moving rats: Fig. 1 shows the experimental setup for neural activities recording and the video captures related animal behavioral task simultaneously.

Fig. 1. The experimental setup (top). Light-color (red virtual ring) was belted up the right forelimb to be recognized the trajectory by video tracking system. The sequence images were captured the rat performing the lever press tasks in return for a reward of water drinking (bottom).

In this paper, we report on the classification of action potentials into a spike wave shape database. We do not intend to classify the waveforms: much work have been already made on this subject, see for instance Van Staveren *et al.* in [26]. But we expect to put in evidence a relationship between the rat posture or motion and the recorded neural activity. So, our research is mostly based on two objectives: phenomenon explanation and phenomenon prediction. This represents one of the current challenges in signal theory research.

2   Material and Methods

2.1   Animal Training and Behavioral Tasks

The study, approved by the Institutional Animal Care and Use Committee at the National Chiao Tung University, was conducted according to the standards established in the Guide for the Care and Use of Laboratory Animals. Four male Wistar rats weighing 250-300 g (BioLASCO Taiwan Corp., Ltd.) were individually housed on a 12 h light/dark cycle, with access to food and water *ad libitum*.

Dataset was collected from the motor cortex of awake animal performing a simple reward task. In this task, male rats (BioLACO Taiwan Co.,Ltd) were trained to press a lever to initiate a trial in return for a water reward. The animals were water restricted 8-hour/day during training and recording session but food were always provided to the animal ad lib every day.

2.2   Chronic Animal Preparation and Neural Ensemble Recording

The animals were anesthetized with pentobarbital (50 mg/kg i.p.) and placed on a standard stereotaxic apparatus (Model 9000, David Kopf, USA). The dura was retracted carefully before the electrode array was implanted. The pairs of 8 microwire electrode arrays (no.15140/13848, 50m in diameter; California Fine Wire Co., USA) are implanted into the layer V of the primary motor cortex (M1). The area related to forelimb movement is located anterior 2-4 mm and lateral 2-4 mm to Bregma. After implantation, the exposed brain should be sealed with dental acrylic and a recovery time of a week is needed.

During the recording sessions, the animal was free to move within the behavior task box (30 cm×30 cm× 60 cm), where rats only pressed the lever via the on right forelimb, and then they received 1-ml water reward as shown in Fig. 1. A Multi-Channel Acquisition Processor (MAP, Plexon Inc., USA) was used to record neural signals. The recorded neural signals were transmitted from the headstage to an amplifier, through a band-pass filter (spike preamp filter: 450-5 kHz; gain: 15,000-20,000), and sampled at 40 kHz per channel. Simultaneously, the animals'behavior was recorded by the video tracking system (CinePlex, Plexon Inc., USA) and examined to ensure that it was consistent for all trials included in a given analysis [26].

3   Data Analysis

3.1   Preprocessing

Neural activity was collected from 400-700ms before to 200-300 ms after lever release for each trail. Data was recorded from 33 channels, action potentials (spikes) crossing manually set thresholds were detected, sorting and the firing rate for each neuron was computed in 33 ms time bins (Fig. 3).

In each channel, the rms noise level is constantly monitored and determines the setting of a level detector to detect spike activity. Each trail was segmented into 19-30 bins approximately. Notice that the real time data processing software reduces the data stream by rejection of data which does not contain bioelectrical activity.

In the following experiment, the data are summarized by a 706×33 matrix $X = (x_{ij})$ and a 706×1 vector $Z = z_j$ where x_{ij} denotes the feature i for the channel j and z_j stands for the class label of the sample i. The class label is 1,2,3 or 4, depending on the trajectory followed by the rat arm in the plane of the video tracking system (Fig. 2).

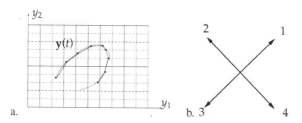

Fig. 2. Arm trajectory in the video tracking system $y_1 y_2$–plane (a). z-label derived from the position vector $\mathbf{y}(t)$(b).

The z-label is obtained from differential geometry. We assume that a Cartesian co-ordinate system has been introduced in $\mathbb{R}^2$. Then, evey point in space can be uniquely determined by the position vector $\mathbf{y} = \mathbf{y}(t) = (y_1(t), y_2(t))$. Therefore the vector $\mathbf{y}(t_2) - \mathbf{y}(t_1)$ gives the direction of the motion and lies in one of the four quadrant of the $y_1 y_2$–plane. The first quadrant is labelled $z = 1$, the second quadrant $z = 2$, etc.

Let $X = (x_{ij})$ the matrix of dataset and $Z = z_j$ the vector of class label.

3.2   Manual Scatterplot Classification

A method for classification is by plotting a selection of 2 or 3 spike features in a scatter diagram. This results in a 2- or 3-D graph with separate groups. The groups can only be assigned when there is enough spacing between the groups. Elliptic shaped areas are drawn around the groups isolating the classes.

3.3   Quantification and Classification of Spike Waveforms

The classes can also be assigned using minimum Mahalanobis distance classification. First a selection of spikes must be manually classified and a set of spikes features must be selected. The so-called 'training-set' must be representative for all waveforms in the measurement. It is created by randomly selecting a number of spikes.

This study aims to compare two different approaches of the classification: a traditional approach which uses LDA, QDA or MDA and which supposes a preliminary selection of features. The second approach uses Sparse Decomposition (SD) proposed by [10]. The traditional approach is briefly reviewed below as well as the principles of SD.

LDA, QDA and MDA. Statistical discriminant analysis methods such as LDA, QDA and MDA arise in a Gaussian mixture model. These methods are concerned with the construction of a statistical decision rule which allows to identify the population mem-bership of an observation. The predicted class is chosen to maximize the posterior class probability given the observation. The main assumption of these methods concerns the distribution of the observations of each class which is Gaussian. We refer to Chapter 4

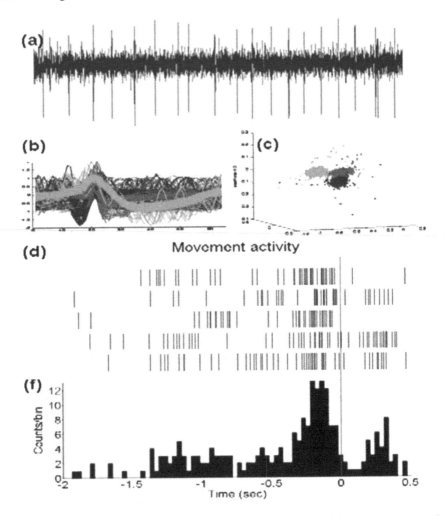

Fig. 3. The raw data of neural recording and an example of spike sorting. (a) Real raw data recording in M1 area. (b) Three individual spikes are detected from the raw data from (a). Visualized result for spike sorting by using principal component analysis. All spike timestamps were displayed for one trail a single neural recording (d) and the firing activity histograms (f) around the time of behavioral task. The bin is 33 ms. The red line denotes that the animal presses the lever.

of [3] for details on LDA and QDA. MDA, developed by Hastie and Tibshirani [16], is a generalization of LDA in which each class is modeled by a mixture of Gaussians. This modelization gives more flexibility in the classification rule than LDA and allows MDA to take into account heterogeneity in a class. Breiman *et al.* [5], MacLachlan and Basford [18] have actually contributed and tested this generative approach on many fields. Let k the number of classes. The proportion of each class in the population is π_i with $i = 1, \ldots, k$. Let π_{ir} the proportion of the subclass $r = 1, \ldots, R_k$ of the class i and $\phi_{i,r}(x, \theta)$ the probabilistic normal density function of the subclass r. We can define the probability of a nuclear x belongs to the class i such as:

$$P(Z = i, X = x) = \pi_i f_i(x) = \pi_i \sum_{r=1}^{R_i} \pi_{ir} \phi(x, \theta_{i,r}), \tag{1}$$

where $\theta_{i,r} = \mu_{i,r}, \sigma$. The π_i estimate by the maximum likelihood stands for the proportion of the class i in the learning set. The parameters Σ, $\mu_{i,r}$ and $\pi_{i,r}$ are estimated with the EM algorithm. At the qth iteration, the parameters are :

$$\pi_{ir}^{(q)} = \frac{\sum_{j=1}^{n} I(z_j = i) p_{jr}^{(q)}}{\sum_{j=1}^{n} I(z_j = i)}, \tag{2}$$

$$\hat{\mu}_{ir} = \frac{\sum_{j=1}^{n} x_j I(z_j = i) p_{jr}^{(q)}}{\sum_{j=1}^{n} I(z_j = i) p_{jr}^{(q)}}, \tag{3}$$

$$\hat{\Sigma} = \frac{\sum_{j=1}^{n} \sum_{r=1}^{R_i} p_{jr} (x_j - \hat{\mu}_{ir})(x_j - \hat{\mu}_{ir})^t}{n}. \tag{4}$$

The rule of Maximum A posteriori predicts the class of an observation x : the observation x belongs to the class which gives it the largest likelihood.

Feature Selection. Density estimation in high dimensions is generally considered to be more difficult – requiring more parameters. In order to be able to use traditional classifiers, it is necessary to summarize into a low dimensional space most of the information content contained in the original high dimensional space. In the literature, different approaches for feature selection have been proposed.A comprehensive overview of many existing methods of feature selection has been written by Dash and Liu [7]. Each variable selection method is made by two main parts: choosing an evaluation function and finding a selection procedure. On the one hand, an evaluation function tries to measure the ability of a feature to distinguish the different class labels. These evaluation functions have been divided into five groups: distance, information, dependance, consistancy measures and classifier error rate. On the other hand, the selection procedure solves the problem of the huge total number of competing variable subsets to be generated (using heuristic function for example). Our approach of the feature selection is typical: we choose an index trying to find a subset of features from the original set that is necessary and sufficient to describe the target. For our problem, we use a distance measure (Wilks'lambda) and a heuristic procedure to choose an optimal subset of features. In each iteration, remaining features yet to be selected (or rejected) are considered for selection (or rejection). If the additional feature improves the Wilks'Lambda then this feature is selected otherwise it is rejected.

A Reminder on Sparse Decomposition. The problem solved by the sparse representation is to search for the most compact representation of a signal in terms of linear combination of atoms in an overcomplete dictionary. Recent developments such as wavelet, ridgelet, curvelet and contourlet transforms are an important incentive for the research on the sparse representation [19]. Compared to methods based on orthonormal transforms or direct time domain processing, sparse representation usually offers better performance with its capacity for efficient signal modelling. In the standard framework of sparse representation, the objective is to reduce the signal reconstruction error with as few number of atoms as possible. On the one hand, discriminative analysis methods, such as LDA, are more suitable for the tasks of classification. On the other hand, discriminative methods are usually sensitive to corruption in signals due to lacking crucial properties for signal reconstruction. We propose here a method of sparse representation for signal classification , which modifies the standard sparse representation framework for signal classification. We first show that replacing the reconstruction error with discrimination power in the objective function of the sparse representation is more suitable for the tasks of classification. When the signal is corrupted, the discriminative methods may fail because little information is contained in discriminative analysis to successfully deal with noise, missing data and outliers. Let the n-dimensional vector $\mathbf{x}$ to be decomposed as a linear combination of the vectors $\mathbf{a}_i, i = 1, \ldots, m$. According to [19], the vectors $\mathbf{a}_i, i = 1, \ldots m$ are called *atoms* and they collectively form a *dictionary* over which the vector $\mathbf{x}$ is to be decomposed. We may write $\mathbf{x} = \sum_{i=1}^{m} s_i \mathbf{a}_i = \mathbf{As}$, where $\mathbf{A} \triangleq [\mathbf{a}_1, \ldots, \mathbf{a}_m]$ is the $n \times m$ dictionary (matrix) and $\mathbf{s} \triangleq (s_1, \ldots, s_m)^T$ is the $m \times 1$ vector of coefficients. If $m > n$, the dictionary is overcomplete, and the decomposition is not necessarily unique. However, the so called "sparse decomposition" (SD), that is, a decomposition with as much zero coefficients as possible has recently found a lot of attention in the literature because of its potential applications in many different areas. For example, it is used in Compressive Sensing (CS) [9], underdetermined Sparse Component Analysis (SCA) and source separation [14], decoding real field codes [6], image deconvolution [12], image denoising [11], electromagnetic imaging and Direction of Arrival (DOA) finding [13], and Face Recognition [27].

The sparse solution of the Underdetermined System of Linear Equations (USLE) $\mathbf{As} = \mathbf{x}$ is useful because it is unique under some conditions: Let $spark(\mathbf{A})$ denote the minimum number of columns of $\mathbf{A}$ which form a linear dependent set [10]. Then, if the USLE:

$$\mathbf{As} = \mathbf{x} \tag{5}$$

has a solution $\mathbf{s}$ with less than $\frac{1}{2}spark(\mathbf{A})$ non-zero components, it is the unique sparsest solution [10,15]. As a special case, if every $n \times n$ sub-matrix of $\mathbf{A}$ is invertible (which is called the Unique Representation Property or URP in [13]), then a solution of (5) with less than $(n + 1)/2$ non-zero elements is the unique sparsest solution.

For finding the sparse solution of (5), one may search for a solution for which the ℓ^0 norm of $\mathbf{s}$, i.e. the number of non-zero components of $\mathbf{s}$, is minimized. This is written as:

$$\text{minimize} \sum_{i=1}^{m} |s_i|^0 \quad \text{subject to} \quad \mathbf{As} = \mathbf{x} \tag{6}$$

Direct solution of this problem needs a combinatorial search and is NP-hard. Consequently, many different algorithms have been proposed in recent years for finding the sparse solution of (5). Some examples are Basis Pursuit (BP) [8], Smoothed ℓ^0 (SL0) [20,21], and FOCUSS [13]. Many of these algorithms, replace the ℓ^0 norm in (6) by another function of $\mathbf{s}$, and solve the problem:

$$\text{minimize} f(\mathbf{s}) \quad \text{subject to} \quad \mathbf{As} = \mathbf{x} \tag{7}$$

For example, in BP, $f(\mathbf{s})$ is the ℓ^1 norm of $\mathbf{s}$ (i.e. $\sum_{i=1}^{m} |s_i|$); and in SL0, $f(\mathbf{s})$ is a smoothed measure of the ℓ^0 norm.

However, up to our best knowledge, in all of the previous works, it is explicitly or implicitly assumed that the dictionary matrix is full-rank. Note that having the URP is a more strict than being full-rank, that is, a matrix which has the URP is full-rank, but a full-rank matrix has not necessarily the URP. Consider however a dictionary $\mathbf{A}$ which is not full-rank (and hence has not the URP), but $spark(\mathbf{A}) > 2$. This dictionary may still be useful for SD applications, because a solution of (5) with less than $\frac{1}{2}spark(\mathbf{A})$ non-zero components is still unique and is the sparsest solution. As an example, $\mathbf{A} = [1, 2, -1, 1; 2, -1, 1, 0; 3, 1, 0, 1]$ is not full-rank (its 3rd row is the sum of its first two rows), but every two of its columns are linearly independent, and hence $spark(\mathbf{A}) = 3$.

On the other hand, for a non-full-rank $\mathbf{A}$, the system (5) does not even necessarily admit a solution (that is, $\mathbf{x}$ cannot be necessarily stated as a linear combination of the atoms $\mathbf{a}_i, i = 1, \ldots, m$. For example, $\mathbf{x} = (1, 2, 3.1)^T$ cannot be stated as a linear combination of the columns of the above mentioned $\mathbf{A}$, because contrary to $\mathbf{A}$, its last row is not the sum of its first two rows. In this case, all the algorithms based on (6) or (7) will fail, because the solution set of (5) is empty. In effect, in this case, the 'sparsest solution' of (5) has not been even defined, because (5) has no solution !

Non-full-rank dictionaries may be encountered in some applications. For example, in SD based classification [27], the idea is to express a new point $\mathbf{x}$ as a sparse linear combination of all data points $\mathbf{a}_i, i = 1, \ldots, m$, and assign to $\mathbf{x}$ the class of the data points $\mathbf{a}_i$ which have more influence on this representation. In this application, if for example, one of the features (one of the components of $\mathbf{a}_i$) can be written as a linear combination of the other features for all the 'data' points $\mathbf{a}_i, i = 1, \ldots, m$, then the dictionary $\mathbf{A}$ is non-full-rank. If this is also true for the new point $\mathbf{x}$, then we are in the case that (5) has solutions but $\mathbf{A}$ is non-full-rank; and if not, then (5) has no solution and our classifier will fail to provide an output (based on most current SD algorithms).

For a non-full-rank overcomplete dictionary, one may propose to simply remove the rows of $\mathbf{A}$ that are dependent to other rows, and obtain a full-rank dictionary. This naive approach is not desirable in many applications. In Compressive Sensing (CS) language, this is like trowing away some of the measurements, which were useful in presence of measurement noise for a better estimation of $\mathbf{s}$ (recall that in simple estimation of the weight of an object, if you have several measurements available, trowing away all but one is not a good estimation method of the weight!).

In the next section, we generalize the definition of sparse decomposition to classification (SDC) and modify the algorithm itself to directly cover non-full-rank dictionaries.

4   Definition of SDC

In linear algebra, when the linear system $\mathbf{As} = \mathbf{x}$ is inconsistent (underdetermined as well as overdetermined), one usually considers a Least Squares (LS) solution, that is, a solution which minimizes $\|\mathbf{As} - \mathbf{x}\|$, where $\|\cdot\|$ stands for the ℓ^2 (Eucidean) norm throughout the paper. Naturally, we define the sparse decomposition as a decomposition $\mathbf{x} \approx s_1\mathbf{a}_1 + \cdots + s_m\mathbf{a}_m = \mathbf{As}$ which has the sparsest $\mathbf{s}$ among all of the minimizers of $\|\mathbf{As} - \mathbf{x}\|$. By a sparse LS solution of (5), we mean the $\mathbf{s} \in \mathcal{S}$ which has the minimum number of non-zero components, that is:

$$\underset{\mathbf{s}}{argmin} \sum_{i=1}^{m} |s_i|^0 \text{ subject to } \|\mathbf{As} - \mathbf{x}\| \text{ is minimized} \tag{8}$$

Note that the constraint $\mathbf{As} = \mathbf{x}$ in (6) has been replaced by $\mathbf{s} \in \mathcal{S}$ in (8). If (5) admits a solution, $\mathcal{S}$ will be the set of solutions of (5), and the above definition is the same as (6).

Replacing the reconstruction error with the discrimination power quantified by the Fisher's discrimination criterion used in the LDA, the objective function (8) that focuses only on classification can be written as:

$$\underset{s}{\arg\min} \sum_{i=1}^{m} |s_i|^0 \text{ subject to } \mathcal{S} : \frac{\Sigma_B}{\Sigma_W} \text{ is minimized} \tag{9}$$

where Σ_B is the 'intra-class distance' and Σ_B is the "inner-class scatter'. Fisher's criterion is motivated by the intuitive idea that the discrimination power is maximized when the spatial distribution of different classes are as far away as possible and the spatial distribution of samples from the same class are as close as possible. Maximizing (9) generates a sparse representation that has a good discrimination power.

The SL0 Algorithm. SL0 algorithm [20,21] is a SD algorithm with two main features: 1) it is very fast, 2) it tries to directly minimize the ℓ^0 norm (and hence does not suffer from replacing ℓ^0 by asymptotic equivalents). The basic idea of SL0 is to use a smooth measure of the ℓ^0 norm ($f(\mathbf{s})$), and solve (7) by steepest descent. To take into account the constraint $\mathbf{As} = \mathbf{x}$, each iteration of (the full-rank) SL0 is composed of:

- Minimization: $\mathbf{s} \leftarrow \mathbf{s} - \mu \nabla f$
- Projection: Project $\mathbf{s}$ onto $\{\mathbf{s} | \mathbf{As} = \mathbf{x}\}$:

$$\mathbf{s} \leftarrow \mathbf{s} - \mathbf{A}^T (\mathbf{A}\mathbf{A}^T)^{-1} (\mathbf{As} - \mathbf{x}) \tag{10}$$

For extending SL0 to non-full-rank dictionaries, the projection step should be modified: $\mathbf{s}$ should be projected onto $\mathcal{S}$ instead of $\{\mathbf{s} | \mathbf{As} = \mathbf{x}\}$. This can be seen from the lemma in [2].

5   Results

In this section, the classification is conducted on the experimental database. An exemple of a scatterplot after finishing the classification is shon in Fig. 4. This classification consisted in a MDA classification (Fig. 4.a) and the application of the SDC algorithm resulting in classes 1,2,3 and 4. Classification with SDC is conducted with the decomposition coefficient (the $\mathbf{x}$ as in equation (8)) as feature and support vector machine (SVM) as classifier.

Table 1. Classification error rates with different level of signal-to-noise ratio (SNR)

Methods	no noise	5 dB	10 dB	20
MDA	**0.02620**	**0.0308**	0.1774	0.2782
SDC	0.02773	0.0523	**0.0720**	**0.1095**

In this experiment, noise is added to the signals to test the robustness of SDC, with increasing level of energy. Table 1 summarizes the classification error rates obtained with different SNR.

Results in Table 1 show that in the case that signals are ideal (noiseless), MDA is the best criterion for classification. This is consistent with the known conclusion that discriminative methods outperform reconstructive methods in classification. However, when the noise is increased the accuracy based on MDA degrades faster than the accuracy base on SDC. This indicates that the signal structures recovered by the standard sparse representation are more robust to noise thus yield less performance degradation.

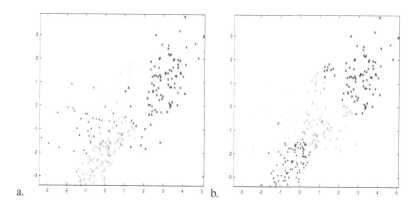

a.   b.

Fig. 4. Expected class assigment for the waveform dataset (a). Class assignments by SDC, visualized w.r.t parameters 1 and 3.

6   Conclusion

In summary, sparse representation for classification (SDC) is proposed. SDC is motivated by the ongoing researches in the area of sparse representation in the signal processing area. SDC incorporates reconstruction properties, discrimination power and sparsity for robust classification. In current implementation of SDC, the weight factors are empirically set to optimize the performance. Approaches to determine optimal values for the weighting factors are being conducted.

It is interesting to compare SDC with the Support Vector Machine. Both SDC and SVM incorporate sparsity and reconstruction error into consideration. For SDC, the two terms are explicitly included into objective function. For SVM, the two terms are included in the Bayesian formula. In SVM, the 'dictionary' used is the collection of values from the 'kernel function'.

References

1. Amirikian, B., Georgopoulus, A.P.: Motor Cortex: Coding and Decoding of Directional Operations. In: The Handbook of Brain Theory and Neural Networks, pp. 690–695. MIT Press, Cambridge (2003)
2. Babaie-Zadeh, M., Vigneron, V., Jutten, V.: Sparse decomposition over non-full-rank dictionaries. In: Proceedings of ICASSP 2009, Taipei, TW (April 2009)
3. Bishop, C.: Pattern recognition and machine learning. Springer, New York (2006)

4. Bodreau, M., Smith, A.M.: Activity in rostal motor cortex in response to predicatel force-pulse pertubations in precision grip task. J. Neurophysiol. 86, 1079–1085 (2005)
5. Breiman, L., Friedman, J.H., Olshen, R.A., Stone, C.J.: Classification and regression trees. Wadsworth International Group, Belmont (1984)
6. Candès, E.J., Tao, T.: Decoding by linear programming. IEEE Transactions Information Theory 51(12), 4203–4215 (2005)
7. Dash, M., Liu, H.: Feature selection for classification. Intelligent Data Analysis 1, 131–156 (1997)
8. Donoho, D.L.: For most large underdetermined systems of linear equations the minimal l^1-norm solution is also the sparsest solution. Technical report (2004)
9. Donoho, D.L.: Compressed sensing. IEEE Transactions on Information Theory 52(4), 1289–1306 (2006)
10. Donoho, D.L., Elad, M.: Maximal sparsity representation via ℓ^1 minimization. The Proc. Nat. Aca. Sci. 100(5), 2197–2202 (2003)
11. Elad, M.: Why simple shrinkage is still relevant for redundant representations? IEEE Transactions on Image Processing 52(12), 5559–5569 (2006)
12. Figueiredo, M.A.T., Nowak, R.D.: An EM algorithm for wavelet-based image restoration. IEEE Transactions on Image Processing 12(8), 906–916 (2003)
13. Gorodnitsky, I.F., Rao, B.D.: Sparse signal reconstruction from limited data using FOCUSS, a re-weighted minimum norm algorithm. IEEE Transactions on Signal Processing 45(3), 600–616 (1997)
14. Gribonval, R., Lesage, S.: A survey of sparse component analysis for blind source separation: principles, perspectives, and new challenges. In: Proceedings of ESANN 2006, pp. 323–330 (April 2006)
15. Gribonval, R., Nielsen, M.: Sparse decompositions in unions of bases. IEEE Trans. Inform. Theory 49(12), 3320–3325 (2003)
16. Hastie, T., Tibshirani, R.: Discriminant analysis by gaussian mixtures. Technical report, AT & T Bell laboratories, Murray Hill, NJ (1994)
17. Krakei, S., Hoffman, D.S., Strick, P.L.: Muscle and movement representation in the primary motor cortex. Science 285, 2136–2139 (1999)
18. MacLachlan, G., Basford, K.: Mixtures models: inference and applications to clustering. Marcel Dekker, New York (1988)
19. Mallat, S., Zhang, Z.: Matching pursuits with time-frequency dictionaries. IEEE Trans. on Signal Proc. 41(12), 3397–3415 (1993)
20. Mohimani, G.H., Babaie-Zadeh, M., Jutten, C.: Fast sparse representation based on smoothed ℓ^0 norm. In: Davies, M.E., James, C.J., Abdallah, S.A., Plumbley, M.D. (eds.) ICA 2007. LNCS, vol. 4666, pp. 389–396. Springer, Heidelberg (2007)
21. Mohimani, H., Babaie-Zadeh, M., Jutten, C.: A fast approach for overcomplete sparse decomposition based on smoothed ℓ^0 norm. Accepted in IEEE Trans. on Signal Processing
22. Morrow, M.M., Miller, L.E.: Prediction of muscle activity by populations of sequentially recorded primary motor cortex neurons. J. Neurophysiol. 89, 1079–1085 (2003)
23. Schwartz, A.B., Taylor, D., Tillery, S.I.H.: Extraction algorithms for cortical control of arm prosthesis. Current opinion in Neurobiology 11, 701–707 (2001)
24. Sergio, L.E., Kalaska, J.F.: Systematic changes in directional tuning of motor cortex cell activity with hand location in the workspace during generation of static isometric forces in constant spatial directions. J. Neurophysiol. 78, 1170–1174 (2005)
25. Taylor, D.M., Tillery, S.I.H., Schwartz, A.B.: Direct cortical control of 3d neuroprosthetic devices. Science 296(7), 1829–1832 (2002)
26. Van Staveren, G.W., Buitenweg, J.R., Heida, T., Ruitten, W.L.C.: Wave shape classification of spontaneaous neural activity in cortical cultures on micro-electrode arrays. In: Proceedings of the second joint EMBS/BMES conference, Houston, TX, USA, October, 23-26 (2002)
27. Wright, J., Yang, A.Y., Ganesh, A., Sastry, S.S., Ma, Y.: Robust face recognition via sparse representation. IEEE Transaction on Pattern Analysis and Machine Intelligence (Accepted, March 2008)

Pareto-Based Multi-output Metamodeling
with Active Learning

Dirk Gorissen, Ivo Couckuyt, Eric Laermans, and Tom Dhaene

Ghent University-IBBT, Dept. of Information Technology (INTEC), Gaston Crommenlaan 8,
9050 Ghent, Belgium

Abstract. When dealing with computationally expensive simulation codes or
process measurement data, global surrogate modeling methods are firmly established as facilitators for design space exploration, sensitivity analysis, visualization and optimization. Popular surrogate model types include neural networks,
support vector machines, and splines. In addition, the cost of each simulation
mandates the use of active learning strategies where data points (simulations)
are selected intelligently and incrementally. When applying surrogate models to
multi-output systems, the hyperparameter optimization problem is typically formulated in a single objective way. The different response outputs are modeled
separately by independent models. Instead, a multi-objective approach would
benefit the domain expert by giving information about output correlation, facilitate the generation of diverse ensembles, and enable automatic model type selection for each output on the fly. This paper outlines a multi-objective approach to
surrogate model generation including its application to two problems.

1 Introduction

Regardless of the rapid advances in High Performance Computing and multi-core architectures, it is rarely feasible to explore a design space using high fidelity computer
simulations. As a result, data based surrogate models (otherwise known as metamodels
or response surface models) have become a standard technique to reduce this computational burden and enable routine tasks such as visualization, design space exploration,
prototyping, sensitivity analysis, and optimization.

It is important to first stress that this paper is concerned with fully reproducing the
simulator behavior with a global model. The use of metamodels to approximate the
costly function for optimization (Metamodel Assisted Optimization) is *not* our goal.
Our objective is to construct a high fidelity approximation model that is as accurate as
possible over the *complete* design space of interest using as few simulation points as
possible (= active learning). This model can then be reused in other stages of the engineering design pipeline, for example as cheap accurate replacement models in design
software packages (e.g., ADS Momentum).

In engineering design simulators are typically modeled on a per-output basis. Each
output is modeled independently using separate models (though possibly sharing the
same data). Instead, the system may be modeled directly using multi-objective algorithms while maintaining the tie-in with active learning (classically a fixed data set is

D. Palmer-Brown et al. (Eds.): EANN 2009, CCIS 43, pp. 389–400, 2009.

chosen up front). This benefits the practitioner by giving information about output correlation, facilitating the generation of diverse ensembles (from the Pareto-optimal set), and enabling the automatic selection of the best model type on the fly for each output without having to resort to multiple runs. The purpose of this paper is to illustrate these concepts by discussing possible use-cases and potential pitfalls.

2 Global Surrogate Modeling

Again we stress that optimization of the simulation output is not the main goal, rather we are concerned with optimization of the surrogate model parameters (hyperparameter optimization) in order to generate accurate global models with a minimum number of data points. Mathematically, the problem is to approximate an unknown multivariate function $f : \mathbb{R}^d \mapsto \mathbb{C}^n$ by finding a suitable function s from an approximation space S such that $s : \mathbb{R}^d \mapsto \mathbb{C}^n \in S$ and s closely resembles f as measured by some criterion $\xi = (\Lambda, \varepsilon, \tau)$. Λ is the generalization estimator, ε the error (or loss) function, and τ is the target value required by the user.

This means that the global surrogate model generation problem (i.e., finding the best approximation $s^* \in S$) for a given set of data points D is defined as

$$s^* = \min_{t \in T} \min_{\theta \in \Theta} \Lambda(\varepsilon, s_{t,\theta}, D) \tag{1}$$

such that $\Lambda(\varepsilon, s_{t,\theta}^*, D) \leqslant \tau$. $s_{t,\theta}$ is the parametrization θ (from a parameter space Θ) of s and $s_{t,\theta}$ is of model type t (from a set of model types T). The first minimization over $t \in T$ is the task of selecting a suitable approximation model type, i.e., a rational function, a neural network, a spline, etc. This is the model type selection problem. In practice, one typically considers only a single $t \in T$, though others may be included for comparison. Then given a particular approximation type t, the task is to find the hyperparameter assignment θ that minimizes the generalization estimator Λ (e.g., determine the optimal order of a polynomial model). This is the hyperparameter optimization problem, though generally both minimization's are simply referred to as the model selection problem. Many implementations of Λ have been described: the holdout, bootstrap, cross validation, Akaike's Information Criterion, etc.

An additional assumption is that f is expensive to compute. Thus the number of function evaluations $f(X)$ needs to be minimized and data points must be selected iteratively, at points where the information gain will be the greatest. This process is referred to as active learning or adaptive sample selection. An important consequence of the adaptive sampling procedure is that the task of finding the best approximation s^* becomes a dynamic problem instead of a static one. Since the optimal model parameters will change as the amount and distribution of data points (D) changes.

3 Multi-objective Modeling

There are different ways to approach the global surrogate modeling problem in a multi-objective manner. The most obvious is to use multiple criteria to drive the hyperparameter optimization instead of a single one. In this case the minimization problem in equation 1 becomes a multi-objective one. This approach is useful because single criteria are inadequate at objectively gaging the quality of an approximation model. This is the so called *"The five percent problem"* [1] which always arises during approximation.

Secondly, a multi-objective approach is also useful if models with multiple outputs are considered. It is not uncommon that a simulation engine has multiple outputs that need to be modeled. Also, many Finite Element packages generate multiple performance values for free. The direct approach is to model each output independently with separate models (possibly sharing the same data). However, it is usually more computationally efficient to approximate the different responses together in the same model. The question then is how to drive the hyperparameter optimization. Instead of simple weighted scalarization (which is usually done) a useful approach is to tackle the problem directly in a multi-objective way. This avoids the weight assignment problem and the resulting multi-objective hyperparameter trace gives information about how the structure of the responses are correlated. This is particularly useful if the hyperparameters can be given physically relevant interpretations.

In addition, in both cases (multi-criteria, multi-output) the final Pareto front enables the generation of diverse ensembles, where the ensemble members consist of the (partial) Pareto-optimal set (see also references in [2]). This way all the information in the front can be used. Rainfall runoff modeling and model calibration in hydrology [3] are examples where this is popular. Models are generated for different output flow components and/or derivative measures and these are then combined into a weighted ensemble or fuzzy committee. Finally, a Pareto based approach to multi-output modeling also allows integration with the automatic surrogate model type selection algorithm described in [4]. This enables automatic selection of the best model type (Kriging, neural networks, support vector machines (SVM), etc.) for each output without having to resort to multiple runs or compromising accuracy. While for this paper we are concerned with the global case, this also applies to the local (optimization) case.

4   Related Work

There is an emerging body of research available on multi-objective hyperparameter optimization and model selection strategies. It is only since the emergence of off-the-shelf algorithms (with NSGA-II being the most popular) that work in this area has taken off. Increasingly authors are proposing multi-objective versions of classical metamodeling methods (e.g., [5]). An extensive and excellent overview of the work in this area is given by Jin et al. in [2]. By far the majority of the cited work uses multi-objective techniques to improve the training of learning methods. Typically an accuracy criterion (such as the validation error) is is used together with some regularization parameter or model complexity measure in order to produce more parsimonious models [6].

Another topic that has been the subject of extensive research is that of multi-objective surrogate based optimization (MOSBO). While initially their use has been constrained to the single objective case, results are increasingly being reported in the multi-objective case. Examples are ParEGO [7], the surrogate based variants of NSGA-II [8] and the work on statistical improvement by Keane et al. [9]. Though the research into MOSBO is still very young, an excellent overview of current research is already available in [10].

Little work seems to have been done on multi-objective multi-output modeling, with only some results for classification problems [11]. The link with active learning has also not yet been explored it seems. A multi-objective approach also enables automatic model type selection, both for the global and local (optimization) case. As Knowles et al. state in [10]: *"Little is known about which types of model accord best with particular features of a landscape and, in any case, very little may be known to guide this choice"*. Thus an algorithm to automatically solve this problem is very useful [12]. This is also noticed by Voutchkov et al. [8] who compare different surrogate models for approximating each objective during optimization. They note that, in theory, their approach allows the use of a different model type for each objective. However, such an approach will still require an a priori model type selection choice and does not allow for dynamic switching of the model type or the use of hybrid models.

5 Problems

We now discuss two problems to illustrate how multi-output problems can be modeled directly using multi-objective algorithms: an analytic test function and a Low Noise Amplifier (LNA). In addition to the results described here, more data and test cases can be found in [1]. For results involving multiple criteria the reader is also referred to [1].

5.1 Analytic Function

To easily illustrate the concepts and potential problems we first take a predefined analytical function with two input parameters x_1, x_2 and two responses y_1, y_2:

$$y_1(x) = (1 - x_1)^2 + 100(x_2 - x_1^2)^2 \tag{2}$$

$$y_2(x) = -20 \cdot \exp\left(-0.2\sqrt{\frac{1}{d} \cdot \sum_{i=1}^{d} x_i^2}\right) - \exp\left(\frac{1}{d} \cdot \sum_{i=1}^{d} \cos(2\pi \cdot x_i)\right) + 20 + e \tag{3}$$

So $f(x_1, x_2) = [y_1, y_2]$ with $x_i \in [-2, 2]$ and $d = 2$. Readers may recognize these two responses as representing the Rosenbrock and Ackley functions, two popular test functions for optimization. Plots of both functions are shown in figure 1.

   We chose this combined function since it is an archetypal example of how two outputs can differ in structure. Thus it should show a clear trade-off in the hyperparameter space when modeled together (i.e., a model that accurately captures y_1 will be poor at capturing y_2 and vica versa).

   It is important to stress that we are **not** interested in optimizing these functions directly (as is usually done), rather we are interested in reproducing them with a regression model (with two inputs and two outputs), using a minimal number of samples. Thus the problem is effectively a dynamic multi-objective optimization problem in the *hyperparameter* space.

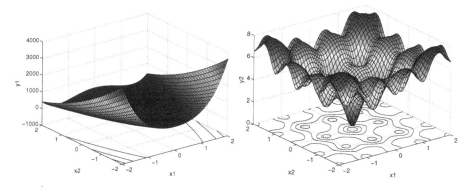

Fig. 1. The Rosenbrock (left) and Ackley (right) functions

5.2   Low Noise Amplifier (LNA)

As second test case we consider an RF circuit block. A LNA is the typical first stage of a receiver, having the main function of providing the gain needed to suppress the noise of subsequent stages, such as a mixer. In addition it has to give negligible distortion to the signal while adding as little noise as possible. The performance figures of a LNA (e.g. voltage gain, linearity, noise figure, etc.) can be determined by means of computer simulations and are functions of the design parameters (e.g. width and length of transistors, bias conditions, values of passive components) [13]. The goal of the design process is to figure out one or more sets of design parameters resulting in a circuit which fulfills the specifications, i.e. constraints given on the performances. More details and results of a modeling study of the noise parameters and admittances (in function of the design parameters) can be found in [14]. For this paper we consider the modeling of the performance figures directly.

We consider the two dimensional case where the power consumption P and third-order linearity $IIP3$ are approximated in function of the transistor width W and the inductance Lm.

6   Experimental Setup

6.1   SUMO-Toolbox

As experimental platform we used the SUrrogate MOdeling (SUMO) Toolbox v6.1.1 The SUMO Toolbox [14] is an adaptive tool that integrates different modeling approaches and implements a fully automated, adaptive global surrogate model construction algorithm. Given a simulation engine the toolbox produces a surrogate model within the time and accuracy constraints set by the user. Different plugins are supported: model types (rational functions, Kriging, splines, etc.), hyperparameter optimization algorithms (PSO, GA, simulated annealing, etc.), active learning (random, error based, density based, etc.), and sample evaluation methods (local, on a cluster or grid). Components can easily be added, removed or replaced by custom implementations.

The toolbox control flow is as follows: Initially, a small initial set of samples is chosen according to some experimental design. Based on this initial set, one or more surrogate models are constructed and their hyperparameters optimized according to a chosen optimization algorithm (e.g., PSO). Models are assigned a score based on one or more measures (e.g., cross validation) and the model parameter optimization continues until no further improvement is possible. The models are then ranked according to their score and new samples are selected based on the best performing models and the behavior of the response (the exact criteria depend on the active learning algorithm used). The hyperparameter optimization process is continued or restarted intelligently and the whole process repeats itself until one of the following three conditions is satisfied: (1) the maximum number of samples has been reached, (2) the maximum allowed time has been exceeded, or (3) the user required accuracy has been met. The SUMO-Toolbox and all the algorithms described here is available from http://www.sumo.intec.ugent.be

6.2   Analytic Function (AF)

In a first use case for this problem the Kriging [15] and NSGA-II plugins were used. The correlation parameters (θ) represent models in the population. Following general practice, the correlation function was set to Gaussian, and a linear regression was used. Starting from an initial Latin Hypercube Design of 24 points, additional points are added each iteration (using a density based active learning algorithm) up to a maximum of 150. The density based algorithm was used since it is shown to work best with Kriging models [16]. The search for good Kriging models (using NSGA-II) occurs between each sampling iteration with a population size of 30. The maximum number of generations between each sampling iteration is also 30.

A second use case of the same problem was done using the automatic model type selection plugin. This algorithm is based on heterogeneous evolution using the GA island model and is able to select the best model type for a given data source. A full discussion of this algorithm and its settings would consume too much space. Such information can be found in [4]. The following model types were included in the evolution: Kriging models, single layer ANNs (based on [17]), Radial Basis Function Neural Networks (RBFNN), Least Squares SVMs (LS-SVM, based on [18]), and Rational functions. Together with the ensemble models (which result from a heterogeneous crossover, e.g., a crossover between a neural network and a rational function), this makes that 6 model types will compete to fit the data. In this case, the multi-objective GA implementation of the Matlab GADS toolbox is used (which is based on NSGA-II). The population size of each model type is set to 10 and the maximum number of generations between each sampling iteration is set to 15. Again, note that the evolution resumes after each sampling iteration. In all cases model selection is done using the Root Relative Square Error (RRSE) on a dense validation set.

6.3   LNA

The same settings were used for this problem except that the sample selection loop was switched off and LS-SVM models were used to fit the data (instead of Kriging). Instead of sampling, a 12^2 full factorial design was used. More extensive results (including sampling and more dimensions) will be presented in a separate publication.

7   Results

7.1   Analytic Function: Use Case 1

Two snapshots of the Pareto front at different number of samples are shown in figure 2. The full Pareto trace (for all number of samples) is shown in figure 3(a). The figures clearly show that the Pareto front changes as the number of samples increase. Thus the multi-objective optimization of the hyperparameters is a dynamic problem, and the Pareto front will change depending on the data. This change can be towards a stricter trade-off (i.e., a less well defined 'elbow' in the front) or towards an easier trade-off (a more defined 'elbow'). What happens will depend on the model type.

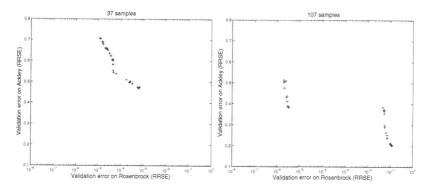

Fig. 2. Two snapshots of the Pareto front (during the model parameter optimization) at different sampling iterations (AF)

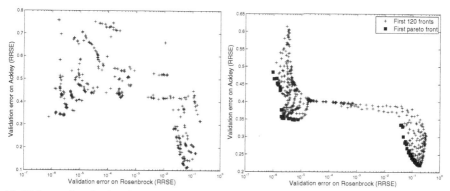

(a) With sample selection (up to 150 samples) (b) Without sampling (brute force hyperparameter search at 124 samples)

Fig. 3. Pareto front search traces (AF)

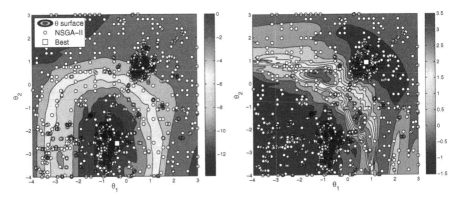

Fig. 4. Kriging θ-surface with the NSGA-II search trace (AF, *left:* Rosenbrock, *right:* Ackley)

From the figure it is also immediately clear that the Rosenbrock output is much easier to approximate than the Ackley output. Strangely though, there seems to be a discontinuity in the front. The Pareto front is split into two parts and as sampling proceeds the algorithm (NSGA-II) oscillates between extending the left front over the right front (or vica versa). The full Pareto trace in figure 3(a) also shows this.

To understand what is causing this behavior, a brute force search of the hyperparameter space was performed for a fixed LHD of 124 sample points. The space of all possible θ parameters was searched on a 100x100 grid with bounds [-4 3] (in log_{10} space) in each dimension. Simultaneously an extensive NSGA-II Kriging run was performed on the same data set for 450 generations. In both cases a dense validation set was used to calculate the accuracy of each model. The combination of both searches (for both outputs) is shown in figure 4 (note that the RRSE is in log scale). The brute force search of the θ-surface also allows the calculation of the true Pareto front (by performing a non-dominated sorting). The resulting Pareto front (and the next 119 fronts, these are shown for clarity) are shown in figure 3(b).

Studying the surfaces in figure 4 reveals what one would expect intuitively: the Rosenbrock output is very smooth and easy to fit, so given sufficient data a large range of θ values will produce an accurate fit. Fitting the Ackley output, on the other hand, requires a much more specific choice of θ to obtain a good fit. In addition both basins of attraction do not overlap, leading to two distinct optima. This means that (confirming intuition) the θ-value that produces a good model for y_1 produces a poor model for y_2 and vice versa. Together with figure 3(b) this explains the separate Pareto fronts seen in figure 2. The high ridge in the Ackley surface makes that there are no compromise solutions on the first front. Any model whose performance on y_2 would lie between the two separate fronts would never perform well enough on y_1 to justify a place on the first front. Thus, the fact that NSGA-II does not find a single united front is due to the structure of the optimization surface and not due to a limitation of NSGA-II itself.

7.2 Analytic Function: Use Case 2

The analytic problem was also tackled with the automatic model type selection algorithm described in [4]. This should enable the automatic identification of the most

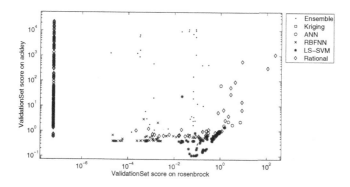

Fig. 5. Heterogeneous Pareto trace (AF)

adequate model type for each output *without* having to perform separate runs. Figure 5 shows the full Pareto search trace for the test function across all sampling iterations.

The figure shows the same general structure as figure 3(a): there is a strong trade-off between both outputs resulting in a gap in the search trace. If we regard the model selection results we find they agree with intuition. The Rosenbrock function is very easily fit with rational functions, and its smooth structure makes for an excellent approximation with almost any degree assignment (illustrated by the straight line at roughly 10^{-7} on the x-axis). However, those same rational models are unable to produce a satisfactory fit on the Ackley function, which is far more sensitive to the choice of hyperparameters. Instead the LS-SVM models perform best, the RBF kernel function matching up nicely with the different 'bumps' of the Ackley function.

Thus, we find the results to agree with intuition. The Rosenbrock function is better approximated with a global model since there are no local non-linearities. While the Ackley function benefits more from a local model, and this is exactly what figure 5 shows. Note the diverse front shown in the figure now allows the generation of a diverse ensemble (further improving accuracy) using standard ensemble techniques (bagging, weighted, etc.).

7.3 LNA: Use Case 1

The analytic function is of course just a synthetic example, but it is useful as an illustration and for pointing out potential problems. When we now turn to the LNA problem we shall see that the same kind of problems can arise when modeling real data. Figure 7(a) shows the search trace of an LS-SVM hyperparameter optimization run using NSGA-II (RBF kernel, optimizing c and σ). Of course the trade-off is not as extreme as with the analytic example but we do note some similarities: one output (P) is significantly easier to model than the other and there is again a gap in the Pareto front. A brute force search of the LS-SVM hyperparameter space (with the NSGA-II results overlaid) confirms this (figure 6). It is also interesting to compare figure 6 with figure 4. Both figures are in line with the authors' experience that optimizing the Kriging θ parameters is typically harder than selecting good values for c, σ. Our experience is that the θ optimization landscape is more rugged, multi-modal, and more sensitive to the

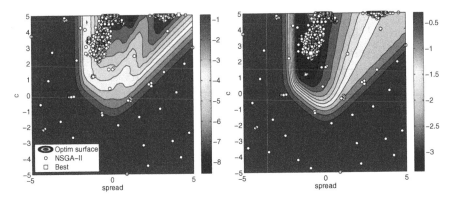

Fig. 6. SVM (c, σ)-surface with the NSGA-II search trace (LNA, *left: P*, *right: IIP3*)

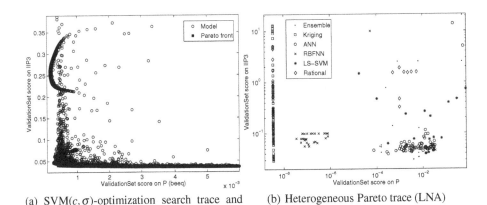

(a) SVM(c, σ)-optimization search trace and   (b) Heterogeneous Pareto trace (LNA)
Pareto front (LNA)

Fig. 7. Pareto front search traces (AF)

data distribution (see also the discussion in [16]). On the other hand, it is usually quite easy to generate an LS-SVM that captures the trends in the data without being too sensitive to the data distribution. An added benefit of SVM type models is that the number of hyperparameters is independent of the problem dimensionality (unlike Kriging).

7.4   LNA: Use Case 2

Finally, we can also apply the automatic model type selection algorithm to the LNA data. The resulting search trace is shown in figure 7(b). Again, note the similarity with figure 5. However, contrary to the previous problem, we see that in this case including different model types can actually alleviate the trade-off in the data. It turns out that while LS-SVM models have problems capturing both outputs accurately, the Kriging models are able to do so, producing a single best model for both outputs. Again the almost straight line at 10^{-8} indicates that the P output is extremely easy to fit, a wide range of hyperparameters will produce a good fit.

8 Conclusion and Future Work

The use of metamodels to aid design space exploration, visualization, etc. has become standard practice among scientists and engineers alike. In order to increase insight and save time when dealing with multi-response systems, the goal of this paper was to illustrate that a multi-objective approach to global surrogate model generation can be useful. This allows multiple outputs to be modeled together, giving information about the trade-off in the hyperparameter space. It further enables selecting the best model type for each output on the fly, permits the generation of diverse ensembles, and the use of multiple criteria. All for roughly the same computational cost as performing multiple independent runs (which is still outweighed by the simulation cost).

While this paper presented some interesting use cases, still much work remains. An important area requiring further investigation is understanding how the iterative sample selection process influences the hyperparameter optimization landscape. For the tests in this paper the authors have simply let the optimization continue from the previous generation. However, some initial tests have shown that an intelligent restart strategy can improve results. Knowledge of how the number and distribution of data points affects the hyperparameter surface would allow for a better tracking of the optimum, reducing the cost further. The influence of noise on the hyperparameter optimization (e.g., neural network topology selection) also remains an issue as is the extension to high dimensional problems (many outputs/criteria). In general, while progress towards dynamic multi-objective optimization has been made, this is a topic that current research in multi-objective surrogate modeling is only just coming to terms with [10].

Acknowledgments

The authors would like to thank Jeroen Croon from the NXP-TSMC Research Center, Device Modeling Department, Eindhoven, The Netherlands for the LNA code.

References

1. Gorissen, D., Couckuyt, I., Dhaene, T.: Multiobjective global surrogate modeling. Technical Report TR-08-08, University of Antwerp, Middelheimlaan 1, 2020 Antwerp, Belgium (2008)
2. Jin, B.Y., Sendhoff: Pareto-based multiobjective machine learning: An overview and case studies. IEEE Transactions on Systems, Man, and Cybernetics, Part C: Applications and Reviews 38(3), 397–415 (2008)
3. Fenicia, F., Solomatine, D.P., Savenije, H.H.G., Matgen, P.: Soft combination of local models in a multi-objective framework. Hydrology and Earth System Sciences Discussions 4(1), 91–123 (2007)
4. Gorissen, D., De Tommasi, L., Croon, J., Dhaene, T.: Automatic model type selection with heterogeneous evolution: An application to rf circuit block modeling. In: Proceedings of the IEEE Congress on Evolutionary Computation, WCCI 2008, Hong Kong(2008)
5. Mierswa, I.: Controlling overfitting with multi-objective support vector machines. In: GECCO 2007: Proceedings of the 9th annual conference on Genetic and evolutionary computation, pp. 1830–1837. ACM Press, New York (2007)

6. Fieldsend, J.E.: Multi-objective supervised learning. In: Knowles, J., Corne, D., Deb, K. (eds.) Multiobjective Problem Solving from Nature From Concepts to Applications. Natural Computing Series. LNCS. Springer, Heidelberg (2008)

7. Knowles, J.: Parego: A hybrid algorithm with on-line landscape approximation for expensive multiobjective optimization problems. IEEE Transactions on Evolutionary Computation 10(1), 50–66 (2006)

8. Voutchkov, I., Keane, A.: Multiobjective Optimization using Surrogates. In: Parmee, I. (ed.) Adaptive Computing in Design and Manufacture 2006. Proceedings of the Seventh International Conference, Bristol, UK, pp. 167–175 (April 2006)

9. Keane, A.J.: Statistical improvement criteria for use in multiobjective design optimization. AIAA Journal 44(4), 879–891 (2006)

10. Knowles, J.D., Nakayama, H.: Meta-modeling in multiobjective optimization. In: Branke, J., Deb, K., Miettinen, K., Słowiński, R. (eds.) Multiobjective Optimization. LNCS, vol. 5252, pp. 245–284. Springer, Heidelberg (2008)

11. Last, M.: Multi-objective classification with info-fuzzy networks. In: Boulicaut, J.-F., Esposito, F., Giannotti, F., Pedreschi, D. (eds.) ECML 2004. LNCS, vol. 3201, pp. 239–249. Springer, Heidelberg (2004)

12. Keys, A.C., Rees, L.P., Greenwood, A.G.: Performance measures for selection of metamodels to be used in simulation optimization. Decision Sciences 33, 31–58 (2007)

13. Lee, T.: The Design of CMOS Radio-Frequency Integrated Circuits, 2nd edn. Cambridge University Press, Cambridge (2003)

14. Gorissen, D., De Tommasi, L., Crombecq, K., Dhaene, T.: Sequential modeling of a low noise amplifier with neural networks and active learning. Neural Computing and Applications 18(5), 485–494 (2009)

15. Lophaven, S.N., Nielsen, H.B., Søndergaard, J.: Aspects of the matlab toolbox DACE. Technical report, Informatics and Mathematical Modelling, Technical University of Denmark, DTU, Richard Petersens Plads, Building 321, DK-2800 Kgs. Lyngby (2002)

16. Gorissen, D., De Tommasi, L., Hendrickx, W., Croon, J., Dhaene, T.: Rf circuit block modeling via kriging surrogates. In: Proceedings of the 17th International Conference on Microwaves, Radar and Wireless Communications, MIKON 2008 (2008)

17. Nørgaard, M., Ravn, O., Hansen, L., Poulsen, N.: The NNSYSID toolbox. In: IEEE International Symposium on Computer-Aided Control Sysstems Design (CACSD), Dearborn, Michigan, USA, pp. 374–379 (1996)

18. Suykens, J.A.K., Gestel, T.V., Brabanter, J.D., Moor, B.D., Vandewalle, J.: Least Squares Support Vector Machines. World Scientific Publishing Co., Pte, Ltd., Singapore (2002)

Isolating Stock Prices Variation with Neural Networks

Chrisina Draganova[1], Andreas Lanitis[2], and Chris Christodoulou[3]

[1] University of East London, 4-6 University Way, London E16 2RD, UK
[2] Department of Multimedia and Graphic Arts, Cyprus University of Technology, 31 Archbishop Kyprianos Street, P. O. Box 50329, 3603 Lemesos, Cyprus
[3] Department of Computer Science, University of Cyprus, 75 Kallipoleos Avenue, P.O. Box 20537, 1678 Nicosia, Cyprus
c.draganova@uel.ac.uk, andreas.lanitis@cut.ac.cy,
cchrist@cs.ucy.ac.cy

Abstract. In this study we aim to define a mapping function that relates the general index value among a set of shares to the prices of individual shares. In more general terms this is problem of defining the relationship between multivariate data distributions and a specific source of variation within these distributions where the source of variation in question represents a quantity of interest related to a particular problem domain. In this respect we aim to learn a complex mapping function that can be used for mapping different values of the quantity of interest to typical novel samples of the distribution. In our investigation we compare the performance of standard neural network based methods like Multilayer Perceptrons (MLPs) and Radial Basis Functions (RBFs) as well as Mixture Density Networks (MDNs) and a latent variable method, the General Topographic Mapping (GTM). According to the results, MLPs and RBFs outperform MDNs and the GTM for this one-to-many mapping problem.

Keywords: Stock Price Prediction, Neural Networks, Multivariate Statistics, One-to-Many Mapping.

1 Introduction

In many problems involving the analysis of multivariate data distributions, it is desirable to isolate specific sources of variation within the distribution, where the sources of variation in question represent a quantity of interest related to the specific problem domain. The isolation of different types of variation within a training set enables the generation of synthetic samples of the distribution given the numerical value of a single type of variation (or data dimension). Usually multiple parameters are required to specify a complete sample in a distribution, thus the process of generating a sample given the value of a simple parameter takes the form of one-to-many mapping. In general one-to-many mapping problems are ill-conditioned, requiring the use of dedicated techniques that use prior knowledge in attempting to formulate an optimized

D. Palmer-Brown et al. (Eds.): EANN 2009, CCIS 43, pp. 401–408, 2009.

mapping function. With our work we aim to investigate the use of different methods for defining a mapping associating a specific source of variation within a distribution and a given representation of this data distribution.

As part of our performance evaluation framework we assess the performance of different one-to-many mapping methods in a case study related to the definition of the relationship between the index value of twenty stocks included in the FTSE 100 UK (www.ftse.com/Indices/UK_Indices/index.jsp) and the daily individual stock prices over a three year time period. We implement and test methods that learn the relationship between the daily general index value and the corresponding individual daily stock prices of twenty of the FTSE 100 UK stocks with largest volume that have available data for at least three consecutive years. Once the mapping is learned we attempt to predict the daily stock prices of each share given the value of the general index. This application can be very useful for predicting the prices of individual share prices based on a given value of the daily index.

As part of our experimental evaluation process we investigate the following neural network-based methods: Multilayer Perceptron (MLP) [1], Radial Basis Functions [2], Mixture Density Networks (MDN) [3, 4] and the non-linear latent variable method Generative Topographic Mapping (GTM) [5]. As a reference benchmark of the prediction accuracy we consider the values of the predicted variables that correspond to the average values over certain intervals of the quantity of interest that we are trying to isolate (the so called Sample Average (SA) method). The SA method provides an easy way to estimate the most typical values of each stock, for a given index value. However, the SA method does not take into account the variance of stock prices hence it cannot be regarded as an optimum method for this problem.

The rest of the paper is organised as follows: in section 2 we present an overview of the relevant literature; in section 3 we describe the case study under investigation, the experiments and give visual and quantitative results and in section 4 we present our conclusions.

2 Literature Review

There exist well-established neural network methods for solving the mapping approximation problem such as the Multilayer Perceptron (MLP) [1] and Radial Basis Functions (RBF) [2]. The aim of the training in these methods is to minimize a sum-of-square error function so that the outputs produced by the trained networks approximate the average of the target data, conditioned on the input vector [3]. It is reported in [4] and [6], that these conditional averages may not provide complete description of the target variables especially for problems in which the mapping to be learned is multi-valued and the aim is to model the conditional probability distributions of the target variables [4]. In our case despite the fact that we have a multi-valued mapping we aim to model the conditional averages of the target data, conditioned on the input that represents a source of variation within this distribution. The idea is that when we change the value of the parameter representing the source of variation in the allowed range, the mapping that is defined will give typical representation of the target parameters exhibiting the isolated source of variation.

Bishop [3, 4] introduces a new class of neural network models called Mixture Density Networks (MDN), which combine a conventional neural network with a mixture density model. The mixture density networks can represent in theory an arbitrary conditional probability distribution, which provides a complete description of target data conditioned on the input vector and may be used to predict the outputs corresponding to new input vectors. Practical applications of feed forward MLP and MDN to the acoustic-to-articulatory mapping inversion problem are considered in [6]. In this paper, it is reported that the performance of the feed-forward MLP is comparable with results of other inversion methods, but that it is limited to modelling points approximating a unimodal Gaussian. In addition, according to [6], the MLP does not give an indication of the variance of the distribution of the target points around the conditional average. In the problems considered in [4] and [6], the modality of the distribution of the target data is known in advance and this is used in selecting the number of the mixture components of the MDN.

Other methods that deal with the problem of mapping inversion and in particular mapping of a space with a smaller dimension to a target space with a higher dimension are based on latent variable models [7]. Latent variables refer to variables that are not directly observed or measured but can be inferred using a mathematical model and the available data from observations. Latent variables are also known as hidden variables or model parameters. The goal of a latent variable model is to find a representation for the distribution of the data in the higher dimensional data space in terms of a number of latent variables forming a smaller dimensional latent variable space. An example of a latent variable model is the well-known factor analysis, which is based on a linear transformation between the latent space and the data space [3]. The Generative Topographic Mapping (GTM) [5] is a non-linear latent variable method using a feed-forward neural network for the mapping of the points in the latent space into the corresponding points in the data space and the parameters of the model are determined using the Expectation-Maximization (EM) algorithm [8]. The practical implementation of the GTM has two potential problems: the dimension of the latent space has to be fixed in advance and the computational cost grows exponentially with the dimension of the latent space [9].

Density networks [10] are probabilistic models similar to the GTM. The relationship between the latent inputs and the observable data is implemented using a multilayer perceptron and trained by Monte Carlo methods. The density networks have been applied to the problem of modelling a protein family [10]. The biggest disadvantage of the density networks is the use of the computer-intensive sampling Monte Carlo methods, which do no not scale well when the dimensionality is increased.

Even though the problem we consider in this paper bear similarities with the problem of sensitivity analysis with respect to neural networks, there are also distinct differences. In sensitivity analysis the significance of a single input feature to the output of a trained neural network is studied by applying that input, while keeping the rest of the inputs fixed and observing how sensitive the output is to that input feature (see for example [11] and references therein). In the problem investigated in this paper, we do not have a trained neural network, but the index based on the values of 20 stocks. Based on our knowledge of the application, i.e., the index, we isolate a specific source of variation and carry out an one-to-many mapping between that isolated source and the model (which is a multivariate data distribution). More specifically, the model refers to all the 20 stock values. This allows us to analyse the variation of the isolated source within the model.

3 Experiments and Results

For the experiments related to stock price prediction described in this paper we have used the historical daily prices available at uk.finance.yahoo.com. Twenty stocks have been selected from those that have the largest volume and that have their daily prices between 16/12/2003 and 10/12/2007. Precisely the set of selected stocks includes: BA, BARC, BLT, BP, BT, CW, FP, HBOS, HSBA, ITV, KGF, LGEN, LLOY, MRW, OML, PRU, RBS, RSA, TSCO and VOD. The daily index values for these 20 stocks have been calculated using the method described in [12] with a starting index point set to 1000.

3.1 Experimental Methodology

We first train neural network models using the MLP with the scaled conjugate gradient algorithm [13], the RBF and MDN methods. The inputs for the neural network model are the numerical values of the daily general index value and the output corresponds to the prices of the 20 stocks. In the MLP model, the network has one input node, one hidden layer with hyperbolic tangent (tanh) activation function and an output layer with linear activation function, since the problem we consider is a regression problem. In the case of RBF similarly to the MLP, the input layer has one node, the output layer has linear outputs and the hidden layer consists of nodes (centres) with Gaussian basis functions. The Gaussian basis function centres and their widths are optimised by treating the basis functions as a mixture model and using the Expectation-Maximisation (EM) algorithm for finding these parameters. The number of hidden nodes in the MLP and RBF networks and the learning rate in the MLP network are set empirically. We also set empirically the number of hidden nodes and kernel functions (mixture components) in the MDN model. Theoretically by choosing a mixture model with a sufficient number of kernel functions and a neural network with a sufficient number of hidden units, the MDN can approximate as closely as desired any conditional density. In the case of discrete multi-valued mappings the number of kernel functions should be at least equal to the maximum number of branches of the mapping. We performed experiments using up to five kernel functions.

The SA method is applied by calculating the average vectors of the 20 stock prices corresponding to the index value in fifty equal subintervals between 715.1 and 1044.8, which are the minimum and the maximum value of the general index.

We have also carried out an experiment for isolating one latent variable using the GTM. The GTM models consist of an RBF non-linear mapping of the latent space density to a mixture of Gaussians in the data space of parameters (20 stock prices). The models are trained using EM algorithm. After training we use the RBF mapping to obtain the parameters corresponding to several values of the latent variable. We show the variation of the latent variable reflected on the stock prices.

3.2 Results

Table 1 represents the quantitative results for each method used, expressed as the mean error between actual and predicted stock prices over the considered period of time. The mean error was calculated using: mean error = $(\Sigma_{i=1,n} \, abs(y_i - a_i))/n$, where y_i is the predicted and a_i is the actual price of the shares, n is the total number days over which the share prices are predicted.

Figure 1 illustrates the graphical results for the actual and the model output prices of one of the stocks - ITV obtained with the SA, MLP, RBF, MDN and GTM methods. These graphical results show the variation of the index value reflected on the prices of the stock in question.

The graphical and quantitative results corresponding to the SA, MLP and RBF models are comparable. The results obtained with the MDN method did not produce better representation of the data which again can be explained with the large dimensionality of the problem. In the case of the GTM method, although the quantitative results are worse than those obtained with the other methods, the graphical results show that the general trend of the actual prices is captured, demonstrating therefore the potential of the GTM method for modeling the distribution of the stock prices in terms of one latent variable.

Table 1. Mean error between actual and predicted prices of the 20 listed shares (see text for details) with different methods

Share	Method				
	MLP	**SA**	**RBF**	**MDN**	**GTM**
BA	45.51	45.94	49.66	57.25	84.40
BARC	40.72	41.36	43.94	58.45	73.96
BLT	143.26	145.12	157.12	191.11	266.71
BP	41.46	41.93	46.16	55.64	74.51
BT	19.07	19.64	24.32	26.09	41.09
CW	14.29	14.68	17.84	20.59	27.12
FP	14.58	14.69	15.01	21.71	22.76
HBOS	67.60	68.22	70.32	99.54	121.99
HSBA	30.85	30.56	31.79	35.03	51.85
ITV	4.73	4.65	5.01	6.57	7.62
KGF	18.51	18.13	19.07	27.97	32.15
LGEN	9.62	9.80	10.88	13.39	17.88
LLOY	27.90	28.40	30.38	35.77	49.97
MRW	22.99	23.45	28.55	31.13	41.92
OML	15.32	15.47	15.65	18.63	27.73
PRU	49.34	50.07	56.36	58.32	93.83
RBS	25.93	26.13	27.70	43.30	44.50
RSA	13.17	13.03	13.69	15.44	28.95
TSCO	28.21	28.16	33.01	37.29	55.30
VOD	7.91	7.65	8.36	11.70	15.01
Total	*32.05*	*32.35*	*35.24*	*43.25*	*58.96*

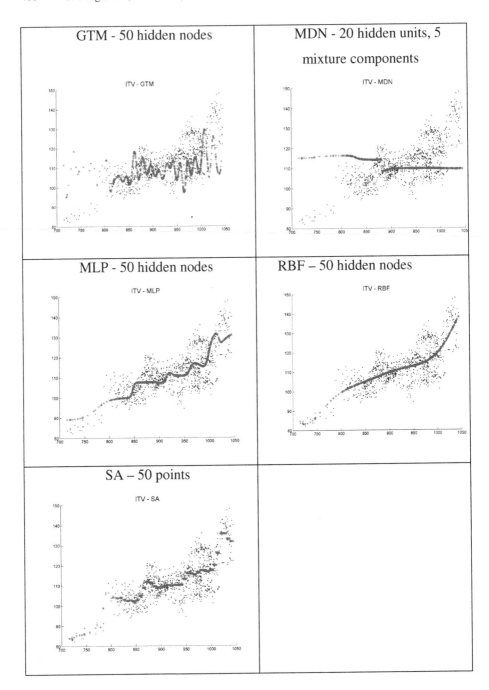

Fig. 1. Sample graphical result for the variation of the index value reflected on the ITV stock price; the solid lines and the scattered dots indicate the predicted and actual stock prices respectively corresponding to the index values

4 Conclusions

In this paper we investigate the use of a number of different techniques in the task of finding the mapping between a specific source of variation within a multivariate data distribution and the multivariate data distribution itself. The source of variation represents a quantity of interest related to a given problem domain. More specifically, we aim to define a mapping function relating the general index value among a set of shares to the prices of individual shares. We look for such mapping which gives a typical representation of the data distribution that exhibits the variation of the specific quantity. In this mapping, the target space has a higher dimension than the input space and for one input value the target output value is not unique. This leads to finding one-to-many multi-valued mapping. More specifically, we investigate several well-known methods used for solving such problems including MLP, RBF, MDN and GTM.

The results of our experiments demonstrate the potential of using neural networks trained with the MLP and RBF methods for isolating sources of variation and generating typical representations of the corresponding data distributions in the considered case study. With the neural network approach we do not make any assumptions about the mapping function; the neural networks are learning the complex mapping between the desired attributes and the parameters related the specific applications. The quantitative results obtained with the MLP and RBF are similar. The best result is achieved with the MLP method. The graphical results obtained with these methods are also similar. The MLP and RBF methods give the conditional averages of the target data conditioned on the input vectors and as expected they do not give a complete description of the target data reported in [4, 6]. For this problem we are addressing it is sufficient to define a mapping that generates typical samples of the data distribution (and not its entire variance) given specific values of the desired source of variation. This makes our results with the MLP and RBF (which are relatively simple methods, compared to MDN and GTM) not only acceptable but quite good for the type of inversion problems we are addressing, compared to the MLP results for the acoustic-to-articulatory inversion mapping reported in [14], [6]. For this one-to-many problem considered and also the problem of reconstructing the same spectrum from different spectral line parameters [4], the entire variance of the distribution is required. It has to be noted also that for the one-to-many problems considered in our paper the training algorithm for the MLP does not have to be modified as suggested by Brouwer [15], resulting in increased complexity. To the best of our knowledge RBFs have not previously been specifically used for one-to-many problems.

The MDN [3, 4] can give a complete description of the target data conditioned on the input vector provided that the number of mixture components is at least equal to the maximum number of branches of the mapping. The experiments carried out in the considered case study demonstrate that in problems for which the modality of the distribution of the target data is very large and not known, the application of the MDN leads to a large number of mixture components and outputs. Therefore it does not provide the desired type of mapping, which explains the poor results obtained with the MDN method. In particular, these results do not show the desired variation of the quantity of interest in the stock prices case study, they do not show complete representation of the target data space.

The GTM [5] is used to map points in the latent variable interval to points in the target data space and our results with this method actually show this mapping of one latent variable to the corresponding data spaces. It has to be noted though, that in order to isolate a specific source of variation, we need to find all latent variables which leads to a problem that is not computationally feasible [9]. In the stock price case study the isolated latent variable might be a different source of variation and not necessarily the desired index value.

The framework presented in the paper can be potentially useful in various applications involving multivariate distributions. With regards to the stock prices application, it is well established that trends of the general index can be easily predicted, unlike price fluctuations at share level. Therefore the ability to infer individual stock prices based on the general index value can be an invaluable tool for efficient portfolio management and prediction of the behavior of individual shares.

References

1. Rumelhart, D.E., Hinton, D.E., Williams, R.J.: Learning representations by back-propagation errors. Nature 323, 533–536 (1986)
2. Powell, M.J.D.: Radial basis functions for multivariable interpolation: A review. In: IMA Conference on Algorithms for the approximation of Functions and Data, pp. 143–167. RMCS, Shrivenham (1985)
3. Bishop, C.M.: Neural Networks for Pattern Recognition. Oxford University Press, New York (1995)
4. Bishop, C.M.: Mixture Density Networks. Technical Report NCRG/94/004, Neural Computing Research Group, Aston University (1994),
   http://research.microsoft.com/~cmbishop/downloads/
   Bishop-NCRG-94-004.ps
5. Bishop, C.M., Svensén, M., Williams, C.K.I.: GTM: The Generative Topographic Mapping. Neural Computation 10(1), 215–234 (1998)
6. Richmond, K.: Mixture Density Networks, Human articulatory data and acoustic-to-articulatory inversion of continuous speech (2001),
   http://www.cstr.ed.ac.uk/downloads/publications/2001/
   Richmond_2001_a.ps
7. Bartholomew, D.J.: Latent Variable Models and Factor Analysis. Charles Griffin & Company Ltd., London (1987)
8. Dempster, A., Laird, N., Rubin, D.: Maximum likelihood from incomplete data via the EM algorithm. Journal of the Royal Statistical Society. Series B 39(1), 1–38 (1977)
9. Carreira-Perpinan, M.A.: One-to-many mappings, continuity constraints and latent variable models. In: Proc. IEE Colloquium on Applied Statistical Pattern Recognition, Birmingham, pp. 14/1–14/6 (1999)
10. MacKay, D.J.C., Gibbs, M.N.: Density networks. In: Proceedings of Society for General Microbiology, Edinburgh (1997)
11. Zeng, X., Yeung, D.S.: A Quantified Sensitivity Measure for Multilayer Perceptron to Input Perturbation. Neural Computation 15, 183–212 (2003)
12. FTSE Guide to UK Calculation Methods,
    http://www.ftse.com/Indices/UK_Indices/Downloads/
    uk_calculation.pdf#
13. Møler, M.: A scaled conjugate gradient algorithm for fast supervised learning. Neural Networks 6(4), 525–533 (1993)
14. Carreira-Perpiñán, M.Á.: Continuous latent variable models for dimensionality reduction and sequential data reconstruction. PhD thesis, Dept. of Computer Science, University of Sheffield, UK (2001),
    http://faculty.ucmerced.edu/mcarreira-perpinan/papers/
    phd-thesis.html
15. Brouwer, R.K.: Feed-forward neural network for one-to-many mappings using fuzzy sets. Neurocomputing 57, 345–360 (2004)

Evolutionary Ranking on Multiple Word Correction Algorithms Using Neural Network Approach

Jun Li, Karim Ouazzane, Yanguo Jing, Hassan Kazemian,
and Richard Boyd

Faculty of Computing, London Metropolitan University,
166-220 Holloway Road, London, N7 8DB, UK
{Jul029,k.ouazzane,y.jing,h.kazemian}@londonmet.ac.uk,
richard.boyd@disabilityessex.org

Abstract. Multiple algorithms have been developed to correct user's typing mistakes. However, an optimum solution is hardly identified among them. Moreover, these solutions rarely produce a single answer or share common results, and the answers may change with time and context. These have led this research to combine some distinct word correction algorithms to produce an optimal prediction based on database updates and neural network learning. In this paper, three distinct typing correction algorithms are integrated as a pilot research. Key factors including Time Change, Context Change and User Feedback are considered. Experimental results show that 57.50% Ranking First Hitting Rate (HR) with the samples of category one and a best Ranking First Hitting Rate of 74.69% within category four are achieved.

Keywords: Neural Network, Metaphone, Levenshtein distance, word 2-gram, Jaro distance, Jaro-Winkler distance, ranking First Hitting Rate.

1 Introduction

Computer users inevitably make typing mistakes. These may be seen as spelling errors, prolong key press and adjacent key press errors etc [1]. Multiple solutions such as Metaphone [2] and n-grams [3] have been developed to correct user's typing mistakes, and each of them may have its unique features. However, an optimum solution is hardly identified among them. Therefore, it is desired to develop a hybrid solution based on combining these technologies, which can put all merits of those distinct solutions together.

Moreover, each function may rarely generate a single answer, let alone multiple functions which may produce a larger list of suggestions. This requires developing an evolutionary and adjustable approach to prioritize the suggestions in this list. Also, the answers may change within different context; and the solutions are also required to evolve based on user's feedbacks. Therefore, this research is motivated by the requirement of combining distinct word correction algorithms and subsequently producing an optimal prediction based on dataset updates and neural network learning process.

D. Palmer-Brown et al. (Eds.): EANN 2009, CCIS 43, pp. 409–418, 2009.

2 Typing Correction Functions

There are many types of errors caused by users, for example, spelling errors, hitting adjacent key and cognitive difficulties. Some efforts have been made based on different technologies such as spell checking, natural language processing and control signals filter. In this paper, a pilot research is carried out and three distinct algorithms (referred to as L.M.T), namely, Levenshtein word distance algorithm [4], Metaphone algorithm, and 2-gram word algorithm are used.

Metaphone is a phonetic algorithm indexing words by their sound, which can be adjusted to correct typing errors. These are two examples,

```
able -> APL
hello-> HL
```

The right side of the arrow is words' phonetic keys. Let's assume that a user intends to type a word 'hello' but mistakenly typed 'hallo' instead, whose phonetic keys are identical. Subsequently, the system is able to index and retrieve possible words from the database based on the phonetic key and present them to a user for selection.

Levenshtein distance is another function that needs to be explored. It is designed based on the calculation of minimum number of operations required to transform one string into another, where an operation is an insertion, deletion, or substitution of a single character, for instance,

```
hello <-> hallo //the string  distance is one
hello <-> all   //the string  distance is three
```

After a comparison with each string stored in the memory, the pair with the least distance can be considered as having the highest similarity, and then the one or the group with the least distance can be presented through the user interface module.

Word 2-gram (i.e. word digram or word bigram) is groups of two consecutive words, and is very commonly used as the basis for simple statistical analysis of text. For instance, given a sentence, 'I am a student', some word 2-gram samples are,

```
I am
am a
```

For example, under an ideal condition, 'am' can be predicted if its predecessor 'I' is typed. Then the predicted word 'am' can be used to make sure that user types it correctly.

Another similar method to Levenshtein distance is Jaro metric [5]. The Jaro distance metric states that, given two strings s_1 and s_2, two characters a_i, b_j from s_1 and s_2 respectively are considered matching only if,

$$i - \frac{\min\left(|s_1|, |s_2|\right)}{2} \le j \le i + \frac{\min\left(|s_1|, |s_2|\right)}{2} . \tag{1}$$

then their distance d is calculated as,

$$d = \frac{1}{3}\left(\frac{|s_1^{'}|}{|s_1|} + \frac{|s_2^{'}|}{|s_2|} + \frac{|s_1^{'}|-t}{|s_1^{'}|}\right) . \tag{2}$$

Where $|s_1^{'}|, |s_2^{'}|$ are the numbers of s_1 matching s_2 and s_2 matching s_1 characters respectively, and t is the number of transpositions.

A variant of Jaro metric uses a prefix scale p, which is the longest common prefix of string s_1 and s_2. Let's define Jaro distance as d, then Jaro-Winkler [6] distance can be defined as,

$$Jaro-Winkler(d + \frac{\max(p,4)}{10} * (1-d)) . \tag{3}$$

The result of the Jaro-Winkler distance metric is normalized into the range of [0, 1]. It is designed and best suited for short strings.

3 Word List Neural Network Ranking and Definitions

As the above solutions rarely produce a single answer or share common results, this implies that a combination will definitely be a more accurate solution. However, it requires a word-list with words priority rather than a single word to be generated. For instance, a user intends to type a word 'hello' but mistakenly typed 'hallo' instead. Let's assume that two functions, namely, Metaphone method and Levenshtein distance are integrated together and the correction results are produced as follows,

```
Metaphone:            'hello', 'hall'
Levenshtein distance: 'hello', 'all', 'allow'
```

Then a words list with 'hello', 'hall', 'all' and 'allow' is made available to the user. It is evident that a ranking algorithm computing each individual's priority is necessary before a word list is presented to a user.

In a real-time interaction, it requires that the word-list priority computation is able to adapt itself timely based on the user behavior and some other factors. In practice, this can be simplified by considering the word-list priority computation as a function of three variables: Time Change, Context Change and User Feedback. Therefore, a ranking algorithm, which is able to learn from user's selection and context changing over time, and subsequently adjust its weights, can be developed. In this research, these three variables are further quantified and represented by frequency increase, word 2-gram statistic and a supervised learning algorithm respectively, and subsequently a novel Word List Ranking neural network model associated with the variables is developed. The definitions introduced below are useful as they are part of the rules which dictate the whole process.

First Rank Conversion Values and First Hitting Rate Definition: In a neural network post processing, if its output follows a 'winner takes all' strategy, that is, the maximum value in output is converted into one and the rest values are converted into zeros, then the converted elements are named as First Rank Conversion Values. Given testing metrics P, target metrics T and testing result metrics R where their numbers of lines and columns are equal and expressed as n, m respectively, then the Hitting Rate is $HR = \{hr_i \mid hr_i = zeros(T - R_i)/n, i \in m\}$, where R_i is the ith Rank Conversion Values of R, $zeros()$ is the function to compute the number of zero vector included in metrics, and the First Hitting Rate is hr_1.

Word-List n-formula Prediction Definition: Let's assume that one has distinct algorithms set $A = \{a_1...a_i...a_n\}$, where $1 \le i \le n$ and i, n are positive integers. To process a sequence s, if there exists a one-to-many mapping $\{s \rightarrow O_i\}$ associated with algorithm a_i between input and output, where $O_i = \{o_i^j \mid 1 \le j \le m_i\}$, o_i^j is a generated sequence from the algorithm, and j, m_i are positive integers, then one has $\sum_{i=1}^{n} m_i$ sequence generated and the sequence set is defined as Word-List. The process based on the use of n algorithms to generate a word-list is called n-formula Prediction.

Word-List Success Prediction Rate Definition: Given a word list generated by several algorithms to correct a wrong typing, if the intended word is in the word list, then it is a Success Prediction. If there is a set of wrong typing, the proportion between the number of Success Prediction and wrong typing is called Word-List Success Prediction Rate (SP Rate). Let's define the number of Success Prediction as o_1 and the number of wrong typing as o_2, then one has $o_1 \le o_2$ and $SP\ Rate = o_1/o_2$.

Simulation Rate Definition: Given natural numbers i, m, n, where $i \le n$ and $m \le n$, let's simulate a testing dataset $p_1...p_i...p_n$ with a trained neural network, and its target dataset $t_1...t_i...t_n$, if output $r_1...r_i...r_n$ has m elements which are $r_i = t_i$, then the Simulation Rate (SM Rate) is m/n. Given Word-List Success Prediction Rate SP and Simulation Rate SM, then the *First Hitting Rate = SP * SM*.

As illustrated above, a word correction function can combine multiple algorithms and all of them produce their self-interpreted results independently, which is the so-called Word-List n-formula Prediction. The results could be rarely similar while a user may require only one of them if Success Prediction is fulfilled. So a functional ranking model will play a major role to present an efficient word list with priority. If one considers the learning factor required by a word list and variability of its related dataset, a neural network model is a good choice with the dataset being constantly updated.

In L.M.T combination, Levenshtein word distance algorithm calculates the similarity between each two words, where all the most similar ones are presented; Metaphone algorithm retrieves words based on phonetic index while word 2-gram algorithm retrieves them based on last typed word index. From the definition of Word-list n-formula prediction, L.M.T correction can be referred to as Word-List

3-formula Prediction. Let's use the example shown below, where the word 'shall' is wrongly typed as '*sahll*'.

```
Tomorrow sahll we go to the park?
```

and assume that a database, which includes a *1*-gram & *2*-gram table, has been initialized by a sentence,

```
Out of your shell! Tomorrow all of us shall
start a new training.
```

Then, L.M.T correction result of word '*sahll*' based on *2*-gram word algorithm is '*all*', the correction results based on Metaphone algorithm are '*shall*' and '*shell*' and the correction results based on Levenshtein word distance algorithm are '*all*' and '*shall*'.

4   Word List Neural Network Ranking Modelling

Let's suppose that, corresponding to every wrong typing, each algorithm generates a maximum of two words in a descending frequency order. Each word is represented by its two features: word frequency and word similarity values. In a real-time database, the word frequency is updated along with user typing. Both, word frequency and word similarity datasets are normalized before the neural network training and testing.

In this paper, a neural network model with *12-3-6* three layer structure is developed as shown in Figure 1, where the number of the input layer neurons is determined by the expression: *Number of Algorithms* (=3) * *Number of Words predicted* (=2) * *Number of Features of each word* (=2). The model is named as word list neural network ranking (WLR) model and BackPropagation algorithm is adopted as its learning algorithm. Each algorithm generates two predictions based on the input, which is a wrongly typed word. Each prediction is represented by its two features, namely, Jaro-Winkler distance and word frequency.

Generally speaking, WLR model is designed to predict a highest ranked word amongst every six recommendations. Then, a ranking issue is converted to a neural network classification question solving issue. At the output layer of WLR model, there is only one neuron fired once at a time. To normalize the difference between the typed word and a predicted word, Jaro-Winkler metric is applied. It normalizes words difference also called words similarity value, into a range of [*0, 1*]. Another parameter: word frequency, is normalized by Normal Probability Density function based on frequencies' mean value and standard deviation.

An application and its related Access database are developed to generate an experimental dataset for WLR model. The related database has been initialized by words' *1* & *2*-gram frequency statistics of a novel - 'Far from the Madding Crowd' [7][8] before the experimental dataset is generated. The database initialization has followed these rules,

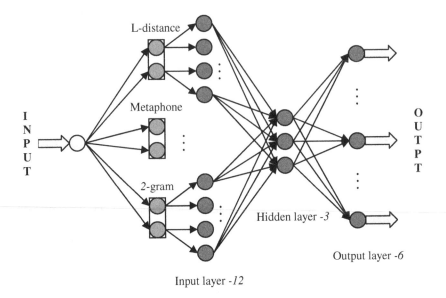

Fig. 1. The circles in blue are neurons of WLR model; the circles in grey are predicted words; the three rectangles represent the three algorithms, namely, Levenshtein distance, Metaphone and 2-gram; the shapes in yellow show the input and the output of WLR model

- *A word is defined as a sequence of alphabets between two separators.*
- *Any symbols are considered as a separator except alphabets.*
- *ALL uppercase are converted into lowercase, e.g. 'If' → 'if', then 'If' is counted as 'if'*
- *Other special cases are not considered. For example, 'read' and 'reading' are considered as two independent words.*

Based on these rules, the word dictionary table and 2-gram dictionary table including their words occurrences are initialized in the Access database. Moreover, for database efficiency purpose, all the 2-gram records whose occurrences are less than two are eliminated. Overall, about *79.10%* of all 2-gram records are eliminated. This will only produce a very limited influence on the performance of WLR model if one considers thousands of repetitive trials in a neural network training and testing. The occurrences of the words' *1 & 2*-gram are kept updated along with user's typing progress (if there is a new 2-gram generated, the 2-gram and its occurrence will be inserted into the database). Therefore, these updated frequencies can well represent a user's temporal typing state captured and stored in a database.

As a simulation to dyslexic's typing, a testing sample [9] is used as the experimental dataset for WLR model as shown below,

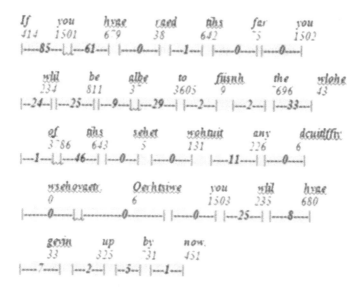

Fig. 2. The numbers in red and black indicate words *1* and 2-gram frequencies respectively

As shown in Figure 2, some words within sentences are wrongly typed, such as 'hvae' (should be 'have') and 'raed' (should be 'read'). The numbers which are right under each word (in red) indicate the frequency of the word after the database initialization. For example, the frequency of the word 'If' is *414* and the frequency of the word 'you' is *1501* in the database. The numbers in black indicate the 2-gram frequency between two consecutive words. For example, the frequency between the first two words 'If' and 'you' is eighty-five, shown as '|---85---|'

Let's assume the frequencies of the words shown above gradually increases in the database while other words are rarely typed. Consequently, the change of other words' frequencies will not have a big effect on the algorithms. Therefore, a simulation can be performed by using the testing dataset which has ignored the influence brought by other words' frequency changes. In this research, *5505* trials of test samples are inserted into the database gradually without considering other words' frequency changes.

Let's define a sampling point as a starting point of sampling in these *5505* trials, and define a sampling step as a gap between two consecutive sampling actions. Twenty five sampling points are set up to collect the three algorithms' prediction results. Only those wrongly typed and completed words are considered at every sampling point. For example, the prediction results for words such as 'hvae' and 'raed' are collected; while the prediction results for right words such as 'if', 'you' and uncompleted words such as 'hva' of 'hvae' are ignored. At each sampling point, the whole dataset are gathered and called a sample. Then, twenty five samples are gathered. The determination of sampling points and sampling step is based on a heuristics method, which shows that the influence of initial frequency updating is essential while further updating influence is waning.

Figure 3 illustrates the sampling procedure, which are classified in four categories [0→5, 10→50, 55→505, 1505→5505]. As illustrated, the influence of frequency updating is waning from one category to another although the sampling steps

416   J. Li et al.

are actually increasing. The four categories are shown in red lines of Figure 3. For instance, five samples have been collected with the frequency being changed from zero to five (i.e. the sampling step is one), and ten samples are collected when the frequency changed from *55* to *505* (i.e. the sampling step is *50*).

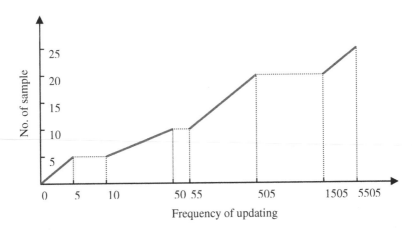

Fig. 3. X-axis refers to the frequency of the whole sample; y-axis refers to the numbers of sampling

The first two subsets of sample one are shown in Figure 4, which lists the predicted results of two mistakenly typed words, which are 'hvae' and 'raed'.

```
Levenshtein word distance       Metaphone              Two-Gram word              output
have 679 0.925000 |hae  3 0.925000 |ve   93 0.583333 |feei  1 0.500000 |are   86 0.527778 |know 65 0.000000 |1  0  0  0  0  0
red  43 0.925000 |read 35 0.925000 |right 65 0.483333 |road 64 0.850000 |been 112 0.500000 |a    27 0.750000 |0  1  0  0  0  0
```

Fig. 4. First line is a comment which marks the three algorithms names and 'output'. The rest are two prediction results based on the three algorithms.

As shown in the columns of Figure 4, each of the three algorithms has generated two predicted words. For instance, Levenshtein word distance algorithm gives two suggestions to the word - 'hvae', which are 'have' and 'hae'. Next to each word, the word's frequency and the similarity values to the target word are displayed. For example, the frequency of the word 'have' is *679* and its similarity to 'hvae' is *0.925*.

The last six columns of Figure 4 clearly show the required output for WLR neural network model. Each of those columns corresponds to one of the words that those three algorithms could generate. If the prediction is true, the corresponding column is set to one, otherwise it is set to zero. For example, the first line of Figure 4 is a prediction for mistakenly typed word 'hvae' while among the six predictions only the first result of Levenshtein word distance algorithm is a correct prediction, therefore the first column of the output is set to one while others are set to zeros. By default, the processing stops at the first '*1*', and the others will be set to zeros. So the output will have a maximum of one '*1*'.

The data shown in Figure 4 still can not be used by WLR model directly, as further pre-processing is required. Therefore the following procedures are applied.

- *Delete the redundancy such as the words of each line.*
- *Normalize all frequencies by applying Normal probability density function*
- *Apply missing data processing rules where it is needed – If some algorithms' prediction results are less than two items, then the frequency and similarity values of the missing items will be set to zeros instead; if none of the algorithms are able to generate results, then this line will be deleted.*

The sampling points are set up according to a heuristic method which analyzes the frequency distribution of the database. For example, the first five frequency updating procedures are considered to be more influential than the case when the frequency changes significantly (e.g. *>1000*). So, the sampling step of the first five is set to one while the rest are sparser.

In this experiment, a vector [*5, 5, 10, 5*] of samples are collected from the four categories and their sampling steps are set to [*1, 10, 50, 1000*]. For example, the first five samples are collected in a step distance of one, the third ten samples are collected in a step distance of fifty.

The dataset is further separated into training dataset [*4, 4, 7, 3*], and testing dataset [*1, 1, 3, 2*]. The post-processing of WLR model follows a 'winner takes all' rule – the neuron which has the biggest value among the six outputs are set to one while others are set to zeros.

After the training process, the Hitting Rates of the testing dataset associated with each category are shown in Figure 5.

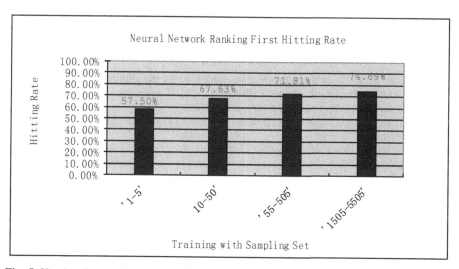

Fig. 5. X-axis refers to the increase of words frequency difference; y-axis refers to the Hitting Rate of WLR model ranking

Figure 5 shows that the samples are separated into four categories based on the step distance of [*1, 10, 50, 1000*]. For example, the first histogram shows a *57.50%* Ranking First Hitting Rate with the samples of category one; the fourth histogram shows a best achievement of *74.69%* Ranking First Hitting Rate with more samples collected between frequency *1505* and *5505* in five separated sampling points. Figure 5 shows an increase of ranking Hitting Rate as words frequency difference and the amount of testing samples increase. This is also partly influenced by the three algorithms previously introduced with learning factors. All the algorithms are adjusting gradually toward a better prediction rate as trials increase.

5 Conclusion

In this paper a hybrid solution based on multiple typing correction algorithms and a Word List Neural Network Ranking model to produce an optimal prediction are presented. Three distinct algorithms, namely, Metaphone, Levenshtein distance and word 2-gram are used in a pilot study. Several key factors including Time Change, Context Change and User Feedback are considered. Experimental results show that 57.50% ranking First Hitting Rate with initial samples is achieved. Further testing with updated samples indicates a best ranking First Hitting Rate of 74.69%. The findings demonstrate that neural network as a learning tool, can provide an optimum solution through combining distinct algorithms to learn and subsequently to adapt to reach a high ranking Hitting Rate performance. This may inspire more researchers to use a similar approach in some other applications.

In practice, an application using WLR theory can be implemented based on propagating rewards to each algorithm and/or word. Currently WLR model adjusts its ranking based on the change of word frequency and similarity. In the future, more parameters such as time element and more typing correction algorithms can be added to achieve a better performance.

Acknowledgments. The research is funded by Disability Essex [10] and Technology Strategy Board [11]. Thanks to Pete Collings and Ray Mckee for helpful advice and discussions.

References

1. Ouazzane, K., Li, J., et al.: A hybrid framework towards the solution for people with disability effectively using computer keyboard. In: IADIS International Conference Intelligent Systems and Agents 2008, pp. 209–212 (2008)
2. Metaphone, http://en.wikipedia.org/wiki/Metaphone (accessed January 23, 2009)
3. N-gram, http://en.wikipedia.org/wiki/N-gram (accessed January 18, 2009)
4. Levenshtein algorithm, http://www.levenshtein.net/ (accessed January 23, 2009)
5. Cohen, W.W., Ravikumar, P., Fienberg, S.: A Comparison of String Distance Metrics for Name-Matching Tasks. In: IIWeb 2003, pp. 73–78 (2003)
6. Jaro-Winkler distance, http://en.wikipedia.org/wiki/Jaro-Winkler (accessed January 23, 2009)
7. Far from the Madding Crowd,
   http://en.wikipedia.org/wiki/Far_from_the_Madding_Crowd (accessed April 15, 2009)
8. Calgary Corpus,
   ftp://ftp.cpsc.ucalgary.ca/pub/projects/text.compression.corpus/text.compression.corpus.tar.Z (accessed January 18, 2009)
9. Davis, M.: Reading jumbled texts,
   http://www.mrc-cbu.cam.ac.uk/~mattd/Cmabrigde/ (accessed January 26, 2009)
10. Disability Essex, http://www.disabilityessex.org (accessed January 18, 2009)
11. Knowledge Transfer Partnership, http://www.ktponline.org.uk/ (accessed January 18, 2009)

Application of Neural Fuzzy Controller for Streaming Video over IEEE 802.15.1*

Guillaume F. Remy and Hassan B. Kazemian

London Metropolitan University
Faculty of Computing, London, United Kingdom

Abstract. This paper is to introduce an application of Artificial Intelligence (AI) to Moving Picture Expert Group-4 (MPEG-4) video compression over IEEE.802.15.1 wireless communication in order to improve quality of picture. 2.4GHz Industrial, Scientific and Medical (ISM) frequency band is used for the IEEE 802.15.1 standard. Due to other wireless frequency devices sharing the same carrier, IEEE 802.15.1 can be affected by noise and interference. The noise and interference create difficulties to determine an accurate real-time transmission rate. MPEG-4 codec is an "object-oriented" compression system and demands a high bandwidth. It is therefore difficult to avoid excessive delay, image quality degradation or data loss during MPEG-4 video transmission over IEEE 802.15.1 standard. Two buffers have been implemented at the input of the IEEE 802.15.1 device and at the output respectively. These buffers are controlled by a rule based fuzzy logic controller at the input and a neural fuzzy controller at the output. These rules manipulate and supervise the flow of video over the IEEE 802.15.1 standard. The computer simulation results illustrate the comparison between a non-AI video transmission over IEEE 802.15.1 and the proposed design, confirming that the applications of intelligent technique improve the image quality and reduce the data loss.

Index Terms: IEEE 802.15.1, neural-fuzzy, rule-based fuzzy, MPEG video.

1 Introduction

Nowadays Moving Picture Expert Group-4 (known as MPEG-4) is broadly used as a video compression technique. Real-time delivery of MPEG-4 is delay-sensitive, as a frame cannot be displayed if its data arrives damaged or late. In practice, many problems can occur and create transmission problem such as noise from other nearby devices. The maximum delay permissible corresponds to the start-up delay accepted by the user. Data may arrive too late for the frame to be displayed, and the result is a poor quality video. Not only data may arrive too late, while retransmission is taking place, other packets may either wait too long at the token bucket of the IEEE 802.15.1 or in extreme cases may overflow the token bucket.

IEEE 802.15.1 standard is a low power radio signal, permits high-quality, high-security, high-speed, voice and data transmission [1]. In IEEE 802.15.1 standard

---

* This work was funded by Engineering and Physical Sciences Research Council (EPSRC), UK. The work was carried out at the London Metropolitan University, UK.

D. Palmer-Brown et al. (Eds.): EANN 2009, CCIS 43, pp. 419–429, 2009.
© Springer-Verlag Berlin Heidelberg 2009

networks, it is really difficult to predict the bandwidth as it is significantly variable, as results of interferences in radio-frequency environments and portability of IEEE 802.15.1 hosts. Furthermore, the maximum bandwidth supported by IEEE 802.15.1 standard wireless is restricted [2].

For multimedia communication, noise and interference on the wireless channel may corrupt the data as it is encoded, decoded and finally screened on a mobile device. This suggests retransmission of corrupted packets should occur. These problems can create delays and data loss.

In [3], Fuzzy Rate Control (FRC) was applied to wireless connection to achieve smooth performance degradation as network load increases. FLC has found applications [4] in routing of broadband network connections, in order to probabilistically avoid shortage of resources. In [5], Fuzzy Logic Control (FLC) has been used to control Bluetooth's default automatic repeat request (ARQ) scheme.

The work in [6] applied FLC to Bluetooth rate control through tandem controllers in an open loop system. In the paper [7], a novel integrated Neuro-Fuzzy (NF) control scheme is utilised to reduce the burstiness and to minimise time delay, data loss and image quality degradation. Of course, deliberate and random packet losses are unsuitable for encoded video and the unbounded delay introduced by Transmission Control Protocol (TCP)'s reliability mechanism also makes it unsuitable for video display. Furthermore in [8], an interesting application of FLC to TCP/IP networks (wired and wireless links) controls the retransmission rate and the RED rate. In all the research carried out above [3-8], it is demonstrated that Fuzzy Logic Controller (FLC) and Neural Fuzzy Controller (NFC) improve the quality of data transmission.

This paper introduces intelligent technique (rule-based fuzzy logic and neural network) to smooth the MPEG-4 data over IEEE 802.15.1 standard. To the best of our knowledge, rule base fuzzy has been already used for similar purpose over Bluetooth, and neural fuzzy scheme like ours has not been applied to increase quality of picture over IEEE 802.15.1 standard. Therefore, the novelty of this design lies in the NFC application to MPEG-4 video transmission over IEEE 802.15.1. The aim of the current work is to minimize the data transmitted during overflow or reduce transmission problem to decrease error at the reception end. The main reason for introducing artificial intelligence is we found that its performance in terms of delivered video quality is better than a conventional scheme, as the tests, in Section 3, illustrate.

The remainder of this paper is organised as follows. Section 2 gives details on fuzzy logic, IEEE 802.15.1 standard and the evaluation methodology. Section 3 contains the results of the evaluation, while Section 4 draws some conclusions.

2 Methodology

This section briefly introduces fuzzy logic controller. In a fuzzy subset, each member is an ordered pair, with the first element of the pair being a member of a set S and the second element being the possibility, in the interval [0, 1], that the member is in the fuzzy subset. This should be compared with a Boolean subset in which every member of a set S is a member of the subset with probability taken from the set {0, 1}, in which a probability of 1 represents certain membership and 0 represents none membership. In a fuzzy subset of "buffer fullness," the possibility that a buffer with a given fullness taken from the set S of fullness may be called high is modeled by a

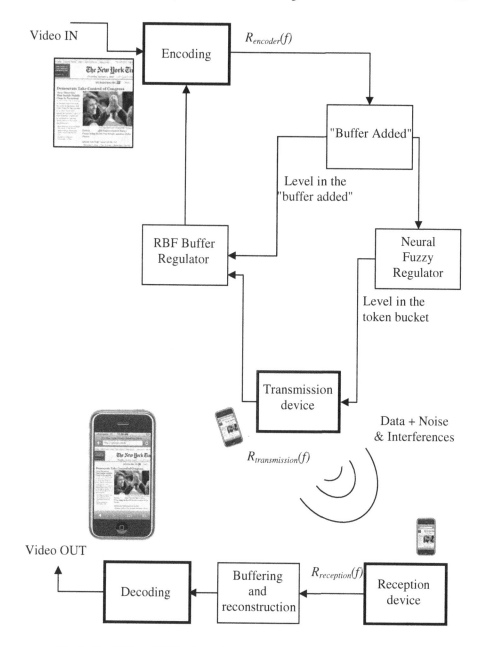

Fig. 1. The RBF and NFC scheme for video transmission over IEEE 802.15.1

membership function, which is the mapping between a data value and possible membership of the subset. Notice that a member of one fuzzy subset can be a member of another fuzzy subset with the same or a different possibility. Membership functions may be combined in fuzzy "if then" rules to make inferences such as if x is high and y is low, then z is normal, in which high, low, and normal are membership functions of the matching fuzzy subsets and x, y, z are names for known data values. In practice, the membership functions are applied to the data values to find the possibility of membership of a fuzzy subset and the possibilities are subsequently combined through defuzzification, which results in a crisp (non-fuzzy) value [9].

IEEE 802.15.1 standard employs variable-sized packets up to a maximum of five frequency-hopping time-slots of 625 microseconds in duration. Every IEEE 802.15.1 standard frame consists of a packet transmitted from a sender node over 1, 3, or 5 timeslots, while a receiver replies with a packet occupying at least one slot, so that each frame has an even number of slots. A source of error bursts is co-channel interference by other wireless sources, including other IEEE 802.15.1 piconets, IEEE 802.11b/g networks, cordless phones, and even microwave ovens. Though this has been alleviated to some extent in version 1.2 of Bluetooth standard by adaptive frequency hopping [10], this is only effective if interferences are not across all or most of the 2.402 to 2.480 GHz unlicensed band. IEEE 802.11b operating in direct sequence spread spectrum mode may occupy a 22 MHz sub-channel (with 30 dB energy attenuation over the central frequency at ± 11 MHz) within the 2.4 GHz Industrial, Scientific and Medical (ISM) band. IEEE 802.11g employs orthogonal frequency division multiplexing (OFDM) to reduce inter-symbol interference but generates similar interference to 802.11b. Issues of interference might arise in apartment blocks with multiple sources occupying the ISM band or when higher-power transmission occurs such as at Wi-Fi hotspots.

In this article only the sending part will be analysed. For the system here, as shown in Fig.1, there is one Fuzzy Logic Controller (FLC), one neural-fuzzy regulator, and an added buffer between the MPEG-4 encoder and the IEEE 802.15.1 standard. IEEE 802.15.1 token bucket is a measure of channel conditions, as was demonstrated in [11]. Token bucket buffer is available to an application via the Host Controller Interface (HCI) presented by an IEEE 802.15.1 standard hardware module to the upper layer software protocol stack. Retransmissions avoid the effect of noise and interference but also cause the master's token bucket queue to grow, with the possibility of packet loss from token bucket overflow.

Therefore, in the Fig. 2 "buffer added" level of membership is presented as used in the ANFIS training, token bucket level of membership is shown in Fig.3. In here, the "buffer added" is full when the variable is very high and the token bucket is full of token (no data in the buffer) when the variable is low. These figures have been produced through MATLAB Fuzzy Logic Toolbox. The inputs were combined according Surgeno Type Fuzzy Controller to produce a single output value. As shown in the Table 1, if the "buffer added" level is very high and the token bucket level is very high (no token left), then big congestion appear and delay occurs as we obtain an overflow. On the other hand, if the "buffer added" level is low (no data in) and the token bucket level is low as well (full of token), then the transmission is very fluid and occurs without problem. In this project the rules are trained by a neural network controller. This controller collects the two inputs (level in the token bucket and level in the "buffer added") and trains them with and Adaptive Neural-based Fuzzy Inference System (ANFIS) training in order to smooth the output (Rtransmission).

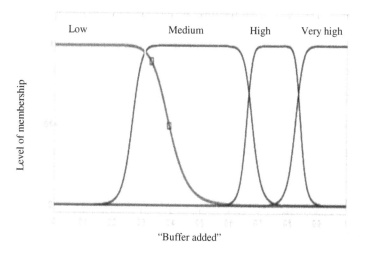

Fig. 2. Membership function plots of the input variable "buffer added" in the Fuzzy inference system

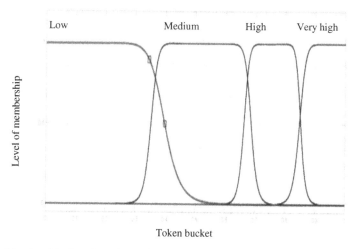

Fig. 3. Membership function plots of the input variable token bucket in the Fuzzy inference system

3 Computer Simulation Results

Synchronous Connection Oriented (SCO) link and Asynchronous Connection-Less (ACL) link have been provided by IEEE 802.15.1 standard [12]. ACL may support up to 724 kb/s download rate and with 57 kb/s upload rate, or symmetrically 434 kb/s in upload and download. In this research, as the video streaming is a one side transmission, ACL link has been selected for video transmission over IEEE 802.15.1. The IEEE 802.15.1 standard specification provides flow control for the HCI. A token bucket is implemented in the IEEE 802.15.1 standard hardware, it controls the flow.

The host can only send data when there is "token" available in the buffer. The data rate during transmission depends on the space available in the token bucket even a rate was set-up at the beginning of the transmission. The noise is generated by a Gaussian noise following the equation:

$$f(x) = \frac{1}{\sigma\sqrt{2\pi}} . e^{-\frac{(x-b)^2}{2\sigma^2}} \tag{1}$$

Where $\frac{1}{\sigma\sqrt{2\pi}}$ is the height of the curve's peak, b is the position of the centre of the peak, and σ controls the width of the "bell".

A video clip called 'X-men' [13] is used to implement the proposed rule based fuzzy and neural fuzzy control scheme and we used from the Group of Pictures (GoP) 50 to the GoP 150 of this clip for the computer simulation. The frame size of the clip is 240 pixels x 180 pixels.

Table 1. "if ·· then" rules

IF token bucket is low AND buffer added is low THEN output is very fluid
IF token bucket is low AND buffer added is medium THEN output is intermediate
IF token bucket is low AND buffer added is high THEN output is fluid
IF token bucket is low AND buffer added is very high THEN output is intermediate
IF token bucket is medium AND buffer added is low THEN output is small intermediate
IF token bucket is medium AND buffer added is medium THEN output is small congestion
IF token bucket is medium AND buffer added is high THEN output is small congestion
IF token bucket is medium AND buffer added is very high THEN output is congestion
IF token bucket is high AND buffer added is low THEN output is fluid
IF token bucket is high AND buffer added is medium THEN output is small congestion
IF token bucket is high AND buffer added is high THEN output is congestion
IF token bucket is high AND buffer added is very high THEN output is congestion
IF token bucket is very high AND buffer added is low THEN output is intermediate
IF token bucket is very high AND buffer added is medium THEN output is congestion
IF token bucket is very high AND buffer added is high THEN output is congestion
IF token bucket is very high AND buffer added is very high THEN output is big congestion

Table 2 compares the numerical results of the testing clip using the Artificial Intelligent system and the No-Artificial Intelligent system for a fixed mean value.

Table 2. Standard deviation comparison between AI system and no-AI system

	AI SYSTEM	NO-AI SYSTEM
Standard deviation for a mean value of 724 Kbps	89.34Kbps	296.57Kbps

In Table 2, if we compare the standard deviation, of a system with and without Artificial Intelligence. We can easily prove that the standard deviation of the intelligent system is much lower than the No-AI system, reducing the burstiness and data loss.

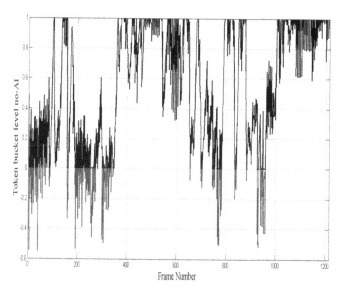

Fig. 4. Token bucket availability for No-AI system X-men GOP 50 to 150 724 Kbps

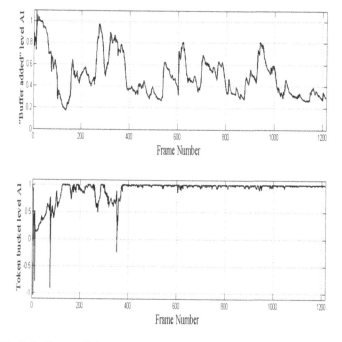

Fig. 5. Buffers availability for AI system X-men GOP 50 to 150 724 Kbps

The following four figures present the MATLAB computer simulation results for 'X-men' clip. Fig. 6 and Fig. 7 have two sets of results respectively the system without Artificial Intelligence (No-AI) and the system with Artificial Intelligence (AI), from the Group of Picture (GoP) 50 to the Group of Picture (GoP) 150 at a mean value of 724 Kbps. On Fig. 6, from above the graphs, it can be seen the departure rate at the entrance of the IEEE 802.15.1 standard buffer (token bucket) (Rencoder), the mean value of the device with noise (Rtransmission), and the superposition of the departure rate at the entrance of the token bucket and the mean value of the device without noise. In Fig 7, from above the graphic shows the departure rate from the IEEE 802.15.1 token bucket, the bit rate of the "buffer added" just after the MPEG-4 encoder, the mean value of the device with noise and the superposition of the departure rate from the token bucket, the bit rate of the "buffer added" just after the MPEG-4 encoder and the mean value of the device without noise.

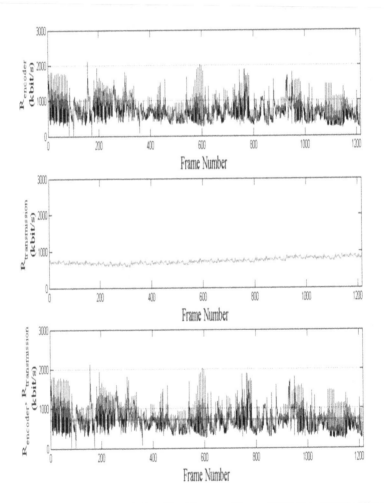

Fig. 6. Rencoder, Rtransmission, No-AI system X-men GOP 50 to150 724 Kbps

The token bucket level is an indication of the storage available in the buffer for data. The token bucket level Fig. 4 shows when the token bucket is overflowed and data are lost. All the data under the value "0" indicates overflow and the token bucket causes data loss. The first graph on top in Fig. 5 indicates the degree of overflow or starvation of the "buffer added". The second graph at the bottom of Fig. 5 outlines the capacity of the token bucket. We can easily see in Fig. 4 the overflow of the token bucket (when the value goes under 0), and in contrast in Fig. 5, the token bucket always has available space between frame 200 and frame 300. For this video clip, the introduction of the RBF and NF control eliminates the data loss altogether, as there is still some storage capacity left in the "buffer added" and the token bucket.

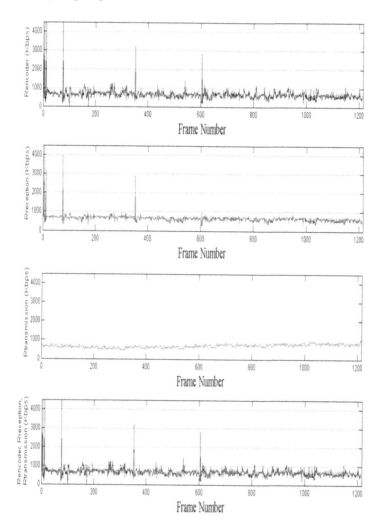

Fig. 7. Rencoder, Rreception, Rtransmission, AI system X-men GOP 50 to 150 724 Kbps

In Fig. 7, the burstiness of the departure bit rate from the Rencoder is reduced on the Rreception signal. Fig. 7 demonstrates that the departure bit rate is much smoother, resulting in reduction in data loss, time delay, and improvement in picture quality over the IEEE 802.15.1 channel. Furthermore, by maintaining a sufficient space available in the token bucket, the novel scheme decreases the standard deviation of the output bit rate and the number of dropped data, resulting in better quality of picture and video stability.

4 Conclusion

This paper applies artificial intelligence techniques to streaming video over IEEE 802.15.1 to improve quality of picture, to reduce time delay and excessive data loss during wireless communication. Neural Fuzzy Controller (NFC) enables a smooth flow of MPEG-4 video data over IEEE 802.15.1 and reduces burstiness, while alleviating the overflow of the token bucket associated with the IEEE 802.15.1 standard. In this research, a significant reduction of the standard deviation of the MPEG-4 video and diminution of the token bucket burstiness have been achieved by using the novel algorithms. These results therefore affect directly the stability of the communication, while providing an improvement in image quality at the receiving end. Finally, as these intelligent techniques are based on efficient algorithms, the proposed design can be applied to delay sensitive applications such as audio, video and multimedia streaming data.

References

1. Haartsen, J.C.: The Bluetooth radio system. IEEE Personal Communications 7(1), 28–36 (2000)
2. http://www.ieee802.org/15/pub/TG1.html
3. Antoniou, P., Pitsillides, A., Vasiliou, V.: Adaptive Feedback Algorithm for Internet Video Streaming based on Fuzzy Rate Control. In: 12th IEEE Symposium on Computers and Communications (ISCC 2007), IEEE Catalog Number: 07EX1898C, Aveiro, Portugal, July 1-4, pp. 219–226 (2007)
4. Vasilakos, A., Ricudis, C., Anagnostakis, K., Pedrycz, W., Pitsillides, A., Gao, X.: Evolutionary - Fuzzy Prediction for Strategic ID-QoS: Routing in Broadband Networks. International Journal of Parallel and Distributed Systems and Networks 4(4), 176–182 (2001)
5. Razavi, R., Fleury, M., Ghanbari, M.: Fuzzy Logic Control of Adaptive ARQ for Video Distribution over a Bluetooth Wireless Link. In: Advances in Multimedia, vol. 2007. Hindawi Publishing Corporation (2007), doi:10.1155/2007/45798, Article ID 45798
6. Kazemian, H.B., Meng, L.: An adaptive control for video transmission over Bluetooth. IEEE Transactions on Fuzzy Systems 14(2), 263–274 (2006)
7. Kazemian, H.B., Chantaraskul, S.: An integrated Neuro-Fuzzy Approach to MPEG Video Transmission in Bluetooth. In: 2007 IEEE Symposium on Computational Intelligence in Image and Signal Processing, CIISP 2007 (2007), IEEE 1-4244-0707-9/07
8. Chrysostomou, C., Pitsillides, A., Rossides, L., Sekercioglu, A.: Fuzzy logic controlled RED: congestion control in TCP/IP differentiated services networks. Soft Computing - A Fusion of Foundations, Methodologies and Applications 8(2), 79–92 (2003)

9. http://plato.stanford.edu/entries/logic-fuzzy
10. Ziemer, R.E., Tranter, W.H.: Principles of Communications – systems, modulation and noise, 5th edn. John Wiley & Sons, Inc., Chichester (2002)
11. Razavi, R., Fleury, M., Ghanbari, M.: Detecting congestion within a Bluetooth piconet: video streaming response. In: London Communications Symposium, London, UK, September 2006, pp. 181–184 (2006)
12. https://www.bluetooth.org/spec/Core_1.2_and Profile_1.2
13. http://www.movie-list.com/nowplaying.shtml

Tracking of the Plasma States in a Nuclear Fusion Device Using SOMs

Massimo Camplani[1], Barbara Cannas[1], Alessandra Fanni[1], Gabriella Pautasso[2], Giuliana Sias[1], Piergiorgio Sonato[3], and The Asdex-Upgrade Team[2]

[1] Electrical and Electronic Engineering Dept., University of Cagliari, Italy
[2] Max-Planck-Institut fur Plasmaphysik, EURATOM Association, Garching, Germany
[3] Consorzio RFX, Associazione EURATOM-ENEA sulla Fusione, Padova, Italy

Abstract. Knowledge discovery consists of finding new knowledge from data bases where dimension, complexity or amount of data is prohibitively large for human observation alone. The Self Organizing Map (SOM) is a powerful neural network method for the analysis and visualization of high-dimensional data. The need for efficient data visualization and clustering is often faced, for instance, in the analysis, monitoring, fault detection, or prediction of various engineering plants. In this paper, the use of a SOM based method for prediction of disruptions in experimental devices for nuclear fusion is investigated. The choice of the SOM size is firstly faced, which heavily affects the performance of the mapping. Then, the ASDEX Upgrade Tokamak high dimensional operational space is mapped onto the 2-dimensional SOM, and, finally, the current process state and its history in time has been visualized as a trajectory on the map, in order to predict the safe or disruptive state of the plasma.

Keywords: Knowledge Discovery, Self Organizing Maps, Tokamak, Disruptions.

1 Introduction

Experimental data are essential information in several fields, such as chemistry, biology, physics, engineering. Utilizing the experimental information for knowledge discovery is a critical task, and the strategy for knowledge discovery usually depends on the contents and characteristics of the data.

As an example, monitoring the state of a plant may generate thousands of multidimensional data points for every experiment performed. A study may consist of many experiments, such that in a single study a huge amount of data points may be generated. Such volumes of data are too large to be analyzed by, e.g., sorting in spreadsheets, or plotting on a single or few graphs, and systematic methods for their inspection are required. In this paper, clustering methods are investigated, for the analysis and organization of large-scale experimental data, and for knowledge extraction in order to map the operational plasma space of a nuclear fusion device.

Tokamak is the most promising device for nuclear fusion. In a Tokamak, the plasma is heated in a ring-shaped vacuum chamber (vessel or torus) and kept away from the vessel walls by applying magnetic fields. The range of 'plasma states' viable

D. Palmer-Brown et al. (Eds.): EANN 2009, CCIS 43, pp. 430–437, 2009.

to a Tokamak is highly restricted by disruptive events: plasma instabilities, usually oscillatory modes, sometimes grow and cause abrupt temperature drops and the termination of an experimentally confined plasma. Stored energy in the plasma is rapidly dumped into the rest of the experimental system (vacuum vessel walls, magnetic coils, etc.).

Hence, one of the most challenging problems in nuclear fusion research consists in the understanding of disruption events. The identification of characteristic regions in the operational space where the plasma undergoes to disruption is crucial for Tokamak development.

In this paper, the Self Organizing Maps (SOMs) are used to map the high-dimensional operational space of ASDEX Upgrade (Axially Symmetric Divertor EXperiment, at present, the Germany's largest fusion device) into a low-dimensional space, while retaining most of the underlying structure in the data.

The produced mapping exploits the similarities of the data by grouping similar data items together, allowing the identification of characteristic regions for plasma scenarios. The operational space consists of 7 plasma parameters from 229 shots carried out during the operation of ASDEX between June 2002 and July 2004.

One of the main drawbacks of unsupervised clustering techniques, as SOMs, is that the number of clusters has to be known a priori. In many practical applications, the optimal number of clusters is not defined and a wrong choice could lead to poor results. In order to overcome this problem, several indexes have been used as a measure of the quality of a clustering partition.

The produced mapping can be used in an off-line manner as a highly visual data exploration tool: it constitutes a display of the operational states of the plasma. For such purpose, the maps are deeply analysed in order to identify *hard* operational limits for plasma scenarios, which lead ultimately to a disruption. The same maps can also be used in an on-line way: the plasma state and its history in time can be visualized as a trajectory on the map. This allows to efficiently tracking the process dynamics. In particular, the SOM facilitates understanding of dynamics so that several variables and their interactions may be inspected simultaneously. During the track of the operation point, location of the point on the map indicates the process state and the possible plasma safe or disruptive phase.

The paper is organised as follows: section 2 recalls the concept of clustering and describes the adopted technique. Section 3 describes the database composition, whereas Section 4 reports the results of the mapping process. In Section 5 the conclusions are drawn.

2 Data Visualization with SOM

Data clustering consists in the organization of a set of items into groups (clusters) that are composed by similar items. Only the information embedded in the input data can be used to identify some aggregation or properties of the patterns.

In this paper, the Self Organizing Map (SOM) [1] has been used. SOM is one of the most famous hard partition clustering techniques. A SOM separates the entire data set in a number of clusters defined a priori, and optimizing a defined criterion function. The method differs from other machine learning techniques because it performs both data clustering and projection. The main advantage of the SOM is in the visualization tools that allow for an intuitive data exploration.

A SOM defines a mapping from the n-dimensional input space X onto a regular (usually two-dimensional) array of K clusters (neurons), preserving the topological properties of the input. This means that points close to each other in the input space are mapped on the same or neighbouring neurons in the output space.

The output neurons are arranged in 2D lattice where a reference vector **w** is associated with every neuron. The ith output neuron represents the ith cluster. Hence, the output of the ith output neuron O_i, $i=1,...,K$, is:

$$O_i = \sum_{j=1}^{n} w_{ij} x_j \tag{1}$$

A competitive learning rule is used, choosing the winner i^* as the output neuron with **w** closest to the input **x**. The learning rule is:

$$\Delta w_{ij} = \eta \Lambda(i, i^*)(x_j - w_{ij}) \quad i = 1,...K; j = 1,...n. \tag{2}$$

The neighbourhood function $\Lambda(i, i^*)$ is equal to 1 for $i=i^*$, and falls off with the distance $\mathbf{d}_{ii*}$ between neurons i and i^* in the output lattice. Thus, neurons close to the winner, as well as the winner itself, have their reference vector updated, while those further away, experience little effect. When training is completed, the reference vectors associated with each neuron define the partitioning of the multidimensional data. A typical choice for $\Lambda(i, i^*)$ is

$$\Lambda(i, i^*) = e^{-d_{ii*}^2 / 2\sigma^2} \tag{3}$$

where σ is a width parameter that is gradually decreased.

When using projections onto 2D lattice, the SOM offers excellent displays for data exploration, useful to approach data analysis visually and intuitively. The visualization techniques mainly used in the experiments are the *Median Distance-Matrix* (D-Matrix) and the *Unified Distance Matrix* (U-matrix) [2].

Moreover, *Operating point and trajectory* [3] can be used to study the behavior of a process in time. The operating point of the process is the reference vector of the current measurement vector. The location of the point corresponding to the current process state can be visualized on the SOM. A sequence of operating points in time forms a trajectory on the map depicting the movement of the operating point.

2.1 Quality Indexes

A common measure used to calculate the precision of the clustering is the average quantization error E_q [4] over the entire data set. Moreover, it is possible to evaluate how well a SOM preserves the data set topology evaluating the topographic error E_t [4]: partitions with a good resolution are characterized by a low value of E_q and a high value of E_t. The Davies-Bouldin (DB) Index [5] and the Dunn (DN) Index [6], are two of the most popular validity indexes. Both indexes are well suited to identify the presence of compact and well separate clusters. The optimum number of clusters corresponds to the minimum of the DB index, whereas the optimum number of clusters corresponds to the maximum of the DN index.

In the present paper, the Davies-Bouldin and Dunn indexes have been modified in order to adapt the indexes to our application. In fact, the main goal of the mapping is the identification of different regions corresponding to different plasma states, and in particular, the safe and unsafe operational regions. The separation of clusters is not required if they contain samples of the same type. In particular, most relevance has been done to minimum distance between disruptive and safe regions with respect to the compactness within each region. Moreover, the modified *DN* index (DN_M) yields a high value if safe region and disruptive regions are well separated.

3 The Data Base Composition

Data for this study consist of signals recorded at ASDEX Upgrade for 229 shots, 149 disruptive and 80 safe. Each pulse has been sampled every 1 ms, resulting in a total number of samples equal to 780 969. The selected shots are in the range 16 000-19 000, corresponding to the experimental campaigns performed between June 2002 and July 2004.

The database is composed by the time series of seven plasma parameters that are relevant to describe the plasma regime during the current flat-top of the discharge. The selected parameters and the corresponding ranges of variations are reported in Table 1.

The overall knowledge of the plasma state can be augmented with an additional information, specifically introduced to indicate the safe or disruptive status of each sample in the shot. In particular, for a disruptive shot, let us define as precursor time t_{PREC} the time instant that discriminates between safe and pre-disruptive phases. On the basis of previous experiences on disruption prediction at ASDEX Upgrade, t_{PREC} has been set equal to t_D - 45ms [7], where t_D is the disruption time.

Table 1. Signals of the Database

SIGNAL	*Acronym*	*Range of variation*
SAFETY FACTOR AT 95% OF THE MAJOR RADIUS	q95	[-6, -2.25]
TOTAL INPUT POWER [MW]	Ptot	[0.3, 14.5]
RADIATED POWER/TOTAL INPUT POWER	Pfrac	[$9.7\ 10^{-3}$, 4]
INTERNAL INDUCTIVITY	li	[0.5, 1.9]
POLOIDAL BETA	βpol	[$5.8\ 10^{-3}$, 2.1]
LOCKED MODE SIGNAL [V]	LM	[0 0.4]
ELECTRON DENSITY/GREENWALD DENSITY	ne$_{Greenwald}$	[$1.1\ 10^{-13}$, $14.5\ 10^{14}$]

The samples of a disruptive shot in the interval [$t_{PREC} \div t_D$] are considered disruptive samples (ds). Moreover, two types of safe samples are considered: the samples of a disruptive shot in the interval [$t_{FLAT-TOP} \div t_{PREC}$] ($ss^d$), and all the samples of a safe shot (ss), where $t_{FLAT-TOP}$ is the flat-top beginning time.

Following this classification, four main categories of clusters can be identified depending on their composition: Empty clusters (ECs), which contain no samples; Disruptive clusters (DCs), which contain disruptive samples ds; Safe clusters (SCs), which contain safe samples ss and ss^d; Mixed Clusters (MCs), which contain both safe and disruptive samples.

4 Results

The first issue to be performed in order to partition the samples in the data base, is a selection of the number of clusters. The number of MCs has to be maintained as limited as possible, in order to obtain a compact representation of the plasma parameter space. To this end the number of clusters should be as large as possible. Nevertheless, using an excessive number of clusters reduces the power of the mapping as a tool to investigate the common features of the patterns belonging to the same cluster.

In the present paper, the number of clusters in the SOM has been chosen in order to optimize the quality indexes introduced in Section 2.1. As previously cited, this choice allows to minimize both the number of samples that belongs to mixed clusters MC, and to limit the number of clusters in the SOM.

A second issue to be performed is the data reduction, in order to balance the number of samples that describes the disruptive phase and that describes the non-disruptive phase. In fact, the number of samples ss and ss^d is much larger with respect to the number of samples ds available in the disruptive phase. Data reduction is performed using SOM applied to each shot as proposed in [8]. As each cluster is supposed to contain samples belonging to similar plasma states, one sample for each safe cluster is retained, whereas all the samples in the disruptive phase are selected. This procedure allows one to automatically select a limited and representative number of samples for the subsequent operational space mapping.

Moreover, SOM uses Euclidian metric to measure distances between data vectors, a third issue to be performed is the scaling of variables, which have been normalized between zero and one.

The composition of the dataset to which the SOM has been applied, was: 38% of ss; 50% of ss^d, and 12% of ds.

Fig. 1 represents the percentage of MCs with respect to the total number of clusters (dashed line) and the percentage of the samples contained in them, with respect to the total number of samples (solid line) versus the total number of clusters. As it can be noticed, both curves saturate for a number of clusters greater then 1421. Thus, further increasing the number of clusters is not useful to reduce the number of MCs.

Fig. 2 shows the trend of the Average quantization error E_q and the Topographic error E_t, versus the total number of clusters. An agreement between the minimization of E_q and the maximization of E_t is achieved for 1421 clusters.

Fig. 3 shows the modified DB index and the modified Dunn index versus the total number of clusters respectively. A good trade-off among the minimization of the DB_M

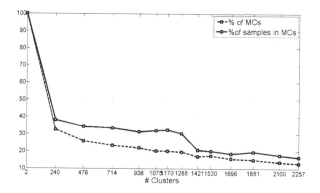

Fig. 1. Percentage of MCs and fraction of the samples in MCs

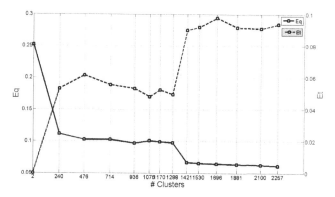

Fig. 2. Average quantization error (continuous line), and Topographic error (dashed line)

index, the maximization of the DN_M index and an acceptable fraction of samples that belong to MCs could be obtained with 1421 clusters. Thus, all three diagrams confirm that is not useful overdraw the map up to 1421 clusters.

The resulting mapping gives: 77% of clusters are SCs; 17% are MCs; 3% are DCs; 3% are empty. Moreover, 20% of samples belong to MCs, whereas the majority of them (78%) is in SCs and only 2% is attributed to DCs.

This means that, as expected, a pre-disruptive phase of 45ms, optimized on the base of all the disruptive shots contained in the dataset, cannot be valid for each of them.

Fig. 4a shows the 2D SOM mapping with 1421 clusters. In this mapping, different colours are assigned to the different type of samples, i.e., ss are blue, ss^d are red, and ds are green. A large safe region (clusters colored of blue and/or red) and a smaller disruptive region (prevalence of green in the clusters) can be clearly identified. Moreover, a transition region, where ss^d and ds coexist, appears between the safe and disruptive regions.In Fig. 4b, the colours of clusters are assigned on the base of the type of cluster, rather than on the samples density. This map clearly highlights the presence of a large safe region (cyan), a disruptive region (green) on the right top side, and a transition region (magenta) formed by MCs, which is at the boundary of the two regions. Empty clusters are white. Therefore, safe and disruptive states of plasma are well separated in the SOM map, provided that an appropriate number of clusters is set.

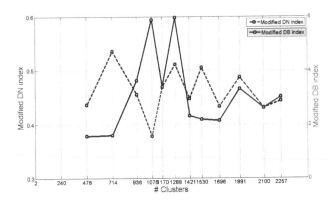

Fig. 3. DB_M (continuous line), and DN_M (dashed line) versus the mapping size

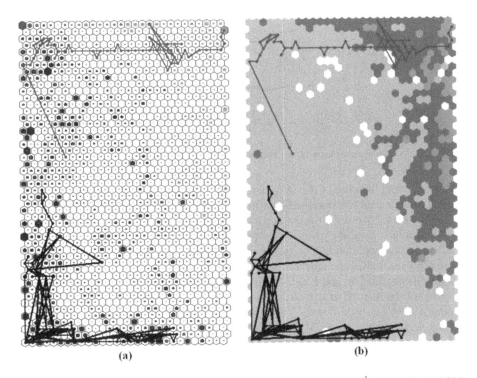

(a)   (b)

Fig. 4. a) SOM coloured on the bases of samples density (blue: ss; red: ss^d; green: ds). b) SOM coloured on the basis of cluster type (cyan: SC; green: DS; magenta: MC; white: EC). Red trajectory tracks a disruptive shot, whereas black trajectory tracks a safe shot.

The SOMs in Fig. 4a and Fig. 4b are used also to display the operational states of the disruptive shots and of the safe shots during the time. The temporal sequence of the samples of a shot forms a trajectory on the map depicting the movement of the operating point. As can be noted, a disruptive shot (red trajectory in the upper side of the maps) starts in a SC cluster containing only ss^d (red cluster in Fig. 4a), crosses mixed clusters, and arrives in a DC cluster. Following the trajectory in the map in Fig. 4b, it is possible to recognize a region (magenta region) where the risk of an imminent disruption is high. The safe shot (black trajectory in the lower side of the maps) starts in a SC cluster containing only ss (blue cluster in Fig.4a), and evolves in the time moving into the safe region (cyan region in Fig. 4b).

5 Conclusions

A strategy to analyze the information encoded in ASDEX Upgrade data base is developed, using the Self-Organizing-Map. It reveals a number of interesting issues for the ASDEX-Upgrade operational space. In particular, the proposed strategy allows us to simply display both the clustering structure of the data corresponding to different plasma states and the temporal sequence of the samples on the map, depicting the movement of the operating point during a shot. Following the trajectory in the map, it is possible to eventually recognize the proximity to an operational region where the risk of an imminent disruption is high.

Acknowledgments

This work was supported by the Euratom Communities under the contract of Association between EURATOM/ENEA. The views and opinions expressed herein do not necessarily reflect those of the European Commission.

References

1. Kohonen, M.T.: Self-Organization and Associative Memory. Springer, New York (1989)
2. Vesanto, F.J., Alhoniemi, E.: Clustering of the self-organizing map. IEEE Transaction on Neural Networks 11(3), 586–600 (2000)
3. Alhoniemi, E.: Unsupervised pattern recognition methods for exploratory analysis of industrial process data. Ph.D. thesis. Helsinki University of Technology, Finland (2002)
4. SomToolbox, Helsinky University of Technology,
   http://www.cis.hut.fi/projects/somtoolbox
5. Davies, D.L., Bouldin, D.W.: A cluster separation measure. IEEE Transaction on Pattern Recognition and Machine Intelligence 1, 224–227 (1979)
6. Dunn, J.: Well separated clusters and optimal fuzzy partitions. Journal of Cybernetics 4, 95–104 (1974)
7. Sias, G.: A Disruption prediction System for ASDEX-Upgrade based on Neural Networks. PhD. dissertation, University of Padova (2007)
8. Camplani, M., Cannas, B., Fanni, A., Pautasso, G., Sias, G., Sonato, P., Zedda, M.: The ASDEX Upgrade Team: Mapping of the ASDEX Upgrade operational space using clustering techniques. In: Proc. of 34th EPS Conference on Plasma Physics and Controlled Fusion, Warsaw, Poland (2007)

An Application of the Occam Factor to Model Order Determination

David J. Booth and Ruth Tye

London Metropolitan University, 166-220 Holloway Road
London N7 8DB, UK

Abstract. Supervised neural networks belong to the class of parameterised non-linear models whose optimum values are determined by a best fit procedure to training data. The problem of over-fitting can occur when more parameters than needed are included in the model. David MacKay's Occam factor deploys a Bayesian approach to the investigation of how model order can be rationally re-strained. This paper uses a case study to show how the Occam factor might be used to discriminate on model order and how it compares with similar indices e.g. Schwarz's Bayesian information criterion and Akaike's information criterion.

Keywords: Over-fitting, Model order, Occam factor, Bayesian statistics.

1 Introduction

An important modeling problem is the inference of a functional relationship f(x,w) between a set of parameters $\underline{w}$ and attribute variables x associated with target values t. This short paper addresses the problem of the determination of both parameter values $\underline{w}$ and model order K for the standard polynomial model denoted H_K. Models that are too simple are unlikely to fit the data adequately, models that are too complex can fit the data and the noise element together and thus fail to generalize when presented with a different data set.

2 The Bayesian Approach

We assume that for model H_K the parameters $\underline{w}$ are based on the data D. The posterior probability for $\underline{w}$, $P(\underline{w} \mid D, H_K)$ is related to the likelihood $P(D \mid \underline{w}, H_K)$, the prior $P(\underline{w} \mid H_K)$ and the evidence $P(D \mid H_K)$ using Bayes' product rule:

$$P(\underline{w} \mid D, H_K) = \frac{P(D \mid \underline{w}, H_K) P(\underline{w} \mid H_K)}{P(D \mid H_K)} \tag{1}$$

The likelihood is calculated from the pdf of $\underline{w}$, the prior is a quantitative expression of our belief about the probability of $\underline{w}$ before the data is presented. The evidence term is derived by integrating (1) over $\underline{w}$.

D. Palmer-Brown et al. (Eds.): EANN 2009, CCIS 43, pp. 438–443, 2009.

3 Formulating the Error Term $E_D(\underline{w})$ for the Simple Polynomial Model

Our model H_K is a sum of powers up to K of the single input variable x
$f(x,\underline{w}) = \sum_{k=0}^{k=K} w_k x^k$. For a data set $D = \{(x_n, t_n), n = 1..N\}$ the accumulated
error is modeled using the sum of the squared errors, $E_D(\underline{w})$

$$E_D(\underline{w}) = \frac{1}{2}\sum_n (t_n - f(x_n,\underline{w}))^2 = \frac{1}{2}(t - X\underline{w})^T (t - X\underline{w})$$

where

$$X = \begin{bmatrix} 1 & x_1 & \cdots & x_1^K \\ 1 & & & \\ \vdots & & & \\ 1 & x_N & & x_N^K \end{bmatrix} \qquad (2)$$

We assume a Gaussian noise model $\mathcal{N}(0, \sigma^2)$ for the error values $(t - X\underline{w})$ which
are assumed to be independent, giving the conditional probability of the data set D,

$$P(D \mid \underline{w}, \sigma^2, H_K) = Z_D \exp(-\frac{E_D(w)}{\sigma^2}), \; Z_D = \left(\frac{1}{2\pi\sigma^2}\right)^{N/2} \qquad (3)$$

The maximum likelihood estimate of the parameters $\underline{w}$ and σ is found by locating the
stationary point of $\log P(D \mid \hat{w}, \sigma^2, H_K)$

$$0 = \nabla_{\underline{w}} \log P(D \mid \hat{w}, \sigma, H_K) \Rightarrow \hat{\underline{w}} = (X^T X)^{-1} X^T t \qquad (4)$$

$$0 = \frac{\partial \log P(D \mid w, \sigma, H)}{\partial \sigma} \Rightarrow \hat{\sigma}^2 = \frac{1}{N}(t - X\hat{\underline{w}})^T (t - X\hat{\underline{w}})$$

Following Ghahramani Z [1], the second order properties are:

$$A = -\nabla\nabla \log P(\hat{\underline{w}} \mid D, H_k) = \frac{X^T X}{\hat{\sigma}^2} \qquad \frac{\partial^2 \log P(D \mid w, \sigma, H)}{\partial \sigma^2} = -\frac{2N}{\hat{\sigma}^2} \qquad (5)$$

Since the Hessian A is positive definite and the second order term in σ is strictly
negative then the point $\hat{\underline{w}}$ and $\hat{\sigma}$ is indeed a maximum. From which the maximized
likelihood of the model data D at $\hat{\underline{w}}$ and $\hat{\sigma}$ is

$$P(D \mid \hat{w}, \hat{\sigma}^2, H_K) = \frac{1}{(\hat{\sigma}\sqrt{2\pi})^N} \exp(-\frac{E_D(\hat{w})}{\sigma^2}) = \frac{1}{(\hat{\sigma}\sqrt{2\pi})^N} \exp(-\frac{N}{2}) \qquad (6)$$

4 Deriving a Second Order Approximation for the Log Posterior

Using Laplace's method, the log of the posterior can also be written as a 2nd order expansion about the maximum with Hessian matrix A. This is justified if the posterior has a high peak about $\hat{\underline{w}}$ which will be the consequence of a reasonably well fitting model.

$$\log P(\hat{\underline{w}}+\underline{\Delta w} \mid D,H_K) \approx \log P(\hat{\underline{w}} \mid D,H_K)+\frac{1}{2}\underline{\Delta w}^T \nabla\nabla \log P(\hat{\underline{w}} \mid D,H_K)\underline{\Delta w}$$

$$P(\underline{w} \mid D,H_K) \approx P(\hat{\underline{w}} \mid D,H_K)\exp(-\frac{1}{2}\underline{\Delta w}^T A\underline{\Delta w}) \tag{7}$$

5 Calculation of the Evidence P (D|H_K)

If we integrate both sides of (1) over $\underline{w}$ then we obtain the marginal likelihood (aka Bayesian evidence) of model H_K

$$P(D \mid H_K) = \int P(D \mid \underline{w},H_i)P(\underline{w} \mid H_K)d\underline{w} \tag{8}$$

Mackay [2], now applies a Box model to estimate the integral.

$$\int P(D \mid \underline{w},H_K)P(\underline{w} \mid H_K)d\underline{w} =$$
$$DEN(P(D \mid \underline{w},H_K)P(\underline{w} \mid H_K))*VOL(P(D \mid \underline{w},H_K)P(\underline{w} \mid H_K)) \tag{9}$$

Where DEN is the probability density at the maximum $\underline{w}$ and VOL is the volume of the box that models the volume enclosed by the probability surface over the parameter space. In the one dimensional case DEN is just the maximum value and VOL is just the width of the box. If $\underline{w}$ is a two dimensional vector then VOL is an area constituting the base of the box. For higher dimensions VOL is a hyper box.

To apply the DEN operator we have assumed that the maximum of the product of it's two terms is the product of the maximum of each term which can be justified by imposing the restriction of a constant prior term in some hyper region about $\hat{\underline{w}}$

$$DEN(P(D \mid \underline{w},H_K)P(\underline{w} \mid H_K)) = P(D \mid \hat{\underline{w}},H_K)P(\hat{\underline{w}} \mid H_K) \tag{10}$$

Expanding VOL in (9) using (1) and a second order expansion of $P(\underline{w} \mid D,H_K)$

$$VOL(P(D \mid \underline{w},H_i)P(\underline{w} \mid H_K)) = VOL(P(D \mid H_K)P(\underline{w} \mid D,H_K)) =$$
$$\int \exp(-\frac{1}{2}\underline{\Delta w}^T A\underline{\Delta w}) = \det{}^{-1/2}(A/2\pi) \tag{11}$$

The final expression for the marginal likelihood is the product of the maximised likelihood and Mackay's occam factor, [2].

$$\int P(D|\underline{w},H_i)P(\underline{w}|H_i)d\underline{w} = \underbrace{P(D|\underline{\hat{w}},H_i)}_{\max imised\ likelihood} \underbrace{P(\underline{\hat{w}}|H_i)\det^{-1/2}(A/2\pi)}_{occam\ factor} \quad (12)$$

The occam factor above is a direct consequence of the application of Bayesian reasoning. It will penalise models with a large number of parameters since the size of the hyper box will force $P(\underline{\hat{w}}|H_i)$ to be low. Secondly an over fitted model will have a sharp narrow peak for the marginal likelihood distribution (9) which corresponds to a large A term and hence a small $\det^{-1/2}(A/2\pi)$. We approach the problem of the prior distribution by making a very simple uniform assumption. In one dimension $P(\underline{\hat{w}}|H_i)$ is of value $1/(\sqrt{(2\pi)}\sigma_w)$ over some interval $[\underline{\hat{w}} +\sqrt{(2\pi)}\sigma_w/2,\ \underline{\hat{w}} - \sqrt{(2\pi)}\sigma_w/2]$ and zero elsewhere. For an K dimensional hyper box the prior will be $(1/\sqrt{(2\pi)}\sigma_w)^K$ giving a final expression for the posterior. This construction gives us a proper prior. The model with the higher value of the posterior would be the preferred model

$$P(D|H_K) = P(D|\underline{\hat{w}},H_K)\frac{\sigma_{wD}}{\left(\sqrt{2\pi}\right)^K \sigma_w^K} = \frac{1}{\left(\hat{\sigma}\sqrt{2\pi}\right)^N}\exp(-\frac{N}{2})\frac{\det^{-1/2}(A)}{\sigma_w^K} \quad (13)$$

Let L be the maximised likelihood $P(D|\underline{\hat{w}},H_K)$. Taking negative logs and introducing a factor of 2 a provisional Occam Information Criterion can defined

$$OIC = -2\ln L + \ln\det(A) + 2K\ln\sigma_w \quad (14)$$

For comparison OIC is compared to the Bayesian Information Criterion, [3] and to the Akaike Information Criterion, [4].

$$BIC = -2\ln L + K\ln N$$

$$AIC = -2\ln L + 2K$$

For all the indices above we would look for a lower value to determine a preferred model. The choice of σ_w is part of the problem of assessing of how probable a particular $\underline{w}$ is prior to any data arriving and does need careful consideration, [2]. In this case we have chosen a value of about twice that of the true coefficients

We have calculated these criteria for four polynomial models of orders 0 to 3. The data set D comprises of a range of points along the x axis and corresponding values from the application of a target quadratic function plus a noise component whose magnitude is a fraction, rv=0.3 of the quadratic's standard deviation, Fig 1.

Figure 2 shows the components needed to compute the information indices. For each model order the standard deviation (sigmahat, upper left), the 'sigwbarD' term is the probability volume that results when the model is applied and corresponds to $\det^{1/2}(A_K)$ term (upper right), the information criterion (lower left) and lastly the posterior probability. We have repeated the analysis for a linear model and show the results in Figure 3. In both cases the Occam Information Criterion does correctly identify the correct order.

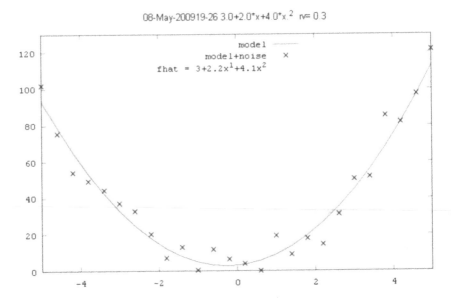

Fig. 1. Quadratic model, generated data and fitted model

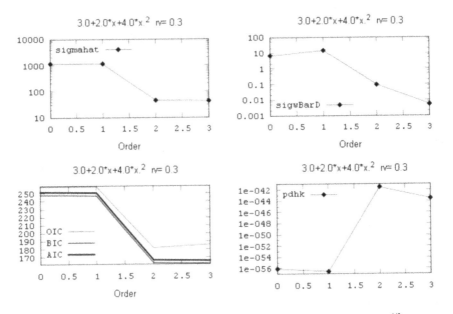

Fig. 2. For each model order top left: the standard deviation, top right: the $\det^{-1/2}(A_K)$ term, lower left: the information criterion and lower right: the posterior probability

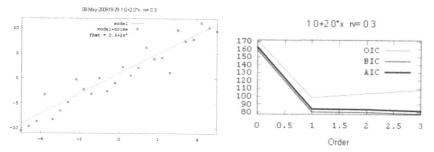

Fig. 3. Left Linear model, generated data and fitted model, right information indices

6 Conclusions and Future Work

This work demonstrates that MacKay's Occam factor and our Occam Information Criterion do correctly discriminate for optimal polynomial models. Our assumptions include: 1) that a polynomial model of order K is a reasonable model, 2) a Gaussian noise model for the error term, 3) a 2nd order expansion for the log posterior, and 4) the prior term for the parameters is constant in some hyper region about the maximum likelihood value. The first two are met by construction and 3) is a logical consequence We need to test assumption 4, in particular the sensitivity of our judgement on the prior probabilities of the coefficients of the model and secondly to the performance under increasingly noisy data. Work is also underway on how effective OIC would be to discriminate between neural network models of differing networks architectures where it is likely that our assumptions are open to challenge. An initial task is the calculation of the Hessian either using direct expressions or using numerical methods since it will be much larger and possibly numerically unstable. Initial results show that the error term $E_D(\underline{w})$ can be very flat around the minimum giving very low eigenvalues for A, [5]. Given that these obstacles can be overcome the OIC should be a useful indicator of optimal model complexity.

References

1. Ghahramani, Z.: Machine Learning 2006 Course Web Page, Department of Engineering. University of Cambridge (2006)
2. MacKay, D.: Information Theory, Inference, and Learning Algorithms. Cambridge University Press, Cambridge (2003)
3. Schwarz, G.: Estimating the Dimension of a Model. Source: Ann. Statist. 6(2), 461–464 (1978)
4. Akaike, H.: A new look at the statistical model identification. IEEE Transactions on Automatic Control 19(6), 716–723 (1974)
5. Bishop, C.M.: Pattern Recognition and Machine Learning. Springer Science and Media LLC (2006)

Use of Data Mining Techniques for Improved Detection of Breast Cancer with Biofield Diagnostic System

S. Vinitha Sree[1], E.Y.K. Ng[1,2], G. Kaw[3], and U. Rajendra Acharya[4]

[1] School of Mechanical and Aerospace Engineering, College of Engineering,
Nanyang Technological University, 50 Nanyang Avenue, Singapore 639798
[2] Adjunct NUH Scientist, Office of Biomedical Research,
National University Hospital, Singapore
[3] Consultant Radiologist, Department of Diagnostic Radiology,
Tan Tock Seng Hospital, Singapore 308433
[4] School of Engineering, Division of ECE, Ngee Ann Polytechnic, Singapore 599489

Abstract. The Biofield Diagnostic System (BDS) is an adjunct breast cancer detection modality that uses recorded skin surface electropotentials for differentiating benign and malignant lesions. The main objective of this paper is to apply data mining techniques to two BDS clinical trial datasets to improve the disease detection accuracy. Both the datasets are pre-processed to remove outliers and are then used for feature selection. Wrapper and filter feature selection techniques are employed and the selected features are used for classification using supervised techniques like Linear Discriminant Analysis (LDA), Quadratic Discriminant Analysis (QDA), Support Vector Machines (SVM) and Back Propagation Neural Network (BPNN). It was observed that the LDA classifier using the feature subset selected via wrapper technique significantly improved the sensitivity and accuracy of one of the datasets. Also, the key observation is that this feature subset reduced the mild subjective interpretation associated with the current prediction methodology of the BDS device, thereby opening up new development avenues for the BDS device.

Keywords: Biofield Diagnostic System, breast cancer detection, data mining, classification, wrapper methods, filter methods.

1 Introduction

The Biofield Diagnostic System (BDS), developed by Biofield Corporation, USA, is based on differentiating cancerous breast lesions from non-cancerous lesions by measuring alterations in skin surface electropotential differences which are known to occur concurrently with abnormal epithelial proliferation associated with breast malignancies [1]. The main requirement of BDS is that it should be used on a patient with a suspicious lesion that has been located by palpation or by using other diagnostic procedures. The device received market approval from the European Union (CE mark) to be an adjunct to physical breast examination or relevant imaging modalities.

D. Palmer-Brown et al. (Eds.): EANN 2009, CCIS 43, pp. 444–452, 2009.

The test is non-invasive, radiation-free demanding less than 15 minutes of examination time. Many preclinical and clinical studies [2-6] have evaluated the efficiency of BDS in detecting breast cancer. These studies have demonstrated high sensitivity values. It has also been established that BDS increases the malignancy detection by up to 20-29% compared to current diagnostic techniques such as mammography and ultrasound [1].

During the BDS test, the patient lies down in a supine position. One sensor is placed over the center of the suspicious lesion on the symptomatic (S) breast (SC) and four additional sensors are placed outside the margins of the lesion - superior to (SU), inferior to (SL), medial to (SI), and lateral to (SO) the sensor placed over the test lesion. U, L, I, O stand for Upper, Lower, Inner and Outer respectively. Two additional sensors (SH and SV) are placed in the center of the two quadrants (horizontal and vertical directions) adjacent to the symptomatic quadrant and one sensor is placed in the quadrant diagonal (SD) to the quadrant with the test lesion SC. This pattern of sensor placement is then replicated in a mirror image form on the asymptomatic (A) breast, whose sensors are labeled as AC, AU, AL, AI, AO, AV, AH and AD accordingly. Two more reference sensors are placed below and between the breasts. Within a few minutes, the sensors measure the skin surface electropotentials (the explanation of the origin of which can be found in [1, 7-10] and the device presents the **BDS index/score**, which is a linear combination of these electropotentials and age [11]. Prior to the test, the doctor determines a Level of Suspicion (LOS) score using the patient's physical examination or mammography or other imaging test reports and patient history. This **prior LOS** can fall into one of these categories: 1-normal, 2-benign tumor/cyst, 3-malignancy cannot be excluded (indeterminate cases), 4-probably malignant and 5-malignant. The BDS score and the prior LOS are then used in an interpretation map which maps a range of the BDS score for a particular prior LOS to a new revised **Post-BDS-LOS** value, which is similar in notation to the prior LOS and represents the final diagnostic prediction for that patient [2]. During the test, the demographic characteristics of the patient such as her menopausal status (Pre/Post), pregnancy history, and family history are also recorded.

Data mining tools have been widely used in the medical diagnostics field. Even though the diagnostic devices predict the presence of disease efficiently, there is a general consensus that mining of the data or images output by these devices would present new and highly significant predictive information that might result in more accurate disease detection. In the case of BDS, the demographic data, the electropotential values, lesion details and the BDS results of the patients form a potentially useful dataset most suitable for data mining tasks. Hence, in this work, we have attempted to apply a proposed data mining framework to two BDS clinical trial datasets to evaluate the usefulness of data mining techniques in improving the current disease prediction accuracy of BDS.

The paper is organized as follows. The description of the datasets and their respective clinical trial results are presented in section 2. Section 3 describes the proposed data mining framework and explains how the various steps were applied to both the datasets. The results obtained using both the datasets are presented and discussed in section 4. Finally, conclusions are given in section 5.

2 Description of the Datasets

This section provides a description of the datasets used in the study and the results of the BDS clinical trials from which the datasets were obtained.

2.1 Dataset 1

The first dataset, henceforth referred to as the *"TTSH dataset"*, was obtained as a result of the BDS clinical trial conducted at Tan Tock Seng Hospital (TTSH), Singapore in 2008. 149 women scheduled for mammography and/or ultrasound tests participated in the study. Of the 149 cases, 53 cases were malignant and 96 were benign. No pre-processing steps were necessary as no outliers/missing data were present. The BDS test conducted on these women resulted in a sensitivity of 96.23%, specificity of 93.80%, and an accuracy of 94.51%. The Area under the Receiver Operating Characteristics (AROC) curve was 0.972. For each patient, the following features were available for further analysis. The encoded data as used in the analysis is given in brackets. Menopausal status (Pre: 0; Post: 1), Parity (Previous pregnancy (no risk):0; Never pregnant (has risk): 1), Family History (No history: 0; Has history: 1), Lesion palpability (Not palpable: 0; Palpable: 1), Lesion location (RUO: 1; RUI: 2; RLO: 3; RLI: 4; RSA: 5; LUO: 6; LUI: 7; LLO: 8; LLI: 9; LSA: 10 where R–Right; L–Left; SA–Sub Areolar), Age, BDS index, Post-BDS-LOS and Class (Benign: 0; Malignant: 1). Since the tested BDS device was meant for clinical purposes, it did not output the 16 sensed electropotential values. Overall, the TTSH dataset had a total of nine features.

2.2 Dataset 2

The second dataset, henceforth referred to as the *"US dataset"*, was from one of the BDS trials conducted in the U.S. back in 1995. The dataset was obtained from the manufacturer. It had a total of 291 cases, out of which 198 were benign and 93 were malignant. The major difference between this dataset and the TTSH dataset is the presence of the 16 sensor values for each patient in addition to the other features listed in section 2.1. This dataset was purposely chosen to study the effect of these additional sensor values on the disease prediction efficiency. It has been reported that the value recorded by the SC sensor for malignant cases must be statistically greater than or equal to mean of SI, SO, SU and SL values and can never be lesser [12]. This fact was used to pick out outliers from the malignant dataset using t-test. Subsequently, the univariate outliers were detected using z-score method. The threshold was kept as ±3 for benign dataset (since sample size was greater than 80) and as ±2.5 for malignant dataset (sample size ≤ 80). The multivariate outliers were then detected using Mahanalobis distance (cases having probability of the distance ≤ 0.001 were considered outliers) [13]. After these data-preprocessing steps, there were 183 benign cases and 58 malignant cases (total of 241 cases) which were used for further analysis. The BDS test results from these 241 cases indicated a sensitivity of 89.6%, specificity of 54.6%, and an accuracy of 63.1%. The AROC was 0.771. In addition to the original electropotential values, ten linear combinations of these electropotentials were formulated and used as features. These Electropotential Differentials (EPDs) are listed in

Table 1. These features were derived so as to reflect the mean electropotentials /electropotential differentials within the symptomatic breast or the differentials between the symptomatic and asymptomatic breasts. Since studies [2, 8] have demonstrated the usefulness of a few such differentials (EPD 3 and EPD 6) in differentiating malignant and benign lesions, only differentials similar to these EPDs were derived in this work. Overall, the US dataset had 35 features.

Table 1. Additional derived features for the US dataset; EPD: Electropotential Differential

Additional derived features
EPD 1 : Mean(symptomatic sensors SC, SU, SL, SI, SO, SV, SH, SD)
EPD 2 : Mean(symptomatic sensors placed in the lesion's quadrant: SC,SU,SI,SL,SO)
EPD 3 : SC-SV
EPD 4 : Within breast differential : 3SC-SD-SH-SV
EPD 5 : Within breast differential : 4SC-SI-SU-SO-SL
EPD 6 : Between breasts differential : EPD 1 - Mean(asymptomatic sensors AC, AU, AL, AI, AO, AV, AH, AD)
EPD 7 : Between breasts differential: EPD 2 - Mean(asymptomatic inv sensors AC,AU,AI,AL,AO)
EPD 8 : EPD 4 - (3AC-AD-AH-AV)
EPD 9 : EPD 5 - (4AC-AI-AU-AO-AL)
EPD 10 : EPD 3 - (AC-AV)

3 Proposed Data Mining Framework

The general framework of data mining comprises of the following key steps: (1) Data pre-processing, (2) Feature selection, wherein good feature subsets are selected using filter and wrapper techniques, (3) Classifier development, wherein the selected feature subsets are used to build classifiers, (4) Evaluation, wherein the built classifiers are evaluated using test data to determine their accuracy in classifying the data. The proposed framework is illustrated in Fig. 1. In the figure, *"Dataset(FeatureSet)"* implies that the classifier was developed/evaluated using the reduced dataset formed by extracting only the selected *FeatureSet* from the *Dataset*. WFS1 stands for Wrapper Feature Subset and the Filter Feature Subsets are denoted by FFS1 and FFS2.

Both the datasets were pre-processed in the manner highlighted in section 2. All the features in the selected dataset were normalized using min-max normalization. Feature selection is the most important task in data mining as a good set of representative features decreases the processing time, and also leads to better generalization and comprehensibility of the mined results. In this work, both filter and wrapper feature selection techniques were employed. Ten fold stratified cross validation resampling technique was used for feature selection. The selected features were used to form a reduced dataset of the original dataset. This reduced dataset was used to build and test classifiers using 10-fold stratified cross validation. That is, the whole dataset was split into 10 parts such that each part contains approximately the same proportion of class samples as the original dataset. Nine parts of the data (learning set) were used for classifier development and the built classifier was evaluated using the remaining one part (test set). This procedure was repeated 10 times using a different part for testing in each case. The 10 test classification performance metrics were then averaged, and the average test performance is declared as the estimate of the true generalization performance.

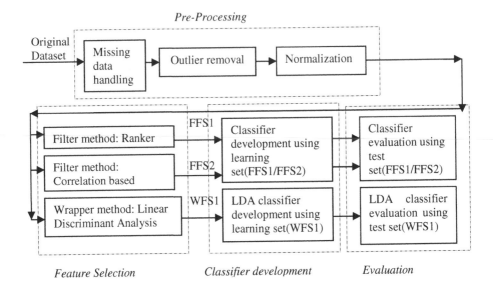

Fig. 1. Proposed data mining framework

The wrapper method uses a classifier to perform classification using every possible feature subset and finally selects the feature subset that gives the best classification accuracy. Since a classifier is built as many times as the number of feature subsets generated, computationally intensive classifiers should not be considered. Hence in this work, the Linear Discriminant Analysis (LDA) classifier based wrapper technique was implemented in Matlab (version 7.4.0.287-R2007a). The search strategy employed was nondeterministic strategy and the search direction was random. At the end of each fold, a rank was assigned to each feature. The top ranked 25% of the features were selected for each dataset. That is, most significant two features were selected to form the reduced TTSH dataset and top eight features were chosen to form the reduced US dataset.

In filter methods, features are first assigned scores based on particular criteria like correlation (Correlation-based Feature Selection (CFS)), χ^2-statistic (Chi-Squared method (CS)) or information measure (Information Gain (IG)/Gain Ratio (GR)). Then they are ranked using methods like best-first, ranker, greedy stepwise etc. The software WEKA was used for this purpose (version 3.6.0) [14]. Again, the top two and top eight features were selected from TTSH and US datasets respectively. The selected features for both datasets using the described feature selection methods are shown in Table 2. It can be observed that all the filter techniques selected similar feature subsets in case of the US dataset.

The reduced datasets formed using the filter subsets were used to build and test the following supervised learning based classifiers - LDA, Quadratic Discriminant

Table 2. Selected feature subsets for both datasets using filter and wrapper methods

Dataset	Wrapper LDA Selected features WFS1	Filter Methods Evaluator: Ranking method	
		CS/IR/GR: Ranker FFS1	CFS: Best First FFS2
TTSH dataset	Post-BDS-LOS, Lesion Palpability	Post-BDS-LOS, Age	Post-BDS-LOS, Lesion Palpability
US dataset	EPD1, EPD2 EPD4, EPD5,SV EPD7, EPD9, SC	EPD4, EPD5, EPD7, EPD8, Post-BDS-LOS, EPD3, SO, SH	EPD4, EPD5, EPD7, EPD8, Post-BDS-LOS, EPD3, SO, SH

Analysis (QDA), Support Vector Machines (SVM) and Back Propagation Neural Network (BPNN). Since wrapper feature sets should be used with their respective classifiers for best results, the reduced dataset formed using WFS1 was used to build a LDA classifier. Classifier performance was evaluated using sensitivity, specificity, accuracy and AROC measures.

4 Results and Discussion

The classification results obtained using both datasets are discussed in this section.

4.1 TTSH Dataset – Classification Results

The LDA/WFS1 classifier recorded a very high sensitivity of 96.23%, specificity of 91.67%, accuracy of 93.29%, and AROC of 0.9406. The results obtained using the filter datasets are presented in Table 3. The BPNN classifier did not result in proper convergence. Looking at the average performance measures over the 10 folds, it is evident that all classifier-feature set combinations (except QDA/FFS2) presented accuracies similar to that given by the LDA/WFS1 classifier.

Table 3. Classifier performance measures for the reduced TTSH datasets formed using the filter subsets FFS1 and FFS2

	PERFORMANCE MEASURES FOR THE TTSH DATASET							
	Sensitivity (%)		Specificity (%)		Accuracy (%)		AROC	
	FFS1	FFS2	FFS1	FFS2	FFS1	FFS2	FFS1	FFS2
LDA	96.23	96.23	91.67	91.67	93.29	93.29	0.94	0.94
QDA	96.23	90.57	91.67	91.67	93.29	91.28	0.94	0.9089
SVM	96.23	96.23	91.67	91.67	93.29	93.29	0.94	0.94

4.2 US Dataset – Classification Results

The LDA/WFS1 classifier recorded a high sensitivity of 93.1%, specificity of 97.81% and very good accuracy of 96.68%. The AROC was 0.9507. The results obtained using the filter subsets are shown in Table 4. It can be seen that all the classifiers demonstrated better results when compared to the original US dataset clinical trial results. In the case of sensitivity and accuracy, QDA recorded the highest values. But, the filter subset based classification results are comparatively lower than the wrapper based results.

Table 4. Classifier performance measures for the reduced US datasets formed using the filter subsets

PERFORMANCE MEASURES FOR THE US DATASET			
	Sensitivity *(%)*	*Specificity* *(%)*	*Accuracy* *(%)*	*AROC*
BPNN	77.59	86.89	84.65	0.8200
LDA	72.41	75.96	75.10	0.7435
QDA	87.93	89.07	88.80	0.8852
SVM	43.10	94.54	82.16	0.6860

4.3 Comparison of Results and Discussion

The TTSH BDS clinical trial resulted in a sensitivity of 96.23%, specificity of 93.80%, and an accuracy of 94.51%. The AROC was 0.972. The US trial, on the other hand, demonstrated a sensitivity of 89.6%, very low specificity of 54.6% and low accuracy (63.1%). The AROC was also only 0.771. It is evident that the BDS device used in the TTSH clinical trial recorded significantly better performance measures when compared to those recorded by the US trial device. The reason for this can be attributed to the fact that both the clinical trials used different interpretation maps to deduce the Post-BDS-LOS from the BDS index and the prior LOS. The TTSH device was an improved version of the device used in the US trial. The reason for including the US dataset was to study the use of the raw independent sensor values in disease detection and also to evaluate the use of classifiers for improving the current very low detection accuracy.

   In the case of the results of the *US dataset*, the best classification results were demonstrated by the LDA/WFS1 combination. It resulted in 93.10% sensitivity, 97.81% specificity, 96.68% accuracy and AROC of 0.9507. These values are significantly very high compared to those observed in the US clinical trial.

   In the case of *TTSH dataset*, the filter and the wrapper datasets' results for the LDA classifier were identical - the sensitivity, specificity, accuracy, and AROC values were 96.23%, 91.67%, 93.29%, and 0.9406 respectively. The use of data mining techniques did not improve the detection accuracy in this case as the results were almost similar to the results obtained using the original TTSH clinical trial dataset.

   Also, it can be observed that the feature subset selected by the wrapper method provides equal or higher accuracy when compared to that given by filter method based subsets. This reinstates the fact that wrapper method is a better feature selection technique to improve a classifier's performance since both learning a classifier and selecting features use the same bias. Another interesting observation is that for both the TTSH and US datasets, the LDA classifier presented the best results. Thus, for BDS datasets, a classifier as simple as LDA seems to be an optimal one.

Considering the US dataset analysis, it can be observed that none of the feature subsets had any of the electropotentials recorded from the asymptomatic side as features. The features depicted either symptomatic sensor values or the difference between symptomatic and asymptomatic electropotentials. This again emphasizes the fact that the surface electropotentials are reliable indicators of the presence or absence of cancer. It can be seen that filter feature subsets had the Post-BDS-LOS as one of the selected features, whereas WFS1 did not have this feature. But among all the feature subsets, WFS1 resulted in the highest sensitivity, specificity, accuracy and AROC. The most important observation here is that even without using the Post-BDS-LOS, LDA/WFS1 is able to classify with clinically acceptable sensitivity and specificity results. Currently, the BDS prediction methodology is dependant on the doctor's subjective prediction of the patient's prior LOS using other modalities and this prior LOS is used to determine the subsequent Post-BDS-LOS. This interpretation methodology makes Post-BDS-LOS parameter a subjective one. Even though the use of the BDS score along with the subjective prior LOS makes BDS a relatively lesser subjective tool compared to mammography or ultrasound, there is still a mild subjective interpretation involved. But, in this study, it has been observed that the LDA/WFS1 classifier demonstrated highly significant detection accuracy without using the subjective Post-BDS-LOS as a feature. Thus, BDS need not be dependant on mammogram or ultrasound or physical examination LOS to calculate Post-BDS-LOS for its predictions. Even without Post-BDS-LOS, using only the original and derived electropotential values, BDS is able to predict the presence of cancer with good accuracy on using classifiers for disease detection. This conclusion justifies the reason why the US dataset was included for analysis in this work.

The TTSH clinical trial has already demonstrated significant accuracy and hence the use of classifiers might not drastically improve the accuracy. That has been demonstrated in this study. The main reason for inclusion of this dataset was to check if a good feature subset could be selected that does not include Post-BDS-LOS. But it is evident that all the feature subsets selected for the TTSH dataset had the Post-BDS-LOS parameter. This proves that without the presence of raw electropotential values as features, the Post-BDS-LOS is a very significant feature. Had the TTSH trial also recorded the raw sensor values, their effect on accuracy could have be studied.

5 Conclusion

Overall, this study has demonstrated that the use of classifiers with carefully selected features does improve the classification accuracy of the BDS device. The most important observation is the fact that a feature subset formed just with original and derived electropotentials is able to provide the best detection accuracy. The LDA/WFS1 technique that gave this best accuracy did not use the subjective Post-BDS-LOS as a feature. This conclusion indicates that there is a possibility of designing BDS as an objective diagnostic device without any dependence on the results of other diagnostic techniques like mammography and ultrasound for interpretation.

Acknowledgements

The authors would like to thank Dr. Michael J. Antonoplos of *Biofield Corp., USA*, for providing the US clinical trial dataset for this work and for comparison with our clinical trial results and for sharing his views and interests on the work. Support and permission to publish these findings granted by Mr. David Bruce Hong of *The MacKay Holdings Limited* is also gratefully acknowledged.

References

1. Biofield Corp., http://www.biofield.com
2. Biofield Diagnostic System, Physician's manual, LBL 024 rev. A
3. Cuzick, J., et al.: Electropotential measurements as a new diagnostic modality for breast cancer. Lancet 352, 359–363 (1998)
4. Dickhaut, M., Schreer, I., Frischbier, H.J., et al.: The value of BBE in the assessment of breast lesions. In: Dixon, J.M. (ed.) Electropotentials in the clinical assessment of breast neoplasia, pp. 63–69. Springer, NY (1996)
5. Fukuda, M., Shimizu, K., Okamoto, N., Arimura, T., Ohta, T., Yamaguchi, S., Faupel, M.L.: Prospective evaluation of skin surface electropotentials in Japanese patients with suspicious breast lesions. Jpn. J. Cancer Res. 87, 1092–1096 (1996)
6. Gatzemeier, W., Cuzick, J., Scelsi, M., Galetti, K., Villani, L., Tinterri, C., Regalo, L., Costa, A.: Correlation between the Biofield diagnostic test and Ki-67 labeling index to identify highly proliferating palpable breast lesions. Breast Cancer Res. Treat. 76, S114 (2002)
7. Davies, R.J.: Underlying mechanism involved in surface electrical potential measurements for the diagnosis of breast cancer: an electrophysiological approach to breast cancer. In: Dixon, J.M. (ed.) Electropotentials in the clinical assessment of breast neoplasia, pp. 4–17. Springer, Heidelberg (1996)
8. Faupel, M., Vanel, D., Barth, V., Davies, R., Fentiman, I.S., Holland, R., Lamarque, J.L., Sacchini, V., Schreer, I.: Electropotential evaluation as a new technique for diagnosing breast lesions. Eur. J. Radiol. 24, 33–38 (1997)
9. Gallager, H.S., Martin, J.E.: Early phases in the development of breast cancer. Cancer 24, 1170–1178 (1969)
10. Goller, D.A., Weidema, W.F., Davies, R.J.: Transmural electrical potential difference as an early marker in colon cancer. Arch. Surg. 121, 345–350 (1986)
11. Sacchini, V., Gatzemeier, W., Costa, A., Merson, M., Bonanni, B., Gennaro, M., Zandonini, G., Gennari, R., Holland, R.R., Schreer, I., Vanel, D.: Utility of biopotentials measured with the Biofield Diagnostic System for distinguishing malignant from benign lesions and proliferative from nonproliferative benign lesions. Breast Cancer Res. Treat. 76, S116 (2002)
12. United States Patent 6351666 - Method and apparatus for sensing and processing biopotentials, http://www.freepatentsonline.com/6351666.html
13. Detecting outliers (2006), http://www.utexas.edu/courses/schwab/sw388r7_spring_2006/SolvingProblems/
14. WEKA data mining software, http://www.cs.waikato.ac.nz/ml/weka/

Clustering of Entropy Topography in Epileptic Electroencephalography

Nadia Mammone*, Giuseppina Inuso, Fabio La Foresta, Mario Versaci, and Francesco C. Morabito

NeuroLab, DIMET, *Mediterranean* University of Reggio Calabria, Italy
{nadia.mammone,giuseppina.inuso,fabio.laforesta,
mario.versaci,morabito}@unirc.it
http://neurolab.unirc.it

Abstract. Epileptic seizures seem to result from an abnormal synchronization of different areas of the brain, as if a kind of recruitment occurred from a critical area towards other areas of the brain, until the brain itself can no longer bear the extent of this recruitment and triggers the seizure in order to reset this abnormal condition. In order to catch these recruitment phenomena, a technique based on entropy is introduced to study the synchronization of the electric activity of neuronal sources in the brain and tested over three EEG dataset from patients affected by partial epilepsy. Entropy showed a very steady spatial distribution and appeared linked to the region of seizure onset. Entropy mapping was compared with the standard power mapping that was much less stable and selective. A SOM based spatial clustering of entropy topography showed that the critical electrodes were coupled together long time before the seizure onset.

Index Terms: Electroencephalography, Renyi's Entropy, Epilepsy, SOM.

1 Introduction

Epilepsy represents one of the most common neurological disorders (about 1% of the world's population). Two-thirds of patients can benefit from antiepileptic drugs and another 8% could benefit from surgery. However, the therapy causes side effects and surgery is not always resolving. No sufficient treatment is currently available for the remaining 25% of patients. The most disabling aspects of the disease lie at the sudden, unforeseen way in which the seizures arise, leading to a high risk of serious injury and a severe feeling of helplessness that has a strong impact on the everyday life of the patient. It is clear that a method capable of explaining and forecasting the occurrence of seizures could significantly improve the therapeutic possibilities, as well as the quality of life of epileptic patients. Epileptic Seizures have been considered sudden and unpredictable events for centuries. A seizure occurs when a massive group of neurons in the cerebral cortex begins to discharge in a highly organized rhythmic pattern then,

---

* Corresponding author.

D. Palmer-Brown et al. (Eds.): EANN 2009, CCIS 43, pp. 453–462, 2009.

for mostly unknown reasons, it develops according to some poorly described dynamics. Nowadays, there is an increasing amount of evidence that seizures might be predictable. In fact, as proved by the results reported by different research groups working on epilepsy, seizures appear not completely random and unpredictable events. Thus it is reasonable to wonder when, where and why these epileptogenic processes start up in the brain and how they result in a seizure. The goal of the scientific community is to predict and control epilepsy, but if we aim to control epilepsy we must understand it first: in fact, discovering the epileptogenic processes would throw a new light on this neurological disease. The cutting edge view of epileptic seizures proposed in this paper, would dramatically upset the standard approaches to epilepsy: seizures would no longer be considered the central point of the diagnostic analysis, but the entire "epileptogenic process" would be explored. The research in this field, in fact, has been focused only on epileptic seizures so far. In our opinion, as long as we focus on the seizure onset and then we try to explain what happened before in a retrospective way, we will not be able to discover the epileptogenic processes and therefore to fully understand and control epilepsy, because seizures are only a partial aspect of a more general problem. Epileptic seizures seem to result from an abnormal synchronization of different areas of the brain, as if a kind of recruitment occurred from a critical area towards other areas of the brain (not necessarily the focus) until, the brain can no longer bear the extent of this recruitment and it triggers the seizure in order to reset this abnormal condition. If this hypothesis is true, we are not allowed to consider the onset zone the sole triggering factor, but the seizure appears to be triggered by a network phenomenon that can involve areas apparently not involved in seizure generation at a standard EEG visual inspection. Seizures had been considered unpredictable and sudden events until a few years ago. The Scientific Community began being interested in epileptic seizure prediction during '70s: some results in literature showed that seizures were likely to be a stage of a more general epileptogenic process rather than an unpredictable and sudden event. Therefore a new hypothesis was proposed: the evolution of brain dynamics towards seizures was assumed to follow this transition: inter-ictal − > pre-ictal − > ictal − > post-ictal state. This emerging hypothesis is still under analysis and many studies have been carried out: most of them have been carried out on intracranial electroencephalographic recordings (IEEG). The processes that start up in the brain and lead to seizure are nowadays mostly unknown: investigating these dynamics from the very beginning, that is minutes and even hours before seizure onset, may throw a new light on epilepsy and upset the standard diagnosis and treatment protocols. Many researchers tried to estimate and localize epileptic sources in the brain, mainly analysing the ictal stage in EEG [1], [2], [3]. If the aim is to detect and follow the epileptogenic processes, in other words to find patterns of epileptic sources activation, a long-time continuous analysis (from either the spatial and temporal point of view) of this output in search of any information about the brain-system might help. In order to understand the development of epileptic seizures, we should understand how this abnormal order affects EEG over the cortex, over time. Thus the point is: how can we measure this order? Thanks to its features, entropy might be the answer, thus here we propose to introduce it to investigate the spatial temporal distribution of order over the cortex. EEG brain topography is a technique that gives a picture of brain activity over the cortex. It consists of plotting the EEG in 2-D maps by color coding EEG features, most commonly the EEG power. EEG topography has been widely used as a tool for investigating the activity of epileptic brains but just for analysing sparse images and not for reconstructing a global picture of the brain behaviour over time [1], [4], [5], [6], [7], [8], [9], [10], [11], [12]. Here we propose to carry out a long-time continuous entropy topography in search of patterns of EEG behaviour. Once entropy is mapped a spatio-temporal SOM based clustering is carried out in order to put in the same cluster the electrodes that share similar entropy levels.

The paper is organized as follows: Section 2 will describe the electroencephalographic potentials generation from neuronal sources and the genesis of epileptic seizures, Section 3 will introduce entropy topography and spatial clustering and Section 4 will report the results.

2   The Electroencephalography and the Neuronal Sources

A single EEG electrode provides estimates of synaptic action averaged over tissue masses containing between roughly 100 million and 1 billion neurons. The space averaging of brain potentials resulting from extracranial recording is a data reduction process forced by current spreading in the head volume conductor. The connection between surface and depth events is thus intimately dependent on the physics of electric field behaviour in biological tissue. The synaptic inputs to a neuron are of two types: those that produce excitatory postsynaptic potentials (EPSP) across the membrane of the target neuron, thereby making it easier for the target neuron to fire an action potential and the inhibitory postsynaptic potentials (IPSP), which act in the opposite manner on the output neuron.

The cortex is believed to be the structure that generates most of the electric potential measured on the scalp. The synaptic action fields can be defined as the numbers of active excitatory and inhibitory synapses per unit volume of tissue at any given time. A seizure occurs when a massive group of neurons in the cerebral cortex begins to discharge in a highly organized rhythmic pattern (increased entrainment). According to one theory, seizures are caused by an imbalance between excitatory and inhibitory neurotransmitters.

Electric and magnetic fields provide large-scale, short-time measures of the modulations of synaptic and action potential fields around their background levels.

Dynamic brain behaviour are conjectured by many neuroscientists to result from the interaction of neurons and assemblies of neurons that form at multiple spatial scales: part of the dynamic behaviour at macroscopic scales may be measured by scalp EEG electrodes. The amplitude of scalp potential depends strongly on the characteristic size of the underlying correlated source patches (amount of source synchronization).

Quantitative methods to recognize spatial-temporal EEG patterns have been studied only sporadically. The technique proposed in this paper aims to describe such spatial-temporal patterns of EEG dynamics. Entropy will be proposed to measure the entrainment of neuronal sources, its topography will give a view of the spatial distribution of the entrainment and spatial clustering will quantify the mutual interactions among neurons.

3   Methodology

3.1   Entropy to Measure "Order" in the Brain

Since we were interested in monitoring the *order* degree of different areas of the brain, in this paper we proposed to map the Renyi's entropy of EEG [13] and we compared it with the mapping of the power of EEG. Entropy can be interpreted as a measure of order and randomness. Given a signal, we can think about its amplitude as a random variable X.

$$H_{R_\alpha}(X) = \frac{1}{1-\alpha} \log \sum_i P^\alpha(X = a_i) \tag{1}$$

For a continuous random variable X, whose probability density function (pdf) is $f_X(x)$, Renyi's entropy is defined as:

$$H_{R_\alpha}(x) = \frac{1}{1-\alpha} \log \int_{-\infty}^{+\infty} f_X^\alpha(x) dx \tag{2}$$

where the order α is a free parameter ($\alpha > 0$ and $\alpha \neq 1$) and P represents the probability. In case of variable with small randomness ("high order"), a few probabilities will be close to one ($P(x = a_i) \to 1$) and most of probabilities will be close to zero ($P(x = a_i) \to 0$): the overall contribution to entropy will be low, because the argument of the logarithm tends to 1. In case of a variable with large randomness ("low order"), the probabilities will be uniformly distributed ($P(x = a_i) \to 1/N, \forall i = 1, ..., N$) and the entropy will be high, because the argument of the logarithm tends to $N^{1-\alpha}$, therefore $H_{R_\alpha}(x) \to logN$.

The order α was set at 2 and the pdf was estimated using kernel estimators [13]. Entropy was estimated, for each channel, within 1sec non-overlapping windows. We arranged the entropy time series of each channel as a row of a matrix (from now on named matrix **X**) whose t column represented the set of the entropy values associated with the n electrodes at a certain window t. EEG power was estimated the same way. For each window, we plotted a 2D map for EEG entropy and a 2D map for EEG power exploiting a function from the toolbox EEGLAB [14]. For each electrode, the corresponding value was plotted, encoding the values according to a continuous color scale going from blue (low values) to red (high values). The range of the scale was fixed, so that we could detect the overall variation over time as well as the local in time entropy spatial distribution. The colour of the spatial points lying between the electrodes are calculated by interpolation, thus a smooth gradation of colors is achieved. Every figure was captured as a frame of a movie and each frame was associated to the corresponding time, so that it was possible to keep the time information and therefore to identify the critical stages while reviewing the movie itself. Movies are available at: http://www.permano.it/research/res-eann2009.php. During the visualization we followed the topographic trends of the most active areas in terms of lowest and highest entropy levels, in other words, we were in search of the electrodes that had been associated with low-entropy or high-entropy values for the longest time and we called these electrodes the "most active".

3.2   Entropy Topography Clustering

In order to quantify the information that we could appreciate visually reviewing the movie and in order to automatize the analysis of the evolution of entropy spatial distribution, a spatio-temporal model was necessary. One neurocomputational approach to time-series analysis, self-organizing maps (SOM), is designed so that certain relationships between features in the input vectors components are reflected in the topology of the resultant clusters weight vectors obtained through unsupervised training using Kohonen based learning [15]. Matrix $\mathbf{X}$ was partitioned into 1min non-overlapping windows. Our aim was to subdivide the head, window by window, into a certain number of areas so that the electrodes sharing similar entropy levels could be clustered together, with particular attention to the areas associated to the extreme entropy levels: the highest entropy region would reflect underlying high random activity, the lowest entropy region would reflect low random activity instead. The areas associated with intermediate entropy would account for "neutral" areas. We decided to subdivide the head into four regions: high entropy area, low entropy area and two intermediate entropy areas.

In other words, we needed the rows (channels) of our matrix $\mathbf{X}$ to be clustered into four clusters, window by window. Therefore, we designed a SOM with 4 processing elements and we implemented it in Matlab®. A SOM was trained over each one of the windows coming up with a spatio-temporal clustering of the channels based on their entropy levels.

In order to quantify the visual review of the movies, as described in Section 3.2, we clustered the electrodes into a low-entropy cluster (LEC), a high-entropy cluster (HEC) and two neutral clusters (NEC1 and NEC2), passing the EEG entropy matrix $\mathbf{X}$ through the SOM.

Window by window, we identified which one of the four processing elements was associated to the LEC, HEC and NEC clusters, so that we could come up with an homogeneous spatio-temporal clustering of the electrodes according to their entropy levels.

Once we had come up with the clustering, we estimated how often an electrode belonged to each one of the clusters. In other words, we estimated how often an electrode was active in terms of either high-entropy and low-entropy: we quantified how often (represented in time % with respect to the overall length of the EEG recording) each electrode belonged to the low-activity cluster, to the high-activity cluster or to the neutral clusters. This way we could define a "membership degree" of an electrode to each cluster.

In order to give a spatial view at a glance of this quantification and to visually associate time% levels with the location of the corresponding electrodes, the quantification is represented with histograms in Figure 3: each subplot is associated with a particular cluster, on the x axis electrodes are represented whereas on y axis time % is represented.

4   Results

4.1   Data Description

The analyzed dataset consists in three EEG recordings with 18 to 20-channels (Figure 1) from three patients affected by frontal lobe epilepsy (A, B) and one patient affected by frontal-temporal intractable epilepsy (C). The EEG was high pass filtered at 0.5Hz and the sampling rate was set at 256Hz for patients A and B, at 200Hz, for patients C. The dataset of patient A and C included one seizure whereas the dataset of patient B included three seizures. Figure 2 shows in 3D some explanatory frames from the movie of patient B: Figure 2.a shows an interictal map whereas Figure 2.b-c-d show the ictal maps of the three seizures.

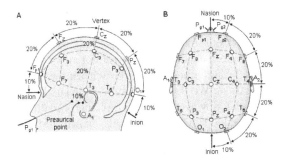

Fig. 1. The international 10-20 system seen from (A) left and (B) above the head. A = Ear lobe, C = central, Pg = nasopharyngeal, P = parietal, F = frontal, Fp = frontal polar, O = occipital.

4.2   Movie Visual Review

The movies of patients A and B showed a very steady spatial distribution of entropy: from the very beginning of the recording a low-entropy area was concentrated in the frontal region, that was the region of seizure onset. The mapping of EEG power was much less stable and selective. The movie of patient C showed a different behaviour, the focal region was associated to high entropy long before the seizure onset.

4.3   Electrodes Clustering

We can point out a very important result: the electrodes belonging to the focal area resulted clustered together from the very beginning and not only during the

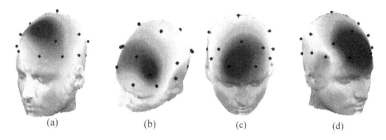

Fig. 2. 3-D plot of entropy spatial distribution of patient B during: (a) interictal stage; (b) first seizure; (c) second seizure; (d) third seizure. The area associated to low entropy (blue) is the epileptogenic focus.

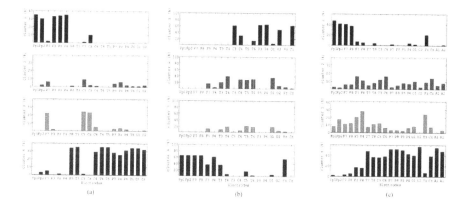

Fig. 3. Quantification of how often (time %) each electrode (x axis) belonged to the low-entropy cluster (cluster 1), to the neutral entropy clusters (clusters 2 and 3) and to the high-entropy cluster (cluster 4) for patient A (a), patient B (b) and patient C (c)

ictal stage. This means that an abnormal coupling among the electrodes that will be involved in seizure development starts before the seizure itself. We will now detail the results patient by patient.

Visually reviewing the movie of patient A, we could realize that the frontal area had been steadily associated to low entropy. Once we applied SOM, we realized that this behaviour impacted on electrode clustering: the electrodes of the frontal areas had been clustered together from the very beginning of the recording and were associated to cluster 1, as we can see from Figure 3. In order to give a deeper look into the evolution of clustering throughout the recording, we analised the trend of cluster 1 visualising the electrodes belonging to it from the beginning of the recording to the time of seizure onset (see Table 1). As we can see from Table 1, cluster 1 is clearly dominated by electrodes Fp1, Fp2, F3, F4 and Fz with an occasional involvement of Cz.

The movie of patient B showed that the frontal area had been steadily associated to low entropy. SOM based clustering showed that the electrodes of the

Table 1. Patient A: Electrodes Belonging to the Critical Cluster

Fp1	Fp1	Fp1	Fp1	Fp1	Fp1	Fp1	Fp1	Fp1	Fp1	Fp1	Fp1	Fp1	Fp1	Fp1	Fp1	Fp1	Fp1	Fp1	Fp1	Fp1	Fp1	Fp1	Fp1	Fp1	Fp1	Fp1	Fp1	Fp1	Fp1	Fp1	Fp1
F7	Fp2	Fp2	Fp2	Fp2	Fp2	Fp2	Fp2	Fp2	Fp2	Fp2	Fp2	Fp2	Fp2	Fp2	F3	Fz	Fp2	Fp2	Fp2	Fp2	Fp2	Fp2	Fp2	Fp2	Fp2	Fp2	Fp2	Fp2	Fp2	Fp2	Fp2
F3	F3	F7	F3	F3	F3	F3	F3	F3	F3	F3	F3	F3	F3	F3	Fz	F4	F3	F3	F3	F3	F3	F3	F3	F3	F3	F3	F3	F3	F3	F3	F3
Fz	Fz	F3	Fz	Fz	Fz	Fz	Fz	Fz	Fz	Fz	Fz	Fz	Fz	Fz	F4	Cz	Fz	Fz	Fz	Fz	Fz	F4	Fz	Fz	Fz	Fz	Fz	Fz	Fz	Fz	Fz
F4	F4	Fz	F4	F4	F4	F4	F4	F4	F4	F4	F4	F4	F4	F4			F4	F4	F4	F4	F4		F4	F4	F4	F4	F4	F4	F4	F4	F4
Cz	Cz	F4	Cz													Cz			Cz		Cz			Cz							
		C3																													
		Cz																													

32	31	30	29	28	27	26	25	24	23	22	21	20	19	18	17	16	15	14	13	12	11	10	9	8	7	6	5	4	3	2	1	0

Time to seizure (min)

Table 2. Patient B: Electrodes Belonging to the Critical Cluster

Fp1	Fp1	Fp1	Fp1	Fp1	Fp1	Fp1	Fp1	Fp1	Fp1	Fp1	Fp1	Fp1	Fp1	Fp1	Fp1	Fp1	Fp1	Fp1	Fp1	Fp1	Fp1	Fp1	Fp1	Fp1	Fp1	Fp1	Fp1	Fp1	Fp1	Fp1	Fp1
Fp2	Fp2	Fp2	Fp2	Fp2	Fp2	Fp2	Fp2	Fp2	Fp2	Fp2	Fp2	Fp2	Fp2	Fp2	Fp2	Fp2	Fp2	Fp2	Fp2	Fp2	Fp2	Fp2	Fp2	Fp2	Fp2	Fp2	Fp2	Fp2	Fp2	Fp2	Fp2
F7	F7	F7	F7	F7	F7	F7	F7	F7	F7	F7	F7	F7	F7	F7	F7	F7	F7	F7	F7	F7	F7	F7	F7	F7	F7	F7	F7	F7	F7	F7	F7
F8	F8	F8	F8	F8	F8	F8	F8	F8	F8	F8	F8	F8	F8	F8	F8	F8	F8	F8	F8	F8	F8	F8	F8	F8	F8	F8	F8	F8	F8	F8	F8
F4	F4	F4	F3	F4	F4	F4	F4	F4	F4	F4	F3	F3	F3	F3	F3		F3	F4	F3	F4	F4	F3	F3	F4	F4	F4	F3	F3	F4	F4	F4
F3	F3	F3	F4	F4	T5		F3	T5	F3	F4	F4	F4	F4	F4		F4	F4	F3	F4	F4	F4		F4	F4	F3	F4	F3	F4	F3	F4	F4
T5	F4	O1	F3	F3		Fz	O1	Fz	Fz	Fz	F3	Fz	F3	F3		F3		Fz	F3	F4	Fz		Fz		Fz	T4	O1	Fz	F4	F4	F3
O1	Fz	Fz	Fz	O1			Fz				Fz		Fz	Fz				Fz						Fz	Fz		T5	T5	Fz	Fz	T6

32	31	30	29	28	27	26	25	24	23	22	21	20	19	18	17	16	15	14	13	12	11	10	9	8	7	6	5	4	3	2	1	0

Time to seizure (min)

Table 3. Patient C: Electrodes Belonging to the Critical Cluster

Fp1	Fp1	Fp1	Fp1	Fp1	Fp1	Fp1	Fp1	Fp1	Fp1	Fp1	Fp1	Fp1	Fp1	Fp1	Fp1	Fp1	Fp1	Fp1	Fp1	Fp1	Fp1	Fp1	Fp1	Fp1	Fp1	Fp1	Fp1	Fp1	Fp1	Fp1	Fp1
Fp2	Fp2	Fp2	Fp2	Fp2	Fp2	Fp2	Fp2	Fp2	Fp2	Fp2	Fp2	Fp2	Fp2	Fp2	Fp2	Fp2	Fp2	Fp2	Fp2	Fp2	Fp2	Fp2	Fp2	Fp2	Fp2	Fp2	Fp2	Fp2	Fp2	Fp2	Fp2
F7	F7	F7	F7	F7	F7	F7	F7	F7	F7	F7	F7	F7	F7	F7	F7	F7	F7	F7	F7	F7	F7	F7	F7	F7	F7	F7	F7	F7	F7	F7	F7
F8	F8	F8	F8	F8	F8	F8	F8	F8	F8	F8	F8		F8	F8	F8	F8	F8	F8	F8	F8	F8	F8	F8	F8	F8	F8	F8	F8	F8	F8	F8
F4	F4	F3	T3	F4		F4	F4	F3	F3	F3	F3	F3	F3		F4		F3	F4		F3	F4	F4	F4	F4	F4	F3	F3	F4	F3	F3	
T3	T3	T4	T5		T5	Fz	F4	F4	F4	F4	F4	F4			F4	Fz	F4	Fz	T3	Fz	Fz	T5	T4	F4	F4						
Fz	Fz	Fz			Fz		Fz	Fz	Fz	Fz	T3				Fz		Fz		O1		Fz	Fz		F3	Fz					F3	F4
							Fz																							Fz	Fz

32	31	30	29	28	27	26	25	24	23	22	21	20	19	18	17	16	15	14	13	12	11	10	9	8	7	6	5	4	3	2	1	0

Time to seizure (min)

frontal areas had been clustered together in cluster 4. The trend of cluster 4 is visualised in Table 2. Cluster 4 is clearly dominated by electrodes Fp1, Fp2, F7, F8, F4 and T3, thus there is some fronto-temporal involvement, compared to the analysis of patient A.

The movie of patient C showed that the frontal area had been steadily associated to high entropy. SOM based clustering showed that the electrodes of the frontal areas had been clustered together in cluster 1. The trend of cluster 1 is visualised in Table 3. Cluster 1 is dominated by electrodes Fp1, Fp2, F7, F8, F4 and F3, there is some fronto-temporal involvement for this patient too.

5   Conclusions

In this paper, the issue of synchronization in the electric activity of neuronal sources in the epileptic brain was addressed. A new Electroencephalography (EEG) mapping, based on entropy, together with a SOM based spatial clustering, was introduced to study this abnormal behaviour of the neuronal sources. Renyi's entropy was proposed to measure the randomness/order of the brain. Three EEG dataset from patients affected by partial epilepsy were analysed. Renyi's Entropy spatial distribution showed a clear relationship with the region of seizure onset. Entropy mapping was compared with the standard power mapping that was much less stable and selective. Thus we can infer that entropy seems to be a possible window on the synchronization of epileptic neural sources. These preliminary results are qualitative, we will pursue a more quantitative description with larger number of experiments and evaluating the statistical significance of the observed behavior. Future research will be devoted to the analysis of the EEG of normal subjects in order to carry out a comparison with the entropy mapping of normal EEG. Furthermore, the analysis will be extended to many other epileptic patients and patterns of entropy activation will be investigated.

Acknowledgements

The authors would like to thank the doctors of the Epilepsy Regional Center of the Riuniti Hospital of Reggio Calabria (Italy) for their insightful comments and suggestions.

References

1. Im, C.-H., Jung, H.-K., Jung, K.-Y., Lee, S.Y.: Reconstruction of continuous and focalized brain functional source images from electroencephalography. IEEE Trans. on Magnetics 43(4), 1709–1712 (2007)
2. Im, C.-H., Lee, C., An, K.-O., Jung, H.-K., Jung, K.-Y., Lee, S.Y.: Precise estimation of brain electrical sources using anatomically constrained area source (acas) localization. IEEE Trans. on Magnetics 43(4), 1713–1716 (2007)
3. Knutsson, E., Hellstrand, E., Schneider, S., Striebel, W.: Multichannel magnetoencephalography for localization of epileptogenic activity in intractable epilepsies. IEEE Trans. on Magnetics 29(6), 3321–3324 (1993)
4. Sackellares, J.C., Iasemidis, L.D., Gilmore, R.L., Roper, S.N.: Clinical application of computed eeg topography. In: Duffy, F.H. (ed.) Topographic Mapping of Brain Electrical Activity, Boston, Butterworths (1986)
5. Nuwer, M.R.: Quantitative eegs. Journal of Clinical Neurophysiology 5, 1–86 (1988)
6. Babiloni, C., Binetti, G., Cassetta, E., Cerboneschi, D., Dal Forno, G., Del Percio, C., Ferreri, F., Ferri, R., Lanuzza, B., Miniussi, C., Oretti, D.V.M., Nobili, F., Pascual-Marqui, R.D., Rodriguez, G., Romani, G., Salinari, S., Tecchio, F., Vitali, P., Zanetti, O., Zappasodi, F., Rossini, P.M.: Mapping distributed sources of cortical rhythms in mild alzheimer's disease. a multicentric eeg study. NeuroImage 22, 57–67 (2004)
7. Miyagi, Y., Morioka, T., Fukui, K., Kawamura, T., Hashiguchi, K., Yoshida, F., Shono, T., Sasaki, T.: Spatio-temporal analysis by voltage topography of ictal electroencephalogram on mr surface anatomy scan for the localization of epileptogenic areas. Minim. Invasive Neurosurg. 48(2), 97–100 (1988)
8. Ebersole, J.S.: Defining epileptic foci: past, present, future. Journal of Clinical Neurophysiology 14, 470–483 (1997)

9. Scherg, M.: From eeg source localization to source imaging. Acta Neurol. Scand. 152, 29–30 (1994)
10. Tekgul, H., Bourgeois, B.F., Gauvreau, K., Bergin, A.M.: Electroencephalography in neonatal seizures: comparison of a reduced and a full 10/20 montage. Pediatr. Neurol. 32(3), 155–161 (2005)
11. Nayak, D., Valentin, A., Alarcon, G., Garcia Seoane, J.J., Brunnhuber, F., Juler, J., Polkey, C.E., Binnie, C.D.: Characteristics of scalp electrical fields associated with deep medial temporal epileptiform discharges. Journal of Clinical Neurophysiology 115(6), 1423–1435 (2004)
12. Skrandies, W., Dralle, D.: Topography of spectral eeg and late vep components in patients with benign rolandic epilepsy of childhood. J. Neural Transm. 111(2), 223–230 (2004)
13. Hild II, K.E., Erdogmus, D., Principe, J.C.: On-line minimum mutual information method for time varying blind source separation. In: 3rd International Conference on Independent Component Analysis And Blind Signal Separation, pp. 126–131 (2001)
14. Delorme, A., Makeig, S.: Eeglab: an open source toolbox for analysis of single-trial eeg dynamics including independent component analysis. Journal of Neuroscience Methods 134, 9–21 (2004)
15. Kohonen, T.: Self-Organizing Maps. Series in Information Sciences, vol. 30. Springer, Heidelberg (1995)

Riverflow Prediction with Artificial Neural Networks

A.W. Jayawardena

International Centre for Water Hazard and Risk Management (ICHARM)
under the auspices of UNESCO, Public Works Research Institute (PWRI)
1-6, Minamihara, Tsukuba, Ibaragi 305-8516, Japan
hrecjaw@hkucc.hku.hk

Abstract. In recent years, Artificial Neural Networks have emerged as a powerful data driven approach of modelling and predicting complex physical and biological systems. The approach has several advantages over other traditional data driven approaches. Particularly among them are the facts that they can be used to model non-linear processes and that they do not require *'a priori'* understanding of the detailed mechanics of the processes involved. Because of the parallel nature of the data processing, the approach is also quite robust and insensitive to noise present in the data. Several riverflow applications of ANN's are presented in this paper.

Keywords: Artificial Neural Networks, Multi-layer Perceptron, Back-propagation, Mekong River, Chao Phraya River, Surma River.

1 Introduction

Flow prediction is an essential component of early warning systems for mitigating flood disasters which account for over 50% of all natural disasters in terms of the human casualties. Various types of approaches can be used for flow prediction with each one having its own pros and cons. When real-time prediction is desired, the mathematical model that is used for flow prediction should be easy to implement, be capable of updating when new data becomes available, and computationally efficient. Data driven types of models have several advantages over distributed type physics-based models in that the former does not require *'a priori'* understanding of the mechanics of the underlying processes. The latter type requires *'a priori'* understanding of the mechanics of the underlying processes as well as a great deal of more spatially and temporally distributed data. The distributed approach is also computationally resource intensive and cannot be easily adopted in a real-time environment. Of the many types of data driven approaches from simple regression models to more complex fuzzy logic type models, the ANN approach has many attractions. They are relatively easy to formulate, insensitive to noise which is inherently present in all types of data, parallel in nature and therefore an error in one element does not cause a major error in the overall system, and can be adopted in a real-time situation.

In the past decade or so, ANNs have been increasingly popular for water resources modelling (e.g. [1], [2]; amongst others). A typical architecture of an ANN consists of an input layer, a hidden layer and an output layer. Inputs are fed into the network at

D. Palmer-Brown et al. (Eds.): EANN 2009, CCIS 43, pp. 463–471, 2009.

the input layer which may have a number of neurons to match the number of input variables. The number of neurons in the hidden layer is usually determined by trial and error whereas the number of neurons in the output layer depends upon the number of output variables expected from the network. Although a nonlinear transfer function has been the traditional choice for the output layer, recent ANN application development shows promise for using a linear transfer function instead (e.g. [3], [4]). Generally, a three layer feed-forward back-propagation neural network with a sigmoid transfer function at the hidden layer and a linear transfer function at the output layer can approximate any function that has finite number of discontinuities, provided that there are a sufficient number of neurons in the hidden layer [5]. Under this situation, the nonlinear properties of simulated processes are contained in the hidden layer.

In this paper, some examples of the application of ANN's in Hydrology relevant to the Asia Pacific Region are illustrated. They include the daily flow predictions in the Mekong River which runs through six countries, the Chao Phraya River of Thailand, and the water level (stage) predictions in Surma River in Bangladesh. These applications have the potential to be used in early warning systems for flood disaster mitigation.

2 Daily Riverflow Prediction

2.1 Mekong River at Pakse Gauging Station, Lao

Mekong River is perhaps the most important and most controversial river in Asia. It originates in China and runs through Myanmar, Lao, Thailand, Cambodia and finally Vietnam before it drains into the South China Sea. The River has a length of about 4620 km and drains a combined area of 795,500 km^2. Using the data at ten discharge gauging stations, an ANN model was constructed to predict the daily discharges at the Pakse gauging station for several lead times. Fig. 1 shows the relative locations of the 10 gauging stations with respective names Chiang Saen (most upstream), Luang Prabang, Chiang Khan, Pa Mong Dam Site, Vientiane, Nong Khai, Nakhon Phanom, Mukdahan, Khong Chiam and Pakse (most downstream). Previous work carried out on river flow and stage prediction include the application of the VIC model [6], and several other studies including the SSARR model for the upper part of the basin, multiple regression models for the lower reach of the delta with overbank flow, and the MIKE-11 for flood mapping in Mekong Delta, carried out by the Mekong River Commission as reported in [7], [8]. The latter works however do not present any results.

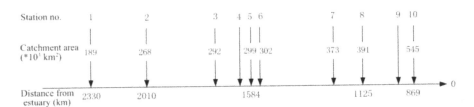

Fig. 1. Schematic diagram of the locations of the gauging stations

The data set used was divided into a training set (May 1, 1972 to December 31, 1983 and January 31, 1989 to December 31, 1989) and a testing set (January 1, 1990 to December 31, 1994) with a total sample size of 6423 (72% for the training set and 28% for the testing set). The training set was chosen so as to include extreme discharges both high and low for better performance of the network. From auto- and cross-correlation analysis it was found that the discharges at Pakse were related to its own time-lagged values, as well as to those at Vientiane, Nong Khai, Nakhon Phanom, Mukdahan and Khong Chiam in decreasing order of magnitude of correlation coefficient.

The ANN model adopted assumed a relationship of the form

$$Q_{t+\gamma}^{10} = F(Q_t^{10}, ... Q_{t-\beta_{10}}^{10}, Q_{t_9}^9, ..., Q_{t_9-\beta_9}^9, ..., Q_{t_2}^2, ..., Q_{t_2-\beta_2}^2, Q_{t_1}^1, ..., Q_{t_1-\beta_1}^1) \qquad (1)$$

where the superscripts refer to the station number, $Q_{t+\gamma}^{10}$ is the discharge prediction with γ-lead time at Pakse, β is the lag time, Q_t^{10} and $Q_{t-\beta_{10}}^{10}$ are the discharges at Pakse at time t and $t-\beta_{10}$, $Q_{t_9}^9$ and $Q_{t_9-\beta_9}^9$ are the discharges at Khong Chiam at time t_9 and $t_9 - \beta_9$, etc. In the model formulation, the inputs and outputs were first normalized to the range (-1, 1) by a linear transformation given by

$$x_i' = low + \frac{(x_i - x_{\min})(up - low)}{(x_{\max} - x_{\min})} \qquad (2)$$

where x_i' denotes the normalized data, x_i is the original data, x_{max} and x_{min} are the maximum and minimum values of the original data, and *low* and *up* are the smallest and largest output values allowed in the network. Although this is not a requirement since the sigmoid transfer function squashes any input to the range (0, 1) but is desirable. With the sigmoid transfer function at the output layer, the model outputs will also be within the range (0, 1). They need to be transformed to the target range for comparison. To avoid computational singularities, the range was limited to (0.1, 0.9). The parameters of the ANN model (momentum term, learning rate, number of neurons in the hidden layer) were estimated by trial and error. The connection weights were estimated by the back-propagation algorithm. A fixed number of 2000 iterations were used as the stopping criterion. Predictions were made with a non-linear output layer as well as a linear output layer, and the latter performed better than the former. Starting with its own time-lagged variables for the discharge at Pakse, time-lagged upstream discharges were added as inputs sequentially. With 7 time-lagged discharges at Pakse, the network performance has reached its optimal value for 1-day led time. Addition of more inputs did not improve the performance significantly. As expected, the reliability of the predictions decreases with increasing lead-time. A representative comparison of the results is given in Fig. 2.

2.2 Principal Component Analysis

When more and more time-lagged variables are added into the input layer the dimensionality of the problem becomes larger and larger without a proportionate contribution to the output. In other words, some input variables tend to become redundant. One way of reducing the dimensionality of the problem is by Principal Component Analysis (PCA). Instead of using the original input variables, a new set of variables which are obtained by a linear combination of the original variables, and which are relatively uncorrelated is used in the network. The relative importance of the new variables is determined on the basis of their variance-covariance structure. By doing so, there is very little loss of information. Usually the first few principal components would be sufficient to describe the relationship adequately. In general, the network with PCA performed better than that without.

2.3   Chao Phraya River at Nakhon Sawan Gauging Station, Thailand

The Chao Phraya River runs through the city of Bangkok and is one of the intensively monitored basins in Asia. The data set used in this study consists of daily discharges at the Nakhon Sawan gauging station for the period April 1978 to March 1994. (Total number of data = 5844). The simulations were carried out by assuming that the predicted discharge values depend only on the 7-days lagged discharge values regardless of the forecasting lead-time. Therefore, the resulting input layer had 7 nodes representing the daily discharges at time levels t, t-1, ..., t-6. A 5-node hidden layer, which gave the best error reduction pattern, was chosen by trial and error. The output layer consisted of only one node, which corresponds to the predicted value. Optimum values of the learning rate parameter and the momentum term, determined by trial and error, for the network with 1-day lead-time were 0.1 and 0.2 respectively. The logistic function was used as the activation function for both hidden and output layers.

The extrapolation capability of the network is enhanced if all the possible extreme events are contained in the training data set. This can be done by using the complete data set during training stage. In this data set however, all the extreme events are distributed within the first 1000 records and the remaining part of the data set does not contain any significant events. Therefore, the first 1000 records were selected for training and the remaining records were selected for evaluating the performance.

Training was carried out in the pattern mode. The weight updating was done after presentation of each sample in the training set to the network. Network was initialised by setting all the synaptic weights to random numbers uniformly distributed in the range of −1.0 to 1.0. The order of presenting the training samples from the training set to the network was also randomised from one epoch to the other. All the input variables as well as the output variables for the training period were linearly normalised to the range of 0.1 to 0.9. These steps were aimed at improving the performance of the back propagation algorithm. The training was stopped after 1000 epochs, chosen by trial and error, in all simulations.

Forecasts with 1-day, 5-days, 7-days, and 10-days to 50-days (in steps of 5-days) lead-times were made to compare their relative performance. The network structure and all other parameters were kept constant during these simulations but the output varied depending on the lead-time. A typical comparison is shown in Fig. 3.

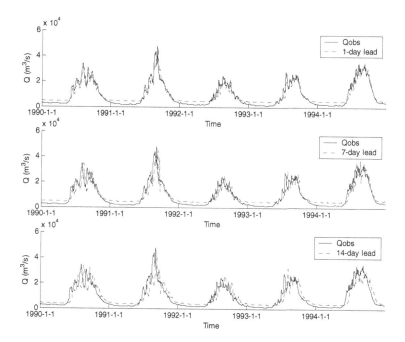

Fig. 2. Comparison of lead-time predictions with the nonlinear output layer using the observed discharge at Pakse (testing period)

Note: Model structure is 14^{III}-3-1 with III indicating inputs $Q_t^{10}, Q_{t-1}^{10}, .., Q_{t-13}^{10}$; 14, 3 and 1 correspond to the numbers of nodes in the input, hidden and output layers.

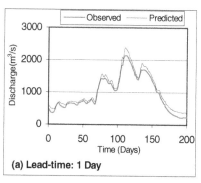

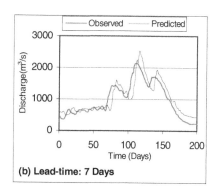

(a) Lead-time: 1 Day   **(b) Lead-time: 7 Days**

Note: Period covered - day 2601 to 2800

Fig. 3. Comparison of observed and predicted discharges at Nakhon Sawan gauging station (Day 1 corresponds to April 1, 1978)

3 Daily Stage Prediction

3.1 Surma River at Sylhet Gauging Station, Bangladesh

Being the most downstream country of several large international rivers that have origins in the high altitudes of the Himalyas, Bangladesh is perennially faced with many different types of natural water-related disasters. In addition to flooding in the flat areas of Bangladesh, the country is also exposed to storm surges with severe consequences. Structural measures to mitigate such damages do not appear to be economically feasible in a developing country like Bangladesh, and therefore non-structural measures such as early warning systems seem to be the only viable option. A major component of an early warning system is a mathematical model that can effectively forecast impending floods in terms of either stage or discharge given the rainfall data as inputs. This application in which ANN's have been used is aimed at forecasting daily water levels at the Sylhet gauging station (24° 42'N; 91° 53'E) across Surma River. The ANN used is of the MLP type with 3 layers: an input layer, a hidden layer and an output layer. The input layer has 4 nodes representing the water level on the previous day, rainfalls on the same day and 2 preceding days. The output layer gives the water level. The data used for training covers the period from August 20, 1980 to December 11, 1989, for validation the period from December 23, 1989 to April 15, 1999, and for application, the period from April 27, 1999 to August 17, 2008. Time series plots of the results using MLP and BP algorithms are shown in Figs. 4-6. Scatter diagrams for training, validation and application data sets give very good correlations with respective R^2 values of 0.9893, 0.9895 and 0.9887. In this study, a momentum term of 0.5 and learning rates ranging from 0.1 to 0.5 have been used.

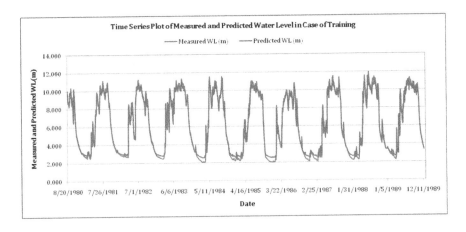

Fig. 4. Time Series Plot of Scaled Measured and Predicted Value of Water Level (Training)

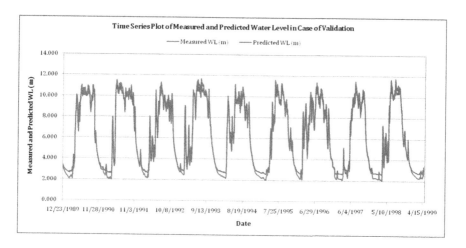

Fig. 5. Time Series Plot of Measured and Predicted Water Level in Case of Validation (Calculated by MLP with BP Algorithm)

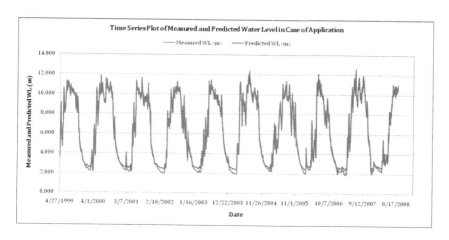

Fig. 6. Time Series Plot of Measured and Predicted Water Level in Case of Application (Calculated by MLP with BP Algorithm)

4 Performance Criteria

The performance of all the models presented in this study were evaluated using four different criteria: root mean square error (RMSE), relative root mean square error with respect to the average observed discharge (RRMSE), Nash-Sutcliffe coefficient of efficiency (NSE) and mean absolute error (MAE). The NSE for all the applications presented in this paper attained values very close the optimal.

5 Concluding Remarks

The applications presented in this paper demonstrate the efficiency and robustness of ANN approach as a tool for flood forecasting. Compared with the distributed approaches which are more resource demanding and yet not as robust, the ANN approach offers a viable alternative which does not require a high level of expertise. The alternative semi-distributed approach which has been used to predict flows in the Mekong and Chao Phraya basins and which required a great deal of additional data have proven to be computationally very expensive. Table 1 illustrates this point. Details of comparisons, as well as the application of the Variable Infiltration Capacity (VIC) model for the same objective and for the same two rivers illustrated in this paper can be found in earlier work of the author [6], [9], and [10].

Table 1. Statistics for grid representation of the Mekong Basin and computation resource consumption for different grid resolutions

Resolution	No. of active grids	Grid mesh dimension	One model evaluation		Optimization	
			VIC	Routing	Evaluations	Total
$2°\times2°$	37	13×8	2 min	5 min	1000	60 hr
$1°\times1°$	113	26×16	5 min	15 min	1000	170 hr
$0.5°\times0.5°$	374	51×31	10 min	1 hr	500	240 hr
$0.25°\times0.25°$	1311	102×61	30 min	3 hr	–	–
$0.125°\times0.125°$	4850	203×121	2 hr	9 hr	–	–

Note: Simulation period for one model evaluation of the VIC model and the linear reservoir routing is for 105 months. Grid mesh dimension is the full dimension of the grid mesh for basin grid representation. As a comparison, 100,000 model evaluations of the lumped SAC-SMA model for the Leaf River watershed use around 20 min on the hpcpower system for 26-month simulation period.

Acknowledgements. The author gratefully acknowledges the contributions made by his former graduate students T.M.K.G. Fernando, Tian Ying and the current graduate student Robin Kumar Biswas.

References

1. Govindaraju, R.S.: Artificial neural networks in hydrology II: Hydrologic applications. J. Hydrol. Engg. 5(2), 24–137 (2000)
2. Maier, H.R., Dandy, G.C.: Neural networks for the prediction and forecasting of water resources variables: a review of modelling issues and applications. Environmental Modelling and Software 15(1), 101–124 (2000)

3. Karul, C., Soyupak, S., Çilesiz, A.F., Akbay, N., Germen, E.: Case studies on the use of neural networks in eutrophication modelling. Ecological Modelling 134(2–3), 145–152 (2000)
4. Safavi, A., Abdollahi, H., Nezhad, M.R.H.: Artificial neural networks for simultaneous spectrophotometric differential kinetic determination of Co(II) and V(IV). Talanta 59(3), 515–523 (2003)
5. Mathworks Inc. Neural Network Toolbox User's Guide. Natick, MA (1998)
6. Jayawardena, A.W., Mahanama, S.P.P.: Meso-scale hydrological modeling: Application to Mekong and Chao Phraya basins. J. Hydrol. Engg. 7(1), 12–26 (2002)
7. Manusthiparom, C., Apirumanekul, C.: Flood forecasting and river monitoring system in the Mekong River basin. In: Proceedings of the Second Southeast Asia Water Forum, Bali, Indonesia, August 29- September 3 (2005)
8. Apirumanekul, C.: Flood forecasting and early warning systems in Mekong River Commission. In: 4th Annual Mekong Flood Forum, Siem Reap, Cambodia, May 18-19 (2006); chapter 2 – Flood Forecasting and Early Warning Systems in Mekong River Commission, pp. 145–151 (2006)
9. Tian, Y.: Macro-scale flow modeling of the Mekong River with spatial variance. A thesis submitted in partial fulfillment of the requirements for the degree of Doctor of Philosophy at The University of Hong Kong, p. 210 (2007)
10. Jayawardena, A.W.: Challenges in Hydrological Modelling – Simplicity vs. Complexity, Keynote Paper. In: Proceedings of the International Conference on Water, Environment, Energy and Society (WEES 2009), New Delhi, India, January 12-16, vol. I, pp. 549–553 (2009)

Applying Snap-Drift Neural Network to Trajectory Data to Identify Road Types: Assessing the Effect of Trajectory Variability

Frank Ekpenyong[1] and Dominic Palmer-Brown[2]

[1] Centre for Geo-Information Studies, University of East London, 4-6 University Way,
London E16 2RD
Tel.: +44(0)2082237243; Fax: +44(0)2082232918
f.u.ekpenyong@uel.ac.uk
[2] Faculty of Computing, London Metropolitan University, 31 Jewry Street,
London EC3N 2EY
Tel.: +44(0)2071334003; Fax: +44(0)2071334682
d.palmer-brown@londonmet.ac.uk

Abstract. Earlier studies have shown that it is feasible to apply ANN to catego-rise user recorded trajectory data such that the travelled road types can be re-vealed. This approach can be used to automatically detect, classify and report new roads and other road related information to GIS map vendor based on a user travel behavior. However, the effect of trajectory variability caused by varying road traffic conditions for the proposed approach was not presented; this is addressed in this paper. The results show that the variability encapsulated within the dataset is important for this approach since it aids the categorisation of the road types. Overall the SDNN achieved categorisation result of about 71% for original dataset and 55% for the variability pruned dataset.

Keywords: Snap-Drift Neural Network, Road network databases, GPS trajec-tory, Trajectory variability, Location-based services.

1 Introduction

Due to increasing traffic density, new roads are being constructed and existing ones modified, thus digital GIS road network databases are rarely up-to-date. There is a need to implement methods that would readily capture road changes, classify feature types and insert them into the database such that users of GIS map-based applications would have up-to-date maps. In [1] a solution that addresses the perceived problem of automated road network by applying Snap-Drift Neural Network on user recorded trajectory data is presented. This solution can be incorporated into Location-Based Service (LBS) applications dependent on Global Positioning System (GPS) and digi-tal Geographical Information System (GIS) road network (e.g. in-car navigation sys-tem), such that when users encounter possible new road segments (departure from the known roads in the database), the on-board device would record such deviations. This data would be processed by the SDNN as discussed in this paper or transferred back

D. Palmer-Brown et al. (Eds.): EANN 2009, CCIS 43, pp. 472–484, 2009.

to the Sat Nav provider and input into their SDNN along with similar track data provided by other service users, to decide whether or not to automatically update (add) the "unknown road" to the road database. In this way, users of applications dependent on road network would be provided with unified and sustainable platform to automatically perform road change update. Also, possible locations of new roads are pinpointed to road database vendors for further investigation. However, only the performance of unsupervised SDNN was assessed. In this paper, the effect of road traffic variability encapsulated in the recorded trajectory data on the SDNN categorisation is assessed. This would further inform on the suitability of using user recorded trajectory data to update road network database.

The following sections of this paper are organised as follows: in Section 2, an overview of related work on trajectory analysis is given. This is followed in Section 3 by an overview of the SDNN. In Section 4, data collection is described followed by presentation of road design parameter derivation from GPS data in section 5. In Section 6, trajectory variability analysis is presented followed by data types in section 7. The results and performance of the SDNN are presented in section 8 and section 9 presents the conclusions.

2 Past Work on Trajectory Analysis

Trajectory data recorded from GPS is an abstraction of the user movement. In the literature this GPS characteristic have been exploited for different studies. Some studies have focused on using users' trajectory data to group their trajectory pattern according to their similarities. For instance, Yanagisawa et al. [2] describe an efficient indexing method for a shape-based similarity search of the trajectory of dynamically changing locations of people and mobile objects. They define a data model of trajectories as lines based on space-time framework, and the similarity between trajectories is defined as the Euclidean distance between these lines. Their approach retrieves trajectories whose shape in a space is similar to the shape of a candidate trajectory from the database. However, the limitation to this approach relates to the use of Euclidean distance within a road network space. Mountain [3] defined three concepts of distance in geographic information system after the work of Laurini and Thompson [4] as Euclidean, Manhattan and Network distance. "The distance in a two dimensional isotropic surface is given by the Euclidean distance. The Manhattan distance assumes movement is restricted to cardinal directions. Network distances acknowledge that movement is constrained to transportation networks and calculates the distance between two points along network edges" [3]. Thus, using Euclidean distance to derive distance with a road network based environment as proposed by [2] is not practical. This concern is also raised in Hwang et al. [5]. They argue that clustering similar trajectories is highly dependent on the definition of distance, the similarity measurements as defined for Euclidean space are inappropriate for road network space and consequently the methods based on Euclidean space are not suitable for trajectory similarity grouping.

In Mountain and Raper [6] a method to automate the discovery and summarisation of a user's spatio-temporal behaviour relying on large archives of recorded spatio-temporal trajectory data is proposed. They were able to extract and visualise trajectory 'episodes' - discreet time periods for which a user's spatio-temporal behaviour was relatively homogenous. Episodes are discovered by analysing breakpoints, which they identified as temporal/spatial jumps and rapid change in direction speed. Other works have focused on using users' trajectory data to predict users' future location by analysing user's historical [7] or current [8] trajectory patterns. These approaches could be used for on-the-fly data pruning in anticipation of a user request in Location based Services (LBS).

GPS trajectory data have not been studied for the purpose of identifying roads segments that might need to be updated. In this study, the concept that trajectory information is an abstraction of user movement is exploited. The characteristics of this movement would in most cases be influenced by the road type or road feature the user is travelling on. A comparison of GPS trail with the actual road map would always reveal some level of conformance between the two data. As such, analysing GPS recorded trajectory data using an Artificial Neural Network should group or classify the trajectory data to reveal the road information or road type the user was travelling. Hence, for a road update problem candidate roads that might need to be updated into a GIS road network database, could be identified and updated. This is the principal concept in this study. Earlier work as shown that the SDNN is able to categorise recorded trajectory data to reflect the travelled roads [1], this papers investigates the effect of road traffic variability encapsulated in recorded trajectory data on the SDNN categorisation.

3 Snap-Drift Neural Networks

Different types of neural networks have been employed in the past for map matching, road extraction purposes and navigational satellite selection [9-12]. In this study the neural network is unsupervised SDNN, developed by Lee and Palmer-Brown [13]. SDNN emerged as an attempt to overcome the limitations of Adaptive Resonance Theory (ART) learning in non-stationary environments where self-organisation needs to take account of periodic or occasional performance feedback [14]. The solution was to swap between two modes of learning (snap and drift) according to performance feedback. The SDNN is able to adapt rapidly in a non-stationary environment where new patterns (new candidate road attributes in this case) are introduced over time. Subsequently, the fully unsupervised version of snap-drift was developed, which swaps its mode after each epoch. The learning process performs a combination of fast, convergent, minimalist learning (snap) and more cautious learning (drift) to capture both precise sub-features in the data and more general holistic features. Snap and drift learning phases are combined within a learning system (Fig. 1) that toggles its learning style between the two modes.

The network is essentially made of 2 parts, F1 and F2 (Fig. 1). F1 has d-wining nodes which provide the input pattern for F2. F1 performs features extraction while F2 carries out feature classification. d is the number of winning nodes in F1 and is optimised. This optimisation is performed empirically by increasing the number of nodes from say 2 upwards until a point when there is no further improvement in performance. Increasing d would increase the number of feature classes. The optimum d-nodes is found when an increase in d does not increase the number of feature classes (no useful features are found).

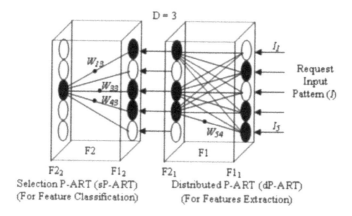

Fig. 1. Snap-Drift Neural Network (SDNN) architecture

On presentation of input data patterns at the input layer F1, the distributed SDNN (dSDNN) will learn to group them according to their features using snap-drift [15]. The neurons whose weight prototypes result in them receiving the highest activations are adapted. Weights are normalised so that in effect only the angle of the weight vector is adapted, meaning that a recognised feature is based on a particular ratio of values, rather than absolute values. The output winning neurons from dSDNN act as input data to the selection SDNN (sSDNN) module for the purpose of feature grouping and this layer is also subject to snap-drift learning.

The learning process is unlike error minimisation and maximum likelihood methods in MLPs and other kinds of neural networks which perform optimization for classification, or equivalents, by for example pushing features in the direction that minimizes error. Such approaches do not have any requirement for the feature to be statistically significant within the input data. In contrast, SDNN toggles its learning mode to find a rich set of features in the data and uses them to group the data into categories. The following is a summary of the steps that occur in SDNN [14]:

Step 1: Initialise parameters: (snap = 1, drift = 0), era = 2000
Step 2: For each epoch (t)
For each input pattern
Step 2.1: Find the D (D = 10) winning nodes at F21 with the largest net input
Step 2.3: Weights of dSDNN adapted according to the alternative learning proce-
          dure: (snap, drift) becomes Inverse(snap, drift) after every successive epoch
Step 3: Process the output pattern of F21 as input pattern of F12
Step 3.1: Find the node at F12 with the largest net input
Step 3.2: Test the threshold condition:
          IF (the net input of the node is greater than the threshold)
THEN
Weights of the sSDNN output node adapted according to the alternative learning
procedure: (α, σ) becomes inverse (snap, drift) after every successive epoch
ELSE

An uncommitted sSDNN output node is selected and its weights are adapted according to the alternative learning procedure: (snap, drift) becomes Inverse(α, σ) after every successive epoch

3.1 The Snap-Drift Algorithm

The snap-drift learning algorithm combines Learning Vector Quantisation (drift) and pattern intersection learning (snap) [16]. The top-down learning of the neural system which is a combination of the two forms of learning is as follows:

$$W_{Ji}^{(new)} = snap(I \cap W_{Ji}^{(old)}) + drift(W_{Ji}^{(old)} + \beta(I - W_{Ji}^{(old)})) \qquad (1)$$

where W_{Ji} = top-down weights vectors; I = binary input vectors, and β = the drift speed constant = 0.5. In successive learning epochs, the learning is toggled between the two modes of learning. When snap = 1 and drift = 0, fast, minimalist (snap) learning is invoked, causing the top-down weights to reach their new asymptote on each input presentation. (1) is simplified as:

$$W_{Ji}^{(new)} = I \cap W_{Ji}^{(old)} \qquad (2)$$

This learns sub-features of patterns. In contrast, when drift=1 and snap = 0, (1) simplifies to:

$$W_{Ji}^{(new)} = W_{Ji}^{(old)} + \beta(1 - W_{Ji}^{(old)}) \qquad (3)$$

which causes a simple form of clustering at a speed determined by β.
The bottom-up learning of the neural system is a normalised version of the top-down learning.

$$W_{iJ}^{(new)} = W_{Ji}^{(new)} / |W_{Ji}^{(new)}| \qquad (4)$$

where $W_{Ji}^{(new)}$ = top-down weights of the network after learning.

Snap-drift is toggled between snap and drift on each successive epoch. This allows strongest clusters (holistic features), sub-features and combination of the two to be captured [14]. The snap-drift algorithm has been used for continuous learning in several applications. In Lee et al. [13], the reinforced version of SDNN is used in the classification of user requests in an active computer network simulation environment whereby the system is able to discover alternative solutions in response to varying performance requirements. The unsupervised version of snap-drift algorithm (without any reinforcement) is used in the analysis and interpretation of data representing interactions between trainee network managers and a simulated network management system and this resulted in the discovery of new patterns of the user behaviour [17]. In Lee and Palmer-Brown [18] SDNN is compared with MLP and proves to be faster and just as effective as the MLP. The unsupervised form of snap-drift algorithm has been used to define unique millisecond features in speech patterns [14]. The SDNN was able to categorise the input patterns according to the general pattern of stammering and non-stammering speech. In [1] the SDNN was use to categorise recorded GPS trajectory to reveal travelled road types. The results showed that SDNN is able to group collected points to reveal travelled road segments. Relying only on the winning node, a grouping accuracy of about 71% is achieved compared to 51% from Learning Vector Quantisation (LVQ). On analysis and further experimentation with the SDNN d-nodes an improved grouping accuracy was achieved, but with a high count of unique d-node combinations.

4 Data Collection

GPS recorded trajectory data were gathered from a 31.2 km drive over a range of road types in London (Fig. 2). The GPS data were collected using Garmin's Etrex Vista with a reported accuracy (95% confidence interval) of 5.8m during the day and 2.1m at nights [19]. The GPS points were collected every 5 seconds. Voice data were also concurrently collected noting road segment characteristics that could affect the recorded data; like stops at junctions, traffic lights, GPS carrier loss and other delays.

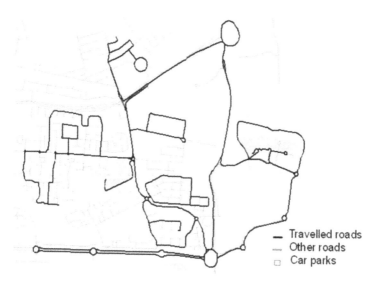

Fig. 2. Travelled routes where GPS trajectory was recorded

The voice data were used to identify collected points features that do not match any road related features from the Ordnance Survey (OS) MasterMap ITN layer [20]. Seven days data (Table 1) was collected to generate data which was used to study the effect of traffic variation on the study approach. One dataset was collected each day and covered different times ranging from 7am till 10pm. Table 1 shows the summaries of the data collection campaign. Day 1 data was collected on a Sunday with less traffic thus the least number of points were recorded, least time journey time and highest average speed. On the contrary, 976 GPS points was recorded on Saturday and it took about an hour and 22 minutes with an average speed of 21Km/h for the trip.

Table 1. Summary of the data trajectory data collection campaign

Trip	Start time	Elapsed time	Total point recorded	Average speed (Km/h)
Day 1: Sunday	07:43:23	00:56:22	679	29
Day 2: Monday	07:57:24	01:08:10	819	25
Day 3: Tuesday	16:19:26	01:06:22	798	26
Day4: Wednesday	09:43:43	01:05:40	789	26
Day 5: Thursday	18:13:16	01:06:01	794	26
Day 6: Friday	20:01:48	01:03:27	763	27
Day 7: Saturday	14:01:35	01:21:17	976	21

It is important to note here that variation in the data collection summary (Tables 1) is due to traffic condition of the travelled roads. For instance, the time window of the Day-7 (Saturday) data collection coincided with when people were out shopping, so much of the points were recorded on traffic delays on road serving shopping malls and at its car parks. To capture these variations, most of the routes were travelled more than once. Using the OS road type naming convention, roundabout features are normally part of other road classes say A roads, local streets or minor roads. But for our purpose we treat roundabout features and points collected at traffic light stops as unique classes considering the fact that we are grouping the trajectory data based on geometry information between successive points. GPS data points were collected from 7 road types and road related features namely; A road, local street, minor road, private road, roundabout, car park, and traffic light.

5   Road Design Parameter Derivation from GPS Trajectory Data

Clearly, GPS information holds significant point-to-point geometrical and topological information to allow the identification of possible roads. However, to decide what a road from such information is, it is imperative that similar road design parameters considered in actual road design are derived from the recorded GPS trail. From a typical GPS reading, the following information could be recorded for each GPS point: date; time; altitude (m); coordinate position (different coordinate systems). Also, the travel direction (^{0}); length (m) and time elapsed are recorded but relative to previous GPS point.

A study of relevant road design manuals showed that road design parameters like the speed; radiuses of horizontal and vertical curvatures are important road design parameters. These parameters where derived from the recorded trajectory data. Other parameters also derived include the acceleration, sinuosity and change in travel direction. The derivations are presented in [1].

6   Trajectory Data Variability Analysis

An individual travelling on a road network trades time for space [21], however, this is influenced by the constraint that limits individual's travel ability. Miller identified the following constraints as: the person's capabilities for trading time for space in movement; access to private and public transport; the need to couple with others at particular locations for given durations; and the ability of public or private authorities to restrict physical presence from some locations in space and time. How different individuals negotiate these constraints while trading time for space influences a road traffic condition. Here the locations of travel pattern variability within the test site are sought with a view to eliminating or reducing the variability. To inform on the effect of traffic variability on the research approach, the performance of SDNN with original data is compared with that with the variability removed or reduced. Point density analysis was carried out to determine clusters (areas of variability) in the recorded dataset. This was achieved using the kernel density implemented in ArcGIS 9 [22].

Fig. 3. Spatial mapping of traffic variability corridors in the study area

In total 18 variability corridors was found in the test site. The result of a spatial summation of all the trajectory variability locations for each data collection day is shown in Fig. 3. This summation reveals likely trajectory variation location (travel delay) in the study area; this is represented by the green shades in the figure. Overall most of the variability was found around traffic lights, road calming measures or temporal obstructions, road entries/exit and on road segment linking shopping malls. Comparison of the total number of recorded GPS points each day with number of recorded GPS points in the mapped variability corridor reveals that these variation represents a significant amount (almost 50%) of the total GPS points recorded each data collection day. For instance, In Day-1 673 GPS points were recorded and 327 of these points reside in the variability corridors. Similarly, in Day-2 819 GPS points were recorded and 428 were found in the variability corridors; in Day-7 976 points were recorded and more than half (563 points) reside in the variability corridors.

6.1 Trajectory Variability Reduction

A method was implemented that reduced the recorded GPS points within the trajectory variability corridor and also preserves the abstraction (geometry) representing the shape of the road segments travelled. The steps are shown below:

1. Calculate the number of points in a trajectory variability corridor (TVC)
2. Calculate the total length covered by the points in the TVC
3. find the speed of last recorded GPS points before the TVC points
4. Using speed-distance formula, where t = 5s, derived the new equidistance where new points would be placed using a constant speed found in 3 above
5. Get the x y coordinate of newly created points.

The number of representative points placed in a TVC depends on the length of points in the TVC and the speed of the last GPS point before the TVC was encountered. Consequently, recorded GPS points in the TVC were replaced with the newly created points and the variables recalculated for the new dataset.

7 Data Types

Two trajectory datasets are presented to the SDNN following the trajectory variability analysis, these are:

1. The original recorded GPS trajectory dataset with its derived variables now referred to as dataset 1
2. The variability pruned GPS Trajectory dataset with its derived variables now referred to as dataset 2, for this dataset there is reduction in the number of data points.

For both datasets 7 variables represented by separate fields in the input vector were presented to the SDNN. These are the speed, rate of acceleration and deceleration, radiuses of horizontal and vertical curvatures, change in direction and sinuosity. The training and test patterns were presented in a random order to the SDNN. This simulates the real world scenarios whereby travelled road patterns measured by GPS-based trajectory data varies depending on the travel speed, geometry of the road and nature of the road such that a given road type might be repeatedly encountered while others are not encountered at all.

8 Results

Fig. 4 shows a comparative plot of the SDNN and dSDNN categorisations across the road patterns for dataset 1. From this plot it can be seen that the SDNN is able to correctly categorised majority of the patterns except for private and minor road patterns. Most of the minor road patterns were categorised as local street patterns by the SDNN and the dSDNN suggesting that the minor road patterns share similar features to the local street patterns. In reality, minor road attributes like the speed limit do overlap with that of local street and as such it is challenging for the SDNN to correctly group features of these road types. Overall an improve categorisation is achieved with the dSDNN. Private road patterns were found on the dSDNN but wrongly generalised by the SDNN as local street and A road patterns. Using a simple winner-take-all, categorisation of 60% is achieved by the SDNN and 71% by the dSDNN.

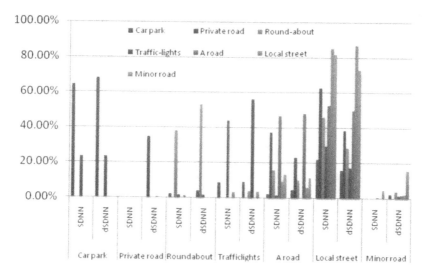

Fig. 4. Comparative plot of SDNN categorisation for the winning (SDNN) and the d-nodes (dSDNN) for dataset 1

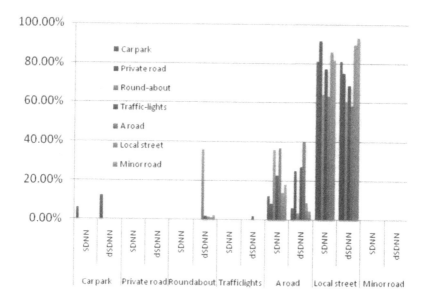

Fig. 5. Comparative plot of SDNN categorisation for the winning (SDNN) and the d-nodes (dSDNN) for dataset 2

Fig. 5 shows a comparative plot of the different categorisation methods across the different road class patterns for variability removed dataset (dataset 2). From this plot it is evident that majority of the road class are categorised as either local street or A road patterns while other patterns are not detected by the SDNN or dSDNN except for the car park patterns. This result shows that variability within the dataset holds significant information that enables the SDNN categorised majority of the dataset to reflect the travelled road types. Hence removing this information results in categorisation where all datasets grouped either as local street or A road patterns. A categorisation accuracy of 50% is achieved by the SDNN and 55% for the dSDNN for this dataset.

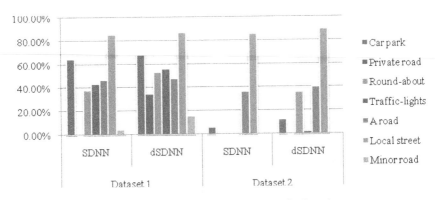

Fig. 6. Plot of distribution of correctly categorised road patterns

Fig. 6 shows a comparative plot of both dataset categorisations across the SDNN and the dSDNN nodes. It can be seen that the categorisation performance of the SDNN (and dSDNN) with the dataset 1 is better than dataset 2 (Fig. 6). For instance, all the road type patterns where detected by the SDNN and dSDNN except for private road for dataset 1. For dataset 2 only car park, local street and A road patterns were detected by the SDNN while only the car park, traffic light, local street and A road patterns were detected by dSDNN. With dataset 2, the traffic lights and roundabout patterns were poorly categorised suggesting that the variability in the recorded dataset is associated to these features; hence removal of the variability also reduces the chance of these patterns being detected by the SDNN. These results also show that without the variability information, most of the roads are categorised as either A roads or Local streets. Overall, it can be seen that the performance of SDNN for the variability pruned dataset (dataset 2) is poor compared to datasets 1. It is concluded that the variability information encapsulated by the recorded trajectory data is useful in grouping the trajectory data to reveal the travelled roads.

9   Conclusion

Unsupervised SDNN is able to categorise (classify) recorded trajectory information such that travelled road information like road type can be revealed. This is because the SDNN adopts a learning process that performs real-time combination of fast, convergent, minimalist learning (snap) and more cautious learning (drift) to capture both precise sub- features in the data and more general holistic features and is able to adapt rapidly in a non-stationary environment where new patterns (new candidate road attributes in this case) are introduced over time. Also, it can be concluded that the variability encapsulated in recorded trajectory caused by varying road traffic conditions is useful for identifying different road types when applying SDNN to trajectory data for the purpose of travelled road class identification. Artificial elimination or reduction of this variability significantly reduces the chances of correctly identifying travel road classes from recorded trajectory data.

Acknowledgment

The authors gratefully acknowledge the Ordnance Survey for provision of MasterMap coverages. All road centreline data in Figure 2 are Crown Copyright.

References

1. Ekpenyong, F., Palmer-Brown, D., Brimicombe, A.: Updating of Road Network Databases: Spatio-temporal Trajectory Grouping Using Snap-Drift Neural Network. In: 10th International Conference on Engineering Applications of Neural Networks, 2007, Thessaloniki, Hellas, August 29–31(2007)
2. Yanagisawa, Y., Akahani, J.-i., Satoh, T.: Shape-based similarity query for trajectory of mobile objects. In: Chen, M.-S., Chrysanthis, P.K., Sloman, M., Zaslavsky, A. (eds.) MDM 2003. LNCS, vol. 2574, pp. 63–77. Springer, Heidelberg (2003)
3. Mountain, D.M.: An investigation of individual spatial behaviour and geographic filters for information retrieval, Department of Information Science, City University, London (2005)
4. Laurini, R., Thompson, D.: Geometries in fundamentals of spatial information systems. Academic Press Ltd., London (1992)
5. Hwang, J.-R., Kang, H.-Y., Li, K.-J.: Spatio-temporal similarity analysis between trajectories on road networks. In: Akoka, J., Liddle, S.W., Song, I.-Y., Bertolotto, M., Comyn-Wattiau, I., van den Heuvel, W.-J., Kolp, M., Trujillo, J., Kop, C., Mayr, H.C. (eds.) ER Workshops 2005. LNCS (LNAI and LNBI), vol. 3770, pp. 280–289. Springer, Heidelberg (2005)
6. Mountain, D.M., Raper, J.: Modelling human spatio-temporal behaviour: A challenge for location-based services. In: 6th International Conference of GeoComputation, University of Queensland, Brisbane, Australia, September 24- 26 (2001),
   http://www.geocomputation.org/2001/papers/mountain.pdf
   (Accessed January 12, 2006)
7. Liu, X., Karimi, H.A.: Location awareness through trajectory prediction. Computers, Environment and Urban Systems 30(6), 741–756 (2006)
8. Brimicombe, A., Li, Y.: Mobile Space-Time Envelopes for Loaction-Based Services. Transactions in GIS 10(1), 5–23 (2006)
9. Barsi, A., Heipke, C., Willrich, F.: Junction Extraction by Artificial Neural Network System - JEANS. International Archives ofPhoto grammetry and Remote Sensing 34, Part 3B, 18–21 (2002)

10. Jwo, D.J., Lai, C.C.: Neural network-based geometry classification for navigation satellite selection. Journal of Navigation 56(2), 291 (2003)
11. Winter, M., Taylor, G.: Modular neural networks for map-matched GPS positioning. In: IEEE Web Information Systems Engineering Workshops (WISEW 2003), December 2003, pp. 106–111 (2003)
12. Jwo, D.J., Lai, C.C.: Neural network-based GPS GDOP approximation and classification. GPS Solutions 11(1), 51–60 (2007)
13. Lee, S.W., Palmer-Brown, D., Roadknight, C.M.: Performance-guided neural network for rapidly self-organising active network management. Neurocomputing 61, 5 (2004)
14. Lee, S.W., Palmer-Brown, D.: Phonetic Feature Discovery in Speech Using *Snap-Drift* Learning. In: Kollias, S.D., Stafylopatis, A., Duch, W., Oja, E. (eds.) ICANN 2006. LNCS, vol. 4132, pp. 952–962. Springer, Heidelberg (2006)
15. Lee, S.W., Palmer-Brown, D.: Phrase recognition using snap-drift learning algorithm. In: The Internation Joint Conference on Neural Neural Networks (IJCNN 2005), Montreal, Canada, July 31 - August 4 (2005)
16. Kohonen, T.: Improved versions of learning vector quantization. In: International Joint Conference on Neural Networks, vol. 1, pp. 545–550 (1990)
17. Donelan, H., Pattinson, C., Palmer-Brown, D.: The Analysis of User Behaviour of a Network Management Training Tool using a Neural Network. Systemics, Cybernetics and Informatics 3(5), 66–72 (2006)
18. Lee, S.W., Palmer-Brown, D.: Phrase Recognition using Snap-Drift Learning Algorithm. In: The International Joint Conference on Neural Networks (IJCNN 2005), Montreal, Canada, July 31- August 4 (2005)
19. Mehaffey, J., Yeazel, J.: Receiver WAAS On and Off, 30-Minute Tests in the Open, Night and Day (2002), http://gpsinformation.net/waas/vista-waas.html (Access on May 10, 2007)
20. Ordnance Survey, OS ITN Layer Dataset. Ordnance Survey, Great Britain (2006)
21. Miller, H.J.: A Measurement Theory for Time Geography. Geographical Analysis 37(1), 17–45 (2005)
22. ArcGIS User's Guide, ESRI, Inc. (2007)

Reputation Prediction in Mobile Ad Hoc Networks Using RBF Neural Networks

Fredric M. Ham[1], Eyosias Yoseph Imana[1], Attila Ondi[2], Richard Ford[2], William Allen[2], and Matthew Reedy[1]

[1] Florida Institute of Technology, Electrical and Computer Engineering Department
[2] Florida Institute of Technology, Computer Sciences Department
{fmh,eyoseph2007,aondi,mreedy}@fit.edu,
{rford,wallen}@cs.fit.edu

Abstract. Security is one of the major challenges in the design and implementation of protocols for mobile *ad hoc* networks (MANETs). 'Cooperation for corporate well-being' is one of the major principles being followed in current research to formulate various security protocols. In such systems, nodes establish trust-based interactions based on their reputation which is determined by node activities in the past. In this paper we propose the use of a Radial Basis Function-Neural Network (RBF-NN) to estimate the reputation of nodes based on their internal attributes as opposed to their observed activity, e.g., packet traffic. This technique is conducive to prediction of the reputation of a node before it portrays any activities, for example, malicious activities that could be potentially predicted before they actually begin. This renders the technique favorable for application in trust-based MANET defense systems to enhance their performance. In this work we were able to achieve an average prediction performance of approximately 91% using an RBF-NN to predict the reputation of the nodes in the MANET.

Keywords: MANET, attribute vector, reputation, radial basis function neural network, ROC curve.

1 Introduction

A Mobile ad-hoc Network (MANET) is a collection of mobile nodes connected to each other through a wireless medium dynamically forming a network without the use of centralized control or existing infrastructure. It is this type of network that is being researched and is projected to be applied in urgent and critical situations like medical emergency, disaster relief and military combat arenas. The sensitivity of such applications and the possible lack of an alternate communications path make these networks attractive to cyber attacks [1]. According to a recent DARPA BAA [2], among all of the cyber threats that exist, one of the most severe is expected to be worms with arbitrary payload, which can saturate and infect MANETs on the order of seconds. Some examples of the mobile devices that can reside at the nodes in the MANET are workstations, sensors, handheld devices, laptop computers, etc.

Maintaining the integrity of the network and each individual node in a MANET is challenging due to the lack of central authority in the network. Preserving this integrity,

D. Palmer-Brown et al. (Eds.): EANN 2009, CCIS 43, pp. 485–494, 2009.

however, is critical for carrying out missions in a military or disaster relief scenario. Although, for example, a computer at a node can be fully tested and configured with the latest "known good" patches (and therefore assumed to be free from malicious code), this known good state might be compromised in the field when a sudden change in the mission or environment could require the users to install new applications or modify certain settings "on-the-fly."

Assuming the presence of a trusted monitoring component in the kernel of the operating system allows a reliable measurement of the state of the computer; however, in itself such a trusted component might be unable to detect the presence of certain malicious code and stop the activities. Moreover, even if the malicious code can be detected, it is often too late: by the time of the detection, the malicious code could have compromised the system in an unknown manner as well as spread to other systems. Also, detecting attacks from remote hosts is useful as those hosts can be isolated, preventing further damage. The mere fact of detecting an attack, however, does not give insight into the vulnerability that allowed the malicious compromise, thus limiting the system to reactive rather than proactive defense.

Intuitively, it should be possible for these trusted components to cooperate and share information about themselves and each other. Correlating the known state of the computers with their reputation relating to malicious activities can be used to identify vulnerable computers based on their states even before they engage in malicious activity, thereby keeping "a step ahead" of the malicious code. As we will demonstrate in this paper, although some compromise is inevitable before the system learns the vulnerable state, the overall damage can be well contained.

In this work we use a Radial-Basis Function Neural Network (RBF-NN) [3] at each node to perform the "step ahead," prediction of the node's reputation, i.e., proactive defense from malicious activities. As explained above, relying on the detection of malicious activity by other nodes or monitoring systems to assign reputations scores is not adequate since malicious code in compromised nodes might wait only a few time steps before it starts attacking other nodes [4]. Moreover, we want to predict a potential compromise before the malicious activity is well underway.

An attribute vector is computed for each network node. The vector contains 10 numeric values related to crucial status indicators of processes and physical characteristics associated with nodal activity. An RBF-NN at each node maps the attribute vector of that particular node to its reputation score so that trustworthiness of each node can be estimated earlier than it would be determined by monitoring the behavior of the node. Figure 1 illustrates a comparison between behavior monitoring systems and our proposed RBF-NN reputation prediction system. In the figure, the node becomes compromised at time step $n = n_k$ and initiates its malicious activities at $n = n_l$ ($n_l > n_k$). Behavior monitoring defense systems detect the compromise at $n=n_l+1$ while the RBF-NN predictor can detect the compromise at $n = n_k + 1$.

The remainder of this paper is organized as follows. Section 2 gives information on related work and Section 3 details how the network and the nodes are modeled in this study. Section 4 gives a brief explanation of the simulation details, including the RBF-NN predictor training. Simulation results are presented in Section 5. Practical considerations that should be considered for implementation of the system that we are proposing is presented in Section 6, and finally, Section 7 concludes the paper with suggestions for future research directions.

Time Step	RBF-NN Predictor System	Behavior Monitoring System	
...			
$n = n_k$	Node becomes compromised	Node becomes compromised	
$n = n_k + 1$	RBF-NN predicts change in the reputation of the node indicating potential attack		Delay of Behavior monitoring system
$n = n_k + 2$	Preventive measures taken to protect the MANET		
...			
...			
...			
$n = n_l$		Node begins malicious activity	
$n = n_l + 1$		Other nodes detect malicious activity and reputation of the node changes	
$n = n_l + 2$		Preventive measures taken to protect the MANET from further attacks	

Fig. 1. Comparison of a RPF-NN predictor system and a behavior monitoring system

2 Related Work

Marti et al. [5] proposed the use of trust-based systems in the selection of next hop for routing. To ascertain the reliability of a node, the node monitors neighboring nodes to determine if it has been cooperative in forwarding other nodes' traffic. If a node has been cooperative, then it has a good reputation and its neighbors are inclined to forward packets to this node. In this way, packets are diverted (or directed away) from misbehaving nodes. Moreover, Buchegger and Le Boudec [6], [7] proposed and analyzed the CONFIDANT protocol, which detects and isolates misbehaving nodes. Similarity, they designed their protocol so that trust relationships and routing decisions are made through experience, which can be observed or reported behavior of other nodes. In [8] the authors analyzed the properties of the mechanisms that mobile nodes use to update and agree on the reputation of other mobile nodes. They suggest that mobile nodes can evaluate the reputation of other nodes based on both direct observation and the reputation propagation algorithm that they formulated in their research. The pitfalls of such systems involve relying on the detection of abnormal and possibly malicious activity to activate the necessary countermeasures, that is, operating reactively as opposed to proactively.

In Zheng et al. [9] it is discussed that malicious code is typically some type of coded computer program and under certain conditions within the computer's hardware and software it can outbreak and infect other code, compromise information and even possibly destroy certain devices within the system. These certain conditions can be modeled by defined combinations of states related to different features at the nodes. Saiman [10] lists features at the nodes that determine the reputation of the nodes within the network. They are classified as *performance metrics evaluation* features and *quantitative trust value* features. In the first category, features have different states assigned with certain numerical values. In the second category, the features are assigned values by mathematical evaluation or physical measurements.

This paper proposes a system which can be applied in any reputation-based or trust-based system. It enhances their performance by being able to determine the reputation value faster than it would be determined by simple behavior monitoring.

3 Modeling the Network

Here we present the assumptions made while building the model of the network. Computers in the network are not homogeneous, i.e. computers can have different settings. Moreover, these settings are not static, but can change as a function of time. There are n distinct settings that describe the state of any given computer in the network. Each setting i can be described by an integer value, s_i between 0 and $s_i^{max} > 1$. The particular value of each setting can be interpreted as a specific version or update of an application or hardware that is represented by that particular setting. The value 0 represents the absence of the corresponding setting (e.g. the application or hardware is not installed on the computer); and the value 1 represents a known "clean" state of the setting. Both states imply that the corresponding setting presents no vulnerability on the computer. At the start of the scenario, all settings have value 0 or 1, representing a "factory install" that is established at the beginning of the mission. All other values represent the fact that the application/hardware has been installed or modified on the computer after it has left the base and initially there exists no information about vulnerabilities that might have been introduced. Once a setting has a value other than 1, it cannot ever change back to a value of 1 (representing the fact that the state of "factory install" cannot be restored in the field).

Some pre-defined (but unknown) value of a particular setting (or a combination of such settings) represents a vulnerable state. When a computer is in a vulnerable state, it might become compromised at a later point in time by malicious code. When such a compromise happens, the machine will eventually behave in a way that is identified as malicious by its peers (i.e. it attacks its peers, causing them damage). Each machine has a reputation value between 0 (fully trusted) and 1 (untrusted) that is assigned to it by its peers. This value is initialized to 0. Once a machine begins attacking its peers, its reputation value (as derived by its peers) begins to increase, until the attack stops (for example, because of a new patch), or the value reaches 1. When a computer is no longer attacking its peers, its reputation value begins to decrease towards 0, representing the "forgiveness" of its peers towards its past behavior.

In order to model the changes in the states of the nodes, the following variables were used. The number of steps for a change on any given computer is determined by a truncated Poisson distribution with mean λ_c and maximum λ_c^{max}; for each change there is a probability, p_v, with a uniform random distribution that reflects a change that introduces a vulnerability. A computer is considered to have changed if at least one of its setting changes. If the change introduces vulnerability, a Poisson distribution with mean λ_m determines the number of time steps it takes for the malicious code to compromise the computer. A compromised computer has probability of $P_f = 0.5$ to improve its reputation (i.e., be "forgiven").

The particular choices for the parameters described above were: $n=10$, for all i: $s_i^{max}=10$, $\lambda_c=5$, $\lambda_c^{max}=10$, $p_v=0.06$, $\lambda_m=3$. There were 50 computers (nodes) in the network, the simulation ran for 1000 time steps, and the reputation values for each compromised machine changed by $\Delta r = 0.05$ per time step. The number of attribute settings set to 1 (as opposed to 0) at the beginning of the simulation was determined by a truncated Poisson distribution with mean 4 and maximum of 10. A heuristic approach was used to select the values of the parameters. There parameters are summarized in Table 1.

Table 1. Simulation Parameters and assumed values

Name	Description	Value
n	Number of distinct settings that describes the state of a node	10
s_i	Possible integer values assigned to setting i	0 - 10
s_i^{max}	Maximum possible value of integer to be assigned to settings	10
λ_c	Mean number of time steps before there will be a change in a computer (Poisson Distribution)	5
λ_m	The mean number of time steps it takes a malicious code to compromise the computer	4
λ_c^{max}	Maximum number of time steps before there will be a change in the computer	10
p_v	Probability that a change is to a vulnerable state	0.06
P_f	Probability that a malicious node moves from a compromised state to a secure one	0.5
Δr	The change in the reputation value of a compromised machine	0.05

4 Simulation Details

From the 50 nodes that were simulated according to the model presented in Section 3, 14 of them were compromised and displayed some level of malicious activity, i.e., their reputation fluctuated between 0 and 1. All of the remaining nodes had reputation of 0(trusted) during the simulation run. Each node had 10 attributes (settings) that can assume any integer from 0 to 10. As previously mentioned, the vector containing the numeric values that represent the "state" of the node is referred to as the attribute vector. The RBF-NN predictor is used to estimate the reputation of a node given the attribute vector at each time step. MATLAB® was used to develop the simulation. The steps involved in training and testing the RBF-NN is now explained. Detailed explanation of RBF-NNs is given in [3].

Let the matrix $\mathbf{A}_i \in Z^{p \times q}$ (a matrix of integers), where $p=1000$ and $q=10$, contain the attribute row vectors generated for i^{th} node for 1000 time steps. The vector $\mathbf{R}_i \in \mathfrak{R}^{p \times 1}$ ($p=1000$) contains the assigned reputation values for each 1000 times steps.

The following steps were followed to train, test and calibrate the neural network:

Step 1: Mean center and variance scale A_i [3].

Step 2: Select 60% of the rows in A_i to be $A_{i,train}$ and 40% to be $A_{i,test}$. Similarity, R_i was decomposed into $R_{n,train}$ and $R_{i,test}$.

Step 3: Train RBF-NN using $A_{i,train}$ with a spread parameter of 300 (determined experimentally).

Step 4: Sort the test data such that $R_{i,test}$ increases monotonically ($A_{i,test}$ should be ordered accordingly). Generate the reputation vector estimate $\hat{R}_{n,test}$ obtained by presenting $A_{i,test}$ to the RBF-NN from step 3.

Step 5: Threshold $R_{i,test}$ so that any value greater than 0.5 is rounded to 1, otherwise set to 0.

Step 6: Use a Receiver Operating Characteristic (ROC) curve [11], developed for each node and based on the simulated data to compute an optimal threshold, TR_i. Specifically, the thresholded reputation test vector $R_{i,test}$ and its estimate, $\hat{R}_{i,test}$, are used in the ROC curve analysis to determine the optimal threshold TR_i associated with the RBF-NN output (this is the calibration step referred to above) in step 3.

Step 7: Threshold $\hat{R}_{i,test}$ such that any value greater than TR_i becomes 1, otherwise it is set to 0.

Step 8: Determine the predictive performance of the RBF-NN by comparing the thresholded vector $\hat{R}_{i,test}$ from step 7 with the thresholded vector $R_{i,test}$ from step 5. The neural network's predictive performance is computed as the percent of the ratio of the correct predictions to the total number of predictions.

Step 9: After performing steps 1 – 8 for all active nodes, the overall Reputation Prediction Performance (RPP) of the MANET is computed as the average of the individual performances of all the nodes.

Assuming a stationary environment, once the neural network is trained on the attributes/reputations of a particular node then a proactive approach can be taken using the RBF-NN to perform a "step ahead" prediction of the node's reputation. This significantly enhances the performance of the wireless network's defense system since it allows prediction of malicious activities before they occur.

For a non-stationary environment, training must be performed periodically "on-the-fly" to update the RBF-NN weights based on the changes in the statistical nature of the environment. The frequency of "re-training" the RBF-NN is determined by how the statistics change within the environment.

5   Simulation Results

As previously mentioned, of the 50 nodes defined in the simulation, only 14 were active in the MANET. From the simulation results, the average RPP achieved was 90.82% using an RBF-NN predictor at each node. The (i) actual node reputation, (ii) the RBF-NN reputation estimate and (iii) the thresholded reputation estimate for nodes 0 and 43 are given in Fig. 2 and 3.

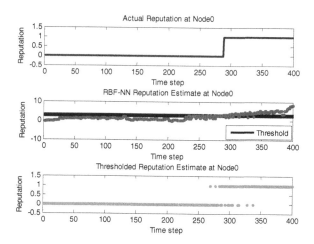

Fig. 2. RBF Estimation for Node 0

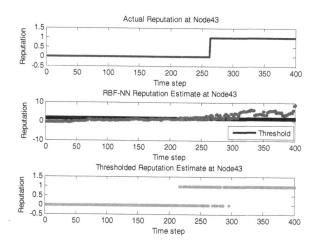

Fig. 3. RBF Estimation for Node 0

Figure 4 shows the ROC curves for nodes 0 and 43. The optimal threshold at the output of the RBF-NN for each node is determined by computing the Euclidean distance from the (0,1) point on the graph to the "knee" of the ROC curve.

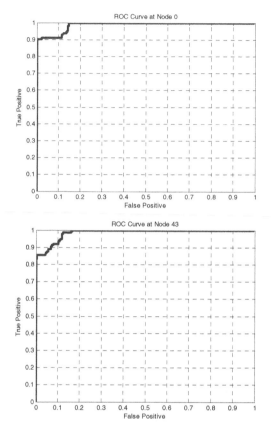

Fig. 4. ROC curves for MANET of node 0 and 43

Figure 5 illustrates the performance of the RBF-NN predictor at the "active" nodes.

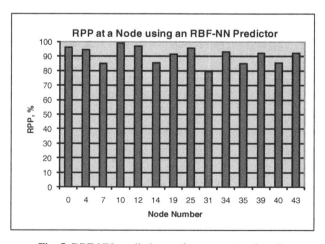

Fig. 5. RBF-NN predictive performance at each node

6   Practical Considerations

There are some practical considerations that must be considered for the successful operation of the proposed system. The first one is the existence of a trusted component in the kernel of the operating system to regularly report the state of the node. This component should be properly secured such that its functionality is not interfered with or obstructed by any software or physical compromise.

The attributes associated with a particular node are definable critical processes and physical characteristics associated with network (MANET) security. Examples are operating system patches, encryption keys, hardware configurations, exposure and location of the node [10]. The numerical assignment of these attributes should be made in such a manner to avoid singularity problems so that the RBF-NN algorithm can perform efficiently. Due to the absence of a central authority, the MANET defense system can be executed in a dynamically assigned node or even a specialized node like a cluster head.

The other issue is reputation scoring. There should be a defined benchmark known at all the nodes to facilitate the assignment of the reputation values for different types of nodal behavior (or activity). In our study, activity either results in an increase in the reputation score or a decrease. A few examples of activities that are normally expected from malicious, misbehaving or suspicious nodes are:

- Packet dropping [10]
- Flooding the MANET with a large number of Route Request (RREQ) [12]
- Sending out incorrectly addressed packets [13]
- A fake Route Reply (RREP) as in a Black Hole attack [14].

It should also be noted that the RBF-NN predictor cannot be operational before it learns the activity and state dynamics of the nodes in the MANET. Moreover, the initial RBF-NN training does not necessarily need to be performed as an integral part of the actual operation of the MANET.

As previously mentioned, a proactive approach to MANET defense as opposed to a reactive approach has the potential to better protect the MANET. For this reason, training of the RBF-NN should be carried out in a laboratory, or possibly during field testing, but in either case before the MANET is deployed for actual operation. It should be noted that the RBF-NN could be updated "on-the-fly" by re-training it adaptively once the actual operation has been initiated.

7   Conclusions and Future Work

We have demonstrated that it is possible to use an RBF-NN to proactively defend the nodes in a MANET by predicting the reputation of the nodes. Accurate prediction can allow time to perform preventative measures so that compromised nodes will not infect or disrupt other nodes in the MANET. Using the simulation approach presented in this paper, an average RPP of 90.82% was achieved using an RBF-NN predictor at each active node.

Our plans for future work include building a more realistic MANET simulator to further test the defense system suggested in this paper. In addition, we plan to investigate the use of a Kalman filter as an N-step predictor capable of estimating the reputation of a node into the future. This would be carried out by predicting the attribute vector out to N time steps, and then using this vector estimate to predict the node's reputation.

Acknowledgments

This research is part of a multi-institutional effort, supported by the Army Research Laboratory via Cooperative Agreement No. W911NF-08-2-0023.

References

1. Abdelhafez, M., Riley, G., Cole, R.G., Phamado, N.: Modeling and Simulations of TCP MANET Worms. In: 21st International Workshop on Principles of Advanced and Distributed Simulation (PADS 2007), pp. 123–130 (2007)
2. TPOC: Ghosh, A.K.: Defense Against Cyber Attacks on Mobile ad hoc Network Systems (MANETs). In: BAA04-18 Proposer Information Pamphlet (PIP), Defense Advanced Research Projects Agency (DARPA) Advanced Technology Office (ATO) (April 2004)
3. Ham, F.M., Kostanic, I.: Principles of Neurocomputing for Science and Engineering. McGraw-Hill, New York (2001)
4. Gordon, S., Howard, F.: Antivirus Software Testing for the new Millennium. In: 23rd National Information Systems security Conference (2000)
5. Marti, S., Giuli, T., Lai, K., Baker, M.: Mitigating routing Misbehavior in Mobile Ad Hoc Networks. In: Proceedings of the Sixth International Conference on Mobile Computeing and Networking, pp. 255–265 (August 2000)
6. Buchegger, S., Le Boudec, J.Y.: Nodes Bearing Grudges: Towards Routing Security, Fairness, and Robustness in Mobile ad hoc Networks. In: 10th Euromicro Workshop and Parallel, Distributed and Network-Based Processing, pp. 403–410 (2002)
7. Buchegger, S., Le Boudec, J.Y.: Performance Analysis of the CONFIDANT Protocol: Co-operation of Nodes–Fairness in Dynamic ad-hoc Networks. In: Proceeding of IEEE/ACM Workshop on Mobile Ad Hoc Networking and Computing, pp. 226–236 (June 2002)
8. Liu, Y., Yang, T.R.: Reputation Propagation and Agreement in Mobile Ad-Hoc Networks. IEEE Wireless Communication and Networking 3, 1510–1515 (2003)
9. Zheng, Z., Yi, L., Jian, L., Chang-xiang, S.: A New Computer Self-immune Model against Malicious Codes. In: First International Symposium on Data, Privacy and E-Commerce, pp. 456–458 (2007)
10. Samian, N., Maarof, M.A., Abd Razac, S.: Towards Identifying Features of Trust in Mobile Ad Hoc Networks. In: Second Asia International Conference on Modeling & Simulation, pp. 271–276 (2008)
11. McDonough, R.N., Whalen, A.D.: Detection of Signals in Noise, 2nd edn. Academic Press, New York (1995)
12. Balakrishnan, V., Varadharajan, V., Tupakula, U., Lucs, P.: TEAM: Trust Enhanced Security Architecture of Mobile Ad-hoc Networks. In: 15th IEEE International Conference on Networks, pp. 182–187 (2007)
13. Stopel, D., Boger, Z., Moskovitch, R., Shahar, Y., Elovici, Y.: Application of Artificial Neural Networks Techniques to Computer Worm Detection. In: International Joint Conference on Neural Networks, pp. 2362–2369 (2006)
14. Deng, H., Li, W., Agrawal, D.P.: Routing Security in Wireless Ad Hoc Networks. IEEE Communications Magazine, 70–75 (October 2002)

Author Index